7th International Conference on Intelligent Textiles & Mass Customisation (ITMC 2019)

IOP Conference Series: Materials Science and Engineering
Volume 827

Marrakech, Morocco
13 - 15 November 2019

Editors:

Mohamed Lahlou
Lieva Van Langenhove
Vladan Koncar
Jacek Mlynarek
Kanji Kajiwara

ISBN: 978-1-7138-3610-0
ISSN: 1757-8981

Printed from e-media with permission by:

Curran Associates, Inc.
57 Morehouse Lane
Red Hook, NY 12571

CURRAN ASSOCIATES INC. proceedings.com

Some format issues inherent in the e-media version may also appear in this print version.

Institute of Physics
Dirac House, Temple Back
Bristol BS1 6BE UK

Phone: 44 1 17 929 7481
Fax: 44 1 17 920 0979

techtracking@iop.org

Additional copies of this publication are available from:

Curran Associates, Inc.
57 Morehouse Lane
Red Hook, NY 12571 USA
Phone: 845-758-0400
Fax: 845-758-2633
Email: curran@proceedings.com
Web: www.proceedings.com

TABLE OF CONTENTS

2019
ITMC
Conference & Smart Textiles Salon
7th Edition of the
International Conference on
Intelligent Textiles
& Mass Customisation
November 13-15, 2019
Marrakech, Morocco
Palm Plaza Hotel
Targeted Topics:
•Smart and functional textiles
•Mass customisation
•Professional clothing
•Advanced manufacturing
•Digital tools
•Nanotechnology
•Design methodologies
•Comfort
•Funding opportunities
•Supply chain management and logistics
•Sustainable production, recycling
•Internet of Things
•Education and training
PROCEEDING
IOP Conference Series:
Materials Science and Engineering
Partners ITMC
GroupeCTTGroup
SHINSHU UNIVERSITY
esith
200 YEARS GHENT UNIVERSITY
ensait
For more information: Phone: +212 5 22 23 41 24, Fax: +212 5 22 23 15 85 , Email: itmc@esith.ac.ma
Website: http://itmc.esith.ac.ma

IOP Conf. Series: Materials Science and Engineering **827** (2020) 011001 doi:10.1088/1757-899X/827/1/011001

Preface

The International Conference on Intelligent Textiles and Mass Customization (ITMC) meets every two years, where the organization rotates among the 5 coordinating countries (Belgium, Canada, France, Japan and Morocco). The 7th edition of the International Conference on Intelligent Textiles and Mass Customization-ITMC 2019 was held at the historical city of Marrakech (MOROCCO), from the 13th to the 15th of November 2019.

The ITMC 2019 conference invited guests from various industries and disciplines related to the textile industry. The purpose of the conference is to explore new ideas, effective solutions and collaborative partnerships for business growth throughout the creation of a beneficial synergy between designers, manufacturers, suppliers and end users of all sectors and making full use of this potential. ITMC conference themes are axed on intelligent textiles and mass customization. For 3 days, inspiring speakers from industries, academies, governments and societies shed the light on new chances and challenges, via global statistics and success stories about cutting edge science and technology. The innovation brought to the table of discussion has bloomed through cooperation, policy, education and training and rose via an outstanding interaction between speakers and participants, which has been assured through new IT tools.

On behalf of the Conference organizers, we would like to thank all the participants coming from compagnies, universities and research institutions of all around the globe for greeting us with their presence and the lively exchange of ideas and experiences at the ITMC2019 Conference. We are looking forward to seeing you again in the next edition of ITMC 2021 in Canada!

IOP Conf. Series: Materials Science and Engineering **827** (2020) 011001 doi:10.1088/1757-899X/827/1/011001

Editorial board

IOP Conf. Series: Materials Science and Engineering **827** (2020) 011001 doi:10.1088/1757-899X/827/1/011001

Keynote Speakers

❖ Keynote 1

"E-textiles for medical and industrial applications"

Prof. Vladan Koncar: Distinguished Professor at ENSAIT (Ecole Nationale Supérieure des Arts et Industries Textiles) textile engineering institute in Roubaix, France.

Abstract: Smart and E textiles play a significant role as well in the European textile sector and assist the textile industry in its transformation into a competitive knowledge driven industry. Numerous materials and systems are available together with devices for sensing and actuation, but they are not compatible with textile substrates or with the textile production processes. They could be transformed into a textile compatible structure or even in a full textile structure. Smart textiles can be defined as textiles that are able to sense and respond to changes in their environment. They are able to detect different signals from the environment (temperature, light intensity, pollution…), to decide how to react and finally to act using various textile based, flexible or miniaturized actuators (textile displays, micro vibrating devices, LED, OLED…). The "decision" can be taken locally in case of embedded electronic devices (textile electronics) to smart textile structures or remotely in case the smart textile is wirelessly connected to clouds containing data base, servers with artificial intelligence software etc. and may be a part of Internet of Things (IoT) concept. This keynote focusses on latest research results within GEMTEX smart and E textiles research team given below:

- PHOS ISTOS, FP7 EU Project in the field of light emitting textiles structures used for a Photo Dynamic Therapy of skin cancer in its initial phase;
- @HEALTH, B2B project, concerning ECG monitoring in real time, with the system embedded into underwear aiming at the development of the predictive medicine concept;
- CONTEXT, ANR, French collaborative scientific project, in the area of textile metamaterials for energy harvesting and data transfer, using high and low frequency (2,4 GHZ and NFC frequency 13,56 MHz) textile antennas; and
- LITEVA, FUI, French industrial project dealing with the new generation of autonomous vehicle dash-boards.

❖ Keynote 2

"Complex 3D Woven Textiles: A Sustainable material"

Prof. Bijoya Kumar Behera: Professor and Head, Textile Technology Department, Indian Institute of Technology, India.

Abstract: Reinforced composite materials open up completely new skylines in an assortment of mechanical applications, compared with traditional materials. They are most useful in structures where the critical issue is how well the final product performs. Their superb properties such as mechanical strength, fatigue resistance and chemical resistance combined with the greater flexibility offer to design near net shape complex structures like solid, hollow, profiled, auxetic and aerodynamic structures result in cost effective solutions in advanced composite industry. Over and above, light weight load bearing structures open up unlimited possibilities of energy saving measures.

In composite making, woven fabrics are the preferred type of reinforcement other than fiber mats.

Woven fabrics are made by joining two or more yarns in an orthogonal framework with great steadiness in the

twist and filling bearings and offer the most noteworthy cover or yarn pressing thickness in connection to fabric thickness. As of now the greater part of the woven fabrics utilized as a part of composites are straightforward developments, for example, plain, satin, twill and basket
weaves. The weaving standard is additionally equipped for making structures that have significant measurement in the thickness heading shaped by layers of fabrics or yarns, termed as the threedimensional (3D) fabrics. A large portion of these structures can be made on the ordinary weaving machines with some modifications. The quick preferences of fabrics with an outstanding thickness measurement incorporate the basic respectability of the woven structure, the fulfilment of the geometric shapes, and volumes that are needed for some end-use applications. At present, both the traditional and uniquely made weaving machines are utilized to make different 3D fabrics, mostly for the structural composite applications in the field of aerospace, marine, automotives, sports, medical, wind energy and civil construction. There have been effective endeavours in growing new weaving gadgets, especially to make 3D woven solid, spacer and profiled structures. These new advances mastermind use of both flat tows and twist yarns in a 3D structure and permit weft yarns to be embedded at diverse levels. Another clear favourable position is that the weaving technology
can make 3D fabrics that are much thicker than the routine innovation.

Efforts are also being made to develop innovative 3D structures with various geometrical configuration. The importance of woven spacer fabrics, their properties and process optimization along with the advantages they carry with them have been studied in great detail. Spacer fabrics are proved to be strong for the structural applications due to their light weight and presence of reinforcement along their thickness direction. They have a promising future in the lightweight application areas and can be the real substitutes for aluminium and other metal alloys. In this way there dependably exists an interest on more creative structures to fulfil the requirement for present lightweight situated innovation. Improvement of complex preforms from a straightforward weaving innovation would dependably be a genuine test for a textile technocrat. Some new structural sandwich composites have also been developed using 3D woven spacer structures for their excellent properties, especially complete molding and good designing.
Regardless of weave architectures, the 3D woven fabrics as composite preforms have an imperative part to play in the improvement of cutting edge material composites. 3D-weaving technology does not only support extremely low cost manufacturing, but also decreases the impact to the environment and contribute to sustainability. In the first place, 3D technology supports the use of more abundant materials like plastics in substitution of metals, as well as less energy intensive manufacturing processes and decreases logistic efforts in the supply chain.

❖ Keynote 3
"Hot and Cold 3D Fabric Design"

Mr. Éric Perlinger: President of the world leader in the manufacture of high performance technical yarns, FilSpec ™, Canada.

Abstract: Since becoming President of FilSpec in late 2015, my direct supports and I have attended many "Smart Textile" academic conferences and we have joined various organisations, such as AFFOA , whose vocation is the development of "Smart Textiles". After attending conferences in Boston, Ghent, Raleigh and Stuttgart, I have been disappointed and somewhat underwhelmed by most of the academic papers. Few discuss "smart textiles" per se. Most consist of presenting "battery-wire-sensor" configurations, which only use the fabric as a superstructure to hold the electronics in place.
For all the media discussion of conductive yarns and acorn effects, few proposition presented fit into our conception of "Smart Textiles". For us, "Smart Textiles" react and interact with the environment in such a way to offer the wearer various desired functions or performances. "Smart Textiles" are more than novel data

collection and electronic stimulation.
FilSpec is a Canadian company based in Sherbrooke, Quebec that manufactures unique yarn solutions for various special applications. We view ourselves not as a yarn spinner, but as an engineering company that works in yarn. We believe that 80% of the performance of any fabric comes from the yarn. Our unique ability lies in the creation of specialty yarns that are converted into multi-functional fabrics.
In an industry suffocating under the weight of commodification, FilSpec has benefitted from phenomenal growth due our unwavering commitment to research, development and specialization.
In the past 3 years, we have nearly doubled the size of the company and developed a multitude of different branded products.
One such product is WarmFil, a yarn that uses the near infra-red rays of the sun to convert them into heat energy, generating a positive delta of 6-10 degrees centigrade (depending on time of day and cloud coverage). WarmFil fits into our definition of a "Smart Textile" as it reacts to its environment (NIR) and uses that reaction to offer a desirable function (heat).
Our first incarnation of a "Smart Textile" inspired us to try to create a fabric that would use part of the WarnFil technology to cool, rather than to heat. One of the properties that WarmFil offer is the ability to evaporate water 5 times faster than regular polyester. As the fibers generate heat they also accelerate evaporation. Our hypothesis is that we can cool the wearer through accelerated evaporation. We call it the "heat-pump effect" and we patented the concept of creating a multi-layer fabric using hydrophobic, hydrophilic and heat generating yarns in order to cool through hyperevaporation.

In September 2018, we started a project with NC State's Wilson College of Textiles to test our patented hypothesis. Students (Claire Rose, Ines Ortiz de Zevallos and Rachel Hall) spent their senior year working on the project and concluded that:
- A low thickness is paramount to allow fabric to have a higher heat loss;
- To be considered a cooling fabric, the fabric must have a high absolute and relative heat loss;
- A faster drying rate is indicative of a faster evaporation rate which showed high heat loss Phase 2 of our research focuses on the optimal fabric configuration to maximize the cooling effect.

❖ Keynote 4

"Digital transformation and open innovation for Railway sector"

Mr. Emmanuel COX: Digital transformation officer, French National Railway Company (SNCF), FRANCE.

Abstract: SNCF Réseau maintains, upgrades and sells access to the French rail network, serving all passenger and freight rail companies. There are basically three fundamental values which should govern the rail system: Safety, Security, and Sustainability (on both environmental and economic terms).
The future challenges for SNCF Réseau are huge:
- Respecting the fundamental values,
- Enhancing the existing capacity fulfilling user demand,
- Increasing the reliability delivering better and consistent quality service of European Rail system,
- Decreasing infrastructure's exploitation cost

To meet the numerous challenges ahead and to limit the impact on environment, the rail sector should increasingly rely on its innovative potential for delivering smart solutions with regards to safety, security, punctuality, availability, accessibility, seamless operation, capacity, connectivity, sustainability and other

IOP Conf. Series: Materials Science and Engineering **827** (2020) 011001 doi:10.1088/1757-899X/827/1/011001

performances. A large digital transformation has started at different levels:

- informing customers and staff,
- training the staff,
- monitoring and automizing processes, deploying IoT on assets for industrial supervision

SNCF Réseau is facing many issues to scale and industrialize IoT deployment over his assets. To cope with this challenge, many initiatives were launched internally to ease its deployment. The conference presents SNCF Réseau's challenges for the future, how the IoT and digital process contributes to transform the company, the leverage for implementation of digital processes and deployment of IoT, the expected impacts. On conclusion, a major initiative will be detailed: the DIGITAL OPEN LAB, a collaborative project that turns predictive maintenance into reality.

❖ Keynote 5

"Lithium ion batteries in a fast-changing world"

Prof. Rachid Yazami: Inventor of graphite anode for lithium-ion batteries, Founding Director of KVI Group, Singapore.

Abstract: Lithium ion batteries (LIB) were first commercialized in Japan in 1991 and since has experience a tremendous market growth to nearly reach about $100b. The commercialization of LIB came 10 years after the discovery by John Goodenough and Rachid Yazami of the lithium cobalt oxide cathode and the graphite anode, respectively. In 1985, Akira Yoshino sat up the first working protype of a LIB cell. LIB captured the news headlines when the 2019 Chemistry Nobel Prize was announced to honor the scientists and engineer who did the pioneer work that led to a success story. LIB owe their popularity owing to outstanding performances as compared to other autonomous power sources including batteries, fuel cells and supercapacitors. Firstly, LIB store the highest amount of energy per weight and volume, close to 270Wh/kg and 700Wh/l. This allows lighter and smaller batteries to be used in mobile electronics and in electro-mobility systems such as electric vehicles. Secondly, LIB provide high power to enable fast charging and discharging, merits that are in very high demand by the end users especially in electric vehicles (EV). Thirdly, depending on cells quality and usage conditions LIB an have a long life providing over 1000 deep charge and discharge cycles. For an EV application it is required a LIB pack to last about 10 years and provide over 200,000 km. Fifthly, LIB are generally safe although we hear here and there some unfortunate thermal incidents and accidents but they are statistically rare. The number of applications of LIB is steadily growing as society is moving toward cleaner environment to cope with the earth global warming issue. Energy storage systems associated with clean energy generation systems such as solar, wind, wave and hydrolytic energy are expected to play a major role in the climate change as we are transiting from greenhouse emitting energy sources to cleaner energy source provided by batteries. Besides well-established markets for LIB of mobile electronics, electro-mobility and energy storage systems, new applications are foreseen in battery imbedded systems such as in textiles. Thin printable and flexible batteries have been developed and tested in e-textile products. This is made possible owing to nanomaterials such as graphene together with polymers as illustrated in the figure below.

❖ Keynote 6

"MASEN Renewable Energy program in Morocco"

Mr. Hicham Bouzekri: Director of R&D and Industrial Integration at the Moroccan Solar Energy Agency (MASEN), Morocco.

Abstract: MASEN is implementing the world leading Moroccan renewable energy strategy to bring the solar, wind and hydro installed capacity to represent 42% of electricity generation in Morocco by 2020 and 52% by 2030. The mission of MASEN in implementing this strategy is complete and covers renewable electricity production industrial integration, local development, research and development.

❖ Keynote 7

"Toward the smart fiber & textile"

Prof. Dr. Eng. Toshihiro Hirai: Fiber Innovation Incubator, Professor at the Faculty of Textile Science and Technology, Shinshu University, Japan.

Abstract: Information technology, such as IoT and AI, is ruling the world in every field of industry. Our life style is also changing from the very basic level. Textile industries are also involved in the trend, and the textile products are expected to be smart in various ways. Many textile companies, in Japan, declared themselves to be chemical companies, and trying to be adoptive to the trend using their technologies. In this presentation, I would like to introduce some of my results on the smart functions of the conventional textile polymers. The functions are electromechanical, electro-optical, and piezoelectric functions. The results suggest that even the conventional polymers can exhibit highly potential "smart" functions, and also suggested that they have good potential as fibers and textiles even in the conventional manner in the AI technology trend. The challenges by Japanese Chemical Fiber Companies will also be introduced.

❖ Keynote 8

"Developments in the Belgian Textile Industry"

Mr. Kris Van Peteghem: Textile and Apparel Advisor at Fedustria, Belgium.

Abstract: As in most western economies, the textiles and apparel industry in Belgium was subjected to a major decline and transformation in the last decades. This presentation will give an overview of the main developments explaining this transformation. In the textiles industry we see 3 different

value chains. That is also how most West-European countries see it: first value chain is textiles for clothing, which we could call the apparel chain; second, textiles used at and in homes, which we call home textiles such as carpets and rugs, furnishing fabrics, terry toweling, household textiles, mattress textiles and so on. And third, there is the value chain of technical textiles.
Technical textiles is a wide range of textiles products with functionalities to meet a technical challenge, such as medical textiles in hospitals and the operation theatre, textiles for the agriculture business, textiles used in cars such as airbags, geotextiles for the road construction, … There are thousands of products in technical textiles. Technical and technological innovation is the main driver for technical textiles.
Basically, the apparel value chain became less and less important in Belgium and Western Europe to make place to the more specialized products in technical textiles. Two elements are important in explaining this transformation. First, the very high manufacturing costs in Western Europe. It pushed away the labour intensive textiles and apparel activities out of Western Europe towards North Africa and the Far East (Asia), certainly the making up of clothing. In a number of cases the weaving activities moved also, certainly to the Far East. In Belgium the clothing activities were replaced by the manufacturing of home textiles, such as carpets, in which Belgium is still one of the most important manufacturers in Europe. And the last 30 years technical textiles came up as a very strong third value chain in Europe. Now is represents almost half of the textiles business in Belgium, whereas home textiles represent 40% and clothing textiles only 10%. Second element is the growing need for new products, which require investment in R&D. New societal demands in the field of health and safety, environment, personal protection and so on creates always new opportunities for technical textiles. It's a growing business.
As mentioned before, the role of innovation and product development as a business strategy is key.
The transformation of the textiles industry also implies new business models and new investment projects, focussing on f.i. flexibility to allow mass customization. The presentation will also explain the necessity of well educated students and the need for vocational training. Otherwise, innovation can not be successfully implemented.

❖ Keynote 9

"Smart textiles: technological review and future perspectives"

Pr. Savvas G. Vassiliadis: Professor at Piraeus University of Applied Sciences, Greece.

Abstract: Smart textiles or e-textiles or electronic textiles are textiles that are characterized by advanced functionality and specifically by two major properties. Namely, their ability to obtain information from their environment and their ability to react based on that information. The main difference between mainstream, functional textiles that have already existed for many years (breathable, flame retardant etc.) and smart textiles is this ability of reaction to stimuli. The idea of textiles that react with their environment, and here textiles are meant to include both garments and a variety of other applications such for example medical applications, is not a new one. Nevertheless, one has only look at one of the first attempts at smart textiles, the MIThril vest from MIT (2003), with its boards and wires suspended on a fabric substrate to immediately understand some of the challenges that someone attempting to integrate electronics and textiles would face. It would take advances in circuit miniaturization, lowering power demands, development of next generation batteries as well as production of conductive textiles (embroidery/ sewing threads, conductive fabrics etc.) to bring us to today where hobbyist, arts/ crafts enthusiasts and fashionist as can use components like the Arduino and Lilypad to create garments with blinking LEDS and dresses that literally light up on the catwalk or at the 2016 MET Gala where actress Claire

Danes wore a Zack Poshen dress incorporating optical fibres. But marker research of current smart textiles products available to customers doesn't come up with as many products as one would expect from the amount of research that is being carried out on the subject. In other words, there is a lot of research in University and Research centre Laboratories but few actual products. Examining those products that are available under "smart textiles", sports applications are prominent. It should be noted that these smart textile sport products are geared towards organized sports or high-end amateurs (e.g. Athos training system, Sensoria tops). Main functions include bio-signal sensors (heartbeat and breathing rate monitors) and accelerometers for posture and movement monitoring. The data collected from the various sensors is aggregated to an electronic system that usually transmits the data to a smartphone or tablet. In these applications, the sensors can and sometimes are textile based and so are the power/ data lines but the communication/ data aggregation modules are electronics that are usually encased in a plastic shell and can be snapped on and off the product. Other commercial applications include health monitors such as baby socks from Owletcare that monitor a baby's bio-signals and the @health system that monitors heart function 24/7. What all these products have in common are the fact that their target groups are very specialized and the solutions used for the integration of the electronics to the textile substrate are the "softest" possible. This means that all the electronics/ power sources are segregated and detachable, the product can sometimes be cleaned with the usual care methods but only in the most gentle of settings (e.g. cold water, immersion only, no spinning/ very light spinning, no dry cleaning). The smart textiles as a widely available consumer product seems still under continuous research and development. It would be worth debating the perceived benefits of investing time and money in researching fully textile electronics when bio-signal monitoring is readily available as a bracelet like device (e.g. Fitbit, Apple watch). The smart textiles must provide corresponding solutions and in the same time must be simple, reliable and competitive to the alternative technologies.

International Scientific Committee

IOP Conf. Series: Materials Science and Engineering **827** (2020) 011001 doi:10.1088/1757-899X/827/1/011001

Organizing Committee

- **Lieva Van LANGENHOVE** University Ghent, Belgium
- **Vladan KONCAR** ENSAIT, France
- **Jacek MLYNAREK** CTT Group, Canada
- **Mohamed LAHLOU** ESITH, Morocco
- **Mustapha HLYAL** ESITH, Morocco
- **Omar CHERKAOUI** ESITH, Morocco
- **Abderrahmane FARHATE** ESITH, Morocco
- **Touria AIT OZZO** ESITH, Morocco
- **Jihane LAHMAR** ESITH, Morocco
- **Hasna RACHADI** ESITH, Morocco
- **Kawtar KOUZMI** ESITH, Morocco
- **Ayoub NADI** ESITH, Morocco
- **Zineb SAMOUH** ESITH, Morocco
- **Ahmed ABED** ESITH, Morocco

Peer review statement

All papers published in this volume of *IOP Conference Series: Materials Science and Engineering* have been peer reviewed through processes administered by the proceedings Editors. Reviews were conducted by expert referees to the professional and scientific standards expected of a proceedings journal published by IOP Publishing.

Oriented membranes processes for facilitated extraction and recovery of some industrial dyes across polymer inclusion membranes containing Chitin as new extractive agent

Y Chaouqi [1, 2,3], M El Bouchti [2], R Ouchn [1,2], Z Habibi[1], O Cherkaoui[2] and M Hlaibi[1]
[1] Laboratoire Génie des Matériaux pour Environnement et Valorisation (GeMEV), Faculté des Sciences Aïn Chock, Université Hassan II, Casablanca, Morocco.
[2] Laboratoire de Recherche sur les Matériaux Textiles (REMTEX), ESITH Casablanca, Morocco.
[3] Author to whom any correspondence should be addressed

chaouqi.youssef@gmail.com

Abstract. In this study, we developed a polymer inclusion membrane (PIM) using the phase inversion technique, based on polymer supports polyvinylidene fluoride (PVDF) and polyvinylpyrrolidone (PVP), with chitin as extractive agent (EA). This PIM was used to study facilitated extraction and recovery processes of Astrazon Red FBL 200%, Blue P3R, Direct Turquoise FBL 400% and BEMACID Yellow N-TF dyes from simulated aqueous textile solutions. To explore the composition, structure and morphology of the developed PIM, both FTIR and SEM techniques were used. To explain the performance of the PIM and describe the mechanism of the process studied for the recovery of dyes from simulated wastewater, kinetic and thermodynamic models were also adopted, which are based on Fick's laws and the saturation law of EA by the substrate (dyes). The overall data indicate that the used EA is effective for theses oriented processes, and the facilitated extraction of dyes is not depending on the pore size but on structures and interactions between the molecules of the EA and the dye.

1.Introduction

To solve the problems related to textile effluents, several studies have investigated the origin of pollution (integrated approach) [1-7] and at the treatment of the final effluents (end-of-pipe approach) [8–10].

In the last few years, the treatment of industrial waste effluents by membrane processes has gained more and more interest because of technical, economical, and political reasons. Technically, the basic technology of process engineering has been developed. Economically, the shortage of water and the increasing cost of auxiliary chemicals and energy have pushed this technology. Politically, the increasing interest of government and people to environmental problems has forced industries to observe severe environmentally safe procedures [11,12].

In the current study, we have elaborated polymer inclusion membranes (PIM) based on polymer supports polyvinylidene fluoride (PVDF) and polyvinyl pyrrolidone (PVP), according to the phase inversion technique, with chitin as extractive agent (EA). These PIM have been used to study facilitated extraction and recovery processes of Astrazon Red FBL 200%, Blue P3R, Direct Turquoise FBL 400% and BEMACID Yellow N-TF dyes from simulated aqueous textile solutions. To explore the composition, structure and morphology of the developed PIM, both FTIR and SEM techniques were used. Kinetic and thermodynamic models were applied [13-18], based on the formation of unstable entity (TS) by the low interactions of the substrate S (dyes) with the immobilized EA (chitin), necessary for the diffusion of the substrate molecules through the membrane phase. These models were used to determine the macroscopic parameters, respectively: initial flux J_0 and permeability P in relation to the performance of the PIM membrane developed for the facilitated extraction process of the dye. Next, Then, we calculated the microscopic parameters, namely the Kass association constants related to the formed entities (S-EA), and the apparent diffusion coefficients D* relating to the diffusion of the substrate S through the membrane phase of the adopted PIM.

2. Materials and methods

The membrane was prepared by phase inversion method, the mixture of PVDF/PVP was completely dissolved in Dimethylformamide as solvent and Chitin as extractive agent and as the water bath step of phase inversion at room temperature.

Extraction experiments were carried out in a cell that contains two compartments, the source phase and the receiving phase, the system is immersed in a thermostatic bath and agitation provided by a multi station magnetic stirring, to follow the extraction evolution of dyes substrate, we took samples from the receiving phase as a function of time, and we used *THERMO SCIENTIFIC* UV adsorption spectroscopy to analyze the samples [13,14].

3. Results and discussion

3.1 Characterization of the membrane:

The Figure 1 shows the infrared spectra of the PVDF/PVP polymer substrate, before/after the addition of EA, the appearance of two new peaks at 1653 cm^{-1} and at 3996 cm^{-1} in the spectrum of the PVDF+PVP+Chitin membrane are attributed respectively to the vibration of C =O carbonyl group of the ester function, and the vibration of O-H groups. Those new peaks indicate that the Chitin extractive agent is inserted correctly in the PVDF+PVP support.

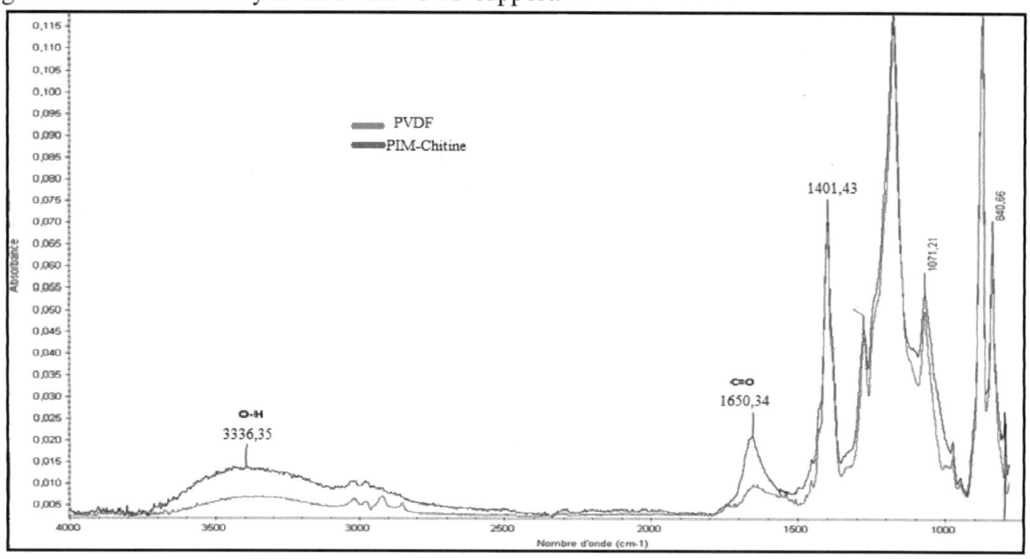

Figure 1. Fourier transform infrared-attenuated total reflection spectra of the membranes, PVDF support and PIM-Chitin.

The SEM image in figure 2 show an asymmetric membranes, and there are homogeneous and contain cavities of different sizes, communicative through pores.

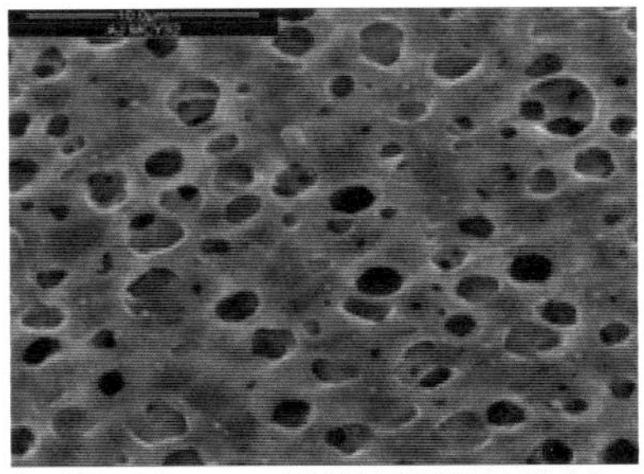

Figure 2. SEM image: PVDF/PVP-Chitin surface.

3.2 Macroscopic and microscopic parameters:

To optimize dyes extraction conditions, various factors affecting the facilitated extraction of the dyes from wastewater, such as the polymer support, the nature and the concentration of the extractive agent (AE), the initial concentration of the substrate, the pH and the temperature of the aqueous phase. All macroscopic (P and J_0) and microscopic (K_{ass} and D^*) parameters are calculated with the thermodynamic model developed by our laboratory [13-18].

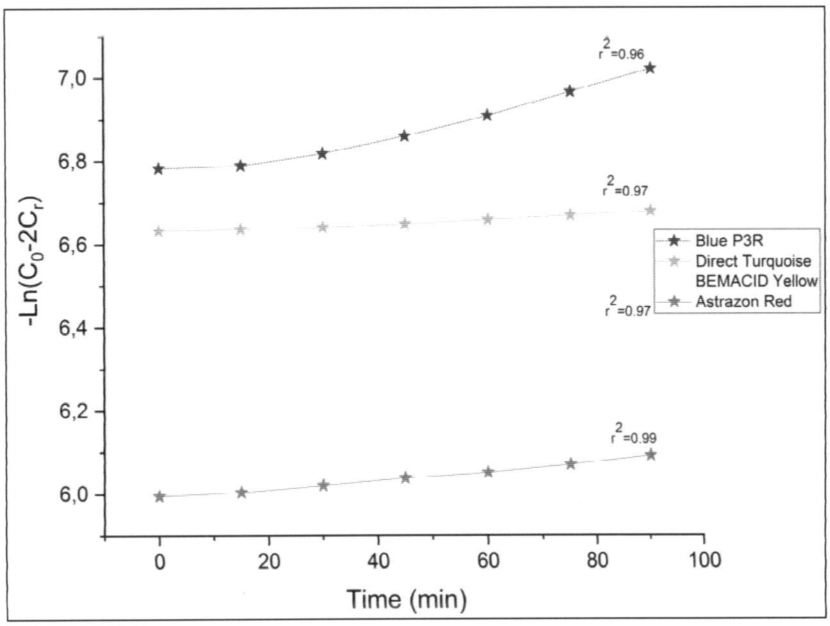

Figure 3. Evolution of the kinetic function -Ln($\mathbf{C_0}$-$\mathbf{2C_r}$) = f(**t**) for the facilitated extraction of Blue P3R dye at different initial substrate concentrations.

$[\text{Chitin}]_0$ = 0.0007035 M, pH = 4, and T = 298 K

Table 1. Effect of initial concentration C_0 on the evolution of P and J_0 parameters for the facilitated extraction process of dyes

Substrate	C_0 (mol L^{-1})	$P\times10^6$ (cm^2 s^{-1})	$J_0\times10^5$ (mmol cm^2 s^{-1})
Blue P3R	0.00113	5.66	0.022
Direct Turquoise FBL 400%	0.00132	0.26	0.0073
BEMACID Yellow N-TF	0.00132	0.103	0.0029
Astrazon Red FBL 200%,	0.0025	1.09	0.0028
pH = 4, and T = 298 K, PIM=PVDF/PVP+Chitin			

The results show a very good performance of the developed membrane for the oriented process of the facilitated extraction of studied dyes.

4. Conclusion

In this work, we prepared by phase inversion method a new PIM membrane based on PVDF+PVP as supports and Chitin as extractive agent. We characterized the obtained membrane by SEM and IR, and we found that insertion of the extractive agent Chitin in the membrane phase of the PIM allows the passage of dyes. On the other hand, we have verified our models and we were able to calculate all parameters related to facilitated extraction process of dyes through the elaborated PIM membrane.

Finally, we have elucidated a mechanism by successive jumps on fixed sites for the studied process through this PIM membrane type, and we can conclude that the elaborated membrane is efficient for a possible application relating to the extraction and the recovery of the studied dyes from the textile effluents.

Acknowledgments

The works in this manuscript are conducted within the framework of a PPR2 project, funded by the Ministry of Higher Education and Scientific Research (MESRSFC) and the National Center of Scientific and Technical Research (CNRST). We are grateful to both organizations for their financial support.

References

[1] Contreras S, Pascual E, Esplugas S, Gim J, Rodrı M 2002 Comparison of different advanced oxidation processes for phenol degradation. *Water Res.* **36** pp.1034–42.

[2] Ghaly AE, Ananthashankar R, Alhattab M, Ramakrishnan V 2014 Chemical Engineering & Process Technology Production , Characterization and Treatment of Textile Effluents : A Critical Review. *J Chem Eng Process Technol.* **5** pp.1–18.

[3] Katheresan V, Kansedo J, Lau SY 2018 Efficiency of Various Recent Wastewater Dye Removal Methods: A Review, *Biochem Pharmacol.*

[4] Al-alwani MAM, Ludin NA, Bakar A, Amir A, Kadhum H, Mukhlus A 2018 Molecular and Biomolecular Spectroscopy Application of dyes extracted from Alternanthera dentata leaves and Musa acuminata bracts as natural sensitizers for dye-sensitized solar cells. Spectrochim Acta Part A *Mol Biomol Spectrosc* **192** pp.487–98.

[5] Liu L, Zhang J, Tan Y, Jiang Y, Hu M, Li S, et al 2014 Rapid decolorization of anthraquinone and triphenylmethane dye using chloroperoxidase : Catalytic mechanism , analysis of products and degradation route *Chem Eng J.* **244** pp.9–18.

[6] Hubbe MA, Hasan SH, Ducoste JJ 2011 Cellulosic Substrates for Removal of Pollutants *BioResources* **6** pp.2161–287.

[7] Lopez A, Ricco G, Ciannarella R, Rozzi A, Di Pinto AC, Passino R 1999 Textile wastewater

reuse: Ozonation of membrane concentrated secondary effluent Water *Sci Technol* **40** pp.99–105.

[8] Forgacs E, Cserháti T, Oros G 2004 Removal of synthetic dyes from wastewaters: A review. *Environ Int.* **30** pp.953–71.

[9] Robinson T, McMullan G, Marchant R, Nigam P 2001 Remediation of dyes in textile effluent: A critical review on current treatment technologies with a proposed alternative *Bioresour Technol.* **77** pp.247–55.

[10] Przystaś W, Zabłocka-Godlewska E, Grabińska-Sota E 2012 Biological Removal of Azo and Triphenylmethane Dyes and Toxicity of Process By-Products. *Water, Air, Soil Pollut.* **223**(4) pp.1581–92.

[11] Askari N, Farhadian M, Razmjou A 2018 Environmental Nanotechnology , Monitoring & Management Simultaneous e ff ects of pH , concentration , pressure on dye removal by a polyamide nano fi lter membrane ; optimization through response surface methodology. *Environ Nanotechnology, Monit Manag* **10** pp.223–30.

[12] Abid MF, Zablouk MA, Abid-alameer AM 2012 Experimental study of dye removal from industrial wastewater by membrane technologies of reverse osmosis and nanofiltration. *Iranian J Environ Health Sci Eng.* **9**(1) p.17.

[13] Kamal O, Eljaddi T, Atmani EH EL, Touarssi I, Lebrun L, Hlaïbi M 2016 Grafted polymer membranes with extractive agents for the extraction process of VO 2 + ions *Polym Adv Technol.* **28** pp.541–48.

[14] El Atmani EH, Benelyamani A, Mouadili H, Tarhouchi S, Majid S, Touaj K, et al 2018 The oriented processes for extraction and recovery of paracetamol compound across different affinity polymer membranes. Parameters and mechanisms *Eur J Pharm Biopharm.* **126** pp.201–10.

[15] Hlaïbi M, Tbeur N, Benjjar A, Kamal O, Lebrun L 2011 Carbohydrate-resorcinarene complexes involved in the facilitated transport of alditols across a supported liquid membrane *J Memb Sci.* **377** pp.231–40.

[16] Benjjar A, Eljaddi T, Kamal O, Touaj K, Lebrun L, Hlaibi M 2014 The development of new supported liquid membranes (SLMs) with agents: Methyl cholate and resorcinarene as carriers for the removal of dichromate ions (Cr2O72−) *J Environ Chem Eng.* **2** pp.503–9.

[17] Hor M, Riad A, Benjjar A, Lebrun L, Hlaïbi M 2010 Technique of supported liquid membranes (SLMs) for the facilitated transport of vanadium ions (VO2+) Parameters and mechanism on the transport process *Desalination* **255**(1-3) pp.188–95.

[18] Chaouqi Y, Ouchn R, Eljaddi T, Jada A, El bouchti M, Cherkaoui O and Hlaibi M 2019 Oriented processes for extraction and recovery of blue P3R dye across hybrid polymer membranes: parameters and mechanism *Journal of Membrane Science and Research* **5** pp.303–9.

IOP Conf. Series: Materials Science and Engineering **827** (2020) 012002 doi:10.1088/1757-899X/827/1/012002

Functional shoe for the detection of walking pattern anomalies

B Malengier[1,5], G B Tseghai[1,2], M C Ciocci[3], D A Mengistie[4] and L Van Langenhove[1]

[1] Department of Materials, Textiles and Chemical Engineering - Centre for Textile Science and Engineering, Faculty of Engineering and Architecture, Ghent University, Ghent,Belgium
[2] Ethiopian Institute of Textile and Fashion Technology, Bahir Dar University, Bahir Dar,Ethiopia
[3] Department of Mathematics: Analysis, Logic, and Discrete Mathematics, Faculty of Sciences,GhentUniversity,Belgium
[4] Division of Organic electronics, Linköping University, Sweden
[5] Author to whom any correspondence should be addressed

Benny.Malengier@UGent.be

Abstract. Analysis of walking patterns can play an important role in the diagnosis of musculoskeletal disorder and detecting anomalies in walking gaits. In this work, we introduce a systematic approach to detect person's walking patterns. A flexible resistive pressure sensor, developed from electro-conductive textile fabric, is non-intrusively integrated in an ordinary shoe together with a time of flight height sensor. The constructed shoe detects both the pressure between shoe and foot and the gap between shoe and ground. The combination of those give a trace of the walking pattern. The shoe should be functional in detecting walking pattern anomalies.

1. Introduction

Several diseases are characterized by problems with walking. One of these is Parkinson's disease (PD), which is a disabling pathology affecting millions of people of all races and cultures globally. The characteristic motor features of the disease include bradykinesia (i.e. slowness of movement), tremor, rigidity (i.e. resistance to externally imposed movements), flexed posture, postural instability and freezing of gait [1]. The primary biochemical abnormality in PD is a deficiency of dopamine due to degeneration of neurons in the *substantia nigra* [2].

For some Parkinson's patients, the lack of dopamine may cause irregular and asymmetrical walking patterns. These are caused by the tailing behind of one leg due to unequal intensity of the neurological signals. In the worst-case scenario, the patient trips and falls as a consequence of the irregular walking pattern. PD affects about 1% of the population over the age of 60 years [3], a figure that is expected to double by 2020 [4].

To detect the irregularity, a 'smart shoe insole' was designed, including sensors capable of analyzing the walking pattern of the wearer. In case of pattern anomaly, the shoe can warn the wearer and further prototyping can show how actuators could be integrated to prevent falling. An alternative application for the shoe could be aiding in avoiding toe-walking in children or adults, by giving an appropriate feedback to help remember to walk in a flat-footed fashion [5]. Such feedback could be for example a vibrating motor, which has already been used to notify people of bad sleeping habits [6]. In figure 1 an example of the difference between a regular and an irregular walking pattern is given. Detecting the irregularity of the step can augment other wearable sensors used for diagnosis and treatment [7].

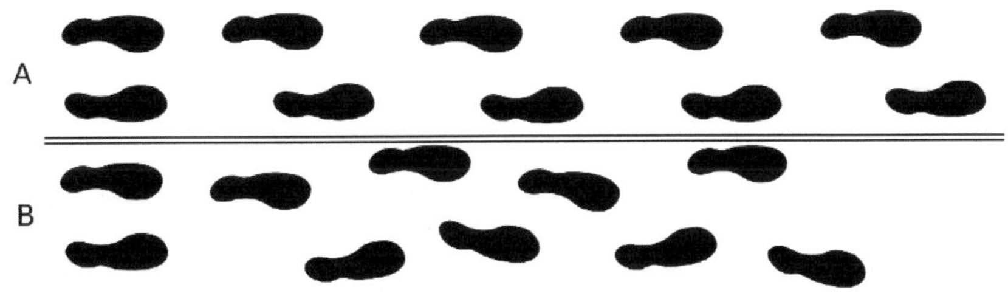

Figure 1. Comparison of a regular walking pattern (top) with an irregular walking pattern (bottom).

2. Material and Methods

Different ways can be used to evaluate a person's walking pattern. In this project, two sensors are considered for integration into a shoe: a textile pressure sensor to detect when the foot exerts pressure onto the shoe, and a distance sensor to measure the distance between the shoe and the ground or the distance to the opposite leg while walking. The registered data provide insight into the wearer's walking pattern, and a baseline for benchmarking can be set up and hence anomalies can be detected. Extra useful information can be collected by an accelerometer which is readily available in several wearables.

Standard thin pressure sensors are common now, but not textile based. They are typically thin film pressure sensors [8], and though easy to use, not ideal to combine with textile material.

A flexible textile pressure sensor was constructed in several ways. A first sensor is constructed by separating two conductive, stainless steel (SS) fabrics (the electrodes) with a non-conductive barrier fabric, and connecting electrical wires to the stainless-steel fabrics by soldering them via connectors as previously developed by some of the authors [9]. This design concept and working principle is shown in figure 2. When enough pressure is applied to the top of this system, surface fibres of the conductive fabric are pushed through the barrier fabric, contacting the conductive layer on the other side. This closes a circuit and thus allows electricity to flow, which can be detected by the electronics connected to the circuit.

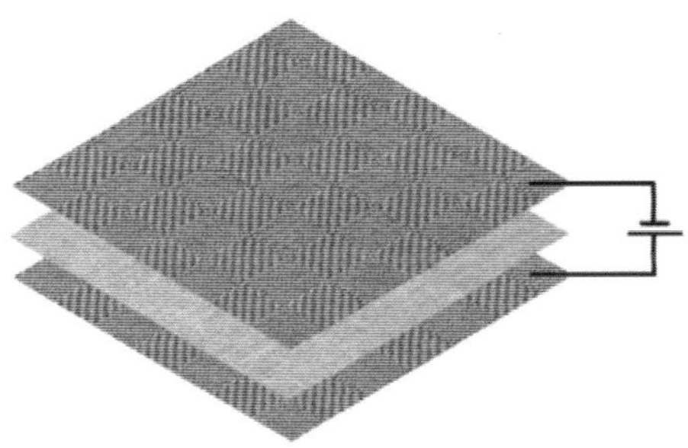

Figure 2. Concept of the pressure sensor: a non-conductive fabric separates the two electrodes which are conductive fabric.

The second and third sensor considered use the same principle but use a commercial non-woven treated with Polypyrrole (PPY-E) of Eeonyx, which acts as a variable resistance sensor under pressure. As this material is difficult to calibrate, using it in the tree layer setup improves the detection as the middle layer creates a binary sensor: it detects pressure only above a certain threshold. If the weight that needs to be detected is larger, a thicker barrier material can be used. PPY-E and PPY-E are used in sensor 2 as electrodes, while in sensor 3 one electrode is SS while the other is E. Moreover, in sensor 3, the barrier material is cotton lightly coated with PPY, see figure 3.

Figure 3. Pressure sensors. Electrodes are, left: SS and SS, middle: E and E, right: SS and E.

Walking creates two main pressure points in a shoe: under the heel where the pressure is highest when the shoe first contacts the ground, and under the ball of the foot where the pressure is highest just before the shoe leaves the ground. For ease of construction the pressure fabric was placed in the heel of the shoe.

As a secondary sensor, a standard distance sensor, was used. We selected the VL53L0X time-of-flight distance sensor. This sensor is laser based and detects distance to surfaces by detecting the reflected wave. We considered two possible manner of placement: directed to the ground or directed towards the opposite leg to detect when it passes.

After a few iterations and analysis of the results we built the final prototype using sensor 3 (SS and E), and with the distance measured as downward as possible. A Wemos D1 mini board is used as data processing unit to read out the sensors. This connects via wifi to a base station, where the results are collected. The pressure sensor is attached to insoles which can be removed and placed in other shoes. Its value is read out with the ADC circuit using a pulldown resistor of 820 kOhm, resulting in ADC values in the range [0, 1023]. A textile connector is used between the insoles and the Wemos board, which is detachable for ease of use. The resulting prototype is given in figure 4.

A drawback of sensors in shoes are the limited place to put the data processing unit and the power source it requires. An alternative solution is to detect walking patterns using accelerometer data from another place on the body, as wearable step counters do attached to the wrist. Ideally for this a place close to the skeleton is used with limited noise from not related movements (like hand waving on the wrist). We put the accelerometer in a belt-bag on the waist as close as possible to the hip bone. We collected data from the accelerometer and analyzed how gait measured via the functional shoe relates to these data. For simplicity a BBC micro:bit was used as it contains an excellent accelerometer.

3. Results and Discussion

The data are validated and used to determine walking pattern irregularity to determine if the approach is viable. The stored ADC values of the textile pressure sensor, and the distance values to the ground are recorded for use in walking pattern analysis. Example processed output is given in figure 5. For normal walking every heel down is clearly registered, resulting in a very good step detection, and allowing steps per minute to be determined. The irregular gait however shows a much more difficult pattern to analyze and extract step data from, however, the clear difference from a normal step can be easily determined. Toe walking on the other hand is clear by a lack of high pressure points.

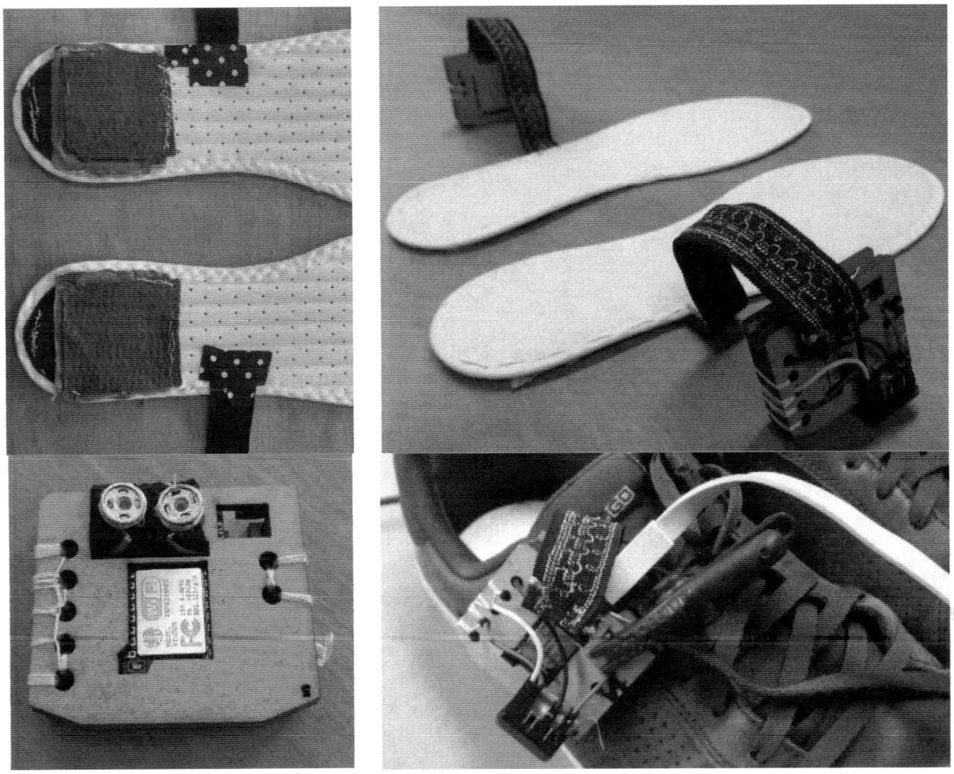

Figure 4. Prototype functional shoe. Left Top: textile pressure sensors at the base of the insole, Right Top: full layout with Wemos board on the side of the shoe, and distance sensor pointing downward, Left Bottom: Wemos board detached, Right Bottom: placement on a shoe for testing.

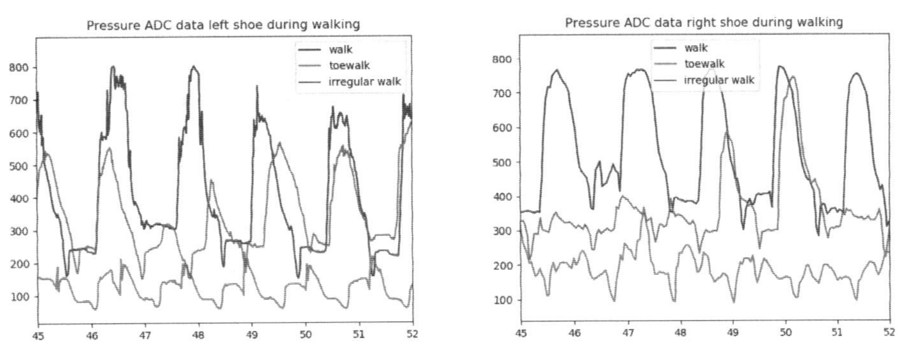

Figure 5. ADC pressure output of left shoe and right shoe showing normal regular walking result, toe walking result and example of irregular gait.

With the data from the left and right shoe, we can clearly determine if the duration of left-foot-down to right-foot-down is equal to the duration of right-foot-down to left-foot-down. Test groups can be set up to compare the measurements from healthy people to those of PD patients in order to validate detection of irregular gait by this method of measurement.

More complex situations would be hard to determine however. Combination with a height sensor has a clear advantage that gaits can be more easily characterized, as is clear from figure 6.

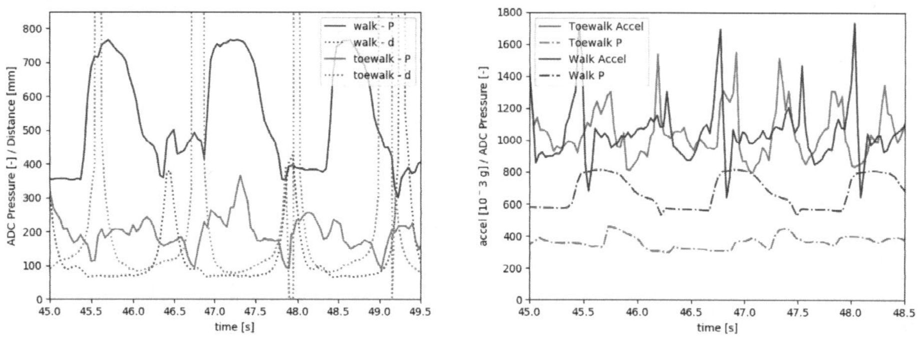

Figure 6. Left: Comparison of pressure data and height measurement at the right shoe. Right: Comparison of acceleration data taken at the right hip and pressure data at the right shoe.

From figure 6 left, we see that first high pressure is detected in normal gait when the shoe is down, while on the lift a peak is found in the distance to the ground. Here it must be noted that due to the rotation of the ankle the height sensor might point to the horizon, leading to infinite distance measure. The height measurement adds little to the pressure data. However, for toe walking where the pressure on the heel hardly changes during walking, the height sensor allows to clearly distinguish steps, augmenting the obtained data from the heel pressure.

In figure 6 right, we compare with acceleration data measured at the right hip. This data is very noisy and would benefit from a smoothing step. Nevertheless, comparing with the pressure data from the heel, it is clear that the peak in acceleration corresponds with the heel down. The following smaller peak in the acceleration data corresponds with the left heel down, which is clearly visible in the acceleration data but almost not present as a feature in the pressure data of the right heel. Again, while the pressure data of the heel makes distinguishing steps for toe-walking almost impossible, the acceleration data shows clear peaks for every foot down, with both left foot down and right foot down peak almost equally large, which was not the case for normal walking.

Our goal of characterizing a 'standard walking pattern' seems feasible using a combination of sensors, though only using one of the three sensors presented should already suffice to know if deviation from the standard walking pattern occurs. Using a textile pressure sensor shows no drawbacks, response times are sufficiently fast and the measurement sufficiently accurate. Next, the data can be further processed. The anomalies would then be detected by software methods. Frequency monitoring would play a big role: how often the wearer steps down onto the ground is the most important factor for an irregular walking pattern, rather than the actual pressure exerted by the foot, or the height the foot is lifted to.

4. Conclusion

We showed that a fabric pressure sensor can be used in step frequency detection, with the data fully matching a distance sensor, and allowing to augment accelerometer data taken at waist height. A further improvement would be to actuate on encountered problems. This would require a constant and real-time comparison of the walking pattern with the aforementioned 'standard walking pattern' by the processing unit. Any significant deviation from this standard pattern would then result in an actuating function of the shoe. Examples of such functions would be the tightening of the shoe grip, vibrations, giving electrical impulses using integrated electrodes that contact the wearer's skin or an app that displays the walking pattern or risk of falling on a smartphone for easy monitoring. Using those actuating functions, the patient would be warned, enabling them to adapt their pace and prevent later problems.

Acknowledgment
The authors would like to express appreciation for the support of the Erasmus+ project Skills4Smartex. This project was funded with support from the European Commission. This publication reflects the views only of the author, and the Commission cannot be held responsible for any use which may be made of the information contained therein.

References

[1] Ferraye M U, Fraix V, Pollak P, Bloem B R, Debû B 2016. The laser-shoe: A new form of continuous ambulatory cueing for patients with Parkinson's disease *Parkinsonism and Related Disorders* **29** pp 127–128

[2] Mostafa S A, Mustapha A, Abed M, Ibraheem R, Arunkumar N, Khanapi M, Ghani A 2019. Examining multiple feature evaluation and classification methods for improving the diagnosis of Parkinson's disease. *Cognitive Systems Research* **54** pp 90–99

[3] Hagell P 2019 Measuring activities of daily living in Parkinson's disease : On a road to nowhere and back again? *Measurement*, **132** pp 109–124

[4] Noyes K, Liu H, Li Y, Holloway R, Dick A W 2006 Economic Burden Associated With Parkinson's Disease on Elderly *Medicare Beneficiaries* **21**(3) pp 362–372

[5] Hirsch G, Wagner B 2004 The natural history of idiopathic toe- walking: a long- term follow-up of fourteen conservatively treated children. *Acta Paediatrica*, **96**(2) pp 196-199

[6] Kahn T 2019 A deep learning model for snoring detection and vibration notification using a smart wearable gadget *Electronics* **8**(9) pp 987

[7] Rovini E, Maremmani C, Cavallo F 2017 How Wearable Sensors Can Support Parkinson's Disease Diagnosis and Treatment: A Systematic Review *Frontiers in neuroscience* **11** p 555

[8] Hollinger A, Wanderley M 2011 Evaluation of Commercial Force-Sensing Resistors *Proc. NIME06*

[9] Tseghai G B, Malengier B, Nigusse A B, Van Langenhove L 2018 Development and evaluation of resistive pressure sensors from electro-conductive textile fabric *Proceedings of the Second International Forum on Textiles for Graduate Students* pp 651-657

Development of super-hydrophobic and stain repellent fabric finish

R Pallye[1] and S Ghosh[2,3]

[1] Ph. D. Fellow, College of Technology, Eastern Michigan University, Michigan, USA
[2] Professor, College of Engineering and Technology, Eastern Michigan University, Michigan, USA
[3] Author to whom any correspondence should be addressed

sghosh@emich.edu

Abstract. This study intended to develop a health and environmental friendly super-hydrophobic textile fabric using a specific Fluoro Silane (Fs) finish. A novel Fs finish was prepared and applied on an acrylic fabric using a pad-dry-cure method. The finished fabric was evaluated for the degree of hydrophobicity, durability and stain repellency. The finished fabric exhibited static water contact angle greater than 170° and received 90 AATCC (4 ISO) rating that is recognized as super-hydrophobicity and this property was maintained even after a 50,000 cycle abrasion test. FTIR analysis identified the characteristic peaks related Si-O-Si and C-F asymmetric stretching bands of the finish on the fabric indicating a robust attachment of the finish on the fabric. Finish fabric did not show any change in appearance or tactile characteristics of the fabric.
Keywords: Acrylic fabric; Contact Angle; Fluoro silane; Super hydrophobic; Stain repellant

1. Introduction

Super-hydrophobic fabric technology evolved from biomimetic technology. This was derived from the lotus leaf phenomenon as researched by many scientists. In the Greek words, "hydro" means water, "philicity" means affinity, and "phobicity" means lack of affinity. So, the term "Super-hydrophobic" means extremely water repellant. When the static water contact angle θ is >150°, the surface is super-hydrophobic [1]. Under such condition, the water drop forms a spherical shape achieving minimum surface area because of the cohesive forces between water molecules. As the water drops fall on the surface, they effortlessly roll off the surface without wetting it at all. Because of very low surface energy of the super-hydrophobic surface, loosely held dirt and soils get easily attached to the rolling drops and are removed. The lotus leaf is one of the best-known natural super hydrophobic surfaces that effectively removes mud as a water drop flows over the surface. A Nano-level hydrophobic natural wax crystal was found to be present on the top of the micro bumps on a lotus leaf, which makes the lotus leaf a self-cleaning as well as strongly hydrophobic [2]. In order to produce super hydrophobic textiles, researchers have used several techniques. *Kawai and H. Nagata* [3] have used soft lithography, X-Ray lithography, electron beam lithography, Nano sphere lithography and photolithography to develop super hydrophobic surfaces. *Zhang et al* [4] and *Chen et al* [5] used the same technique by creating silicon Nano-pillar arrays. A Phase separation technique was used to prepare Super- hydrophobic surfaces. *Polizos et al* [6] obtained a hierarchical porous structure through interconnection of different dimensions of pores. They used Polyethylene glycol (PEG) and Polydimethyl siloxane (PDMS) in 1:1 ratio in the phase separation, to create a surface that had a water contact angle of 160°. *Bao et al* [7] used chemical vapor deposition method, where a substrate was created in the form of non-volatile film due to reaction of the gaseous reactants. Metal oxide nanoparticles like ZnO, Al_2O_3 and Fe_3O_4 etc., were used to fabricate super hydrophobic surfaces. Several researchers have used sol-gel and other chemical to produce super-hydrophobic surfaces. *Sahoo and Kandasubramanian* [8] used organic, inorganic materials and metal alkoxides for sol-gel techniques. *Hoefnagels et al.* [9] produced super-hydrophobic cotton fabric using silica particles. They reported that WCA was 155°. *Ivanova et al.* [10] fabricated super-hydrophobic cotton fabric by altering the fabric structure using Fluoro silane (Fs) material. *Xue et al* [11] used amino and epoxy functionalized silica nanoparticles along with stearic acid and 1H, 1H, 2H, 2H-perfluorodecyl-trichlorosilane on cotton fabric to create super-hydrophobic characteristics and that they claimed was 170^0 WCA. *Meng et al.*, [12] *Das and De, Meng et al.* [13] used long chain fluorinated compounds on cotton fabric to create super-hydrophobic characteristics.

The objective of this study was to develop a Super-Hydrophobic, stain repellant fabric finish synthesizing a fluoro-silane compound and apply to acrylic fabrics to create super-hydrophobic properties. However, this finish can be applied to polyester, cotton, wool, nylon, silk and blend fabrics. This silane has been rated as 0 health hazard and 0 physical hazard by US HMIS and NFPA agencies [14] and is extremely durable. The process is straight forward and commercialization friendly.

2. Experimental Methods

2.1 *Materials*

This finish was applied on plain weave fabric produced from acrylonitrile and 4-5% Methyl-methacrylate as copolymer. Fabric weight was 8.24 oz. per sq. yd. having construction 25 x 32 (ends and picks per inch).

2.2. *Finish preparation and application*

Reagent-grade solvent butanol, Disodium hydrogen phosphate, 72% Sulphuric Acid were obtained from Sigma Aldrich. A specific Fluoro silane (Fs) was obtained from Matrix Scientific. All chemicals were used without further purification. Pre-determined amount of Fluoro silane was added to the reagent-grade solvent butanol and stirred continuously at room temperature at 350 RPM. A small amount of the catalyst, disodium hydrogen phosphate, was added to facilitate the addition of the finish to the prepared acrylic specimens. A few drops of Sulphuric acid were added slowly to the mixture while stirring to maintain the very low pH of 2, to control the condensation reaction. The required amount of additional butanol was added and stirring was continued. The solution was stirred continuously for one hour and was placed in the refrigerator for twenty-four hours prior to application. Care was taken to evade rapid condensation reaction and precipitation by carefully monitoring pH and condensation time. Finish was applied on to the acrylic fabric sample using a pad-dry-cure method. Prior to the padding process, finish was force sprayed on the fabric to ensure even application on the fabric surface. Treated fabric was dried at 60° C for15 minutes and then cured at 165°-170° C for 15 minutes in an oven. A schematic diagram of the application process is illustrated in Figure.1

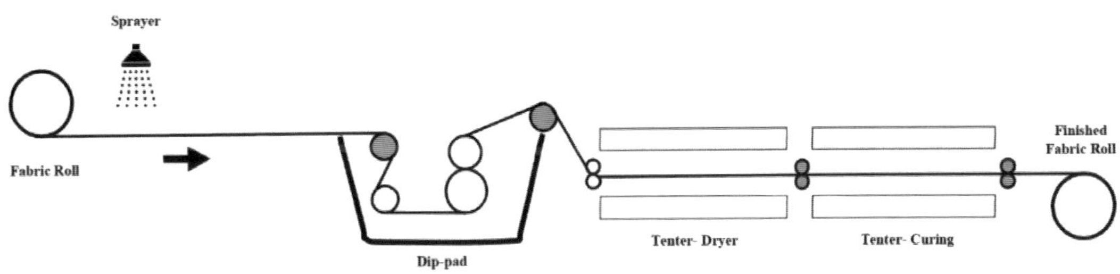

Figure 1. The Super-hydrophobic finish application process.

2.3. *Water Contact Angle (WAC) measurement*

The Static Contact Angle Measurement is one of the measures of hydrophobicity. The Device by First Ten Angstrom (FTA-200) was used for measuring the static contact angle. Images for contact angle measurement were captured using Amscope MU853B 14MP high-speed digital camera, attached to the device set up. Five readings were taken from different areas of the same sample and analyzed by the computer plugin Low Bond Asymmetric Drop-Shape Analysis (LBDSA) [15] available with the open-sourced JAVA application Image R. In the LBADSA method, the theoretical profile of the drop is not fitted to a certain drop contour, instead optimized according to an image energy approach. The measurement is based on perturbation solution of the axisymmetric Laplace equation. During the fitting process, complete pixel information is used. The use of this global model result in highly accurate contact angles and has an edge over the other methods, when a precise accurate contour detection is difficult because of un-sharp or noisy boundaries (as observed in the textile fabrics).

2.4. *Water Repellency: Spray test*

AATCC Test Method 22 (ISO 4920) was used to determine the degree of hydrophobicity of the fabric after the application of the finish. In this method, water is sprayed against the taut surface of a test specimen, under controlled conditions, that produces a wetted pattern whose size depends on the relative repellency of the fabric. The evaluation is accomplished by comparing the wetted pattern with a given standard chart. Rating ranges from 0 to 100 according to AATCC where a 0 to 5 ratting is given according to ISO. A 100 (5) rating means that there was no sticking or wetting of the upper surface, while a 0 rating indicates complete wetting of the whole upper surface.

2.5 *Subjective Visual Evaluation*

Industries generally follow a popular subjective visual observation method for evaluating super-hydrophobic behavior of fabrics. It consists of putting drops of water on the surface of the fabric and observing whether the water drops adhere to and wet the surface of the fabric or not. In this process, water drops poured on to the fabric and it is observed whether the drops take stable spherical shapes, and whether the drops roll over the surface without any residue when the fabric surface is tilted or moved.

2.6 *Abrasion resistance*

The durability of the finish was measured using 50,000 cycles by ASTM method D4966-12 on a Martindale Abrasion tester. Before and after abrasion, the specimens' weight was measured and any change in contact angles (WCA) was evaluated.

2.7 *Fourier Transform Infrared (FTIR) Spectroscopic Analysis*

Attachment of the Fluoro Silane on the acrylonitrile fabric was examined using FTIR spectroscopy. Transmission scans of the fabric before and after the super-hydrophobic finish treatment were produced. Expected bands (peaks) arising from the application of the finish were evaluated. Furthermore, the spectra before and after the abrasion were analyzed to evaluate the durability of the finish on the fabric sample. Secondary derivative transformation of spectra was performed, using a Perkin Elmer software to bring them to a common baseline and to remove some of the noise that may arise from the art effects. This method facilitates the comparison between two or more spectra. The first derivative of a spectrum is simply the measure of the slope of the spectral curve at every point. The slope of the curve; however, is not affected by the baseline offsets in the spectra. Second derivative ignores the offset and it is not affected by any linear "tilt" that may exist in the data. As a result, second derivative spectrum is a very effective method for removing both the base line offset and the slope from the spectrum [16]. It has been observed that second derivative spectra allow us to better understand the overlapping broad bands that often exist in complex material like polymeric fibers spectra. From the second derivative spectra, appropriate bands' peak is evaluated.

2.8. *Stain Repellency Evaluation*

Stain repellency test against ketchup, mustard, and jam were performed on the fabric surface separately. The test specimens were left for 15 minutes after the application of the staining substances. Then, the fabric was rinsed with normal tap water and dried. The post-washing samples were rated for stain using AATCC Test - 130 Stain Release Replica where 5 is no stain and 0 is maximum staining of the specimens.

3. Results and Discussion

Acrylic fabric specimens were treated with Fluoro silane finish and cured at an elevated temperature. Acid group present in the copolymer were expected to get attached with the silane through a condensation reaction under acidic condition. The fabric containing COO- groups was immersed in the Fluoro silane solution containing triethoxy silane. As a consequence, plenty of silane groups reacted with COO- groups on the substrate followed by curing which resulted in a robust attachment of finish on the fabric surface.

Fluorinated methyl groups are less reactive and possess lower surface energy compared to the usual hydrophobic -CH_3 groups, owing to the presence of very strong C-F bonds [17, 18]. Consequently, this finish is expected to provide very high hydrophobicity while it does not attach to the soil strongly. It is important to consider that silica based, and other hydrophobic coatings are not stable enough for day - to-day commercial uses [12]. This Fluorosilane finish; however, is very sustainable and durable for practical uses. This fluorine complex was chosen because US health and environmental agencies like OSHA and NFPA rated the compound as 0 for health and physical effects [14].

WCA is considered a good indicator of hydrophobicity of a fabric. The fabric is recognized as super hydrophobic, when WCA exceeds 150°. In this investigation, treated acrylic fabric was tested for static contact angle at 5 different locations. As seen in Figure 2, WAC values of treated fabric varies from177.5 to 175.4 degrees which is significantly greater than 150° confirming a very high level of super-hydrophobicity of the test specimens. Another practical super-hydrophobicity test on these Fs treated acrylic specimens was carried out using AATCC 22 spray test that is most commonly used by the industry. In this study, Fs treated specimens received >=90 rating where a rating of >=80 is considered as a super hydrophobic substrate. After 15 minutes, the specimen was tilted, and it was observed that the liquid drops rolled off the fabric leaving the specimen fabric completely dry and without any stain.

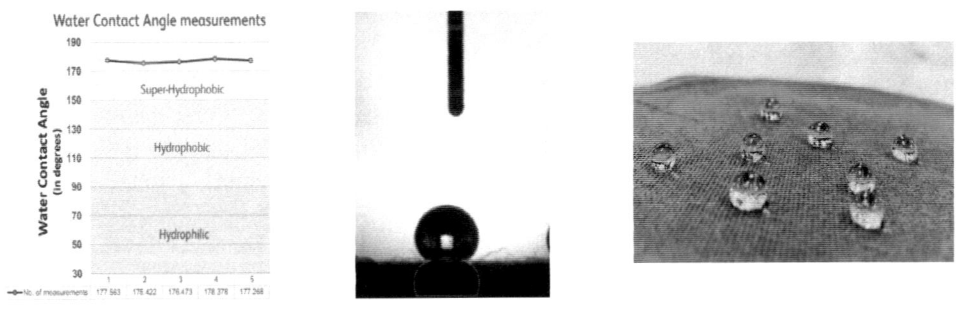

Figure 2. Contact angle values of the Fs treated test specimens and an illustration of the water drop on the fabric specimen and the liquid drops on the treated specimen after 15 minutes.

FTIR study was performed to understand the chemical bonding of Fs finish on the acrylic fabrics. Second derivative of the transmission spectra of Fs treated, and untreated fabric specimen is shown in Figure 3. The peaks that appeared at 724 $cm^{-1}$and at 996 cm^{-1} are associated with C-F bond [19] found on the Fs treated specimen which was not seen on the untreated acrylic fabric. The peaks at 852 cm^{-1} and at 1094 cm^{-1} arising from asymmetric stretching vibration of Si-O-Si [18] were found with high intensity on the treated fabric spectra. These bands on the treated samples support that the Fs finish attachment on the acrylic fabric.

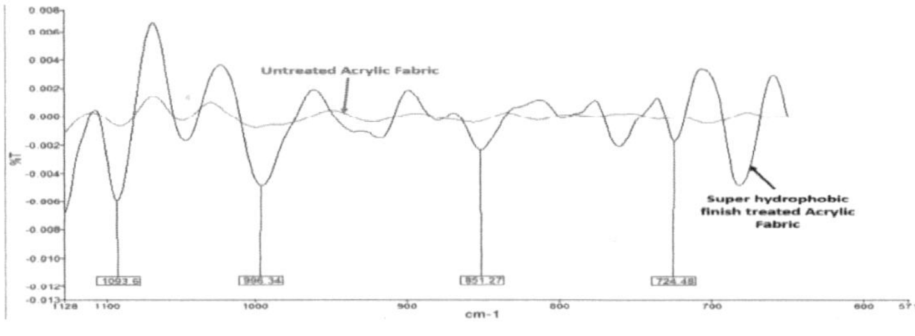

Figure 3. The 2nd Derivative FTIR spectra of the untreated and treated with Fs finish fabrics.

The durability of the Fs finish on the fabric was determined by abrasion testing using 50,000 cycles of abrasion on the fabric surface on a Martindale Abrasion Tester. Post 50,000 cycles of abrasion fabric did not show any changes in surface appearance and had no loss in super hydrophobic finish applied to the fabric. The weight loss percent was 0.22%, which is attributed to insignificant amount of unattached finish. The abraded fabric did not show any changes in surface appearance and in tactile (hand) properties.

FTIR scan, Figure 4 exhibited the same peaks at 724 cm^{-1}, 852 cm^{-1}, 996 cm^{-1} and 1094 cm^{-1} associated with the Fs finish on the fabric before and after abrasion.

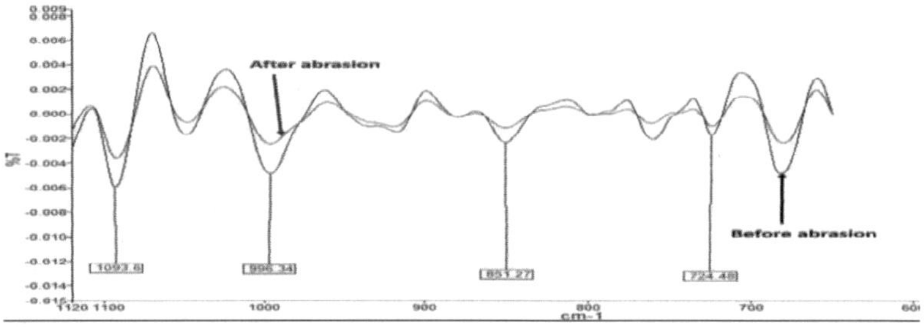

Figure 4. FTIR scan of the Fs finish treated fabric before and after abrasion.

The contact angle of the specimen remained the same value (above 170°) before and after abrasion. So the contact angle didn't change before and after 50,000 cycles of abrasion.

For the stain repellency test, ketchup, mustard, and jam were applied to the fabric surface separately and left for 15 minutes. Then, the fabric was rinsed with normal tap water and dried. No residue and stain was observed on the washed fabric; therefore, a 5 rating was given according to the Stain Release Replica AATCC Test 130.

4. Conclusion

This study developed a Super-hydrophobic acrylic fabric using a specific Fs finish. The treated fabric did not show any changes in the surface appearance and in tactile property. Experimental data showed that super-hydrophobic property of the fabric was maintained even after undergoing 50,000 abrasion cycles. The spherical drops of liquid rolled off the fabric surface without wetting or adhering to the fabric. The treated fabric had a water contact angle >=170°, even after 50,000 abrasion cycles, implying the finish was durable. Water Repellency: spray test provided 90 (AATCC Rating) which is the commonly accepted value by industry for a super-hydrophobic fabric. No residue and stain was observed on the washed fabric during the stain-repellency test against Ketchup, Mustard and Jam. FTIR data identified the presence of Fs finish on both the treated fabric before and after abrasion. It is concluded from the experimental data that the Fs finish used on the acrylic fabric provided a durable super hydrophobic characteristic. A unique feature of this finish and the method for the super-hydrophobic fabric is its suitability for the commercial process and cost.

References

[1] Zhao H and Ras H A 2014 Definitions for hydrophilicity, hydrophobicity, and superhydrophobicity: getting the basics right *The J. Physical Chemistry Letters* pp 686–688

[2] Barthlott W and Neinhuis C 1997 Purity of the sacred lotus, or escape from contamination in biological surfaces *Planta* **202**(1) pp 1-8

[3] Kawai A and Nagata H 1994 Wetting behavior of liquid on geometrical rough surface formed by photolithography *Japanese J. Applied Phys.* **33**(9A), p.L1283

[4] Zhang X, Zhang J, Ren Z, Li X, Zhang X, Difu Z, Wang T, Tian T and Yang B 2010 Modulating two-dimensional non-close-packed colloidal crystal arrays by deformable soft lithography *Langmuir* **26**(4) pp 2930-36

[5] Chen J K, Qui J Q, Fan S K, Kuo S W, Ko F H, Chih Wei C and Chang F C 2012 Using colloid lithography to fabricate silicon nanopillar arrays on silicon substrates *J. Colloid and Interface Science* **367**(1),pp 40-48

[6] Polizos G, Tuncer E, Qiu X, Aytug T, Michelle Kidder K, Messman J M and Sauers I 2014 From hydrophobic to superhydrophobic and super hydrophilic siloxanes by thermal treatment *Langmuir* **30**(44) pp.13235-43

[7] Bao X M, Cui J F, Sun H X, Liang W D, Zhu Z Q, An J, Yang B P, La P Q and Li A 2014 Facile preparation of super-hydrophobic surfaces based on metal oxide nanoparticles *Applied Surface Science* **303** pp 473–480

[8] Sahoo B N and Kandasubramanian B 2014 Recent progress in fabrication and characterization of hierarchical biomimetic super-hydrophobic *Royal Society of Chemistry Advances* **4**(42) pp 22053-93

[9] Hoefnagles H F, Wu D, With G and Ming W 2007 Biomimetic superhydrophobic and highly oleophobic cotton textiles *Langmuir* **23**(26) pp 13158-63

[10] Ivanova N A, Lavretsky A K, Ivanova N A and Zaretskaya A K 2011 Imparting waterproof properties to cotton surface *Protection of Metals and Physical Chemistry of Surfaces* **47**(3) pp 369-71

[11] Xu C H, Jia ST, Zhang J and Tian LQ 2009 Super-hydrophobic surfaces on cotton textiles by complex coating of silica nanoparticles and hydrophobization *Thin Solid Films* **517**(16) pp 4593-98

[12] Das I and De G 2015 Zirconium based super-hydrophobic coatings on cotton fabrics exhibiting excellent durability for versatile use *Scientific Reports* **5** p.18503.

[13] Meng S, Ye Y, Mansouri J and Chen V 2015 Crystallization behavior of salts during membrane distillation with hydrophobic and super-hydrophobic capillary membranes *J. Membrane Science* **473** pp 165-176

[14] Sigma-Aldrich Corporation 2015 Triethoxy-1H,1H,2H,2H-perfluorodecylsilane Safety data sheet Version 4.4 *Aldrich* 658758 (Print Date 02/03/2018)

[15] Stalder, Aurélien F, Melchior T, Müller M, Sage D, Blu T and Unser M 2010 Low-bond axisymmetric drop shape analysis for surface tension and contact angle measurements of sessile drops *Colloids and Surfaces A: Physicochemical and Engineering Aspects* **364**(1-3) pp 72-81

[16] Workman J and Springsteen AW 1998 *Applied Spectroscopy- A compact reference for practitioners* (New York: Academic Press)

[17] Barriet D and Lee T R 2003 Fluorinated self-assembled monolayers, structure and interfacial properties *Curr. Opin. Colloid interface Sci.* **8**(3) pp 236-242

[18] Tsibouklis J and Nevell T G 2003 Ultra-low surface energy polymers: the molecular design requirements *Adv. Mater* **15**(7-8) pp.647-650

[19] Brassard J, Sarkar D K and Perron J 2012 Fluorine Based Superhydrophobic coatings *Applied Sciences* **2**(2) pp 453-464

IOP Conf. Series: Materials Science and Engineering **827** (2020) 012004 doi:10.1088/1757-899X/827/1/012004

Hybridization of game theory and ridesharing to optimize reverse logistics of healthcare textiles

Mustapha Ahlaqqach [1, 2,3], J Benhra[1], S Mouatassim[1] and S Lamrani[1]

[1] LRI, OSIL Team, HASSANII University, Casablanca, Morocco
[2] CELOG-ESITH, Casablanca, Morocco
[3]Author to whom any correspondence should be addressed

Ahlaqqach@esith.ac.ma

Abstract. The reverse logistic of Healthcare Textile has become nowadays a major concern in Morocco. Consequently, it imposes to authorities and professionals the obligation to perform a sustainable reverse logistic model. Through this study, we are going to hybridize game theory and ridesharing to optimize a multi-objective reverse logistics of healthcare textiles. We opted for the Genetic algorithm to face the complexity of the ridesharing model. For cost allocation, the allocation based on volume give best results compared to Alternative Cost Avoided Method. Besides, it allows a higher gain for players with the most important activity.

1. Introduction

Healthcare Textiles (HT) reverse logistics, as a source of contamination, increasingly affects the quality of life of citizens[1]. Therefore, a considerable amount of recommendations has been cited in the GIZ report to limit the impact of this contamination[2]. These recommendations have not been successfully implemented in Morocco[3]. Thus, clinics use their own transport to transport the used HT to laundries. This situation has led to excessive transport costs that are beginning to weigh on the cash flow of these clinics. Then, while waiting for a logistic operator, allowing the launch of the project of outsourcing, we aim through this study to provide an adequate solution to the clinics located in the city of Casablanca.

Ridesharing (RS) as a solution to minimize transportation costs and respect for the environment has recently gained prominence on social networks. This success has encouraged researchers to study the implementation of RS in various fields [4-5-6]. Indeed, this solution can reduce transportation costs, but it is necessary to build homogeneous coalitions and distribute costs equitably among the clinics that make up each coalition. Our objective is to study the best way to implement RS in the case of HT logistics. Thus, the second section presents the proposed solution and outline the state of the art of the RS and game theory (GT). The third part exposes the model formulation of our projected solution. The remaining part of the paper carries out a series of experiments and results discussion.

2. Problem definition and state of art:

The motivation of our work is to exploit the benefits of RS to ensure a temporary solution for clinics that reduces their transport costs, while minimizing the risk of contamination from HT. The latter is fundamental for the clinics involved in our study. Indeed, some clinics are unambitious to participate in high-risk coalitions. The origin of this potential risk is associated with the transport of HT, where the possibility of contamination exists between HT coming from different clinics. Therefore, we will take in consideration the main criteria, stated in interviews with clinic managers, to avoid such high-risk coalitions. RS is defined as a mode of transportation in which several travelers share a vehicle for a trip and split the trip costs[7]. several research studies have shown that RS can bring many environmental and economic benefits, including reducing traffic congestion and carbon emissions[8-9]. In the literature, RS is often formulated as an optimization problem aimed at maximizing customer coverage, maximizing driver income and many other objectives. The latter are subject to constraints from the origin-destination pairs, the time windows (TW) and so on. The complexity of these models is supported by meta-heuristics such as genetic algorithms [10] or other algorithms like greedy-matching[11]. For a better implementation of the RS solution in our case study, we propose the model illustrated in figure 1, where our model select the starting clinic corresponding to the driver. Then, it will constitute a coalition of clinics and corresponding laundries to ensure both the collection and deposit of HT.

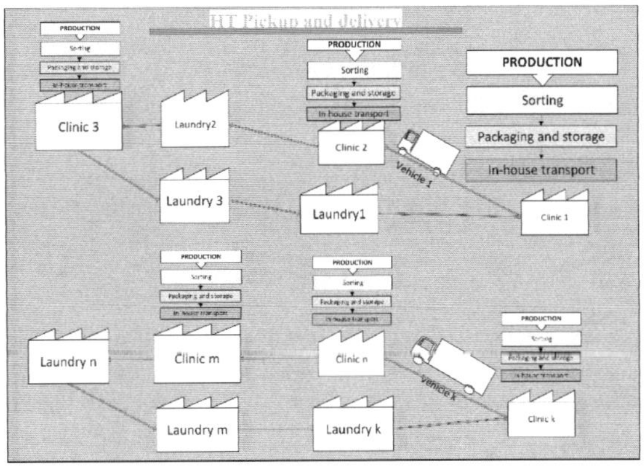

Figure 1. Proposed RS CVRPPDTW Scheme for the HT Logistics Management.

As shown in the figure 1. our model try to set up routings, with picking-up of HT (case of clinics) or delivering of HT (case of laundries), using heterogeneous capacitated vehicle, with requirements in terms of TW. Consequently, one can conclude that we face a Heterogeneous Capacitated Vehicle Routing Problem Pickup and Delivery with TW (HCVRPPDTW). Once coalitions are formed, the other challenge is to distribute the costs fairly among the coalition members, also ensuring an economic gain for each member to encourage them to join the coalition. The main problem in cooperation is cost allocation [12], more than 40 different cost allocation methods have been used in the literature on collaborative transportation[13], and each solution concept satisfy a number of characteristics. However, no concept fulfils all criteria listed in literature. A simple approach for cost allocation is to use a proportional allocation that can be based on the overall volume or weight of the products transported[14]. Moreover, other methods are commonly used in the case of joint distribution and RS like Shapley value and methods based on separable and non-separable cost as ACAM[15]. According to [16], ACAM and Volume can provide an interesting result in the case of transportation problem, insuring individual rationality. Therefore, we will use these methods to allocate costs.

Roughly speaking our case is a RS multi-objective case (cost, coalition risk) with HCVRPPDTW. The present study fills a gap in the literature by adding some particularities compared to the existing models: Multi-objective (MO) model, Heterogeneous fleet (HF), Multiple Depot (MD), Strict TW, Hazardous Materials (HT), PD, RS and GT (Allocation cost).

3. Model Formulation

3.1. HFCVRPPDTW model:

The HCVRPPDTW can be defined on a directed graph G = (N, A), where: N = {0, …, n-1, n, ..., 2n-1, 2n,…, 3n-1} is the set of nodes, A is the set of arcs where n= the number of clinics = the number of laundries = the number of virtual depots and K is a set of heterogeneous vehicles.

Decision variables:

x_{ijk}: a binary variable (BV) equal to 1 if arc (i, j) is used by vehicle k, 0 otherwise, $(i,j) \in N, k \in K$
Z_1: total cost of routing (Variable cost + fixed cost of used vehicles)
The model that govern the CVRPPDTW is the same as presented in [17].

3.2. *Risk Calculation:*

As aforementioned above, based on interviews with clinics, we have specified the list of criteria for assessing the risk of contamination. Then, we applied weighting vote method [18] to established a risk matrix, which give the risk contamination when two clinics are on the same coalition. The total Z_2 risk calculation is calculated with the same objective-function and constraints as presented in [17].
We denote α, β as the Weighting parameters of cost function and risk function respectively.
We express our aggregate multi-objective function as follows:

$$Minimize\ \alpha * Z_1 + \beta * Z_2 \tag{1}$$

3.3. *RS constraints:*

The aim of our model is to limit the number of vehicle used by clinics. Roughly speaking, some clinics will engaged their vehicle to transport their HT and those of other clinics taking into account transport cost and coalition preference. Thus, we will enhance our model with constraints linked to RS. Let:
u_i: a CV which represent node potentials indicating the visit order of node i in the tour

$$x_{0ik} = x_{i+n0k} \qquad i \in \{n, \ldots, 2n-1\}, k \in K \tag{2}$$

$$x_{0i+nk} = x_{i0k} = 0 \qquad i \in \{n, \ldots, 2n-1\}, k \in K \tag{3}$$

$$\textstyle\sum_{j=0}^{3n-1} x_{ijk} = \sum_{p=0}^{3n-1} x_{i+njk} \qquad i \in \{n, \ldots, 2n-1\}, k \in K \tag{4}$$

$$u_i - u_j \le (1 - x_{ijk}) * n - 1 \quad i, j \in \{n, \ldots, 3n-1\}, i \ne j, k \in K \tag{5}$$

$$u_i \le u_{i+n} \qquad i \in \{n, \ldots, 2n-1\} \tag{6}$$

$$u_i \ge 0 \qquad i \in \{n, \ldots, 3n-1\} \tag{7}$$

Equation (2) make sure that a tour begins from the clinic and ends at his destination. However, constraint (3) prevent starting from laundries or ending at clinics. Constraint (4) ensure the use of the same vehicle by origin and destination. Constraints (5) and (6) eliminate sub tours and ensure that each clinic precedes its corresponding laundry. Finally, constraint (7) defines the non-negativity of u_i.

3.4. *Cost allocation methods :*

3.4.1. *Alternative Cost Avoided Method (ACAM).*

The Alternative Cost Avoided Method uses the weights $w_i = \frac{[C(\{i\}) - [C(N) - C(N \setminus \{i\})]]}{[C(N) - \sum_j C(N) - C(N \setminus \{j\})]}$, expressing savings that are made for each participant by joining the grand coalition instead of operating alone.

$$\varphi_i(C) = C(N) - C(N \setminus \{i\}) + w_i * [\,C(N) - \textstyle\sum_i (C(N) - C(N \setminus \{j\}))]] \tag{8}$$

3.4.2. *Allocation based on volume.*

Volume weights is based on distributing the total cost of the grand coalition, $C(N)$, among the participants according to a volume or a cost weighted measure.

$$\varphi_i(C) = \frac{C(\{i\})}{\sum_j C(\{j\})} C(N) \tag{9}$$

For choosing an adequate allocation method, it should be decided which axioms and characteristics are relevant for the case at hand. No cost allocation method fulfills all listed characteristics.

3.5. *Data of computational study:*

IOP Conf. Series: Materials Science and Engineering **827** (2020) 012004 doi:10.1088/1757-899X/827/1/012004

We study 18 clinics disseminated in the whole region of Casablanca. The HT generated historical data of each clinic show the stability of HT's boxes generated. Consequently, we consider the demand as the average number of daily HT's boxes generated. We have a set of heterogeneous vehicle. Each clinic has its own vehicle and we get from the clinic the fixed and variable costs.

4. Solution methodology

The resolution by the exact approach has shown its limitations because the research space is too large, especially in the case of the MO problem[19]. Based on a very broad field of studies, the genetic algorithm (GA) is a promising approximation algorithm that has addressed hard problems in recent decades [20]. For this reason, we propose the GA to deal with this case. The next section will show experimentation of the chosen GA. Then, we will use GT in order to allocate cost and ensure adequate profit and individual rationality for each clinic.

4.1. Heuristic Approach:

4.1.1. Genetic Algorithm. The GA is one of the best known evolutionary optimization techniques, this Meta-heuristic covers several optimization problems such as VRPTW [21]. *Alba et al. 2004* proposed a cellular Genetic Algorithm (cGA) to solve CVRP [22]. The interesting results from the operators of cGA will guide our choice towards the Alba's operators. The proposed GA and the case studied are coded and implemented on the platform HeuristicLab 3.3.15.15587[23] and experiments are carried out on an Intel® CORE(TM) i5-4200 CPU with a 2.3 Ghz processor and 4 GB installed memory RAM.

4.1.2. Validation and analysis of the approach. The benchmarking is the best way to validate the meta-heuristic proposed in the case of RS HCVRPPDTW. However, this is the first work, to our knowledge, that dealt with multi-objective RS in case of HCVRPPDTW. Nonetheless, we can generate our GA for small instances to compare the results with the Exact Approach (EA). The benchmark results, exposed in table1, show a little difference between EA and GA, however the execution time (ET) in GA is very interesting. These results encouraged us to justify the acceptability of a proposed GA's performance and to perform the experimentation of the GA in large instances. To do so, we carried out 50 tests for real instance. The GA, gave the best solutions within a reasonable time. The table 2 represents the best results from GA. The RS constraints chose clinics 5, 13 and 7 as a starting point. Therefore, their vehicles will be used to cover the collection from clinics and delivery to laundries according to the sequences specified in the table 2. The tuning experiments showed that there is a compromise range between cost and risk. Consequently, we can perform RS as an alternative solution to manage the logistic of HT.

Table 1. Average results of tested GA.

Algorithm	Objective Function	GAP to Global optimum	ET (Minutes)
EA	684		23:49.0
GA	686.5	0.36%	01:05.1

Table 2. Best result coming from real instance experiments.

	Pickup and Delivery Sequence	Cost [c]	Risk
Routing 1	C[a]8 C14 L[b]14 C12 C15 L12 L15 C16 L16 L8	1066.9	87
Routing 2	C13 C11 C7 C2 L2 L7 C4 C1 L4 C18 C17 L11 L18 L17 C13	1318.3	416
Routing 3	C10 C5 C3 L3 L5 C6 L6 C9 L9 L10	1287.1	105
	Total	3672.3	608

[a] C: Clinics
[b] L: Laundry
[c] MAD: Moroccan Dirhams

4.2. Cost allocation

The stand-alone cost of each clinic and the cost of every coalition (Routing 1, 2, 3) generated by the heuristic approach are used to share the total cost according to Volume method. As basis for computation of ACAM, we use the heuristic approach to solve the transportation problem for each coalition when one clinic is missing. This will allow us to calculate the marginal cost of each clinic. In this section, cost refers to normalization between cost (MAD) and risk introduced before.

Results show that ACAM and volume respect the individual rationality in the case of coalition 1 and coalition 3 and there is a certain similarity between allocated costs. While for coalition 2, only the volume respect the individual rationality, since for the ACAM, at least one participant is worse in this allocation comparing to its standing alone. Collaboration in the case of clinics allows optimizing considerably the total cost. The profit made by each clinic and by each coalition is very significant.

Table 3. Comparison between ACAM and Volume.

	Clinics	Stand alone	ACAM	Profit	Volume	Profit
Coalition 1						
	8	151.5	11.0	140.5	36.6	114.9
	14	182.5	52.2	130.2	44.1	138.4
	12	182.6	58.1	124.6	44.1	138.5
	15	226.3	48.6	177.7	54.7	171.6
	16	225.5	64.0	161.5	54.5	171.0
	Sum	968.4	234	734.4	234	734.4
Coalition 2	Clinics	Stand alone	ACAM	Profit	Volume	Profit
	13	183.1	87.6	95.5	77.3	105.8
	11	182.8	78.6	104.2	77.2	105.6
	7	151.6	39.9	111.7	64.0	87.6
	2	76.5	55.2	21.3	32.3	44.2
	4	107.3	55.4	52.0	45.3	62.0
	1	76.9	122.0	-45.1	32.4	44.4
	18	301.4	56.6	244.9	127.2	174.2
	Sum	1306.1	551.3	754.7	551.3	754.7
Coalition 3	Clinics	Stand alone	ACAM	Profit	Volume	Profit
	10	182.6	94.9	87.8	72.2	110.4
	5	136.3	52.3	84.0	53.9	82.4
	3	106.4	48.5	57.9	42.1	64.3
	6	136.4	45.3	91.0	53.9	82.4
	9	152.2	41.3	110.9	60.2	92.0
	Sum	713.9	282.3	339.6	282.3	339.6

5. Conclusion

This research sheds new light on reverse logistics in the case of HT, by proposing a RS MO model of a very complicated case of HCVRPPDTW. Our objective was to support medical instances to find a solution to their sustainable reverse logistics. The affinities between the clinics led us to integrate the RS and the GT to set up homogeneous coalition between clinics and to share properly their transport costs. The simulation coming from the GA helped us to seek a compromise between the main objectives studied. For cost allocation, the practical case studied confirms that the ACAM does not guarantee individual rationality in coalition 2, since in the case of its creation one of the clinics pays more than he would have paid by standing alone. The Volume gives a solution that respects individual rationality and allows a higher gain for players with the most important activity. As perspective of this work, we suggest a deep study of coalition stability.

6. References

[1] Ahlaqqach M, Benhra J, Mouatassim S and Lamrani S 2018 Smart City Through the Control of the Healthcare Textiles Transport *SSRN Electron. J.*

[2] SWEEPNET 2014 *Country report on the solid waste management in Kingdom of Morocco*

[3] Ahlaqqach M, Benhra J and Mouatassim S 2017 Optimization of routing for the collection and delivery of medical waste passing through a common warehouse *Logistique Manag.* **25** 25–33
[4] Sun H, Wang H and Wan Z 2019 Model and analysis of labor supply for ride-sharing platforms in the presence of sample self-selection and endogeneity *Transp. Res. Part B Methodol.* **125** 76–93
[5] Olsson L E, Maier R and Friman M 2019 Why Do They Ride with Others? Meta-Analysis of Factors Influencing Travelers to Carpool *Sustainability* **11** 1–16
[6] Skok W and Baker S 2019 Evaluating the impact of Uber on London's taxi service: A critical review of the literature *Knowl. Process Manag.* **26** 3–9
[7] Sánchez D, Martínez S and Domingo-Ferrer J 2016 Co-utile P2P ridesharing via decentralization and reputation management *Transp. Res. Part C Emerg. Technol.* **73** 147–66
[8] Caulfield B 2009 Estimating the environmental benefits of ride-sharing: A case study of Dublin *Transp. Res. Part D Transp. Environ.* **14** 527–31
[9] Ahlaqqach M, Benhra J, Mouatassim S and Lamrani S 2019 Modeling and Optimization of a Multi-Objective Ridesharing Problem in the Case of Medical Waste *Int. J. Recent Technol. Eng.* **8** 1911–8
[10] Guo Y, Goncalves G and Hsu T 2013 Retraced article: A Multi-agent Based Self-adaptive Genetic Algorithm for the Long-term Car Pooling Problem *J. Math. Model. Algorithms Oper. Res.* **12** 45 66
[11] Jiang W, Dominguez C R, Zhang P, Shen M and Zhang L 2018 Large-scale nationwide ridesharing system: A case study of Chunyun *Int. J. Transp. Sci. Technol.* **7** 45–59
[12] D'Andrea A 2017 Blood Components as Joint Products: A Literature Review of Cost-Allocation Methods *Int. J. Bus. Manag.* **12** 46
[13] Guajardo M and Rönnqvist M 2016 A review on cost allocation methods in collaborative transportation *Int. Trans. Oper. Res.* **23** 371–92
[14] Verdonck L, Beullens P, Caris A, Ramaekers K and Janssens G K 2016 Analysis of collaborative savings and cost allocation techniques for the cooperative carrier facility location problem *J. Oper. Res. Soc.* **67** 853–71
[15] Frisk M, Göthe-Lundgren M, Jörnsten K and Rönnqvist M 2010 Cost allocation in collaborative forest transportation *Eur. J. Oper. Res.* **205** 448–58
[16] Salma M, Jamal B, Sabry A H and Mustapha A 2018 An Analysis of Biform Game Strategies for Supply Planning in a Full Distribution Network: Case of a Bottling Company *SSRN Electron. J.*
[17] Mouatassim S, Ahlaqqach M, Benhra J and Eloualidi M 2016 Model based on hybridized game theory to optimize logistics case of blood supply chain *Int. J. Comput. Appl.* **145(15)** 37–48
[18] Zhu X, Song Q and Jia Z 2010 A Weighted Voting-Based Associative Classification Algorithm *Comput. J.* **53** 786–801
[19] Ahlaqqach M, Benhra J, Mouatassim S and Lamrani S 2019 Ruin and Recreate Approach Applied to Multi-objective Optimization of HHCVRPTW in the Case of Medical Waste Collection *Indian J. Sci. Technol.* **12**
[20] Liu R, Xie X, Augusto V and Rodriguez C 2013 Heuristic algorithms for a vehicle routing problem with simultaneous delivery and pickup and time windows in home health care *Eur. J. Oper. Res.* **230** 475–86
[21] Thangiah S R, Nygard K E and Juell P L 1991 GIDEON: a genetic algorithm system for vehicle routing with time windows *The Seventh IEEE Conference on Artificial Intelligence Application* vol i (IEEE Comput. Soc. Press) pp 322–8
[22] Alba E and Dorronsoro B 2004 Solving the Vehicle Routing Problem by Using Cellular Genetic Algorithms (Springer, Berlin, Heidelberg) pp 11–20
[23] Wagner S, Kronberger G, Beham A, Kommenda M, Scheibenpflug A, Pitzer E, Vonolfen S, Kofler M, Winkler S, Dorfer V and Affenzeller M 2014 Architecture and Design of the

HeuristicLab Optimization Environment (Springer, Heidelberg) pp 197–261

Influent product and process parameters on the mechanical behaviour of 3D warp interlock fabrics made with E-glass yarns

A Kececi[1], F Boussu[1,2], D Soulat[1]
[1] ENSAIT, GEMTEX - Laboratoire de Génie et Matériaux Textiles, F-59000 Lille, France.
[2] Author to whom any correspondence should be addressed

francois.boussu@ensait.fr

Abstract. 3D weaving process has been more and more used as reinforcement for composite materials. However, 3D weaving applications are recent and the mechanical properties of the 3D warp interlock fabrics are not well known. Tensile tests have been performed both in the warp and weft directions of the fabrics, indeed they lead to different behaviour regarding the architecture. Results from experiments have been statistically analysed using PLS regression. This study led to correlations between fabrics parameters and mechanical behaviour. Moreover, it revealed that fabrics parameters such as the binding type, the yarns crimp and the picks density have influence on the quasi-static mechanical behaviour.
Keywords: 3-D reinforcement; Mechanical properties; Weaving; E-glass fibres; Para-aramid fibres.

1. Introduction

Textile architectures used as fibrous reinforcement for composite material can be declined into various structures depending on their initial process. Planar or tubular parts can be easily obtained from large scale production processes as flat weaving and knitting, braiding and tubular knitting, non-woven and non-crimp fabrics [1,2]. However, the choice of the suited textile architecture, according to the several process and product parameters as well as the manufacturing costs and quality, seems to be quite a challenge [3,4]. To solve this issue, a better knowledge of the textile material may help to find the suited architecture for the dedicated final application [5,6]. Among all these textile materials, 3D fabrics and, especially the 3D warp interlock fabrics, have been largely described as a promising solution for many applications due to their specific consolidation mode into the thickness [7]. Due to the presence of the linking warp yarns inside the thickness of the 3D warp interlock fabric architecture [8], they tend to increase the quasi-static intra-ply delamination resistance [9-13] and high speed impact resistance [14-19].

Despite all of these existing research results, the optimized combination of product and process parameters of a 3D warp interlock fabric, responding to given specifications of the final composite material solution, still remains a challenge. One of the reasons that could explain this difficulty may lie in the lack of experimental and modelling knowledge of this specific 3D textile reinforcement. Moreover, clear understanding of the influence of product and process parameters, considered alone or combined together, has not yet been revealed.

Thus, the objective of this paper is to provide additional knowledge on mechanical behaviour of different 3D warp interlock fabric architectures made with different types of yarn's raw material, especially E-glass and para-aramid yarns. An experimental characterization of these different 3D woven architectures has been achieved to measure the influence of product parameters on their mechanical behaviour in tensile and bending. Then, a statistical analysis will help us to highlight the influent product and process parameters according to their different mechanical behaviours.

2. Material and Methods

Several woven architectures have been considered to measure their influence on the mechanical behaviour of 3D warp interlock fabrics.

2.1. Types of raw materials

It has been decided to use the E-glass EC9 900 Tex (see Table 1).

Table 1. Yarns parameters.

Type of yarn	Provider (Country)	Nominal linear density (Tex)	Maximal breaking force (N)	Elongation at maximal breaking force (%)	Tenacity (N/Tex)	Tenacity (MPa)
E-glass	PPG (US)	900	308.0 ± 48.1	2.7 ± 0.4	0.96	2526

2.2. List of architectures

All these architectures have been produced on the same dobby loom in order to keep constant the process parameters; as the end density to 10 warp yarns/cm. The chosen architectures aim at reflecting the existing classification of 3D warp interlock fabrics [11] with Angle (A) or Orthogonal (O) binding warp yarns types and Layer-to-layer (L) or Through-the-thickness (T) linking warp yarns paths. A total of 3 layers of weft yarns for the 3D warp interlock fabrics without stuffer warp yarns and 4 layers of weft yarns for the 3D warp interlock fabrics with stuffer warp yarns have been chosen. Different well-known weave diagrams as Twill, Satin and Basket have been selected to measure their influence on their mechanical behaviour.

3. Results and discussion

According to the obtained results based on different physical and mechanical characteristics of the 3D warp interlock fabrics, explanations of material behaviour could be linked to the different product parameters. Assuming that the number of tests done is sufficient, the use of a statistic tool could help us to more deeply detect these product parameters influences on the material behaviour of the different 3D warp interlock fabrics.

3.1. Introduction to PLS Regression

The PLS Regression [20][21] has been used in this study in order to find potential links between fabric properties and tensile behaviour that have not necessarily been revealed in the previous analysis. Moreover, the low number of samples can lead to high standard deviation, so these results have to be compared with global analysis. For each architecture, the explanatory variables that are assumed independent are: the binding step, the binding depth, the picks density, the binding warp yarns average crimp, the stuffer warp yarns average crimp, the weft yarns average crimp, the binding angle A, the binding angle O, the binding path L, the binding path T.

3.2. Results with PLS Regression

Figure 1 shows results for E-Glass fabrics. Maximal load and strain at maximal load have been analysed with PLS Regression. These graphs show the fabrics characteristics that are the most influent on a mechanical parameter. It can increase or decrease the value of this parameter. For instance, regarding maximal load in the weft direction, picks density has the highest influence on maximal load increase, whereas O binding angle has the highest influence on maximal load decrease.

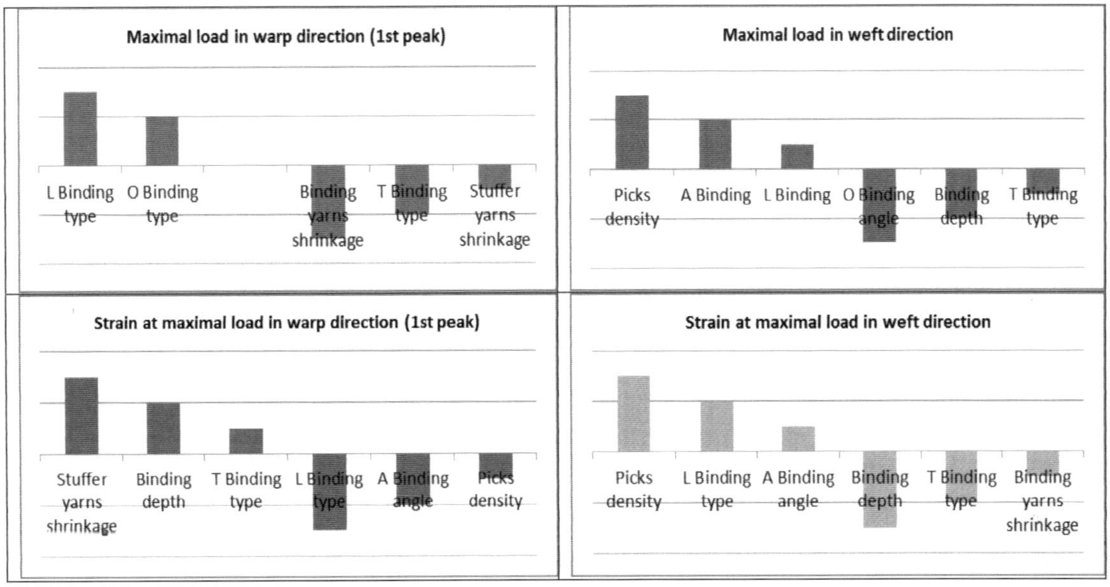

Figure 1. Influent parameters on tensile behaviour for E-glass fabrics.

Considering the obtained results, the main influent parameters as the pick density in the weft direction and stuffer warp yarns crimp in the warp direction are highlighted regarding their link to the mechanical behaviour of 3D warp interlock fabrics made with E-glass yarns. However, second order influent parameters as: the binding depth, O and A binding type, binding warp crimp, L and T binding path may explain the difference of mechanical performance of 3D warp interlock fabrics made with E-glass yarns. For instance, for the strain at maximum load in the warp direction (for the first peak), more the binding depth value is, higher the strain value will be. This binding depth reflects the capacity of the binding warp yarn to be "stored" in the thickness of the 3D woven architecture which could help the material to more elongate during tensile test and increase the global strain value. On the contrary, more the binding depth value is high, less the maximum load, the strain at maximum load will be in the weft direction. This reflects a resulted inverse behaviour in the warp and weft directions of the 3D warp interlock fabrics.

By the same, more the binding path of 3D warp interlock fabric is close to L binding, less the strain value at maximal load value will be. This L binding path reflects also the capacity of the yarn to be layer to layer linked in the thickness of the 3D warp interlock fabrics and then tend to be more "straight" compared to T binding (through-the thickness) which consume more length of binding yarn inside the 3D woven structure.

4. Conclusion

The main objective of this study aims at providing more knowledge on 3D woven structures used as fibrous reinforcement for composite material, especially here the 3D warp interlock fabrics. Several architectures, based on the existing classification into four main clusters as O-L, O-T, A-L and A-T woven structures, have been designed and produced with the same E-glass yarns. Several physical and mechanical parameters have been chosen to characterize these 3D warp interlock fabrics both in the warp and weft directions as: the bending rigidity, the warp and weft yarn crimp, the strain at maximum load and the maximum load values. A deeper analysis, done with PLS regression statistical tool, have revealed second order increasing and decreasing influent parameters, whom mechanical behaviour of 3D warp interlock fabric depends on.

Future works will be engaged to test the same types of 3D warp interlock fabrics with other raw materials types. Additionally, bias tests and forming tests will be also studied to check the material behaviours of 3D warp interlock fabrics to provide complementary knowledge on their moulding capacity for complex 3D shape of composite materials.

References

[1] Kamiya R, Cheeseman B A, Popper P and Chou T W 2000 Some recent advances in the fabrication and design of three-dimensional textile preforms: A review. *Composites Science and Technology,* **60**(1), pp.33–47.

[2] Wang P, Legrand X and Soulat D 2017 Three- Dimentional Textile Preform Using Advanced Textile Technologies for Composite Manufacturing. *Textiles for Advanced Applications* 6th ch, London, UK: IntechOpen, pp.161-189

[3] Mouritz A P 2011, Three- dimensional (3D) fibre reinforcements for composites, *P. B. T.-C. R. for O. P.*, Woodhead Publishing pp.157–199.

[4] Saleh M N, Lubineau G, Potluri P, Withers P J and Soutis C 2016, Micro-mechanics based damage mechanics for 3D orthogonal woven composites: Experiment and numerical modelling, *Composite Structures*, **156**, pp.115–124.

[5] Boussu F, Dufour C, Veyet F and Lefebvre M 2015, Weaving processes for composites manufacture, *In Advances in Composites Manufacturing and Process Design* 55–78, Elsevier.

[6] Chen X, Taylor L W and Tsai L J 2011 An overview on fabrication of three-dimensional woven textile preforms for composites, *Textile Research Journal*, **81**(9), pp.932–944.

[7] Cristian I, Boussu F and Nauman S 2011 Interesting parameters of 3D warp interlock fabrics influencing the mechanical properties of the final composite structures, *10th Anniversary World Textile Conference AUTEX 2010*, Vilnius, Lithuania **26**(3), p.113.

[8] Gude M, Hornig A, Hufenbach W, Langkamp A, Petrinic N and Wiegand J 2006, Delamination Behaviour of 3D-Textile Reinforced Composites - Experimental and Numerical Approaches, *Journal of Nondestructive Testing*, **11**(12).

[9] Nauman S, Boussu F, Cristian I and Koncar V 2009. Impact of 3D woven structure onto the high performance yarn properties. *2nd Conference on Intelligent Textiles and Mass Customisation*, Casablanca, Morocco.

[10] Tong L, Mouritz A P and Bannister M K 2002, 3D Fibre Reinforced Polymer Composites, *Elsevier.*

[11] Hu J 2008 *3-D Fibrous Assemblies: Properties, Applications and Modelling of Three-Dimensional Textile Structures,* Elsevier pp1-32.

[12] Sheng S Z and Hoa S Van 2003, Composite Materials Modeling of 3D Angle Interlock Woven Fabric Composites, *Journal of Thermoplastic Composite Materials*, ***16***, 45–58.

[13] Mouritz A P, Bannister M K, Falzon P J and Leong K H 1999, Review of applications for advanced three-dimensional fibre textile composites. *Composites Part A: Applied Science and Manufacturing*, **30**(12), pp.1445–61.

[14] Lapeyronnie P, Le Grognec P, Binetruy C and Boussu F 2010, Angle-interlock reinforcements : Weaving and the mechanical properties of composites, *JEC Composites*, **58** pp.58–59.

[15] Tan P, Tong L and Steven G P 1999, Micromechanics models for mechanical and thermomechanical properties of 3D through-the-thickness angle interlock woven composites, *Composites Part A: Applied Science and Manufacturing*, **30**(5), pp.637–648.

[16] Tsai K H, Chiu C H and Wu T H 2000, Fatigue behavior of 3D multi-layer angle interlock woven composite plates, *Composites Science and Technology*, **60**(2) pp.241–248.

[17] Baucom, J N and Zikry M A 2003, Evolution of failure mechanisms in 2D and 3D woven composite systems under quasi-static perforation, *Journal of Composite Materials*, **37**(18) pp.1651–74.

[18] Tanzawa Y, Watanabe N and Ishikawa T 1999, Interlaminar fracture toughness of 3-D orthogonal interlocked fabric composites, *Composites Science and Technology*, **59**(8), 1261–70.

[19] Brandt J, Drechsler K and Arendts F J 1996 Mechanical performance of composites based on various three-dimensional woven-fibre preforms, *Composites Science and Technology*, **56**(3), 381–386.

[20] Binard C 2012. *Introduction à la regression PLS*. Groupe de travail PLS.

[21] Tenenhaus M 1998, *La régression PLS, théorie et pratique Paris*, Ed. Technip.

Core-shell particles based on porous carbon@Fe_3O_4 for efficient removal of dyes from textile effluents

Zakaria Anfar[1, 2, 3], Abdellah Ait El Fakir[1,2], Amane Jada[2,3,4], Noureddine ElAlem[1]
[1]Laboratoire Matériaux et Environnement (LMM), Ibn Zohr University, Agadir, 8000, Morocco.
[2]Institut de Sciences Des Matériaux De Mulhouse (IS2M-UMR 7361 CNRS - UHA), Mulhouse-France
[3]Université de Strasbourg, Strasbourg, F-67081, France.
[4] Author to whom any correspondence should be addressed

amane.jada@uha.fr

Abstract. The removal of textile toxic contaminants is an essential process to provide clean water to the increasing population rate. It requires a design of a selective material to adsorb dyes pollutants with stability, effectiveness and reusability. Regeneration efficiency and recovery extent (RCE) are one of the critical factors to measure the performance of the adsorption process. In this study, magnetic Core-shell beads based on porous carbon@Fe_3O_4 NPs with high specific surface area were successfully synthesized by simple Co-precipitation method. The structural, textural, chemical and morphological proprieties of the prepared materials were studied by XRD, FTIR, Raman spectroscopy, nitrogen adsorption-desorption, XPS, SEM and TEM. The adsorption kinetic and isotherm data were well fitted to the pseudo-second-order and Langmuir models. Magnetic particles exhibited an excellent ability to adsorb methylene blue (MB) from aqueous solutions with maximum MB adsorption capacity of 180.42 mg/g (porous carbon@Fe_3O_4-NPs powder) and 49.66 mg/g (magnetic Core-shell beads based on porous carbon@Fe_3O_4-NPs).
The synthesized core shell adsorbent showed good magnetic response, high regeneration efficiency with excellent recovery extent. We believe that in the near future these new magnetic Core-shell beads based on porous carbon@Fe3O4-NPs can be used as an alternative to solve the separation problem involving the material pollutant interfaces and can be commercialized as a benchmark adsorbent for the removal of toxic dyes.

1. Introduction

In recent years, nano-Fe_3O_4 has received great attention as a new kind of functional materials for potential applications especially in the field of wastewater treatment. Porous carbon have also been considered excellent candidates for wastewater treatment [1]. However, numerous studies have been conducted to develop porous carbon with large specific surface area and interesting physicochemical characteristics [2]. In the present work, magnetic Core-shell beads based on porous carbon (PC)@Fe_3O_4-NPs with high specific surface area were successfully synthesized by simple Co- precipitation method.

2. Material and methods

The core-shell particles used in this work were prepared in three stages. The first step is to prepare the porous carbon then the iron nanoparticles (Fe_3O_4) have been impregnated by co-precipitation method and finally core-shell magnetic particles were formed by using Ca^{2+} ions with alginate polymer. X-ray photoelectrons spectroscopy (XPS) was carried out using Thermo Fisher Scientific ESCALAB 250Xi X-ray photoelectron spectroscopy system equipped with Al Kα X-ray source = 1486.7 eV for chemical states of synthesized PC@Fe_3O_4. Thereafter, kinetics (adsorbate-adsorbent contact time ranging from 0 to 120 min) and equilibrium isotherms (dye amount ranging from 5 to 90 mg/L of MB) adsorption experiments were carried out using batch method. The remaining MB amount after adsorption was determined by using UV-spectrophotometer JENWAY 6705 at maximum wavelength, λ_{max} = 664 nm. PC@Fe_3O_4 and Core-shell beads based on PC@Fe_3O_4-NPs reusability were performed under optimum conditions using 20 mL of HCl (0.1 M).

3. Results and discussions

Chemical states of the elements in the surface of the PC@Fe_3O_4 were observed using X-ray photoelectron spectroscopy (XPS) (Figure 1). The XPS spectrum of C 1s of PC is shown in Figure. 1b. The peaks at 283.0, 284, 286.2, 287.8, and 288.6 eV were attributed, respectively, to the bonds C- Fe, C-C/C=C, C-O, C=O, and O-C=O [3]. The deconvolution of the O1s illustrated in Figure. 1c indicates three crests at, respectively, 530 eV (C-OH), 532 eV (C=O), and 533 eV attributed to the sp^2- bonded carbon atoms[4]. Similar C 1s and O 1s peaks were also observed after the functionalization of PC_350 by Fe_3O_4-NPs, as can be seen, respectively, in Figures 1d and 1e. From Figures 1a and 1f the binary peaks shown in the range around 700 and 724 eV are typical of, respectively, Fe $2p_{3/2}$ and Fe $2p_{1/2}$ [5–7]. The high determination XPS spectra revealing the existence of both Fe^{2+} and Fe^{3+} species in the Fe_3O_4 NPs. The Fe 2p (Figure 1) peaks of PC@Fe_3O_4 observed at 710, 712 and 717 eV could be assigned, respectively, to Fe^{2+}, Fe^{3+} in octahedral coordination (Fe^{3+} (o)), and Fe^{3+} in tetrahedral coordination (Fe^{3+} (t)) of Fe_3O_4 [3,8,9]. Thus, the XPS data confirm the formation of Fe_3O_4-NPs via the co-precipitation synthesis route and the development of Fe_3O_4 nanoparticles at the PC@Fe_3O_4 surface. The prepared magnetic particles were found to exhibit excellent ability to adsorb methylene blue (MB) from aqueous solutions with maximum MB adsorption capacity of 180.42 mg/g (porous carbon@Fe_3O_4-NPs powder) and 49.66 mg/g (Core-shell beads based on porous carbon@Fe_3O_4-NPs). The adsorption removal of MB decreases as the number of regeneration cycles increases. After four cycles, 93.02% and 80.19% of regeneration efficiency of the regenerated porous carbon@Fe_3O_4-NPs and Core-shell beads based on porous carbon@Fe_3O_4-NPs were still achieved, respectively. In addition, the recovery extents of PC, PC@Fe_3O_4 and Core-shell beads based on porous carbon@Fe_3O_4- NPs were examined and presented in Figure 2b. After the five filtration cycles, RCE were found to be 96%, 84%, and 40% mg for Core-shell beads based on porous carbon@Fe_3O_4-NPs, PC@Fe_3O_4 and PC, respectively, indicating that Fe_3O_4 and alginate maintain the relatively of extents recovery of PC. Practically in the case of Core-shell beads based on porous carbon@Fe_3O_4-NPs adsorbents, almost Core-shell particles were recovered.

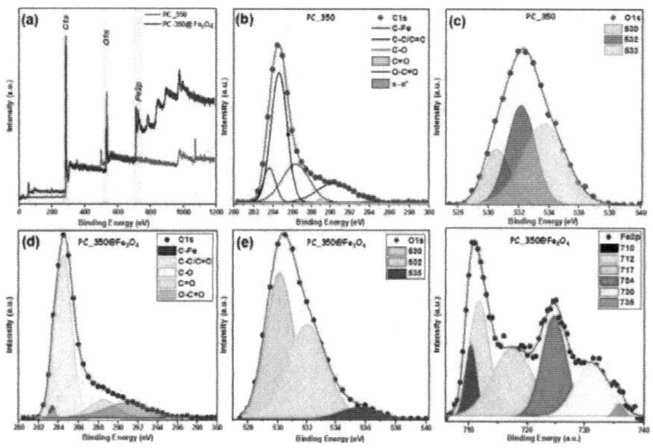

Figure 1. a) XPS patterns of PC and $PC@Fe_3O_4$ and Normalized XPS spectra b) C 1s of PC c) O 1s of PC d) C 1s of $PC@Fe_3O_4$ e) O 1s of $PC@Fe_3O_4$ and f) Fe 2p of $PC@Fe_3O_4$.

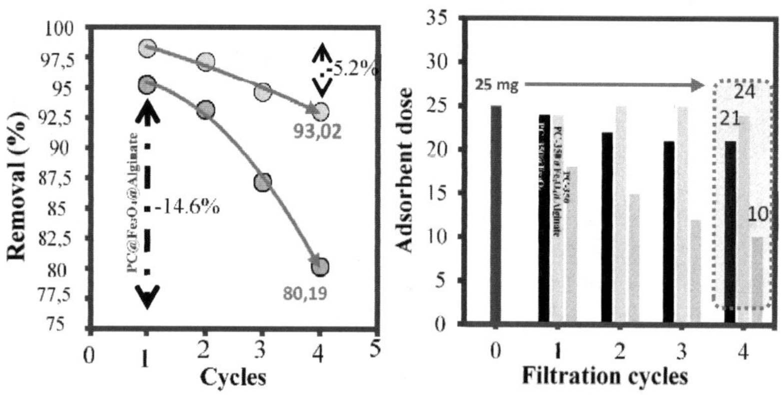

Figure 2. a) Regeneration efficiency and b) Recovery extents (RCE) of the adsorbents.

4. Conclusion

The synthesized core shell adsorbent showed good magnetic response, high regeneration efficiency with excellent recovery extent. It is likely believable that these new magnetic Core-shell beads based on porous carbon@Fe_3O_4-NPs can be used, in the near future, as alternatives to solve the separation problem involving the material pollutant interfaces and can be marketed as a benchmark adsorbent for the toxic pollutants removal.

Acknowledgment

This work was supported by MNEFPESRS, CNRST, SCAC, IFM, Institute of Materials Science Mulhouse - University Haute Alsace and the Materials and Environment Laboratory (LME) - University Ibn Zohr in the framework Franco-Moroccan cooperation, research projects CEDocs under No. 08/00057. We thank the Mascir Foundation for their support.

References

[1] Hashemian S, Salari K and Yazdi ZA 2014 *Journal of Industrial and Engineering Chemistry* **20** pp 1892–1900.

[2] Ait Ahsaine H, Zbair M, Anfar Z, Naciri Y, El haouti R, El Alem N and Ezahri M 2018 *Materials Today Chemistry* **8** pp 121-132

[3] Huang Y, Tang J, Gai L, Gong Y, Guan H and He R 2017 Chemical Engineering Journal

[4] Zbair M, Anfar Z, Ait Ahsaine H, El Alem N and Ezahri M 2018 *Journal of Environmental Management* **206** pp 383-397

[5] Zhou L, Shao Y, Liu J, Ye Z, Zhang H, Ma J, Jia Y, Gao W and Li Y 2014 *ACS Applied Materials*

& Interfaces **6** pp 7275–7285.
[6] Veerakumar P, Panneer Muthuselvam I, Te Hung C, Lin KC, Chou FC and Bin Liu S 2016*, ACS Sustainable Chemistry and Engineering* **4** pp 6772–6782.
[7] Fan H, Niu R, Duan J, Liu W and Shen W 2016 *ACS Applied Materials and Interfaces* **8** pp 19475–19483.
[8] Yang J, Zhao Y, Ma S, Zhu B, Zhang J, Zheng C 2016 *Environ. Sci. Technol* **50**(21) pp 12040 12047
[9] Geng L, Zhang M, Zhang W, Jia M, Yan W and Liu G 2015, Catal. Sci. Technol **5** pp 3097–3102.

Novel biohybrid aerogel composites based on cellulosic and cobalt metallic nanoparticles: efficient and recyclable catalysts for green reduction reactions

Nouaamane EL Idrissi[1,2], Aicha EL Mouden[1], Hamid Kaddami[1] and Larbi Belachemi[1]

[1] Laboratory of Organometallic and Macromolecular Chemistry-Composite Materials, Department of Chemical Sciences, Faculty of Sciences and Technologies, Cadi Ayyad University Marrakech, Morocco

[2] Author to whom any correspondence should be addressed

nouaamane.elidrissi@edu.uca.ac.ma

Abstract. In this study, a cellulose nanomaterial (Cellulose nanofiber, CNF) was employed as solid support and stabilizing agent for the highly active cobalt nanoparticles via in situ green and facile synthesis as an environmentally benign supported catalyst. The process of preparation involves the reduction of cobalt precursors on the surface of CNF in ambient conditions with the presence of $NaBH_4$. The structure of the synthesized organic-inorganic hybrid Co/CNF was characterized by field emission scanning electron microscopy, energy dispersive spectroscopy and Fourier transform infrared spectroscopy. The hybrid aerogel nanocomposite Co/CNF demonstrated excellent catalytic activity for the reduction of 4-nitrophenol (4-NP) to 4-aminophenol (4-AP) in water with $NaBH_4$. Even with a very low amount of catalyst and was found to be good enough to achieve 100% reduction of 4-NP with a higher reaction rate (in 8 min). The heterogeneous nanocomposite catalyst was easily isolated from the reaction mixture by simple filtration and reused 5 times without significant loss of its catalytic activity.

1. Introduction

There is a growing need for environmentally benign, economic and highly active catalysts. Supported catalyst can played an important role in chemical reactions and environmental remediation. In recent years, investigations on supported heterogeneous catalysts become one of the major focuses of modern worldwide research in catalysis. This research has been directed towards the exploration of sustainable and ecological resources, in other words green chemistry.

Cellulose as the most abundant biopolymers in the world has attracted considerable attention as biomaterial support for catalytic systems due to their unique properties, their renewability, biocompatibility, biodegradablity, non-toxicicity, low-cost and environmental benefits. [1-3]

On the other hand, minimizing the size of solid heterogeneous catalysts to the nano-scale increases the specific surface area and the availability of active sites which improves the catalytic performances. The development of nano-sized systems as supported heterogeneous catalysts makes it possible to improve the reactivity and the selectivity of the catalysts compared to the non-supported and “homogeneous” ones [4].

Metal and metal oxide nanoparticles have attracted tremendous interest in catalytic applications in many important industrial processes, like those for the production of fine chemicals and pharmaceuticals

[5] because of their versatile catalytic activities and high surface areas [6–9]. However, metal nanoparticles tend to agglomerate because of their high surface energy which reduces their catalytic performances. Moreover, there are no practical ways to recover the metal nanoparticles from the reaction medium, which presents an additional challenge that limits the usage of metal nanoparticles materials in catalysis. The effective approach to overcome these problems is immobilization of metal nanoparticles on different materials [10].

In this work, we developed a novel bio-hybrid aerogel based on cellulosic nanofibers and cobalt Metallic particles (Co/CNF) and investigated the catalytic performances of this nanocomposite material for the green reduction reaction of nitrophenol.

2. Material and Methods

2.1. Chemicals
Details about the materials and chemicals used for the TEMPO-catalyzed extraction of CNF can be found in a previously published work [10]. Cobalt (II) sulphate heptahydrate [$CoSO_4.7H_2O$], Sodium borohydride $NaBH_4$, 4-Nitrophenol (4-NP) and all solvents were obtained from Sigma-Aldrich and used as supplied without any purification.

2.2. Characterization techniques
FTIR spectra were recorded by BRUKER-Vertex 70. SEM was measured with TESCAN – VEGA3. EDX measurement was performed using AMETEK, EDAX- OCTANE PLUS. UV-Visible used for the kinetic study is a Shimadzu UV–vis spectrometer (UV-2450).

2.3. Preparation of the CNF
The rachis of date palm tree was used in this work as the original source of cellulose. Cellulose Nanofiber CNF was extracted following the procedure well described in our previous work [11].

2.4. Preparation of Co/CNF aerogels
Highly active cobalt nanoparticles supported by porous aerogels cellulose nanofiber composites (Co/CNF) were prepared by an environmentally benign method. The aerogel preparation begins with the formulation of a matrix based on cellulose nanofiber and $CoSO_4$. These two components were mixed by magnetic agitation and 5 minutes sonication to better homogenize the dispersion. After that, the composite mixture was treated with sodium tetrahydroboride $NaBH_4$ in distilled water and finally lyophilized as described in the Figure 1 below

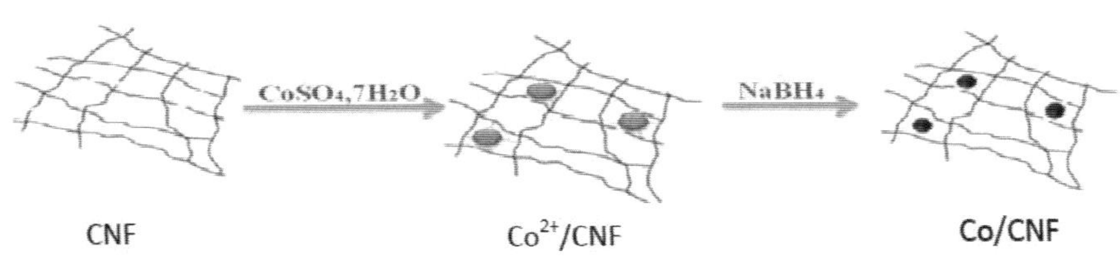

Figure 1. Preparation of Co/CNF aerogels.

2.5. Catalytic Reduction of para-nitrophenol

First, a solution of para-nitrophenol (4-NP) (0.5 g / l) is prepared, then 10 ml are taken from this solution and 10 equivalents of $NaBH_4$ with 5 mol% of Co/CNF are added with stirring magnetic at room temperature. The yellow color of the mixture decreases gradually and disappears after 8 minutes according to the following reaction (figure 2). The kinetic was followed using UV visible spectroscopy.

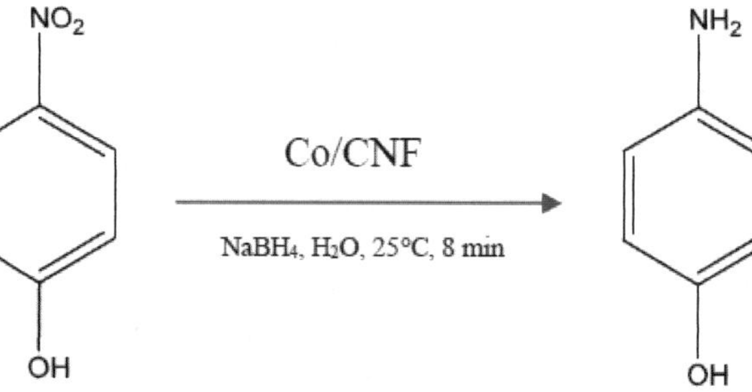

Figure 2. Catalytic Reduction of para-nitrophenol.

3. Results and discussion

3.1. Aerogel Characterization

The new recyclable and environmentally friendly hybrid aerogels composites catalysts Co/CNF was characterized by mean of FTIR spectroscopy, scanning electron microscopy (SEM) and elementary analysis (EDX). Figure 3 shows the SEM picture of pure CNF aerogel taken as reference and that of the Co/CNF aerogel composite. The SEM micrograph of cross section of the aerogel made of pure CNF (Figure 3-A) shows that the CNF are organized in films. The chemical groups carried by the CNF (OH and COOH due to the oxidation) allow H-bounding and permit the cellulose nanofibers to interact and form such a uniform structure [12]. Similar structure is observed for the aerogel composite Co/CNF (Figure 3-B). However, when zooming on the films (Figure 3-B), one can observe metal particles grafted on the films surfaces. At these magnifications, particles of about 1µm are detected but also few aggregates of more than 10 µm could be observed. But, in general, on can confirm that a good dispersion of cobalt particles is obtained. EDX analyses (Figure 3- C and D) confirm that these particles are composed of cobalt. At that level the type of interaction between the CNF film and the metal particle is of high importance. SEM analyses cannot give information about this inquiry. However, in our approach the cobalt particles are fixed by the hydrogen bonds to the CNF surface. It seems that the carboxylate groups at the CNF surface play an emulsifying effect to control the size and insure good dispersion of cobalt particles in the aerogel.

IOP Conf. Series: Materials Science and Engineering **827** (2020) 012007 doi:10.1088/1757-899X/827/1/012007

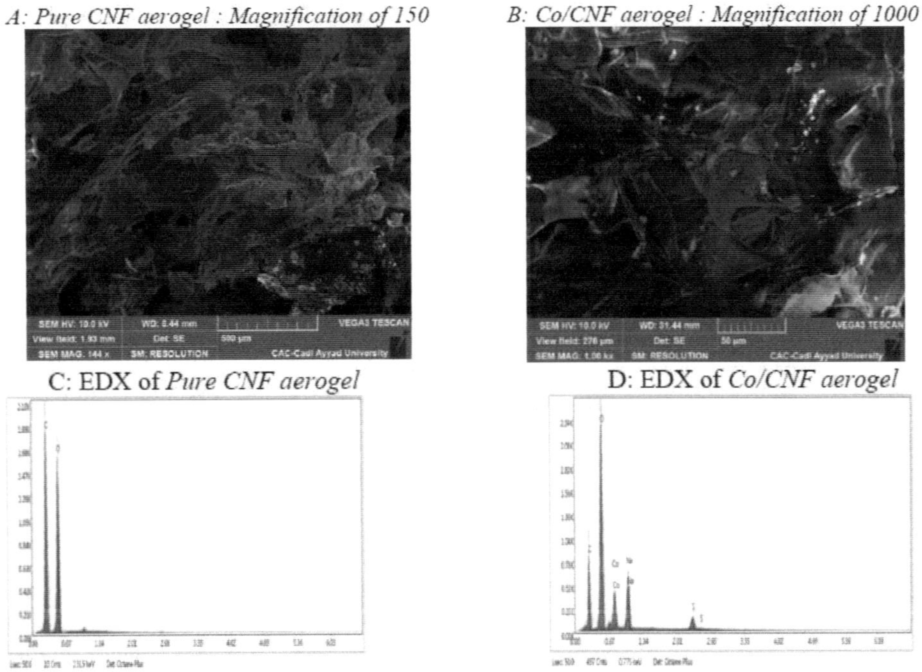

Figure 3. MEB and EDX characterizations of the CNF and Co/CNF composite aerogels.

Figure 4 present the FTIR spectra of the Co/CNF composite aerogel and that of the pure CNF aerogel taken as reference. Some characteristic bands related to the physical and chemical structure are the hydrogen-bonded OH stretching at ca. 4000–2995 cm^{-1}, the OH bending of adsorbed water at 1635 or 1638 cm−1, the CH stretching at 2900 cm^{-1}, the HCH and OCH in-plane bending vibrations at 1430 cm^{-1}, the CH deformation vibration at 1375 cm^{-1}, the stretching vibration of C=O from the free COOH band near 1737 cm^{-1}, the C–OH out-of-plane bending mode at 668 cm^{-1}. Bands at 4000–2995 cm^{-1}, 2900 cm^{-1}, 1430 cm^{-1}, 1375 cm^{-1}, and 900 cm^{-1} are especially sensitive to the state of the crystalline and amorphous regions. While comparing the two spectra, one can observe that the stretching vibration of C=O from the free COOH band near 1737 cm^{-1} disappears from the FTIR spectrum of the composite. On the other hand the intensity of the band of C–OH out-of-plane bending mode at 668 cm^{-1} strongly increases. These two observations might confirm that the COOH and OH groups of cellulose are involved in strong interactions with cobalt particles, which helped to ensure their stabilization and their good dispersion in cellulose matrix.

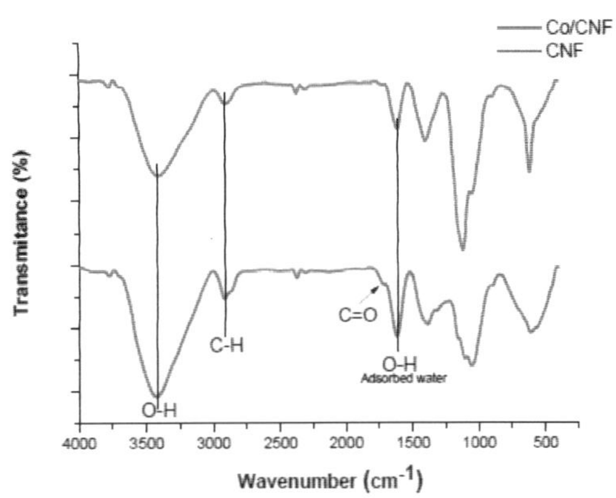

Figure 4. FTIR spectra of CNF and Cobalt/CNF composite aerogels.

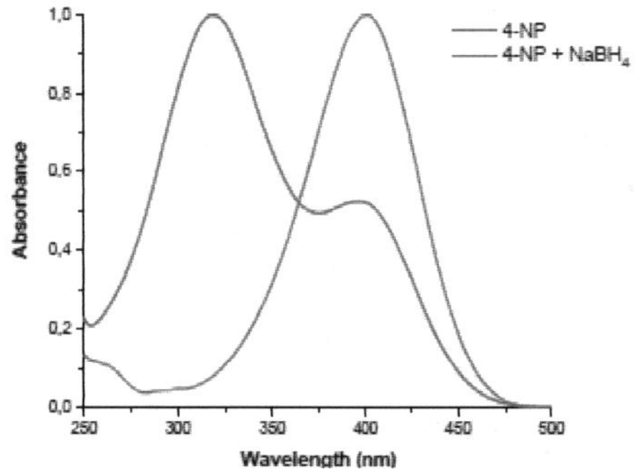

Figure 5. UV-Vis spectrum of p-nitrophenol without and with $NaBH_4$.

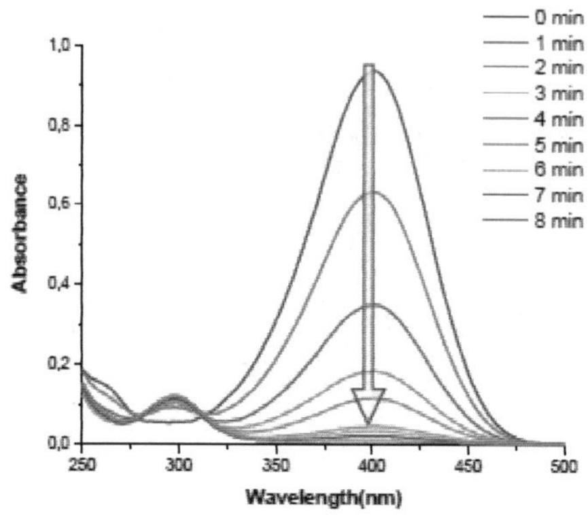

Figure 6. Results of the kinetic study. The UV-visible spectra of Co/CNF.

3.2. Catalytic Activity

Initially, blank reactions without any catalyst were carried out; the results are shown in Figure 5. The p-NP demonstrated an intense UV absorption band at about 315 nm. After adding $NaBH_4$, it was found that the absorption band was shifted towards 400 nm due to the formation of nitrophenolate ions.

The Co/CNF demonstrated excellent catalytic activity towards the reduction reaction of para-nitrophenol to para-aminophenol in water. To our delight, even a very low amount of catalyst was also found to be good enough to achieve 100 % reduction of para-nitrophenol with a higher reaction rate (8 min). Results of kinetic study of para-nitrophenol reduction with (Co/CNF) catalyst are shown in Figure 6. The high reaction rate could be explained by the immobilization of nanoparticles on a cellulosic catalytic support which increases surface area and makes cobalt metal nanoparticles more available and more accessible to the reagents in the reaction medium, and therefore increases the catalytic performance of the catalyst.

3.3. Recyclability Study

Co/CNF catalyst can be easily separated from the reaction mixture and reused with 100% reduction efficiency for five cycles. This catalyst has very interesting catalytic potential applications in eco-friendly processing and in heterogeneous catalysis and reduction reactions according to green chemistry.

Another important advantage of heterogeneous catalysis is the ease of separation from the reaction medium by a simple filtration and its reuse for another reduction reactions. In our case, after filtering the mixture at the end of the reduction reaction and washing the catalyst several times with distilled water, the catalyst was recycled several times to highlight its recyclability. The results of this recyclability study are presented in figure 7.

These experiments were performed to test the catalyst recyclability. The results of the recyclability tests show that the regenerated catalyst has a good catalytic activity with a total conversion of reagent even at the 5th reduction cycle.

The heterogeneous catalyst Co/CNF possesses a high catalytic activity in the reduction reaction of para-nitrophenol to para-aminophenol by $NaBH_4$, taken as a model reaction.

On the other hand, from these results (figure 7) we can see that this reaction is of order 1; so the reaction rate is dependent of the concentration of the reactive.

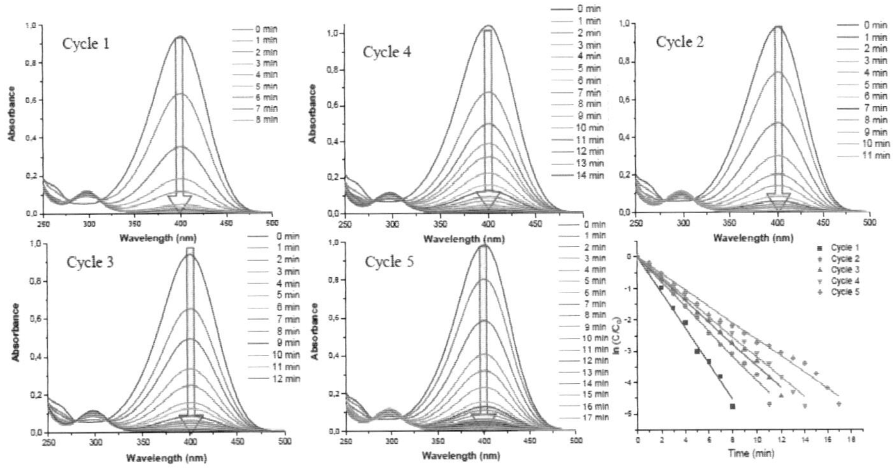

Figure 7. Recyclability study of Co/CNF Catalyst.

4. Conclusion

We successfully synthesized a recyclable heterogeneous catalyst Co/CNF based on cellulose nanofibers (CNF) used as an ecologically catalytic support.

The microstructure investigations showed that Co nanoparticles are homogeneously dispersed in the walls of the CNF based aerogel. This heterogeneous catalyst shows a high catalytic activity towards the reduction reaction of para-nitrophenol with a 100% yield and was easy to recover and recycle. Moreover, it showed excellent reusability for five successive cycles with 100% reduction efficiency.

References

[1] Eisa W H, Abdelgawad A and Orlando J. Rojas 2018 *ACS Sustainable Chemistry & Engineering*, pp. 1-37
[2] Gopiraman M, Deng D, Saravanamoorthy S, Chung M and Kim I S 2018 *RSC Advances* **8** pp. 3014-23
[3] Gu j, Hu C, Zhang W, Dichiara A 2018 *Applied Catalysis B: Environmental* pp. 1-25
[4] Mondal A, Adhikary B and Mukherjee D 2017 *The Bulletin of Materials Science* pp. 321–8.
[5] Choo G C Y, Miyamura H and Kobayashi S 2015 *Chem Sci*, **6** (3), pp.1719-27.
[6] Mondal P,Sinha A,Salam N, Roy A S, Jana N R and Islam S M 2013 *RSC Adv*. **3**, pp.5615–23.
[7] Alvarez R, Diaz-Alvarez A E, Crochet P and Cadierno V 2013 *RSC Adv*., **3**, p. 5889.
[8] Hu M, Jiang J, Bu F, Cheng X, Lin C and Zeng Y 2012 *RSC Adv*. **2**, pp. 4782–4786.
[9] Cao Y W, Jin R C and Mirkin C A 2002 *Science* **297**, pp. 1536.
[10] Sadanand V, Rajini N, Satyanarayana B, and Rajulu A V 2016, *Int. J. Polym. Anal. Charact.* **21**, p.408.
[11] Benhamou K, Dufresne A, Magnin A, Mortha G and Kaddami H 2014 ,*Carbohydrate Polymers*, **99**, pp. 74– 83.
[12] Bendahou D, Bendahou A, Seantier B, Grohens Y and Kaddami H 2015, *Industrial Crops and Products*,**65**, pp. 374-38

IOP Conf. Series: Materials Science and Engineering **827** (2020) 012008 doi:10.1088/1757-899X/827/1/012008

Synthesis of $BaCO_3$ particles tailored by carboxylated cellulose fibers

A Jada[1,2,4], S Ladhari[1, 2] and K Jradi [3]
[1] Institut de Sciences Des Matériaux De Mulhouse (IS2M-UMR 7361 CNRS- UHA), Mulhouse-France
[2] Université de Strasbourg, Strasbourg, F-67081, France.
[3] Centre for Research on Lignocellulosic Materials. University of Quebec at Trois Rivieres, 3351, boulevard des forges. Trois –Riviéres, QC, G9A 5H7
[4] Author to whom any correspondence should be addressed

amane.jada@uha.fr

Abstract. The present work deals with the use of carboxylated cellulose fibers, thereafter called Cellulose Nano Fibers (CNF), as eco-friendly materials and sustainable resources for remediation and valuation, of metal ions present in contaminated water. Barium divalent cation (Ba^{2+}) is among the most common heavy metal pollutants found in water such as arsenic, cadmium, chromium, copper, nickel, lead and mercury. Thus, CNFs containing carboxylate groups were first prepared by the oxidation of the cellulose fibers using the NaOCl-NaBr-TEMPO system, and thereafter, they were used as adsorbent for Ba^{2+} ions removal from $BaCl_2$ aqueous solutions. Such Ba^{2+} removal was evidenced by the decrease of both the CNF surface charge and the CNF absorbance, upon the addition of the metal divalent cation, $BaCl_2$ to the CNF aqueous dispersions. It is shown that the interaction occurring between the fiber carboxylic groups and the positively charged metal ions, Ba^{2+}, occurs via electrostatic interactions and leads to the CNFs –Ba^{2+} complex formation. In the second part of this work, we investigated the effect of the CNFs on the crystallogenesis of barium carbonate particles, $BaCO_3$. Thus, the CNFs aqueous solution was first added to barium chloride $BaCl_2$ solution, and the mixture was stirred for a period of time (complexation time), prior the addition of sodium carbonate (Na_2CO_3) solution. The precipitation of barium carbonate particles in the presence of CNFs occurred at ambient temperature, in a water-jacketed Pyrex glass vessel. Finally, the resulting Barium carbonate- CNFs composites were examined by various methods such as Scanning Electronic Microscopy (SEM), and X-ray Diffraction. From the SEM data, two $BaCO_3$ morphologies were observed, dendritic and bent or rods particles, respectively, in the absence and in the presence of the CNFs.

1. Introduction

Barium carbonate ($BaCO_3$) is a mineral used in various applications such as in the production of optical glass, ceramic, barite, pigment and as a precursor for magnetic ferrites and ferroelectric materials [1-6]. At Standard Ambient Pressure and Temperature (T=298.15 K (25 °C), Pressure, P= 1 bar), witherite is the stable form of $BaCO_3$ and has the aragonite structure with space group Pmcn.

Several methods have been used to prepare $BaCO_3$ particles. Hence, Li et al., [7] have used, water in oil microemulsion, as crystallization medium to prepare $BaCO_3$ nanobelts and nanorods. These authors found that the $BaCO_3$ synthesis was function of main three parameters which are the reactant concentration, the surfactant and the aging time.

Thongtem et al., [8] prepared Strontium carbonate ($SrCO_3$) and Barium carbonate ($BaCO_3$) nanoparticles by sonochemical method using strontium nitrate or barium nitrate and sodium carbonate as starting materials at 80 °C, and for 1–5 h. *Xu and Xue* [9] used hydrothermal method to prepare uniform hexagonal pencil-like $BaCO_3$ whisker, by the chemical reaction of barium chloride and urea. In the present work, barium carbonate ($BaCO_3$) particles were prepared from the mixture of $BaCl_2$ and Na_2CO_3 on the carboxylated cellulose nanofiber as support. The aim of this study is to show how modified cellulose nanofiber (CNF), specifically CNF bearing carboxylate groups, can be used as biomaterials for remediation and valuation, of metal ions such as Barium cations, present in contaminated water, with an ultimate interest in developing Barium carbonate-CNFs composites for various applications

2. Material and methods

The barium chloride salt, $BaCl_2$ (Rectapur), was used as received without further purification. The cellulose nanofiber used as additive in the $BaCO_3$ precipitation medium were obtained by TEMPO (2,2,6,6-tetramethylpiperidine-1-oxyl radical)-mediated oxidation and successive mild disintegration in water of native wood celluloses, [10]. This procedure is known to introduce carboxylate moieties on the cellulose structure and to preserve the micrometric length of the cellulose nanofibers (including both the crystalline and amorphous domains). The resulting individual cellulose nanofibers (CNF) have 3–4 nm wide and 30 nm to few microns in length, i.e. the nanofibers have aspect ratios greater than 100.

2.1. Determination of the CNF carboxylate groups amount and assessment of Interaction between Ba^{2+} and CNF carboxylate groups

The Cellulose Nano Fibers (CNF), in their oxidized form, were titrated by cationic surfactant, cetyltrimethylammonium chloride, CTACl, and the resulting Streaming Induced Potential, SIP, was measured by Müteck titrator (particle charge detection, PCD, Müteck instrument), to determine the carboxylate group's content present in the CNF, following a similar method as described elsewhere [11]. The same SIP measurement method, was also used to evaluate the interaction occurring between the CNF carboxylate groups and the Ba^{2+} cations. Thus, the fiber aqueous dispersion was placed in the measuring cylindrical cell and then was titrated with standardized $BaCl_2$ aqueous solution.

In order to assess the electrostatic interaction occurring in aqueous solution between the fiber carboxylic groups and the positively charged metal ions (Barium (II) cations), the UV spectrophotometry method, in the wavelengths range λ= 220-280 nm was used to measure the absorbance of the cellulose nanofibers, CNF, aqueous dispersions in the absence and in the presence of increasing $BaCl_2$ salt amounts

2.2. Preparation of barium carbonate particle on the CNF support

Barium carbonate particles were prepared in the absence and in the presence of cellulose Nanofibers CNF in solutions super saturated with $BaCl_2$ and Na_2CO_3, at 20°C.

3. Results and discussions

The impact of $BaCl_2$ amount on the CNF aqueous dispersion absorbance is important and will help us to highlight the interaction mechanism occurring between the fiber carboxylic groups and the positively charged Barium cations. Figure 1 shows the absorbance variation versus the wavelength, λ, in the range λ=220-280 nm, for CNF aqueous dispersions, in the absence and in the presence of various $BaCl_2$ amounts.

3.1. Absorbance measurements

As can be seen in Figure 1, in all cases, the increase of the absorbance when λ decreases from 280 to 220 nm is due to the contribution from the Rayleigh scattering which increases proportion to particle size and smaller light wavelengths. Further, the absorbance of the CNF aqueous dispersions decrease substantially as the $BaCl_2$ amount, expressed in moles, increases. Such decrease in the absorbance is due to the neutralization of the CNF carboxylate groups (COO^-) by the barium cations Ba^{2+}, leading hence

IOP Conf. Series: Materials Science and Engineering **827** (2020) 012008 doi:10.1088/1757-899X/827/1/012008

to less ordered CNF structure in the aqueous solution, and likely to the settling of the CNF-Ba complexes.

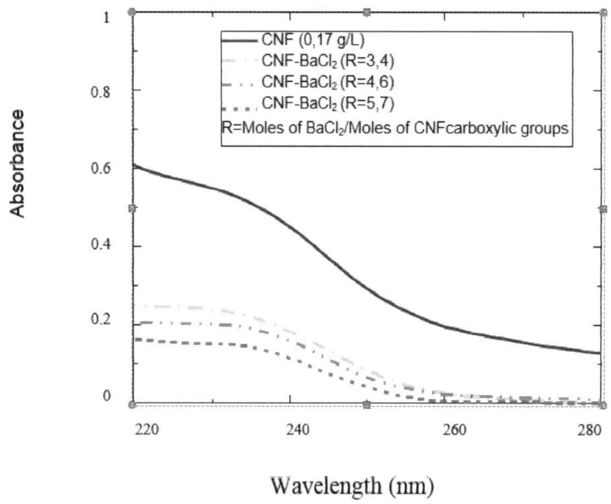

Figure 1. Absorbance of CNF aqueous dispersions, in the absence and in the presence of various $BaCl_2$ amount.

3.2. Streaming induced Potential

Figure 2 presents the variation of Streaming Induced Potential (SIP) of CNFs versus the amount of added divalent $BaCl_2$ salt, at constant initial salt concentration = 0.1 M. The curve in Fig. 2, shows that an increase in $BaCl_2$ amount leads to a decrease in the magnitude of the CNFs initial SIP value, and a plateau SIP value is reached at higher salt concentration.

As can be observed in the figure, diluted solution of Ba^{2+} divalent cations, exhibits an important decrease in the magnitude of the CNFs surface charge, without the solid surface charge sign reversal. Note that before any addition of $BaCl_2$ salt, the CNFs are initially negatively charged owing to the ionization of the carboxylate groups present on their surfaces. Thus, the addition of Barium ions leads to the formation of complexes between the Ba^{2+} cations and the CNF carboxylate groups reducing hence, the magnitudes of the CNFs surface charge and surface potential.

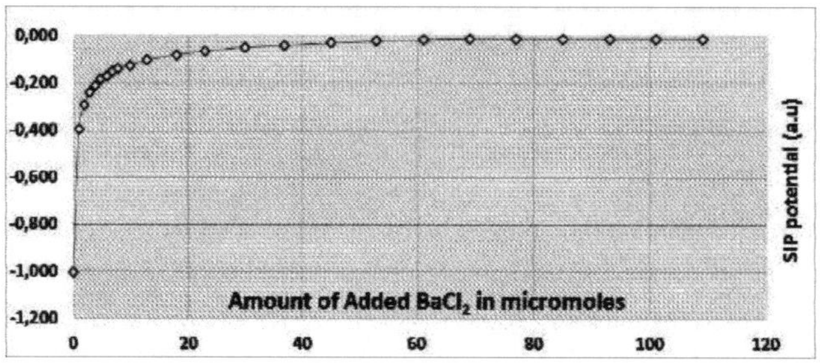

Figure 2. Titration of the CNF (1.74 mg) by BaCl2 (0.1 M) aqueous solution.

3.3. Morphology and crystalline structure of barium carbonate particles as prepared in the presence of CNFs

Figure 3 presented the SEM micrographs of the CNF in the absence of $BaCO_3$ (Figure 3 a), the synthesized $BaCO_3$ in the absence of CNF (Figure 3b, dendritic structure), and the $BaCO_3$ (having belt and/or rod shapes) prepared on the CNF surface Figure 3c). The results obtained showed that CNFs form complexes with barium ions which modify their electrical charges, their textures and morphologies.

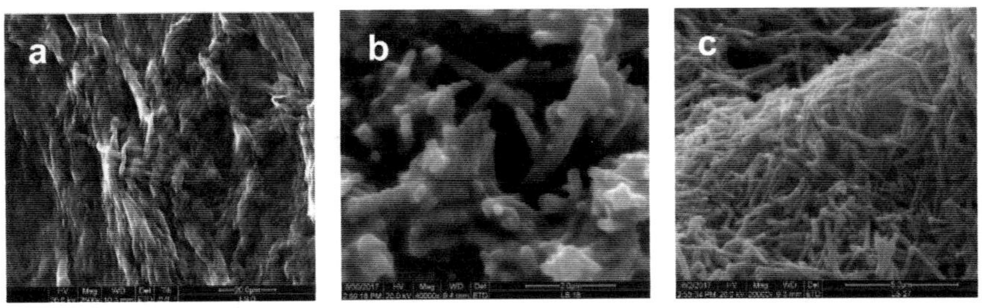

Figure 3. SEM images of (a) CNF; (b) BaCO3 (dendritic structure) and (c) CNF- BaCO3 (belt and/or rod structures) composite.

Furthermore, in the Barium carbonate-CNF composites, the barium carbonate ($BaCO_3$) particles are deposited on the surface of the CNF. The barium ions adsorbed on the surface of the cellulose fiber are transformed into barium carbonate particles with a rather narrow size distribution.

4. Conclusion

In this work, it was shown that Cellulose Nanofibers can be used as adsorbent for Ba^{2+} cations removal from water, and as support for the conversion of the CNFs $-Ba^{2+}$ complexes, formed in situ, to metallic carbonate, $BaCO_3$ particles, anchored on the CNF surface. The final $BaCO_3$-CNF composite obtained is less toxic than Ba^{2+}, and it has many chemical and industrial applications.

References

[1] Jiao JQ, Liu X, Gao W, Wang CW, Feng HJ, Zhao XL, Chen LP 2010 *Materials Research Bulletin* **45** pp 181–185.
[2] Melero-Garcia E, Santisteban-Bailon R, Garcia-Ruiz JM 2009 *Crystal Growth and Design* **9** pp 4730–4734.
[3] Balamurugan K, Kumar NH, Ramachandran B, Rao MSR, Chelvane J A, Santhosh P N 2009 *Solid State Communications* **49** pp 884–887.
[4] Ni YH, Zhang HY, Hong JM, Zhang L, Wei XW 2008 *Journal of Crystal Growth* **310** pp 4460–4467.
[5] Homeijer SJ, Olszta MJ, Barrett RA, Gower LB 2008 *Journal of Crystal Growth* **310** pp 2938–2945.
[6] Buscaglia MT, Buscaglia V, Alessio R 2007 *Chemistry of Materials* **19** pp 711–718.
[7] Li L, Chu Y, Liu Y, Dong L, Huo L and Yang F 2006 *Mater. Lett* **60** pp 2138-2142
[8] Thongtem T, Tipcompor N, Phuruangrat A and Thongtem S 2010 *Mater. Lett* **64** pp 510-512
[9] Xu J, and Xue D 2006 *J. Phys. Chem. Solids* **67** pp 1427-1431
[10] Saito T, Nishiyama Y, Putaux JL, Vignon M, Isogai A 2006 *Biomacromolecules* **7** pp 1687-1691
[11] Wäsche R, Steinborn G 1997 *Deutsche KeramischeGesellschaft DKG* **74** pp 235-239.

A Comparative study between TiO_2 and ZnO photocatalysis: Photocatalytic degradation of textile dye

Meriem Kouhail [1,2,4], Kaoutar Elberouhi[3], Zakia Elahmadi [1], Abbes Benayada [1], Said Gmouh[3].

[1] Mohammed-V University Rabat-EMI, Laboratory of Applied Geophysics, Geotechnics, Engineering Geology, and the Environment, Mohammedia School of Engineering, Rabat, Morocco.

[2] LEC -ESITH, Laboratory of Expertise and Control - Higher School of Textile and Clothing Industries, Casablanca, Morocco.

[3] Laboratory of Engineering and Materials LIMAT, University Hassan II de Casablanca, Morocco.

[4] Author to whom any correspondence should be addressed

kouhail@esith.ac.ma / mkouhail@gmail.com.

Abstract. The present study involves the photocatalytic degradation of three dyes: Reactive Bézactiv yellow (RBY); Direct Triamine yellow86 (DY86); and Methyl Orange (OM) with different structure and containing different substitute groups. The photocatalytic degradation has been investigated using two types of synthesis semiconductors namely titanium dioxide (TiO_2) and zinc oxide (ZnO), as photocatalyst, in aqueous solution under solar irradiation.

The synthesized nanoparticles were characterized by: Fourier Transform Infrared (FTIR), X-ray diffraction (XRD), UV visible diffuse reflectance spectroscopy and scanning electron microscopy (SEM). The progress of photocatalytic degradation of the commercial dyes, has been observed by monitoring the change in substrate concentration of the compound employing High-Performance Liquid Chromatography (HPLC). The photodegradation rate was determined for each experiment and the highest values were observed for ZnO.

Keywords: Comparative; Photodegradation; Solar light; Textile dye; TiO2; ZnO.

1. Introduction:

Many industries such as textile, plastics, paper, and pulp generate streams of waste effluents, which contain a considerable amount of organic dyes. When these compounds are discharged to the main water bodies without any prior treatment, they can cause havoc to the ecological balance in the environment, as these molecules have carcinogenic and mutagenic properties towards aquatic organisms and thus pose threat to human life at the end of the food chain [1].

Various chemical and physical processes, such as chemical precipitation and separation of pollutants, coagulation, electrocoagulation [2], elimination by adsorption on activated carbon, etc. are applied for color removal from textile effluents. One difficulty with these methods is that they are not destructive but only transfer the contamination from one phase to another, therefore, a new and different kind of pollution is faced which calls for further treatment [2].

In particular, photocatalysis is a more promising tool. Moreover, photocatalysis can be used to cause redox transformations and decompose a dye molecule. The use of photosensitive semiconductors such as TiO_2, ZnO, Fe_2O_3, CdS, ZnS and V_2O_5 has been reported in the literature for their use in reducing the color of the dye solutions due to their environmental-friendly benefits in saving resources such as water, energy, chemicals, and other cleaning materials [3].

The main aim of this present work is to study the photocatalytic degradation of different textile dyes with different molecular structures like reactive dye, direct dye, and the methyl orange in the presence of synthesized ZnO and TiO_2 by Solar light(SL). X-ray diffraction (XRD), MEB, UV-vis and FT-IR were used to characterize these catalysts.

2. Material and Methods

2.1. Material

Zinc chloride, NaOH is a reactive agent to synthesize ZnO nanoparticles; the titanium n-butoxide $Bi_4Ti_3O_{12}$ and ethanol is a reactive agent to synthesize TiO_2 were purchased from Sigma-Aldrich.
Reactive Bezactiv Yellow (RBY), Direct Triamine yellow86 (DY86) are azo dyes used in the industry for the dyeing of cotton and cellulose fiber; and Methyl Orange (OM) is an indicator.

2.2. Synthesis

TiO_2 nanoparticles were synthesized by a sol-gel method using titanium n-butoxide $Bi_4Ti_3O_{12}$ as precursors. ZnO nanoparticles were synthesized by direct precipitation method using zinc chloride and NaOH as precursors.

2.3. Characterization

The powder X-ray diffractograms (XRD) of the ZnO and TiO_2 samples, were recorded and analysed to determine their crystal structure and lattice parameters. The average particle size of each metal oxide was estimated using the Debye–Scherrer formula. The particle morphology was analyzed through scanning electron microscopy (SEM) imaging and the Fourier transform infrared (FTIR) was recorded on VETEX70, the spectrum of the precursor was between 400 and 4,000 cm^{-1} and diffuses reflectance. The optical properties of the catalyst were analysed by diffuse reflectance UV–Vis spectroscopy using a Shimadzu UV–Vis spectrophotometer with an integrating sphere and a spectral reflectance standard over a wavelength range of 200–800 nm.

2.4. Methods

The photocatalytic degradation of dyes by nanopowders was studied in TiO_2 and ZnO suspension under solar irradiation between 11.00 a.m. to 3 p.m. The average intensity of sunlight during this period is 78000 Lux unit measured with a lux metre Testo 545; every experiment is recorded in the same conditions (Contact time, pH; anions and intensity irradiations).
Initially, the nanopowder was added to the dye solution to form a suspension. Before irradiation, the suspension was magnetically stirred for 30 min in the dark to reach an adsorption-desorption equilibrium. After given time intervals, the suspensions were filtered to remove the photocatalysts and aliquot of the residual solutions were transferred to HPLC measurement. The extent of removal of the dye, in terms of the values of residual concentration $\mathbf{C_r}$ of dye, has been calculated using the following equation:

IOP Conf. Series: Materials Science and Engineering **827** (2020) 012009 doi:10.1088/1757-899X/827/1/012009

$$Cr = \frac{Ct}{Ci} \quad (1)$$

Where C_i is the initial concentration of dye (ppm) and C_t is the final concentration of dye (ppm) at a given time.

3. Results and discussion

3.1 Characterization

3.1.1 Scanning Electron Microscopy (SEM). The morphology of TiO_2 and ZnO, presented in figure 1 and 2 respectively, shows mono-dispersed nanoparticles, uniformly distributed and spherical shape. The larger particles in these figures may be attributed to the aggregates of the smaller particles.

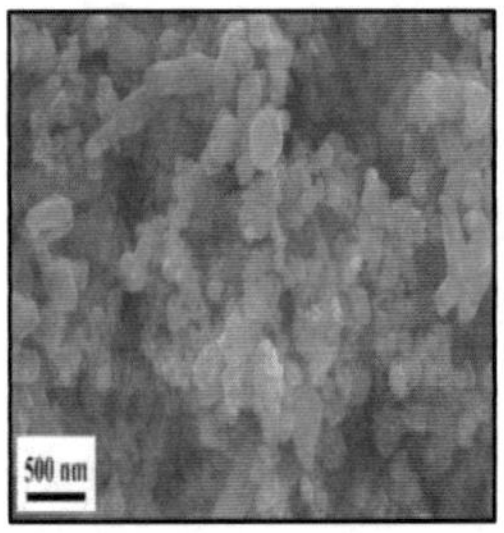

Figure 1. SEM of TiO_2

Figure 2. SEM of ZnO.

3.1.2 DRX: TiO_2 support structure. The diffraction peaks of TiO_2 (2θ=25.25°, 2θ=37.8°, 2θ=47.9°, 2θ=53.59° and 2θ=62.36 °) which are the typical characteristic peaks of anatase TiO_2. The diffraction data were in good agreement with JCPDS files # 21-1272; with a nanoparticle size 12 nm.
The diffraction peaks of ZnO (31.75 °, 34.44 °, 36.25 °, 47.54 °, 56.55 °, 62.87 °, 66.38 °, 67.91 °, 69.05 °, and 72.61 °) can be indexed to those of hexagonal wurtzite ZnO and these values correspond to the file (JCPDS Card No.00-005-0664); with a nanoparticle size 43 nm. [4] [5].

3.1.3 FT-IR analysis. The FT-IR spectrum of TiO2 showed various characteristic peaks an absorption band at 3434.88cm^{-1} is related to stretching vibration of hydroxyl (OH); the other peaks at 1634 cm^{-1} were indicated to stretching of titanium carboxylate, which formed from TTIP and ethanol as precursors only the strong absorption band between 501 and 444 cm^{-1} was remained, which attributed to formed of TiO_2 nanoparticles[6][7].

The FT-IR spectra of the synthesized ZnO particles. The peaks at 599cm^{-1} are related to the stretching vibrations of Zn-O bonds. The peak at 3436cm^{-1} indicates the presence of –OH residue[8][9].

3.1.4 UV–Visible absorption spectrum. ZnO possess a higher absorption than TiO_2 in UV region about 400 nm and TiO_2 spectrum possess an intense absorbance band at 360 nm, indicating a higher efficiency for the generation of electron-hole pairs than TiO_2. . Excess absorbance in ZnO requires more exposure for the valence band electrons, which may be possible to create more vacant anionic sites on the catalyst surface.

The more photogenerated electron hole pairs together with their much higher mobility in the microscale ZnO than in TiO_2 suggest that ZnO would show higher photocatalytic activity than TiO_2[8] [9].

3.2 Photocatalytic activity

The experimental data, of different photocatalytic degradation of three dyes are shown in figure 3.

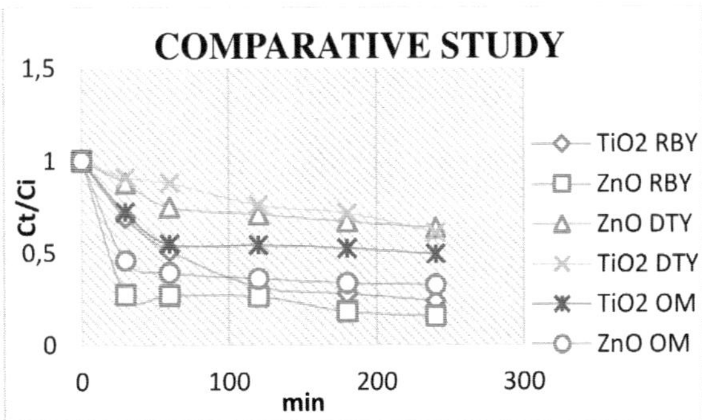

Figure 3. Degradation photocatalytic oxidation with TiO_2 and ZnO.

The three dyes (RBY, DY86, and OM), has a different efficiency, of degradation in the presence of two ZnO and TiO_2. The degradation rate of the dyes, follow the order RBY> OM> DTY.

It can be seen also that ZnO showed a better degradation efficiency than that TiO_2 suggesting that ZnO absorbs a larger fraction of the solar spectrum and absorption of more light quanta than TiO_2 [10].

The degradation of all dye occurs at a faster rate with the solar light in the first 30 min (first step) and became slow after 30 min of irradiation (second step) and more slow or constant after 60 min (third step).

The results of the first step, can be attributed to the availability of a larger number of active sites on the surface of the photocatalyst, leading to an increase in the number of hydroxyl and superoxide radicals; but during the second step, the degradation became slow; this result is probably due to the formation of intermediates and its competitiveness, with parent dye molecules in the photocatalytic degradation process, and/or to the difficulty in converting the N-atoms, of dye into oxidized nitrogen compounds, another reason for this slow kinetics is the slow reaction of short-chain aliphatics with OH radicals [11].

3.3 Chromatogram analysis

To improve the formation of intermediate products; we analyse the HPLC chromatogram of dyes, at 0 min and 240 min.

In figure 4, we observed that the surface area of the major peak at (t=0min) was reduced for all the three dyes and new peaks were appeared with a new retention times in the case of RBY and DY86; confirming the degradation of dye and/or the transformation of parent dye on an intermediates products.

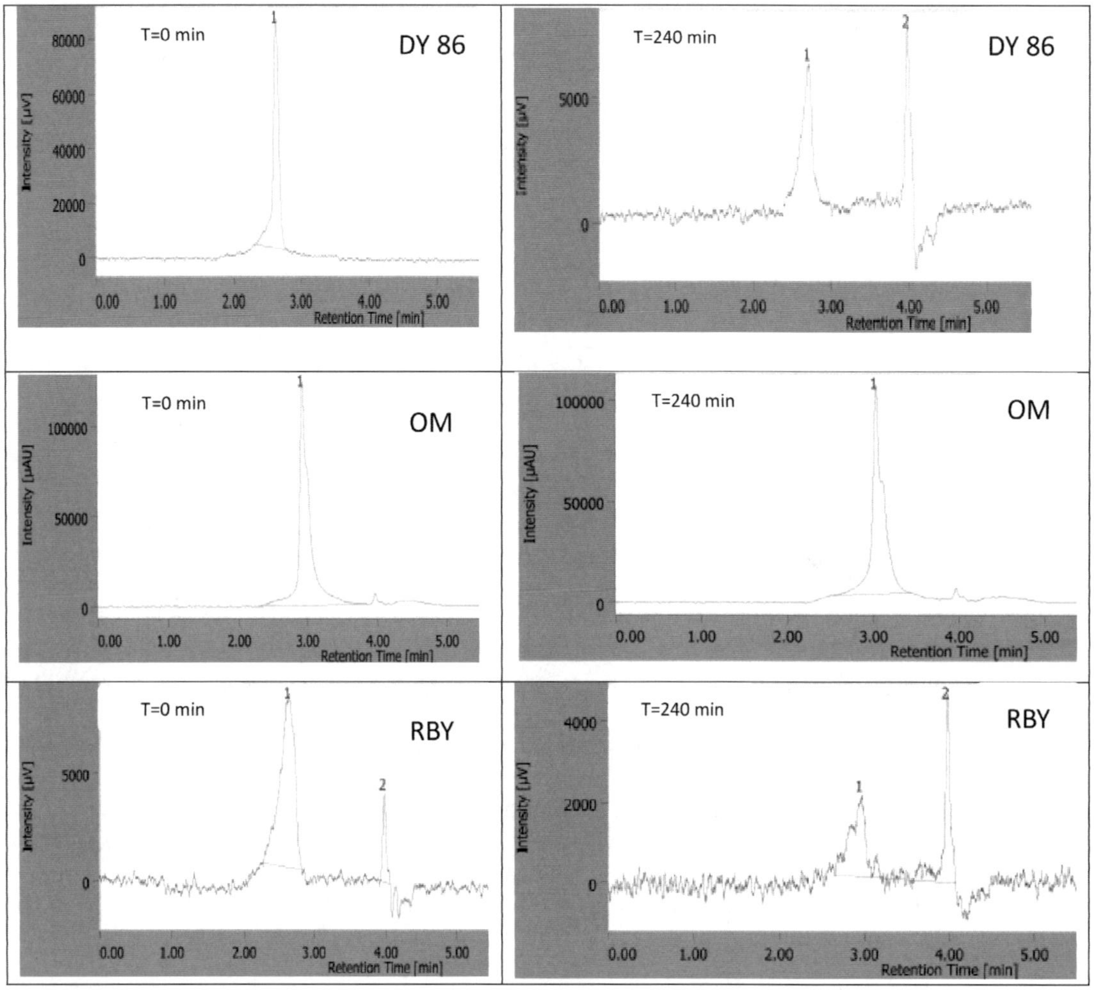

Figure 4. HPLC diagram at 0 min (a) and 240 min (b).

4. Conclusion

This study demonstrates that certain commercial dyes with different structure and different substitute groups can be decolorized and degraded by solar photocatalysis. The results of degradation are not the same is depends on the nature of the dye and the nature of the catalyst.

The formation of intermediates and its competitiveness, with parent dye molecules slow down the photocatalytic degradation process.

Reference

[1] Muthirulan P, Meenakshisundararam M and Kannan N, 2013. Beneficial role of ZnO photocatalyst supported with porous activated carbon for the mineralization of alizarin cyanin green dye in aqueous solution. *Journal of advanced research*, **4**(6), pp.479-484.

[2] Daneshvar N, Salari D and Khataee A R, 2004. Photocatalytic degradation of azo dye acid red 14 in water on ZnO as an alternative catalyst to TiO2. *Journal of photochemistry and photobiology A: chemistry*, **162**(2-3), pp.317-322.

[3] Rauf M A, Meetani M A and Hisaindee S, 2011. An overview on the photocatalytic degradation of azo dyes in the presence of TiO2 doped with selective transition metals. *Desalination*, **276**(1-3), pp.13-27.

[4] Satyanarayana Talam, Srinivasa Rao Karumuri, and Nagarjuna Gunnam,2012, Synthesis, characterization, and spectroscopic properties of ZnO nanoparticles. *ISRN Nanotechnology*,**2012**.

[5] Vijayalakhsmi R and Rajendran V,2012. Synthesis and characterization of nano -TiO2 via different method, *Archives of applied science research* ,**4**(2). pp.1183-90.

[6] Ba-Abbad M M, Kadhum A A H, Mohamad A B, Takriff M S and Sopian K, 2012. Synthesis and Catalytic Activity of TiO2 Nanoparticles for Photochemical Oxidation of Concentrated Chlorophenols under Direct Solar Radiation , *Int J Electrochem Sci*, **7**, p. 18, 2012.

[7] Al-Amin M , Dey S C, Rashid T U and Shamsuddin S, 2016,Solar Assisted Photocatalytic Degradation of Reactive Azo Dyes in Presence of Anatase Titanium Dioxide *International Journal of Latest Research in Engineering and Technology*,**2**(3) p. 9.

[8] Li Y, Xie W, Hu X, Shen G, Zhou X, Xiang Y, Zhao X and Fang P, 2010. Comparison of dye photodegradation and its coupling with light-to-electricity conversion over TiO2 and ZnO. *Langmuir*, ***26***(1), pp.591-597.

[9] Djouadi D,Meddouri M and Chelouche A, 2014.Structural and optical characterizations of ZnO aerogel nanopowder synthesized from zinc acetate ethanolic solution, *Opt. Mater.*,**37**, p. 567□571.

[10] Chakrabarti S and Dutta B, 2004,Photocatalytic degradation of model textile dyes in wastewater using ZnO as semiconductor catalyst, *J. Hazard. Mater.*, **112**(3), p. 269□278.

[11] Sakthivela S, Neppolianb B, Shankar M V,Arabindoob B, Palanichamyb M and Murugesanb V,2003. Solar photocatalytic degradation of azo dye: comparison of photocatalytic efficiency of ZnO and TiO2, *Solar Energy Materials & Solar Cells* **77**(1), pp 65–82.

IOP Conf. Series: Materials Science and Engineering **827** (2020) 012010 doi:10.1088/1757-899X/827/1/012010

Value added creation with advanced mass-customized textile labels

M. Barteld[1,3], R. Gebhardt[1], F. Siegel[1], U. Portsch[2]

[1] Sächsisches Textilforschungsinstitut e.V., Chemnitz, Germany
[2] Qualitype GmbH, Dresden, Germany
[3] Author to whom any correspondence should be addressed

marco.barteld@stfi.de

Abstract. Process-oriented value-added creation in textile networks for mass customization tries to close the backlog and the research gap for mass customization in textile networks by using IT-based security labels according to the research projects "Profund" and "TechProTex - Etikett". Special attention will be paid to the complex product identification and tracking to both B2B and B2C markets. By using the developed platform, the cycles for the new development of a product are significantly reduced. New labels printed with various identification features and QR Code are available in four security levels. The identification is counterfeit-proof, customer-friendly as well as the basis for future-oriented cross-company textile value chains.

1. Initial point

The importance of counterfeit products and brands is growing worldwide. Such products are having increasingly serious consequences, especially for small and medium-sized companies (SMEs), such as costly proofs of complaints or product liability, loss of image and loss of turnover. The marketability of SMEs in the textile chain thus depends, on the one hand, on the reliable differentiation between original and fake goods. On the other hand, both the market and the more and more automated production environment are oriented towards digitally networked, customer-integrated and smart value creation with security requirements at a high level.

Especially in the case of process-oriented mass customization in textile networks, SMEs must establish, demonstrate and ensure trusting cooperation with each other in the B2B sector. For this purpose intelligent framework conditions and tools that can be easily integrated into the textile chain are needed [1-2]. At the same time customers are an increasing part of textile manufacturing and sales processes. Technical textiles are not only becoming smarter but are also more transparent and informative, thus promoting so-called customer integration competence.

The development of textile based materials in combination with IT-based technologies for a new material protecting label is seen as a sustainable solution. This label was developed as a combination with digital-based identification and information scenarios with individual, counterfeit-proof markings with suitable printing technologies within the research project "TechProTex - Etikett". Individual, integrable security features were combined to form a marking that is highly resistant to counterfeiting in the long term. The paper refers to a label service concept with scalable security levels that can be adapted to different application areas in the context of a future-oriented textile label. This label is considered in the context of mass customization and envisioned SME cooperation from the research project "Profund".

Included are basically visible and invisible functions, in particular directly readable information such as brand, manufacturer, size or maintenance instructions and additional data that can either be accessed directly via a QR code or encrypted for an authorized user group, such as batch or tracking number, material origin or production details. As a reference object, an online platform with product information was developed. Manufacturers, dealers and customers can virtually zoom in on the details of the textile product and get more information about it. It is assisted in a targeted manner and integrated into the textile life cycle.

The increasing demand for ecologically and fairly produced textiles and the knowledge of their origin can also be channelled through the new label and transferred into a format that is very effective in the media. In the context of social digitization, successful textile manufacturers and service providers need new tools both to visualize their products and to occupy their social reputation in terms of transparency, credibility and ethical responsibility.

2. Approach

The structuring and realization of the future-oriented label is based on the individual development modules within the framework of customer-specific ready-to-wear production, printing, labelling technologies, process modelling, data management and logistics.

A different label structuring was created by process engineering, modelling and development of the connection between textile material and securing marking particles. The basis was a digital printing ink that could be used in the label as well as pastes for screen and transfer printing. The printing technology was adapted to the finishing of various flat linen products or to the finishing of contract textiles with innovative, customizable labels and integrated into the textile manufacturing process.

Process modelling led to a coded, copy-protected and controlled information deposit as well as to the provision of necessary interfaces, algorithms and visualizations. For example, it was planned to make the information to be read more flexible in various national languages. By incorporating this flexibility into the data solution, the textile labels can be produced cost-effectively in larger quantities. The prerequisite for this is the use of interfaces between printing systems, merchandise management systems, readout units and data management.

The question of an economic vision to organize and configure textile value creation was analyzed both in a corporate network and in the B2C sector. Requirements and information structures at the manufacturing, customer and sales levels have been classified and converted into security levels that can be implemented by the label.

3. Results

3.1. Mass-customized textile labels

Invisible security features (anti-counterfeiting) were added to the label and the textile product, both in terms of printing technology (screen printing and micro valve printing) and textile technology, in particular in terms of textile surface formation and ready-made clothing. The printing of the security features was realized by developing a special ink for application by means of a micro valve nozzle and two pastes suitable for screen and transfer printing. In order to achieve the security-compliant and IT-functional distribution, these security inks and printing pastes were applied to the respective textile surface in defined geometries using screen or digital printing processes.

It was possible to produce the marking spots using the direct printing process and, based on this, the complete label. Two formulations based on polyurethane and polyvinyl chloride were developed. Figure 1 shows an example of a section of a white polyester fabric screen-printed under UV radiation. The layer quality is sufficient for use in corporate wear.

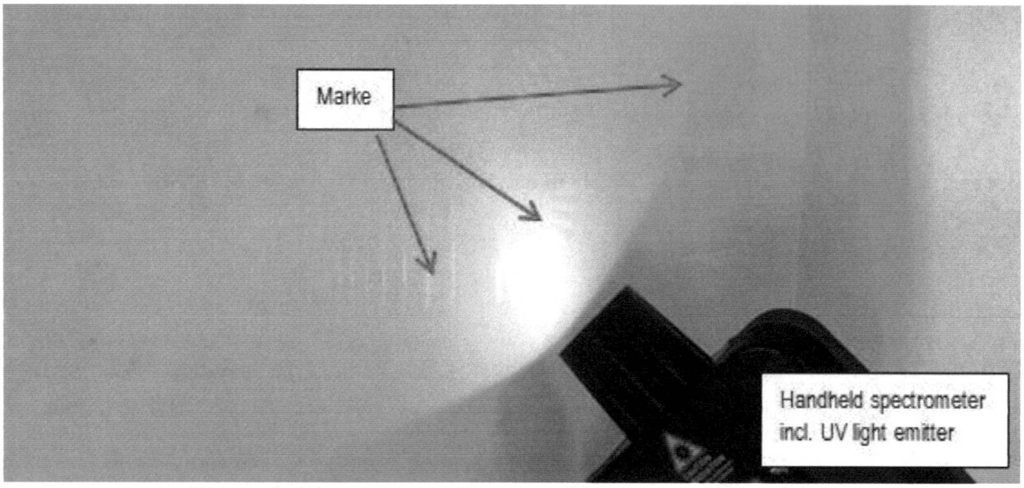

Figure 1. Direct screen-printed marking on a polyester fabric.

As result, different demonstrators were produced. Figure 2 shows a printed label with QR code, security feature and the logos of all partners involved. The label information can either be accessed directly via a QR code or selected (e.g. batch or tracking number, material origin or production details) or made available in encrypted form to an authorized user group.

Figure 2. Printed label demonstrator.

Four security levels have been implemented.

- BASIC (Additional information without identification): Open accessible additional information without identification, ohne key, without additional security features.
- ADVANCED (Additional information with identification): Open accessible additional information, information with access control, e.g. special information on batches, 2D QR code identifies up to batch level, simple PIN (by manufacturer) on product/label as key for access.
- LOWTECH - Identification with individual code and marking characteristic, e.g. batch no.: same as level 2, PIN has a higher security architecture, e.g. 2 parts, simple visible or invisible marking, identification without much effort.
- HIGHTECH - High security level with identification, code and more than one security feature: same as level 3, PIN with even higher security architecture, highly secured - MasterPin, complex marking as integrated high-tech marking, technology or specialist knowledge required for identification.

Non-destructive reading of the marking is ensured by a hand-held reader that is individually adjusted to the exact luminescence frequency of the printed label. Commercially available sensors can be used (e.g. Tailor-Spec as a mini spectrometer variant), which can also be used as an offline system or in the industrial workflows and have production speeds of up to 100 per second. Identification is carried out automatically according to the key-lock principle. End devices can be connected via a USB connection and the reading of the label marking can thus be coupled with an existing enterprise resource planning system.

In order to be able to provide the desired information content of the label for each individual customer at each security level, data from different areas of the textile chain as well as from the different actor groups were considered. In addition to the usual consumer information, additional information such as the origin of individual textile components, ecological footprint, collection recommendations or other technical details on textile construction are of interest. Information on the article number, relevant certificates, quality gradations and even error messages are also important for retailers. On the other hand, the industrial or manufacturer side, , is more interested in test values, serial numbers and information on the routing of the textile chain or on product and brand piracy. In total, the most important 150 items of information were worked out in grouped data categories, of which approx. 85 % were transferred to information technology structuring.

The large number of label variants and individual conceptual views that can be implemented in terms of textile, information and market technology have been combined and addressed to form a standard for mass-customized labels (figure 3). The individual elements of the information are aligned to the target groups in blocks and are subordinated to the four security levels. This results in a roadmap for each label variant, depending on the specific application.

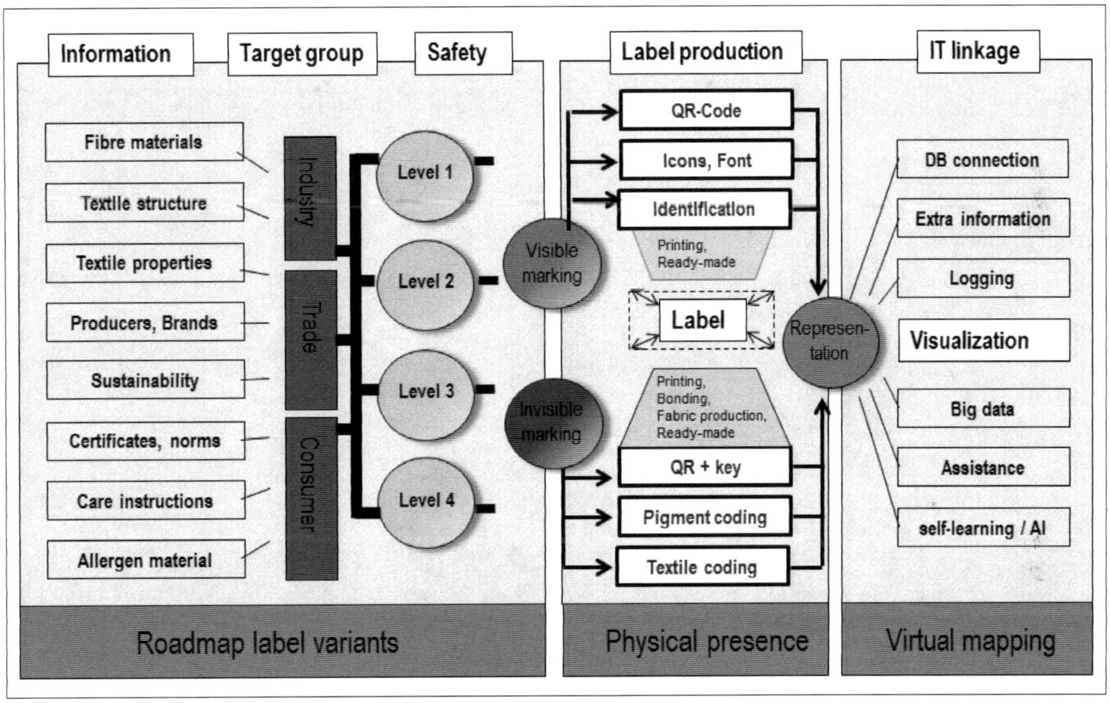

Figure 3. Mass-customized label standard.

Depending on the intended visibility, the textile technical implementation takes place on the one hand with QR code, symbols, fonts and textile labels and on the other hand with QR code including key, pigment coding or textile coding. In principle, the physically produced label is always accompanied by its virtual image. In addition to the actually embodied product characteristics, information on proper laundry care and trade information, this enables the storage and visualization of further data levels. The mass-customized label with its QR code is understood here as a bridge to a comprehensive information pool (portal) to which further scenarios can be linked in the future, such as assistance systems, big data or self-learning modules. Entering the virtual image is controlled by the different key strengths and positions in the security levels.

3.2. Complex product platform

The safe and smart textile label acts as an advantageous customer-specific bridge between textiles, platform and textile chain. In various scenarios, the mass-customized label connects the user with the information to be provided. On the basis of mass-customized labels and MC-specific issues [1-3], a

Proof of Concept (PoC) has been developed which shows how IT-based security labels can be implemented along the Co-Creation (CC) process.

The MC idea is reflected in the PoC by the fact that a platform is proposed which can be used by any number of providers for value creation process coordination (similar to a social network for value creation processes). The CC concept is taken into account, as the customers of the network also have access to the platform in order to be integrated into the CC process and to increase the quality of value creation. The PoC addresses the issue of process modularity by storing the solutions as configurations of a production process instance. The knowledge base is populated with existing company data. It can be further extended by newly created solutions on the platform, thus creating a learning system [4].

With the customer-specific product presentation, the label information can either be called up or selected directly via a QR code, for example via batch or tracking number, material origin or production details. Encrypted access to product data or settings is also possible for an authorized user group. This makes the textile product clearly identifiable, traceable and customer-friendly. As a reference object, an online platform with product information has been developed. The demonstrator refers to the developed platform (Figure 4).

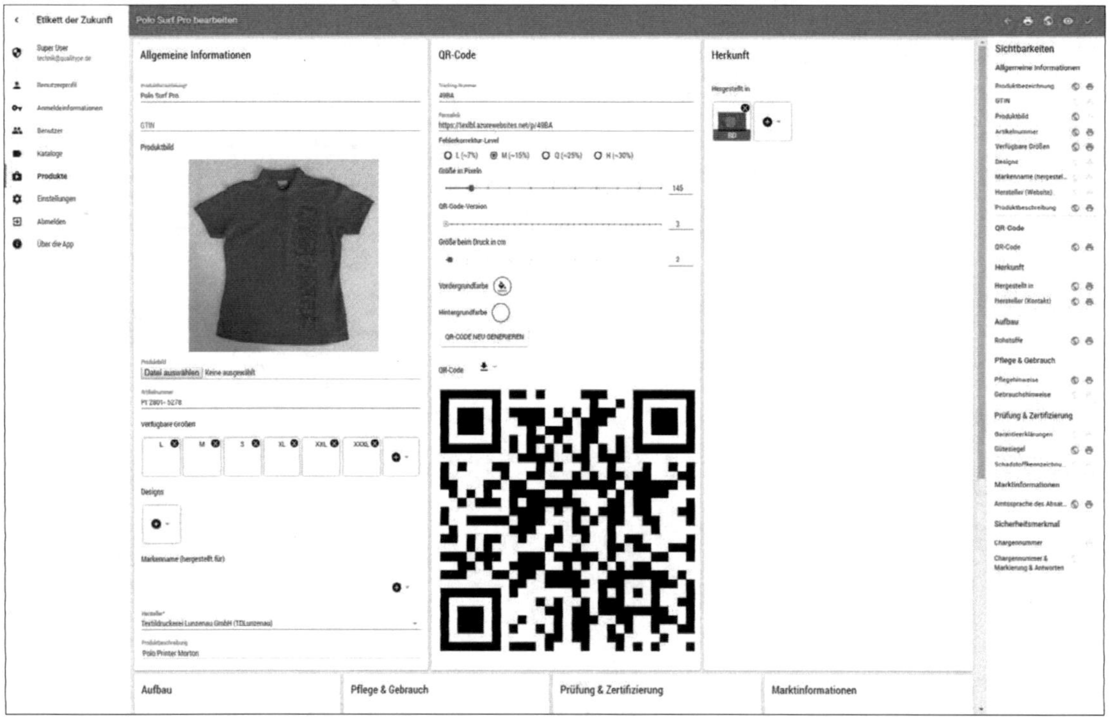

Figure 4. Product information linked and readable with the label of TDL (Textildruckerei Lunzenau GmbH, Penig, Germany) - Backend view.

In an initial pilot phase, interested textile companies and fashion labels were able to apply the mass-customized label to their articles and show customers individual information, pictures or data of the production history, depending on the security level. The origin of individual textile components and the ecological footprint of the whole textile chain are playing an increasingly important role. The rising demand for ecologically and fairly produced textiles and the knowledge of their origin can be channelled through value added creation with advanced mass-customized textile labels and transferred into a format that is very effective in the media. In the context of social digitalization, successful textile manufacturers and service providers need new tools both to visualize their products and to demonstrate their social reputation in terms of transparency, credibility and ethical responsibility.

IOP Conf. Series: Materials Science and Engineering **827** (2020) 012010 doi:10.1088/1757-899X/827/1/012010

4. Conclusions

The mass-customized label provides secure and unambiguous identification and traceability of textile products from the moment they are used. It supports a realignment of strategic management and innovative business models for digitally networked value creation. The label accompanies service-centric business models, co-creation of complex products through smart standardization and improvement of customer integration competence, and MC-oriented optimization of value creation processes. As an IT artefact, the label thus becomes part of both the marking and traceability solution in interaction with the underlying IT platform and a cross-company textile value chain. As a link between textile and clothing products and a future-oriented information and development data platform close to textile applications, it creates the possibility of maintaining a digital twin. A connection to different platforms (qualitype label, Profund etc.) is possible. Interested textile manufacturers or fashion companies can easily order an individual label with a specific QR code with the desired security level and, thus, offer their customers informative added value.

Acknowledgment

This work was supported by the futureTEX project PROFUND and by the ZIM project TechProTex - Etikett.

Bundesministerium für Bildung und Forschung

zwanzig20 PARTNERSCHAFT FÜR INNOVATION

PROFUND futureTEX

ZIM Zentrales Innovationsprogramm Mittelstand

We would like to thank the Federal Ministry of Education and Research (BMBF) for the funding as well as the Project Management Organization Jülich (Research Centre Jülich GmbH as Project Management Organization of BMBF) for cooperation and support. Special thanks for the support of the label project go to the project management organization ZIM-Netzwerkförderung VDI/VDE Innovation + Technik GmbH. Many thanks also to the Network Technical Product Protection in the Textile Chain (TechProTex), which supports research and development projects for marker technologies to protect against product and brand piracy.

References

[1] Gebhardt R and Grafmüller L and Barteld M and Mosig T 2016 Mass customized technical textiles – challenges to the textile industry for tomorrow *XXIV International Congress IFATCC* Pardubice 76-85

[2] Grafmüller L K and Habicht H 2016 The value of mass-customized products. Exploring its peculiarities for business customers, *7th International Conference on Mass Customization and Personalization in Central Europe, Novi Sad*

[3] Wache H and Dinter B and Kollwitz C 2019 Analytics Use Cases for Mass Customization–A Process-based Approach for Systematic Discovery *52nd Hawaii International Conference on System Sciences, Maui*

[4] Barteld M and Gebhardt R and Hönigsberg S and Dinter B 2019 Process-oriented value creation in textile networks for mass customization in SMEs *19th World Textile Conference on Textiles at the Crossroads, Ghent, Belgium*

Polypyrrole coated hematite sand (HS@Ppy) as a new catalyst for degradation of organic dye based on sulfate radicals

A Ait El Fakir[1, 2], Z Anfar[1, 2, 3], A Jada[2, 3,4], N. El Alem[1]
[1] Laboratoire Matériaux et Environnement (LME), Université Ibn Zohr, Agadir, 8000, Morocco.
[2]Institut de Sciences Des Matériaux De Mulhouse (IS2M-UMR 7361 CNRS - UHA), France.
[3] Université de Strasbourg, Strasbourg, F-67081, France.
[4] Author to whom any correspondence should be addressed

amane.jada@uha.fr

Abstract. The present work deals with the degradation in water of organic dye such as Orange G by using sulphate radicals as oxidant. Thus, a new catalyst based on polypyrrole coated hematite sand particles, (HS@Ppy) was first prepared. Thereafter, sulphate radicals were produced by activating the persulfate (PS) agent in the presence of the HS@Ppy catalyst. The resulted HS@Ppy/PS system was finally mixed with Orange G aqueous solutions for dye degradation. The data indicate that polypyrrole shell coating amount in the HS@Ppy particles has a great influence on the Orange G degradation. The HS@Ppy /PS system exhibited the highest OG removal rate (the removal rate is 98 % after 15 minutes of reaction) as compared to the various systems: PS alone, HS@Ppy alone, or HS/PS alone.

1. Introduction

The discharge of wastewater into the environment, including dyes, is considered as a toxicological and aesthetic problem. Indeed, many industries (textiles, paper, leather, etc.) continue to generate various pollutants and toxic azo dyes [1,2]. Most of them are non-biodegradable, toxic and potentially carcinogenic compared to the previous work on the oxidation of organic pollutants by sulphate radicals $SO^{-\bullet}$ [3-5]. In this present research, our objective is to find efficient and less expensive catalysts for the depollution of wastewater by the valorization of natural resources. In this sense, a natural and magnetic hematite sand (HS) catalyst has been used as a (persulfate) PS activating agent. This study provides enhanced catalytic properties of sand hematite as a new catalyst for the in situ remediation of waters contaminated with organic pollutants. Besides, this method is considered to be less expensive in terms of operating costs.

2. Material and Methods

The hematite sand used in this work was taken from the Taghazout beach, 23 km from the city of Agadir Morocco. The pyrrole monomer (Aldrich) was distilled before use. The preparation of our composite material (polypyrrole coated hematite sand (HS@Ppy)) was carried out by a simple polymerization with a known monomer / oxidant ratio. The physico-chemical properties of HS@Ppy were determined by using various structural, microstructural, and spectroscopic techniques such as XRD, FX, MEB/EDS, FTIR, and ATG. The catalytic degradation of dye by the HS@Ppy / PS system was carried out in a 100 mL beaker containing 50 mL of solution of dye at room temperature and with constant stirring. Catalyst

(1 g.L^{-1}) and PS ($Na_2S_2O_8$) concentration was (45 mg.L^{-1}) were added to the reaction.

3. Results and discussions

Thermogravimetric analysis (TGA) under air atmosphere was employed to quantify the coating amount of Ppy shell (Figure 1). For HS@Ppy, the weight rise region was displaced by an obvious weight loss stage in 200-500 °C resulting from the decomposition of polymer shell. The total mass losses (W_T) are 22%, 18%, 16% and 10% for HS@Ppy-4 to HS@Ppy-1 respectively. The ratio of Ppy shell (W_{Ppy}) was calculated by Equation (1):

$$\mathbf{W_{Ppy} = W_T - W_{H2O}} \quad (1)$$

Where W_{H2O} the ratio of H_2O (<150 °C).

The results suggested the ratio of polypyrrole to vary from 8% to 20%. The influence of shell coating amount on Orange G degradation was investigated. HS@Ppy-3 exhibited the highest OG removal rate, and was used in the following experiments (Figure 2a). The catalytic performance of our catalysts has been shown in Figure 2b. It can clearly be seen that the HS@Ppy / PS system exhibits high fading efficiency with respect to the OG toxic molecule (the removal rate is 98 % after 15 minutes of reaction). In the absence of PS, the kill rate of OG was approximately 10 % after 18 min. This is probably due to the adsorption capacity of HS@Ppy / PS and that the PS was not able to degrade the OG directly without activating agent and without light irradiation.

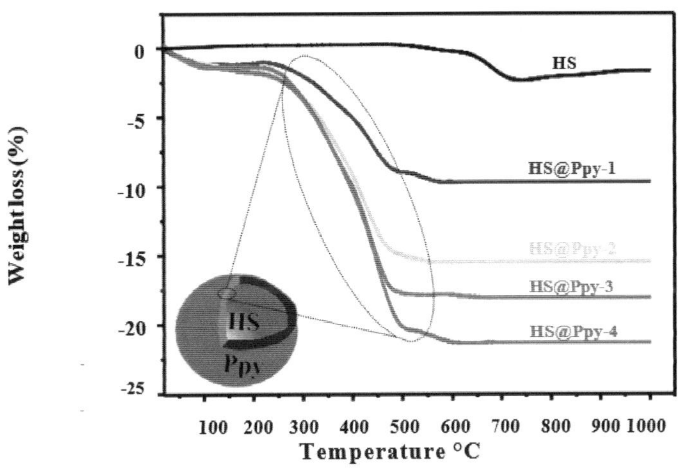

Figure 1. TG curves of HS, HS@Ppy-1, HS@Ppy-2, HS@Ppy-3 and HS@Ppy-4.

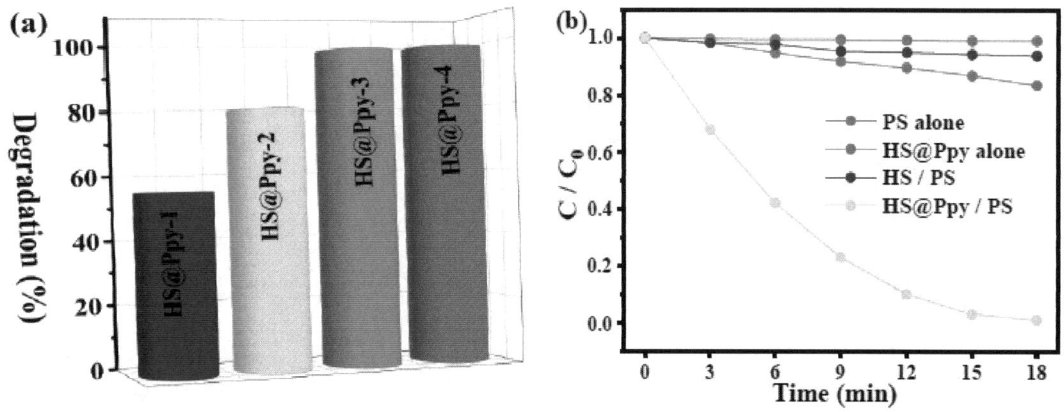

Figure 2. **(a)** Orange G degradation with HS@Ppy-1, HS@Ppy-2, HS@Ppy-3 and HS@Ppy-4. **(b)** Orange G degradation by various catalytic PS systems. Conditions: [PS] = 2 mM, [HS@Ppy] = 1 g.L^{-1}, [OG] = 45 mg/L, room temperature, ambient pH.

4. Conclusion

This study proposes a way for the preparation of a new natural catalyst based on hematite sand. The prepared polypyrrol coated hematite sand compounds were found very effective in activating the persulfate and producing large amount of sulfate radicals to degrade Orange G molecules.

References

[1] Han D, Wan J, Ma Y, Wang Y, Huang M, Chen Y, Li D, Guan Z and Li Y 2014 *Chem. Eng. J* **256** pp 316–323
[2] Cai M, Su J, Lian G, Wei X, Dong C, Zhang H, Jin M, and Wei Z 2016 *Ultrason. Sonochem* **31** pp 193–200
[3] Hu P, Su H, Chen Z, Yu C, Li Q, Zhou B, Alvarez P J J and Long M 2017 *Environ. Sci Technol* **51** pp 11288–11296
[4] Shi P, Su R, Wan F, Zhu M, Li D, and Xu S 2012 *Appl. Catal. B Environ* **123–124** pp 265–272
[5] Zhu Y, Yue M, Natarajan V, Kong L, Ma L, Zhang Y, Zhao Q and Zhan J 2018 *RSC Adv* **8** pp 14879–14887

Improved piezoelectric properties of PLA/PZT hybrid composite films

N Chakhchaoui [1,2,6], R Frahan [3], A Eddiai [3], M Meddad [4], O Cherkaoui [1], Y Boughaleb [2], L Van Langenhove [5]

[1] REMTEX Laboratory, Higher School of Textile and Clothing Industries (ESITH), Casablanca, Morocco

[2] BGIM Laboratory, Higher Normal School (ENS), Hassan II University, Casablanca, Morocco

[3] Laboratory of Physics of Condensed Matter (LPMC), Faculty of Sciences Ben M'Sik, Hassan II University, Casablanca, Morocco

[4] LAS Laboratory of Setif, Mohamed el Bachir el Ibrahimi BBA University, Algeria

[5] Centre for Textile Science and Engineering, Department of Materials, Textiles and Chemical Engineering, Ghent University, Belgium

[6] Author to whom any correspondence should be addressed

chakhchaoui.nabil@gmail.com

Abstract. Piezoelectric polymer-ceramic composites are promising materials for Sensing, Wireless electronics and actuating applications. We report the fabrication of highly piezoelectric biocompatible films containing lead zirconate titanate (PZT) ferroelectric ceramic particles dispersed in poly lactic acid (PLA) with different volume fractions using a solvent cast technique. The properties of the piezoelectric polymer-ceramic films were investigated by Fourier transform infrared spectrometry (FTIR) and field emission scanning electron microscopy (FE-SEM). In the FTIR spectra appear a large number of absorption bands which are attributed to the phases from PLA matrix confirming the total embedding of PZT filler into the matrix. The SEM results showed a good distribution of fillers in the matrix. We find that the added PZT imposes a significant effect on the α–β phase transformation. Our finding can lead to extraordinary enhancement of piezoelectric properties for the PLA/PZT composite films.
Keywords: Piezoelectric polymer-ceramic, PLA, PZT, solvent cast technique.

1. Introduction

Polylactic acid (PLA) remains the leader of electroactive polymers which are used in 43% of the scientific work in the field of recovery and conversion of electromechanical energy [1-6]. This polymer has the advantage of having a better mechanical strength, thermal and ease of implementation compared to other polymers [7]. Although it can present three primary crystalline phases (α, β and γ) according to its conformation in helices and symmetries of different lattices. These forms appear under thermal or mechanical constraints, the β phase is the most common and interesting one since it is the only one that can present a strong piezoelectric response in the PLA.

Adding fillers to prepared PLA polymer composites represents a practical way to enhance β-phase content in PLA films. In the few decades, several nanofillers, such as carbon nanotube (CNT), graphene oxide and graphene, were used as improving agents of the β –phase [8–10]. Recently, there have been many studies to directly relate the phase variation induced by carbon nano-fillers with enhanced piezoelectric properties. For example, *Wu et al.* [8] added the CNF and CNT into PVDF to improve the piezoelectricity of PVDF. It was observed that the fraction of β -phase improved significantly in CNT or CNF films, which was confirmed by X-ray diffraction (XRD). *Lee et al.* [11] PVDF/CNT composite films were prepared by solution blending method. It was confirmed by FTIR that the β -phase improved with the presence of CNT.

To the best of our knowledge, there is no work about the preparation of PLA/PZT films by the solvent cast method. Thus, in the present work, the PLA/PZT composite were prepared using this technique with different volume fractions. These samples were characterized by Fourier Transform Infrared Spectroscopy (FTIR) and scanning electron microscopy (SEM). We added micro particles into polylactic acid (PLA) to prepare composite films of exceptional piezoelectricity. Based on the survey on crystal structures, we clarified the roles played by PZT in phase transformation.

2. Experimental

2.1 Materials

PLA (Nature Works In geo 3100 HP) with a melt flow index of 24 g/10 min (at 210°C and 2.16 kg) was used as matrix of the composite.

Lead zirconate titanate (PZT) was supplied by Ferropem (Denmark) with an average diameter between 1 and 10 µm. This micro particles exhibit a curie temperature of 235 °C and a piezoelectric coefficient d_{33} of 600 pC/N.

2.2. Preparation of the PLA / PZT composites

PLA/PZT composites is prepared by the solvent cast method. A measured quantity of PLA is dissolved in THF (Tetrahydrofuran) by stirring and heating in the microwave oven until it dissolves. The PZT is dispersed completely in the polymer PLA solution. The prepared solution PLA/PZT/THF is spread at room temperature on a glass plate using a stainless steel and is then heated in the oven at 66 °C for 15 min. When the solvent evaporates completely, PLA/PZT composite films are obtained. The process is repeated with different volume fractions (ranging from 0.1, 0.3, 0.5, 0.8, 1 to 2wt.%).

2.3 Characterization of PLA/PZT composites

The chemical structure analysis of micro particles, PLA and its composites, was performed using a Fourier transformation infrared spectrometer (FT-IR, Nicolet iS10, spectra from 500 to 4000 cm^{-1}). The morphological observations of PLA/PZT composites were performed with scanning electron microscopy (SEM, SH-5000P-EDS), operating at an acceleration voltage of 15 kV. The samples' surfaces were coated under vacuum with carbon for SEM observation.

3. Results and discussion

3.1 SEM analysis of PLA/PZT composite films

Figure 1 shows the surface morphology of PLA/PZT composites with 0.5 and 1 volume fraction. Furthermore, on the level of 200 micrometers, the composites display an homogeneous and identical microstructure. When looking at the 1% volume fraction PZT image, it becomes apparent that the microstructure is also homogeneous at the micrometer level. The analysis of microstructure of the 1% volume fraction PZT composite indicates the good adhesion of polymer matrix with the particles of PZT ceramic. We conclude that the ceramic particles are homogeneously dispersed in the polymer matrix and are adhering well to it.

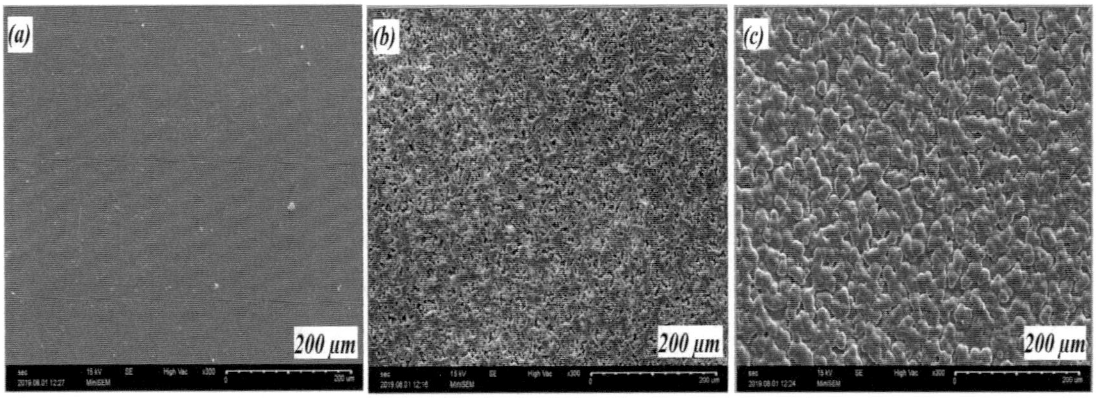

Figure 1. SEM micrographs of thin films hybrids nanocomposites of (A) PLA neat, (B) PLA/0.5 wt % PZT, (C) PLA/1wt % PZT.

Thus, we conclude the presence of uniform micro particles dispersion without agglomeration in the polymer matrix for lower filler volume fraction.

3.2 FT-IR analysis of PLA/PZT composites films

The samples were characterized by FTIR, and the corresponding spectra in the 800-1500 cm^{-1} region are presented in figure 2. From this figure, it is obvious that PZT effect the phase transformation significantly in initial crystallization process. Indeed, an appearance of the piezoelectric β crystalline phase at 1212 is 872 cm^{-1} was observed in all PLA / PZT nanocomposite films with different volume fractions of PZT ceramic, which remains important in energy harvesting. Consequently, the addition of a small percent of PZT micro fillers (0.1 wt %) in the PLA matrix, it was possible to obtain the β-phase in the crystallization.

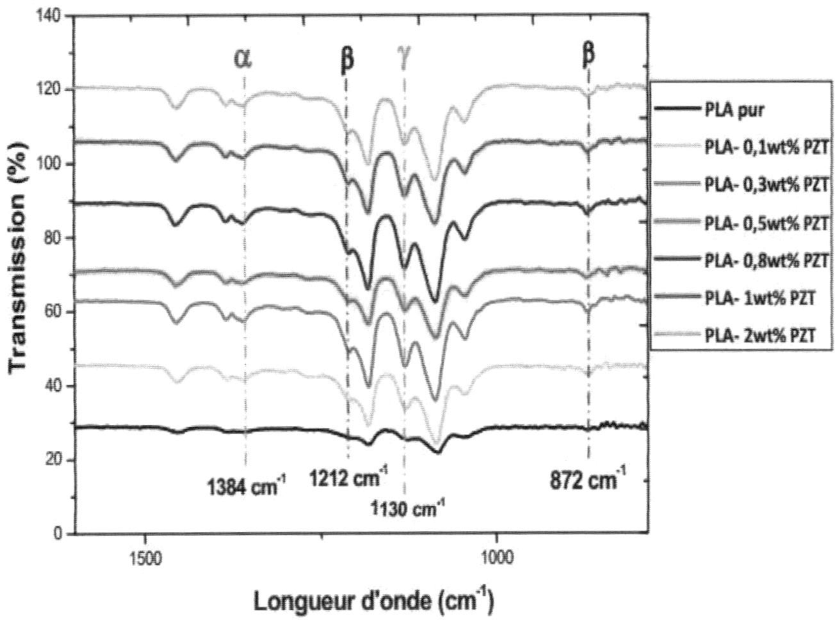

Figure 2. FTIR analysis of PLA/PZT nanocomposite.

We can see that the crystalline phase of the resulting polymer matrix is not disturbed during the addition of PZT ceramic. We can say that we combined in these film nanocomposites, the flexibility of a polymer matrix with the piezoelectric character of the ceramic particle of PZT. Also, the absorption peak intensities of charge materials increase proportionally with the charge content volume in the PLA. This indicates that both polymer matrix and the charge materials have retained their original chemical composition after the dispersion process.

4. Conclusion

PLA polymer and PZT fillers can be mixed in conditions which enables the formation of a homogeneous solution in THF, resulting in the formation of a stable solution at room temperature. By casting solutions of PLA and PLA-PZT mixture on glass plates, high quality and flexible PZT-filled PVDF nanocomposite films have been produced. Due to the high compatibility between PZT and PLA, PZT fillers are well dispersed and distributed within the PLA matrix.

The properties of the piezoceramic filler before and after dispersion in solvent and embedding into the PVDF polymer matrix were investigated. The FTIR spectrum indicated the total embedding of PZT filler into the matrix, by keeping the absorption bands characteristics for the α – phase of PLA. It was observed in the SEM images that the PZT filler kept their morphology and their size at the same nanometric range after its incorporation in PVDF matrix.

It was observed that PZT influenced the phase formation significantly during the preparation. The properties enhancements are directly related to the strong and specific interfacial interaction that results in the adsorption of macromolecular chains of PLA onto the PZT surface.

The PLA/PZT composites open new opportunities to develop thin films for technological applications such as: energy harvesting, sensors and actuators.

References

[1] Priya S and Inman D 2009 *Energy Harvesting Technologies* (New York: Springer)

[2] Eddiai A, Meddad M, Farhana R, Chakhchaoui N, Mazroui M, Rguiti M and Guyomar D 2019 Using PVDF piezoelectric polymers to maximize power harvested by mechanical structure *Superlattices and Microstructures* **135** pp.106-189.

[3] Chakhchaoui N, Jaouani H, Ennamiri H, Eddiai A, Hajjaji A, Meddad M, Van Langenhove L and Boughalen Y 2018. Modeling and analysis of the effect of substrate on the flexible piezoelectric films for kinetic energy harvesting from textiles Jounal of composite materilas **53** pp.3349-61.

[4] Sodano H A, Inman D J and Park G,2005. Generation and storage of electricity from power harvesting devices. J Intell Mater Syst Struct..**16** pp.67–75.

[5] Farhan R, Rguiti, Eddiai A, Mazroui M, Meddad M and Courtois C 2019 Evaluation of performance of polyamide/ lead zirconate titanate composite for energy harvesters and actuators *Journal of Composite Materials* **53** 345–352.

[6] Eddiai A, Meddad M, Mazroui M and al. 2016,Strain effects on an electrostrictive polymer composite for power harvesting: experiments and modeling *Polym. Adv. Technol.* **27** , pp. 677–684.

[7] Ning W, Xingxiang Z, Xiaofei M 2008 Jianming Fang, Influence of carbon black on the properties of plasticized poly(lactic acid*). Composites Polymer Degradation and Stability* **93** pp. 1044–52.

[8] Wu L, Yuan W, Nakamura T and al. 2013 Enhancement of PVDF's piezoelectricity by VGCF and MWNT Adv. *Compos. Mater*. **22**, pp.49–63.

[9] Layek R K, Samant S, Chatterje D P and Nandi A K 2010 Physical and mechanical properties of poly (methyl methacrylate) -functionalized graphene/ poly (vinylidine fluoride) nanocomposites: Piezoelectric β polymorph formation Polymer 51 5846–5856.

[10] El Achaby M, Arrakhiz F Z, Vaudreuil S, Essassi E M and Qaiss A 2012 Piezoelectric β-polymorph formation and properties enhancement in graphene oxide – PVDF nanocomposite films *Appl. Surf. Sci.* **258** pp. 7668–77.

[11] Lee J S, Kim G H, Kim W N, Oh K H, Kim H T, Hwang S S and Hong S M 2008 Crystal structure and ferroelectric properties of poly (vinylidene fluoride)-carbon nano tube nanocomposite film *Mol. Cryst. Liq. Cryst.* **491**, pp.247–254.

IOP Conf. Series: Materials Science and Engineering **827** (2020) 012013 doi:10.1088/1757-899X/827/1/012013

Preparation and characterization of a new low-cost polyacrylonitrile adsorbent

Hayat Bouchoum[1,2,5], Mehdi El Bouchti[1], Amane Jada[3,4], Mohamed Tahiri[2], and Omar Cherkaoui[1]

[1] Laboratory REMTEX, Higher school of textile and clothing industries, Casablanca, Morocco.
[2]Hassan II University, faculty of science Ain Chok, Laboratory of geosciences, Casablanca, Morocco.
[3]Univ Haute Alsace, CNRS, IS2M, UMR7361, F-68100 Mulhouse, France
[4]Univ Strasbourg, Strasbourg, France
[5] Author to whom any correspondence should be addressed

bouchoum.hayat@gmail.com

Abstract. Amidoximated polyacrylonitrile fibers (AO-PANF) was synthesized using hydroxylamine chelating agent to prepare an effective adsorbent for heavy metal and dyes removal. This study is the continuation of an earlier study and focuses on the characterization of the adsorbent. The thermal properties of adsorbent were evaluated using TGA and DSC analysis. The mechanical properties of the adsorbent were also been investigated. Characterization results reveal an improvement of thermal stabilities due to the modification process using hydroxylamine chelating agent. However, a reduction in mechanical properties was showed.

1. Introduction

Adsorption technology is considered to be the most effective and economical method to remove pollutants from aqueous solutions due to its high removal percentage, simple process and adsorbent recovery [1, 2]. The efficiency of adsorption process and economic sustainability depends mainly on the nature of the adsorbents used to remove pollutants from aqueous solutions [3, 4]. The surface modification is an important step to prepare an effective adsorbent for pollutants removal.

Polyacrylonitrile is the most polymeric material used for water purification. It is a commercial and low-cost polymer material with good mechanical, chemical and thermal stability as well as a good resistance to corrosion and solvents [1]. It has been used to produce variety of products such as activated carbon [5], nanofibers [1], composite materials [6], protective materials [7], heavy metals adsorbents [8], conductive materials [9] and others. Polyacrylonitrile fibers (PANF) is characterized by a large surface area and short diffusion distance [10], facilitating the adsorption of metals ions from aqueous solutions. It contain hydroxyl (OH) and amino groups (NH_2) in their backbone making it a good adsorbent for water purification, heavy metals and dyes removal from aqueous solutions [11].

Chemical modification of polyacrylonitrile fibers (PANF) surface with hydroxylamine, diethylenediamine [10], ethylenediamine, triethylenetetramine [8], carboxyl and imidazole [1] improve their properties and enhance the adsorption efficiency of the material. Among these chelating agents, hydroxylamine is promising, containing nitrogen and oxygen atoms, which have a strong affinity towards heavy metals ions. Nitrile groups containing on polyacrylonitrile surface can be easily transformed onto a variety of functional groups (carboxyl, amine, amidoxime, phosphoric, etc…).

In our previous study, we confirmed the capacity of the adsorbent to remove hexavalent chromium from aqueous solutions. In this study, we investigated the mechanical and thermal properties of the adsorbent before and after the modification process.

2. Materials and methods

2.1. Materials

Commercial polyacrylonitrile fibers (PANF) was cut, washed with distillated water and then dried in an oven. Hydroxylamine Hydrochloride, with a purity >99%, was purchased from Scharlab. NaOH and HCl were used to adjust the pH of the solution.

2.2 Surface modification

The surface modification of AO-PANF by one step reaction was done according to the procedure in our previous researches [12]. The modification consists of the conversion of the nitrile groups on PANF surface onto amidoximes groups, using Hydroxylamine Hydrochloride while respecting the cost of production. After the treatment, the adsorbents were separated from the solution by filtration, washed and dried for one day at 60°C. The modification of PANF using 0.15 M of hydroxylamine for 60 min at 70°C was found to produce a high conversion rate of nitrile groups of 33.7 % and acceptable mechanical properties. The FTIR spectroscopy, SEM analysis and calculating the conversion of nitrile groupswere also studied in our previous work [12].

2.3 Samples analysis

2.3.1 Mechanical properties. The mechanical properties of the samples were investigated by measuring the elongation at break and breaking strength using an electronic tensile tester (LUDWIC mpk) according to NF EN ISO 2062. The samples were dried before the measurement for 1 H at 40°C to remove the water molecules.

2.3.2 Thermal analysis. Weight change with temperature was determined by thermogravimetric analysis (TGA) using Cahn Versa Therm analyzer under air atmosphere (25 ml.min^{-1}) from room temperature to 600°C with a heating rate of 10°C.min^{-1}. As a complementary to TGA analysis, differential scanning calorimetry (DSC) was performed to study the thermal behaviors of samples using Setaram DSC Evo 131 analyzer under nitrogen atmosphere 30 ml.min $^{-1}$ with 10 mg of samples in a temperature range of 40°C to 400°C at a heating rate of 10°C.min^{-1}.

3. Results and discussions

3.1. Thermogravimetric analysis

The weight loss of PANF, AO-PANF and AO-PANF-Cu (II) complexes during stabilization process was determined using thermogravimetric analysis (TGA-DTG) and is shown in (Figure 1). The dTg curves (Figure 1-b) demonstrate better the degradation steps of the samples.

Raw PANF show two degradation steps (Figure 1-a). The first degradation step with a small weight loss was observed between 50°C and 120°C due to the release of moisture [13], which was also observed for AO-PANF and AO-PANF- Cu (II) complexes. The onset degradation of PANF was observed occurring from 290 to 430°C with the main weight loss of 32.46 %. The degradation can be attributed to the release of nitrogenous gases from the cyclization, dehydrogenation or cross-linking reaction of nitrile groups containing on PANF surface [14]. Beyond 430°C, the thermal degradation of PANF backbone chains occurs [13]. The results is in accordance with previous studies [15].

Compared to PANF, the decomposition temperature of AO-PANF starts at a lower temperature of 197°C and continues up to 310°C with a weight loss of 9.81% due to the decomposition of grafted functional groups on AO-PANF surface [16]. The reduction is due to the presence of amidoximes group on AO-PANF surface indicating that it was beneficial for further cyclization.

The last degradation step occurs between 380°C and 470°C, the main weight loss of 9.14 % was observed due to the thermal decomposition and carbonization process of AO-PANF. The residual weight at 600°C was 63% suggesting that AO-PANF is a suitable support for industrial wastewater treatment.

The residual weight of AO-PANF is 63% higher than PANF (59.9%) demonstrating the improvement of thermal stability of samples using hydroxylamine chelating agents.

After adsorption of copper, a significant change was observed on the adsorbent with a rapid thermal decomposition of 33.59% between 390°C and 520°C. The results evince that the AO-PANF-Cu (II) complexes metals adsorbed onto AO-PANF become less stable with a yield residual weight of 30%.

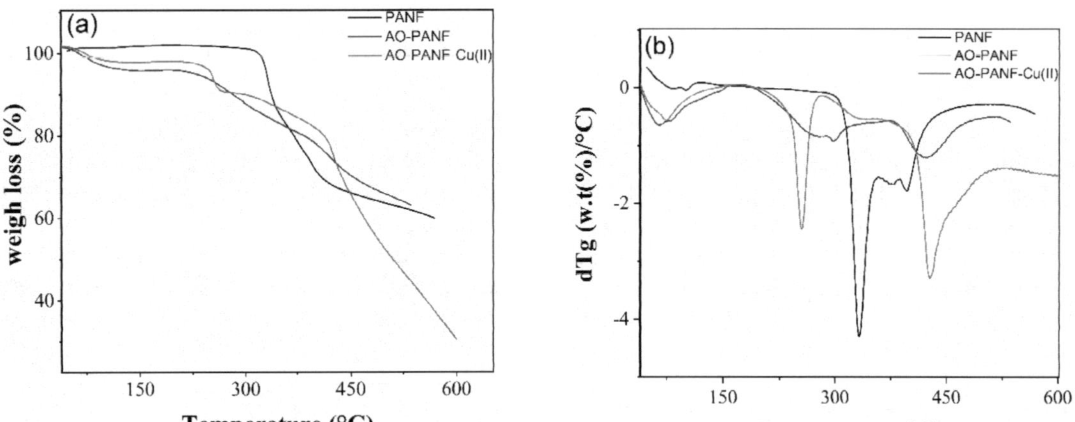

Figure 1. TG (a) and dTg curves (b) of PANF, AO-PANF (C(NH_2OH) = 0.15 M) and AO-PANF-Cu (II) complexes (C (Cu II) = 15 mg L^{-1}) in air atmosphere.

3.2. Differential scanning calorimetry (DSC)

The thermal properties of raw PANF and AO-PANF was achieved using a DSC analyzer from a single heating process and the results is shown in Figure 2. The highly exothermic peaks of PAN and AO-PANF, in nitrogen atmosphere, were observed at 316°C and 263°C, respectively. This reaction indicates the cyclization reactions of nitrile groups containing in PANF during the thermal process [17] and confirms the results obtained in thermogravimetric analysis section. This results is in agreement with the previous study [18]. The shift of exothermic peak for AO-PANF at lower temperatures indicated that the modification with hydroxylamine hydrochloride affects the cyclization mechanism of PANF in which the reaction starts easily by ionic mechanism. The amount of heat liberated during this process was different. The raw PANF compared to AO-PANF had a strong and sharp exothermic peak with an enthalpy of 50.442 J.g^{-1}, whereas AO-PANF demonstrated a broader peak with an enthalpy of 52.64 J.g^{-1}. The reduction of intensity of cyclization reaction for AO-PANF can reduce the structural defect formed in the final carbon material [19]. Additionally, AO-PANF showed an endothermic peak at 103°C which can be due to the removal of moisture generated by the addition of functional groups on PANF surface making it strongly hydrophilic [18]. In PANF, no endothermic peaks was observed suggesting that it is not melt at this heating rate [15].

IOP Conf. Series: Materials Science and Engineering **827** (2020) 012013 doi:10.1088/1757-899X/827/1/012013

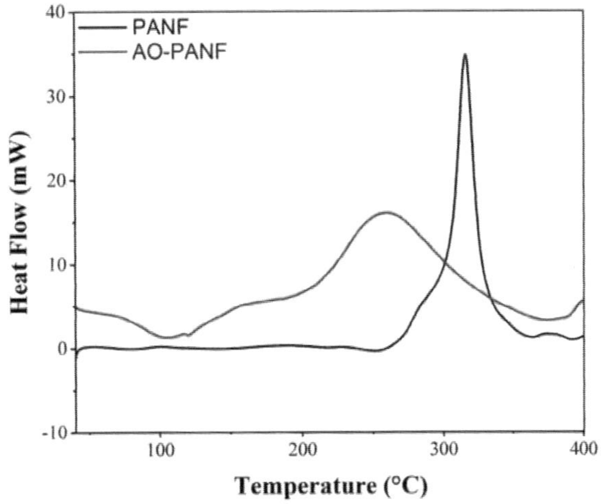

Figure 2. DSC curves of raw PANF and AO-PANF ($C(NH_2OH) = 0.15$ M) in nitrogen atmosphere.

3.3. Mechanical analysis

The mechanical properties were investigated in order to study the effect of the modification on the mechanical properties of the adsorbent. Previous works mentioned that the treatment with amidoximes groups tends to reduce the mechanical strength of original PANF [20, 21]. Table 1 regroups the mechanical parameters of PANF and AO-PANF. From Table 1, a comparison between raw PANF and AO-PANF showed a decrease in mechanical properties, making AO-PANF less resistive. These results were in accordance with the morphological structures observed previously by SEM analysis [12]. This can be due to the amidoximation process which generates a low crystallinity of AO-PANF [18].

Table 1: Mechanical parameters of PANF and AO-PANF ($C(NH_2OH) = 0.15$ M)

Samples	Tensile Strength (MPa)	Strain ε (%)	Tensile Modulus (MPa)
PANF	482	26.31	29.47
AO-PANF	143.37	21.54	13.68

4. Conclusions

This study presented a synthesis procedure of an adsorbent compound of polyacrylonitrile fiber using hydroxylamine chelating agent. The thermal and mechanical properties of the adsorbent before and after modification process has also been investigated. The capacity of AO-PANF to remove heavy metals ions from aqueous solutions was confirmed in our previous studies. According to the results, in the thermal analysis, the comparison between raw PANF and AO-PANF showed an improvement of thermal stabilities due to the use of hydroxylamine chelating agents. However, AO-PANF showed a reduction on the mechanical strength. This can be due to the low crystallinity of AO-PANF.

References

[1] Ndayambaje G, et al., 2016 Adsorption of nickel (II) on polyacrylonitrile nanofiber modified with 2- (2′-pyridyl) imidazole. *Chemical Engineering Journal* **284**: p. 1106-1116.

[2] Hassan A and R Bulánek 2019. Preparation and characterization of thiosemicarbazide functionalized graphene oxide as nanoadsorbent sheets for removal of lead cations. *International Journal of Environmental Science and Technology,* **16**(10) p. 6207-6216.

[3] Weng C H, et al.,2007 Adsorption characteristics of copper (II) onto spent activated clay. *Separation and Purification Technology* **54**(2) p. 187-197.

[4] Ghariani, B., et al., 2019 Porous heat-treated fungal biomass: preparation, characterization and application for removal of textile dyes from aqueous solutions. *Journal of Porous Materials* p. 1-14.

[5] Arbab S and A Zeinolebadi 2013 A procedure for precise determination of thermal stabilization reactions in carbon fiber precursors. *Polymer degradation and stability,* **98**(12) p. 2537-2545.

[6] Yavuz M, et al. 2008 An economic removal of Cu2+ and Cr3+ on the new adsorbents: Pumice and polyacrylonitrile/pumice composite. *Chemical Engineering Journal* **137**(3) p. 453- 461.

[7] Kharaghani D, et al.,2018 Preparation and in-vitro assessment of hierarchal organized antibacterial breath mask based on polyacrylonitrile/silver (PAN/AgNPs) nanofiber. *Nanomaterials,* **8**(7): p. 461.

[8] Soltanzadeh M, G Kiani and Khataee A 2014 Adsorptive capacity of polyacrylonitrile modified with triethylenetetramine for removal of copper and cadmium ions from aqueous solutions. *Environmental Progress & Sustainable Energy* **33**(4) p. 1139-1147.

[9] Koutsonas S 2017 Electrical conductivity of degraded polyacrylonitrile powder by microwave irradiation for supercapacitor devices or other mobile applications. *Materials Letters* **193** p. 203-205.

[10] Abdouss M et al., 2012 Fabrication of chelating diethylenetriaminated pan micro and nano fibers for heavy metal removal. *Chemical Industry and Chemical Engineering Quarterly* **18**(1) p. 27-34.

[11] Ibrahim A G, et al. 2019 Chitosan‑g‑maleic acid for effective removal of copper and nickel ions from their solutions. Int J Biol Macromol, **121** p. 1287-1294.

[12] Bouchoum H, et al. 2019 Synthesis of amidoximated polyacrylonitrile fibers and its use as adsorbent for Cr (VI) ions removal from aqueous solutions. *Environmental Progress & Sustainable Energy.*

[13] Cheng, Y., et al., 2019 Polyamine and amidoxime groups modified bifunctional polyacrylonitrile- based ion exchange fibers for highly efficient extraction of U (VI) from real uranium mine water. *Chemical Engineering Journal.*

[14] Chaúque E F, et al.2016 Modification of electrospun polyacrylonitrile nanofibers with EDTA for the removal of Cd and Cr ions from water effluents. *Applied Surface Science* **369** p. 19-28.

[15] Mapazi O et al. 2018 High temperature thermochromic polydiacetylene supported on polyacrylonitrile nanofibers. *Polymer* **149** p. 106-116.

[16] Chen C et al., 2019 Preparation and Performance of Aminated Polyacrylonitrile Nanofibers for Highly Efficient Copper Ion Removal. *Colloids and Surfaces A: Physicochemical and Engineering Aspects.*

[17] Kim S S and Lee J 2014 Antibacterial activity of polyacrylonitrile–chitosan electrospun nanofibers. *Carbohydr Polym,* **102** p. 231-237.

[18] Han, Z., Y. Dong, and S. Dong, 2010 Comparative study on the mechanical and thermal properties of two different modified PAN fibers and their Fe complexes. *Materials & Design,* **31**(6) p. 2784-2789.

[19] Korobeinyk A V, Whitby R L and Mikhalovsky S V 2012High temperature oxidative resistance of polyacrylonitrile-methylmethacrylate copolymer powder converting to a carbonized

monolith. *European Polymer Journal* **48**(1) p. 97-104.

[20] Xu G., et al.2018 Highly selective and efficient adsorption of Hg2+ by a recyclable aminophosphonic acid functionalized polyacrylonitrile fiber. *Journal of Hazardous Materials* **344** p. 679-688.

[21] Wu D, et al. 2018 Electrospun blend nanofiber membrane consisting of polyurethane, amidoxime polyarcylonitrile, and β-cyclodextrin as high-performance carrier/support for efficient and reusable immobilization of laccase. *Chemical Engineering Journal* **331** p. 517-526.

IOP Conf. Series: Materials Science and Engineering **827** (2020) 012014 doi:10.1088/1757-899X/827/1/012014

Applications of smart textiles in occupational health and safety

Patricia I. Dolez[1,3], Justine Decaens[2], Thibaut Buns[2], Dominic Lachapelle[2] and Olivier Vermeersch[2]

[1] Department of Human Ecology, University of Alberta, Edmonton, AB, T6G 2N1, Canada (formerly with CTT Group, Canada)
[2] CTT Group, St-Hyacinthe, QC, J2S 1H9, Canada
[3] Author to whom any correspondence should be addressed

pdolez@ualberta.ca

Abstract. Smart textiles can detect, react, and adapt to various stimuli; they offer promising perspectives in occupation health and safety (OH&S). The objective of this study was to identify technologies, solutions, and products based on smart textiles and flexible materials that could have an application in OH&S and provide a response to some of the current needs. The collection of information included three aspects: 1) technologies, solutions, and products involving smart textiles and flexible materials found in the literature over the period 2000-2016; 2) issues in OH&S associated with traditional textiles and flexible materials; and 3) current or foreseeable problems associated with the use of smart textiles and flexible materials in OH&S. Issues with traditional textiles and flexible materials in OH&S were cross-matched with technologies, solutions and products relevant to smart textiles and flexible materials. This allowed us to propose short, mid, and long-term developments that could provide a response to some of the current needs in OH&S. The analysis shows that smart textiles and flexible materials offer a promising response to current challenges observed with PPE used in OH&S. With manufacturing and R&D capabilities available in various countries, the feasibility of development of these solutions is very high.

1. Introduction

While the use of electricity in illuminating and animated garments and accessories can be traced back to the end of the 19th century [1], the development of smart textiles only really began in the mid 80's with the invention of wearable computers [2] and manufacturing equipment allowing the incorporation of optic fibers into textiles [3]. The first commercial products incorporating smart textiles reached the market in the years 2000 with applications mainly targeting vital sign monitoring [4].

Smart textiles can detect, react, and adapt to various stimuli, e.g., electrical, magnetic, thermal, mechanical, chemical, and optical [5]. These three levels of functionalities define the three categories of smart materials [6]: passive with only sensing capability; active with both sensor and actuator functionalities; and ultra-intelligent, i.e. able to sense, react and adapt to stimuli. Smart textiles offer promising perspectives in occupation health and safety (OH&S), with embedded geolocation, monitoring of people's physiological condition, heating and cooling, wireless communication, and energy harvesting for instance.

Three generations of smart textiles can be identified [6]. In the first generation, of which examples can already be seen on the market, active functions are added at the final product assembly stage. For the second generation, smart components are embedded in the textile during its production. The third generation of smart textiles, on which researchers are still working and which is not expected to reach the market before 10 years, will hold the active parts within their fibers and yarns; they will be integrated at the extrusion or spinning stages.

The objective of this study was to identify technologies, solutions, and products based on smart textiles and flexible materials that could have an application in OH&S and provide a response to some of the current needs in OH&S. The research also involved determining current or foreseeable problems associated with the use of smart textiles and flexible materials in OH&S.

2. Methods

The collection of information involved three aspects. First, we analyzed the technical and scientific literature to identify technologies, solutions, and products involving smart textiles and flexible materials over the period 2000-2016. In total, more than 500 papers in journals and conference proceedings, patents, articles in textile news bulletins, and technical datasheets on manufacturer and distributor websites relevant to the subject were identified. The information gathered included a short description of the technology, solution or product; the nature of the stimulus; the type of system (device or material); the level of technology readiness; existing and suggested applications; and the limitations indicated. A list of 184 companies involved in smart textiles was also prepared with some information including the type of product (clothing; accessories; component; software) and a description of the technology; this list comprises more than 60 Canadian manufacturers.

The second aspect investigated dealt with the identification of issues in OH&S associated with traditional textiles and flexible materials, i.e. whose properties and performance are fixed and cannot adapt to the conditions encountered. Information was obtained from prevention counsellors in various professional activity sectors in Quebec, Canada (automotive; construction; farming; fisheries; food transformation; foundries; manufacturing of transportation equipment and machines; metal, electric, apparel & printing; mining; municipal affairs; provincial administration; pulp and paper; social affairs; and textile & hosiery). Interviews were also conducted with OH&S researchers, sales representatives of companies manufacturing personal protective equipment (PPE) used in OH&S, product development managers at protective textile manufacturers, and procurement officers in organizations using PPE. For each issue put forward, information was obtained about the activity sector(s) concerned, the category and number of workers affected, the solution currently implemented if any, and the presence of other constraints (e.g. related to cost or conditions of the work environment).

All these resource-persons were also questioned about the third aspect of the study related to current or foreseeable problems associated with the use of smart textiles and flexible materials in OH&S. They were asked if they had been in contact or aware of products used in OH&S that contained smart textiles and flexible materials; if positive, a short description of the product, activity sector and use conditions, and problems identified were noted. Input from experts at the Canadian Department of National Defence was obtained too. Finally, a brainstorming session with the research team was organized to complete the list of current or foreseeable problems associated with the use of smart textiles and flexible materials in OH&S based on their analysis of the relevant scientific and technical literature.

3. Results and discussion

3.1. Technologies, solutions, and products based on smart textiles and flexible materials that could have an application in OH&S

Three main categories of technologies, solutions, and products relevant to smart textiles and flexible materials were identified: sensors, actuators, and materials/components. A complete list of the technologies, solutions, and products identified in the study can be found in [7].

Examples of textile-based sensors include temperature, heat flux, pressure, strain, gas, pH, chemical, humidity, obstacle, proximity and location sensors as well as textile electrodes. Their input signal or stimulus is the main characteristic of interest. Sensors responding to a mechanical and electrical stimulus represent 59 and 23%, respectively, of the sensor-type technologies, solutions, and products identified

(Figure 1). A few sensor-type products designed for OH&S applications were identified, for instance a smart shirt collecting vital signs that was used by Canadian astronaut David Saint-Jacques during his mission aboard the International Space Station in 2018-19 [8]; protective gloves that change color when sensing toxic chemicals in the air [9]; smart trousers that detect when a chain saw gets too close and switch it off [10]; an instrumented glove that captures the motion of the user's hand and fingers and can be used for the collection of ergonomic ground data in the workplace [11]; and a lifeline system for firefighters that stores tactical information such as location of doors and victims [12].

Actuators can be defined by their output signal. They include heating fibers, antibacterial fabrics, textile displays, self-cleaning surfaces, a self-actuated defibrillator integrated in a garment, and photovoltaic fibers. Thermal, optical and power actuators each account for about a quarter of the actuator-type technologies, solutions, and products identified (Figure 1). A few actuator-type products designed for OH&S applications were also identified. They include a chemical protective garment that provides on-demand cooling [13]; antibacterial medical protective garments that can be recharged by washing with chlorine bleach [14]; surgical gowns with self-adjustable breathability [15]; and an active vest designed for caregivers that prevents incorrect movements when lifting heavy loads [16].

Finally, materials and components relevant to smart textiles and flexible materials comprise various polymers, carbon-based materials, metals, inks, connectors, and antennas.

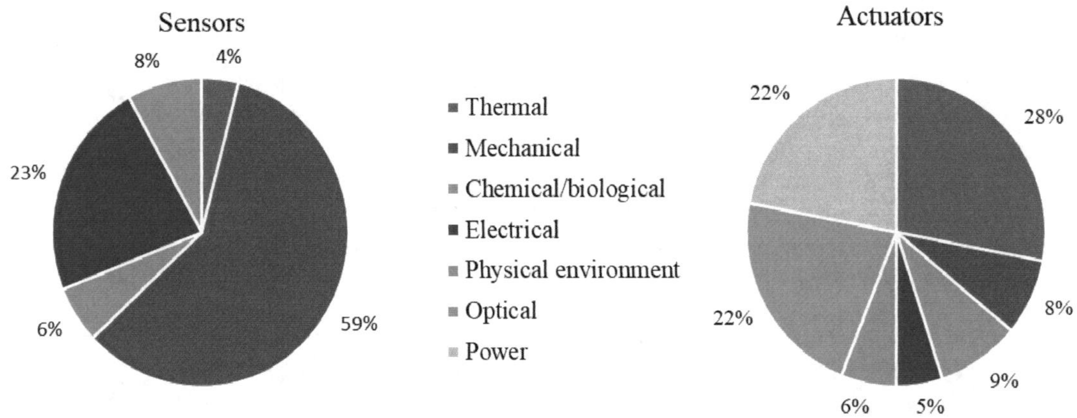

Figure 1. Distribution of technologies, solutions, and products relevant to smart textiles and flexible materials as a function of the stimulus for sensors (left) and output signal for actuators (right).

3.2. Issues in OH&S associated with traditional textiles and flexible materials

The same type of classification was used to analyze the information obtained in terms of issues in OH&S associated with traditional textiles and flexible materials. The complete list of issues mentioned by the resource-persons interviewed can be found in [7].

Sensors represent about 25% of the issues reported. Among them, 31% involve a stimulus related to the physical environment (Figure 2). Issues associated with a mechanical, electrical or chemical stimulus each account for 20% of the sensor-type issues. Needs include for instance chemical sensors embedded in lab coats to detect chemical exposures; temperature sensors monitoring the conditions inside the protective clothing to avoid heat stress; safety harnesses equipped with mechanical sensors to alert other workers in case of a fall from height; and vital sign sensors combined with embedded geolocation systems for people working alone in remote locations or in confined environments.

In the case of actuators, the most frequent needs reported involve a thermal output (32%) and a mechanical output (31%) (Figure 2). Examples of actuator-type needs include thermoregulating systems for highly protective or fully encapsulating garments; protective gloves whose level of protection and dexterity vary based on the task; protective gloves and clothing which neutralize the biological agent before it reaches the skin in case of cut or needle puncture; and protective garments increasing their breathability in case of high intensity activity or when going from a cold to a warm area.

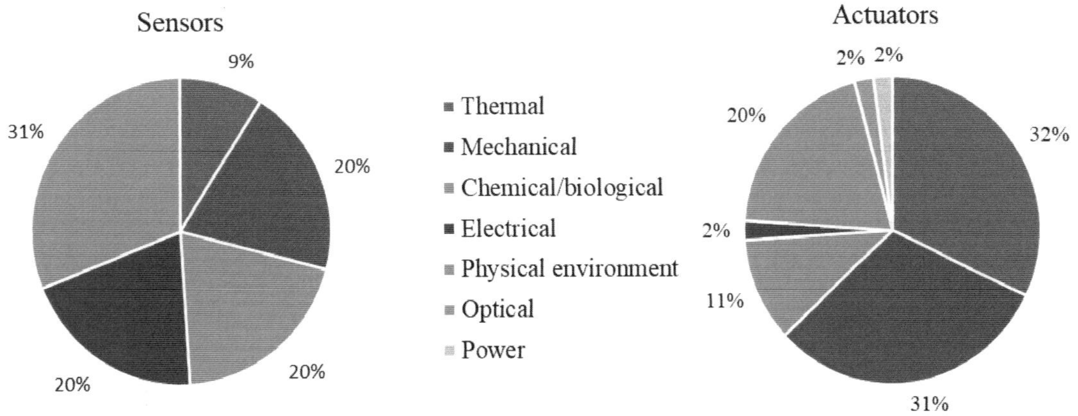

Figure 2. Distribution of issues in OH&S associated with traditional textiles and flexible materials as a function of the stimulus for sensors (left) and output signal for actuators (right).

3.3. Current or foreseeable problems associated with the use of smart textiles and flexible materials in OH&S

A series of current or foreseeable problems associated with the use of smart textiles and flexible materials in OH&S have been identified and grouped into eight categories [7]:

- *Lack of maturity of the technologies associated with smart textiles.* Current systems still need improvements in terms of durability (resistance to sweat, repeated launderings, wear, etc.), reliability (accuracy, repeatability, and reproducibility) and energy efficiency. In addition, energy harvesting systems are still far from reaching the market. Thermal signature is also a challenge for military applications.
- *Potential negative health and safety outcomes.* Concerns have been raised both in terms of manufacturing, with particles of all sizes entering or generated by the process, as well as use. This includes the potential toxicity of nanoparticles and metals, effects of electromagnetic fields, accidental electric shock, and inability to activate the emergency shut -off device in case of a malfunctioning smart protective glove. An important point also regards ethics, with the question of information management and personal data safety.
- *Potential interference and incompatibility issues.* Responses are needed regarding the risk of interference of wearable electronics with communication systems and other electronic systems such as pacemakers. In addition, the presence of electrically conductive components in PPE is not compatible with requirements for electric shock and electrical arc protection. Internal and external environmental conditions such as profuse sweat and rain may also interfere with smart textile proper operation.
- *Gap in the relevant knowledge, standards, and test methods.* Needs include an in-depth understanding of new product operation and the generation of data about device safety and behavior over time for instance. The smart textile industry also misses strategies for managing improper use. Finally, the lack of appropriate standard test methods and regulatory framework may generate risks for the user. This delay in background research is often observed with new technologies with a strong commercial growth potential that benefit from large investments.
- *Issues with accessibility.* The accessibility of products based on smart textiles and flexible materials in OH&S is limited by the cost of the product itself, which can be traced in some instances to the high price of the functionalizing component/material, as well as of accessories necessary for its operation, for instance chip-reading gates for tracking the use of safety harnesses.
- *Issues with use and care.* A major limitation to smart textile use is related to batteries: their weight and bulkiness as well as the charging process. Care is also a huge challenge, especially for protective

clothing. To be relevant for use in OH&S, smart textiles have to resist wear and severe environmental conditions such as high moisture for rain gear and high temperature for firefighter protective clothing.

- *Lack of validation of the real value for the user.* The adoption of smart textiles and flexible materials in OH&S is conditional to demonstrating that benefits outweigh additional costs and potential negative side effects. This analysis should involve experts in all aspects relevant to the smart product application.
- *Challenges related to end of life.* Smart textiles present the same challenges as traditional e-waste [17]: their amount is increasing worldwide and they contain problematic as well as valuable substances. They also raise additional issues due to the fact that they are more integrated and in larger number, have a shorter life span, and are often disposed of with their batteries. In particular, it is critical to ensure that the incineration of domestic waste performed by some countries to reduce landfilling does not create new hazards for health and the environment because of the presence of smart textiles.

3.4. Short, mid, and long-term developments proposed to respond to current needs in OH&S using smart textiles and flexible materials

The issues in OH&S reported with traditional textiles and flexible materials were cross-matched with the technologies, solutions and products relevant to smart textiles and flexible materials identified to propose short, mid, and long-term developments that could provide a response to some of the current needs in OH&S.

For instance, a short-term response to the need for thermoregulation in protective garments could rely on commercial products for athletes to warn workers when their vital signs point towards thermal stress onset and on existing cooling/heating systems with an external power source. In the mid-term, existing technologies such as phase change materials could be integrated as removable components in the garment. Ultimately, the solution would be completely embedded in the fabric, yarns or fibers; allow long-term thermoregulation; and be resistant to extreme work conditions, wear, and laundering.

The detection of contaminants could take advantage of existing technologies such as chemical-sensitive dyes in the short-term. In a next phase, textiles would be developed to detect and filter toxic contaminants. Ultimately, they would neutralize them.

In terms of musculoskeletal support, short-term solutions could rely on commercially available customized exoskeletons. Mid-term developments would use active technologies such as electrostimulation and shape memory textiles. In the long run, textiles will adjust their shape and stiffness depending on the mechanical constraints.

The complete list of development projects proposed for smart textiles and flexible materials in OH&S can be found in [7].

4. Conclusion

The analysis conducted as part of this study shows that smart textiles and flexible materials offer a promising response to current challenges observed with protective clothing and PPE used in OH&S. A series of short, mid, and long-term developments are proposed based on a cross analysis between issues in OH&S reported with traditional textiles and flexible materials and technologies, solutions, and products relevant to smart textiles and flexible materials identified in the technical and scientific literature.

With manufacturing and R&D capabilities available in various countries, including more than 60 companies and several research groups involved in smart textiles and flexible materials in Canada, the feasibility of the corresponding development projects is very high. This is especially true for short and mid-term proposals as they rely primarily on the optimization of existing products and technologies to improve their reliability, durability, and ease of operation.

Acknowledgments

This study has benefited from the support of the Institut de recherche Robert-Sauvé en santé et en sécurité du travail (IRSST) [Project Number: 2016-0028]. The authors also want to thank the OH&S community and textile industry for the information shared. They also wish to acknowledge the contribution to the study of Mohamed Ameur Arfaoui, Léa Bisson, Kévin Cloutier, and Stevan Roué of CTT Group, and François Ouellet of IRSST.

References

[1] Marvin C 1990 *When Old Technologies Were New: Thinking About Electric Communication in the Late Nineteenth Century* (New York, NY: Oxford University Press)
[2] Post R, Orth M, Russo P and Gershenfeld N 2000 E-broidery: design and fabrication of textile-based computing *IBM Syst. J.* **39** (3-4) 840–60
[3] Schwar R C and Wainwright H L 1998 *Apparatus for implanting optical fibers in fabric panels* Patent 5738753 (Alexandria, VA: U.S. Patent and Trademark Office)
[4] Lymberis A and Paradiso R 2008 Smart fabrics and interactive textile enabling wearable personal applications: R&D state of the art and future challenges *Conf. Proc. IEEE Eng. Med. Biol. Soc. (Vancouver, BC)* 5270-3
[5] Tao X (ed) 2001 *Smart fibres, fabrics and clothing* (Cambridge, UK: Woodhead Publishing / Boca Raton, FL: CRC Press)
[6] Begriche A and Lachapelle D 2010 Intelligent textiles for protection – Overview and technologies *Soldier Systems Technology Roadmap – Soldier survivability/sustainability/mobility Workshop* (Ottawa, ON)
[7] Dolez P, Decaens J, Buns T, Lachapelle D, Vermeersch O and Mlynarek J 2018 *Analyse du potentiel d'application des textiles intelligents en santé et en sécurité au travail* (Montreal, QC: Institut de recherche Robert-Sauvé en santé et en sécurité du travail)
[8] Carre Technologies inc 2019 HexoSkin smart shirts www.hexoskin.com
[9] Trupp S 2013 Colour-changing gloves for toxic-substance detection *Smart Textiles and Nanotechnology* **2013** (9) 11
[10] Riedl R-M 2011 Cutting edge smart fabrics *Smart Textiles and Nanotechnology* **2011** (5) 4-5
[11] CyberGlove Systems 2017 CyberGlove III www.cyberglovesystems.com
[12] Siegl V 2012 Smart textiles for firefighters *Smart Textiles and Nanotechnology* **2012** (1) 7
[13] Philips A 2015 Cool suit for chemical protection *Smart Textiles and Nanotechnology* **2015** (12) 5
[14] Uniforms-i 2019 Medical uniforms get smarter, work harder uniforms-i.com
[15] Cardinal Health 2018 SmartGown™ Surgical Gowns www.cardinalhealth.com
[16] Weigelt G 2015 Power vest *Fraunhofer Research News* **2015** (3) 4
[17] Köhler A R, Hilty L M and Bakker C 2011 Prospective Impacts of Electronic Textiles on Recycling and Disposal *Journal of Industrial Ecology* **15** (4) 496-511.

IOP Conf. Series: Materials Science and Engineering **827** (2020) 012015 doi:10.1088/1757-899X/827/1/012015

100% PVDF 3D textiles structures to improve energy harvesting

A Talbourdet[1], C Cochrane[1,2], F Rault[1], G Lemort[1], C Campagne[1], and E Devaux[1]
[1]ENSAIT, GEMTEX - Laboratoire de Génie et Matériaux Textiles, F-59000 Lille, France
[2] Author to whom any correspondence should be addressed

Cedric.cochrane@ensait.fr

Abstract. With a final view to prototyping a textile energy harvesting system, piezoelectric textile structures based on 100% poly(vinylidene fluoride) (PVDF) were developed and characterized. Multifilaments of 246 tex were produced by melt spinning. The mechanical stretching during the process provides PVDF fibers with an optimal β-phase ratio (97%). Some studies have already been carried out on piezoelectric PVDF-based structures as films or textiles. The goal of the study is the investigation of the differences between 2D and 3D woven fabrics structures from piezoelectric PVDF multifilament yarns. The textile structures were poled after the weaving process, and a maximum output voltage of 2.3 V was observed on an angle - through-the-thickness interlock (interlock 3D structure) under compression by DMA tests. Energy harvesting is optimized in a 3D interlock thanks to the stresses of the multifilaments in the thickness. This finding has led to the design of an inner sole prototype from a knitting structure and another structure with piezoelectric fibers outside the plane of the fabric. The prototype is able to harvest energy and the results are consistent with the measurement realized with DMA under dynamic compression close to walking.

1. Introduction

One of the future challenges in the e-textile field is about the management of energy. Certainly, electronic systems are becoming less electricity consuming (with BLE, bluetooth Low energy for instance) and batteries are more quickly rechargeable but the energy sources are often non-renewable and non-wearable. In the area of wearable e-textile, the movements of the human body could be an interesting renewable source of mechanical energy that could be captured, converted and stored. Since the discovery of the piezoelectric character of poly(vinylidene fluoride) (PVDF) in 1969 by *Kawai* [1], this polymer is widely used to develop energy harvesting systems. To reach a flexible system, films, fibers and nanofibers based on PVDF or its co-polymers may be processed [2-4]. But more recently, more complex structures, 3D and/or including electrodes, have emerged. For instance, *Soin et al.* [5] have produced a 3D knitting structure made of PVDF monofilament and silver-plated PA6.6 wire acting as electrodes within the textile structure.

2. Material and Methods

Commercial spinning PVDF (Kynar®705 supplied by Arkema) was used to produce multifilaments on a melt-spinning pilot (Spinboy I from Busschaert Engineering, Belgium). The spinning rate is controlled by a metering pump and allows to obtain by adjusting the speed of the alimentation roll (R1) and drawing roll (R2), multifilament of 246 tex. A previous study [6] has shown that a draw ratio (R2/R1) above 4

IOP Conf. Series: Materials Science and Engineering **827** (2020) 012015 doi:10.1088/1757-899X/827/1/012015

allowed to have an optimal β-phase ratio (Piezoelectric phase, 97 % of the total crystal phase). In order to ensure the smooth running of the weaving, a twist of 25 rpm is applied to the 246 tex multifilaments.

Two weaves were chosen: (i) a plain weave (2D structure) and (ii) an angle - through-the-thickness diagonal warp interlock (3D structure). In 3D fabric structures, the third dimension is in the thickness layer. This third direction allows developing the mechanical properties in the thickness in comparison with the conventional 2D structures [7]. These structures, shown in Table 1, were made on a hand loom (ARM B60, 20 pins/cm, 24 frames).

Table 1. Structures and pictures of plain weave and angle - through-the-thickness interlock.

	Weave repeating unit	3D diagram	Picture (× 1.2)
2D Plain Weave		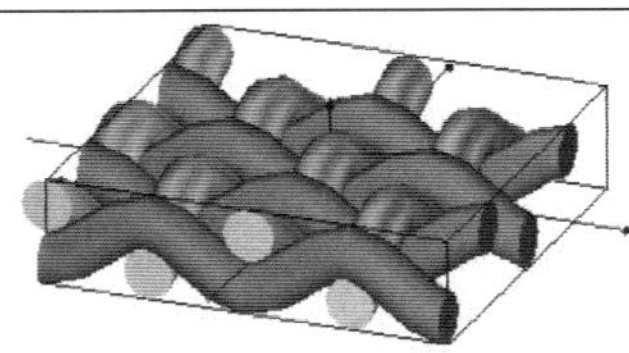	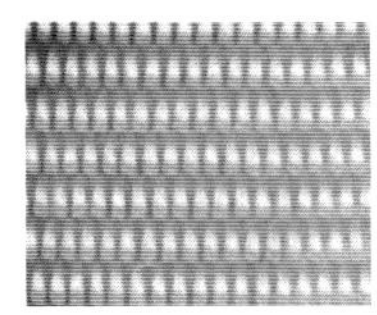
3D Warp Interlock fabric		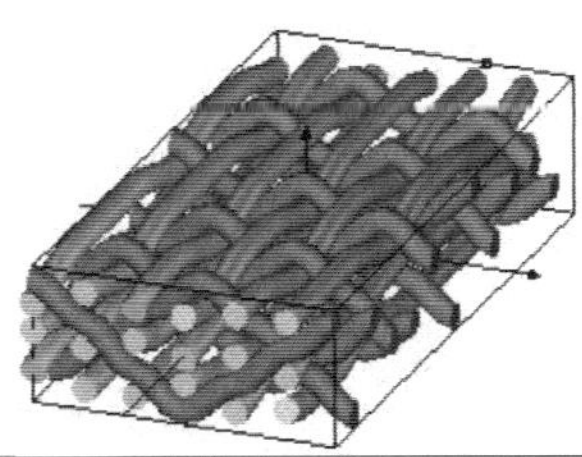	

The use of piezoelectric material requires an electrical polarization step, which is essential to align and reorient the macroscopic dipoles. The polarization is carried out by contact by entrapping the textile structure between two circular copper electrodes (surface area of 9 cm^2) connected to a voltage generator. The structures were polarized in an oven set at 90°C [8]. All textiles structures were polarized under the same field of 2.7 ± 0.1 $V.\mu m^{-1}$ during 15 min. The setup is shown in figure 1.

Figure 1. Setup used for contact polarisation.

For the piezoelectric characterization, the textile was subjected to a dynamic compression strain perpendicular to the structure using a DMTA equipment (DMA TA Instrument Q800), with a plate-to-plate setup. Samples were sandwiched between circular copper electrodes and compression pans (surface area of 1.77 cm^2). The electrodes are connected to an electrical circuit comprising a 1 MΩ resistor and a Keithley multi-meter (3706A) for data acquisition. The force applied by the DMA is between 1 and 5 N for a rate of deformation of 5%–60% and a frequency of 100 Hz, for 10 consecutive cycles. In the case of characterization in order to simulate walking, a force of 18 N, a rate of deformation 60% and a frequency 4.4 Hz are applied. This frequency corresponds to a speed of 10.1 m/s [9].

IOP Conf. Series: Materials Science and Engineering **827** (2020) 012015 doi:10.1088/1757-899X/827/1/012015

3. Results

Table 2 shows the characteristics of the 100% PVDF 2D plain weave and 3D Warp Interlock fabric. As expected regarding the structure, the 3D fabric is more than two times thicker and heavier than plain weave.

Table 2. Main characteristics of 100% PVDF 2D and 3D woven.

	2D Plain Weave	3D Warp Interlock fabric
Thickness (mm)	1,08 ± 0,07	2,38 ± 0,11
Basis weight (g/m²)	785 ± 0,3	1790 ± 4
Theoretical porosity (%)	59	57

Figure 2 shows output voltage and power produce by 2D and 3D fabric during dynamic compression under DMTA. The output voltage is the maximum value of recorded voltage, RMS is calculated from all recorded data and harvested energy is calculated according to measurement time and electrodes size.

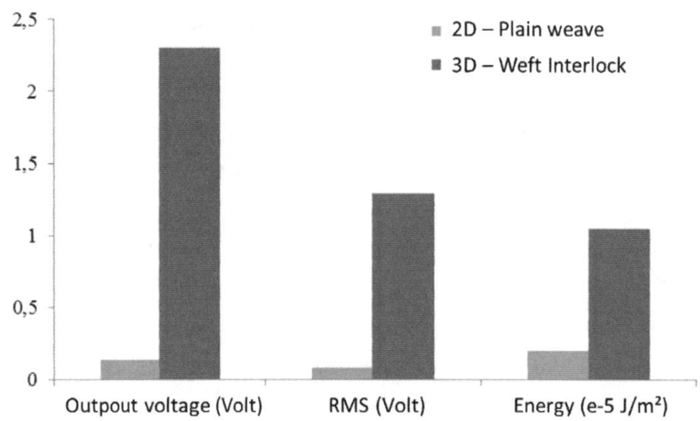

Figure 2. Output voltage (left and center) and power (right) for 2D plain weave and 3D Warp Interlock fabric under compression.

In all cases, 3D Warp Interlock fabric displays better performance for energy harvesting;more than twice the performance of the 2D plain weave. However, this cannot be only explained by the difference between the thickness of fabrics. Thus, as expected, the structure and placement of PVDF filaments in this one are very important. It seems essential to have fibers outside the plane of the fabric. Consequently, a prototype in the form of an inner sole has been designed to convert the energy of walking into electricity. For this prototype, we have tried another textile structure with fibers outside the plane of the fabric, a knitting (rib 1x1, 2.7 mm of thickness, 1319 g/m² and 72% of theoretical porosity). In addition, flexibles electrodes have been developed. Conductive polymers composite (CPC) based on PVDF and 10 wt.-% CNT was mixed by twin-screw extruder and rod with a bulk conductivity of 60 S/m is produced for 3D-printing. The rod is deposited on each side of the PVDF woven with a geometrical shape enabling certain flexibility for the whole system (woven + electrodes). Figure 3 shows, on (a), the prototype of the inner sole for the energy harvesting. On (b), the graph shows the output voltage measured on the capacitor (1 µF) linked to different samples included the prototype of the inner sole.

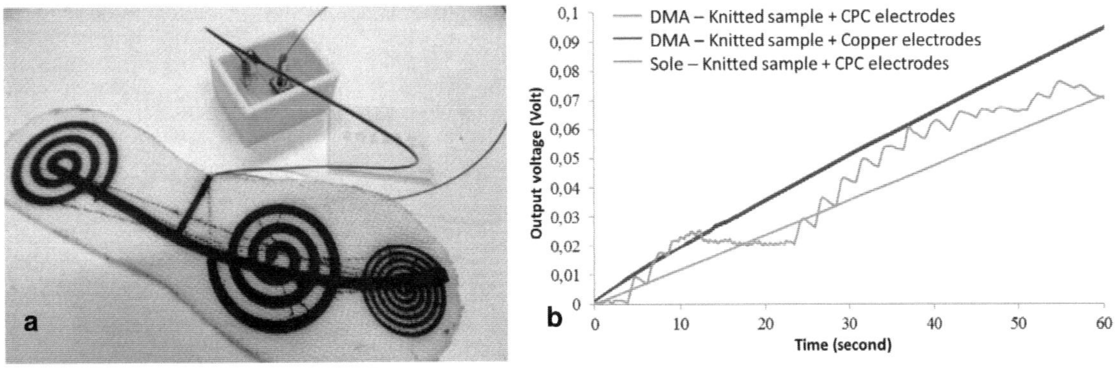

Figure 3. (a) Knitted piezoelectric inner sole and 3D printed CPC electrodes, (b) output voltage for samples under DMA (4.4 Hz, 18 N) and for the inner sol in walk condition.

The prototype is able to harvest energy and the results are consistent with measurement realized with DMA under dynamic compression close to walking.

4. Conclusion

In this study, we compared 100% PVDF 2D and 3D woven structures from the energy harvesting point of view. DMA measurements under compression have demonstrated that 3D structure (Warp Interlock fabric) harvested more energy even though it is thicker. In fact, the textiles structures with fibers outside the plane of the fabric are the most performant in harvesting energy. Thus, the output voltage is 2.3 V for 3D structure opposing to 16 times less voltage for the 2D structure. Finally, based on 100% PVDF Woven structure, we have designed a functional prototype of inner sole able to harvest energy from the walk.

Acknowledgments

Authors are grateful for financial support from BPI France and Arkema to Autonotex Project.

References

[1] Kawai H, 1969. The piezoelectricity of poly (vinylidene fluoride). *Japanese journal of applied physics*, **8**(7), p.975..

[2] Ahn Y, Song S and Yun K.S, 2015. Woven flexible textile structure for wearable power-generating tactile sensor array. *Smart materials and structures*, **24**(7), p.075002.

[3] Fan K, Chang J, Pedrycz W, Liu Z and Zhu Y, 2015. A nonlinear piezoelectric energy harvester for various mechanical motions. *Applied Physics Letters*, **106**(22), p.223902..

[4] Shin S.E, Ko Y.J and Bae D, 2016. Mechanical and thermal properties of nanocarbon-reinforced aluminum matrix composites at elevated temperatures. *Composites Part B: Engineering*, **106**, pp.66-73.

[5] Soin N, Shah T.H, Anand S.C, Geng J, Pornwannachai W, Mandal P, Reid D, Sharma S, Hadimani R.L, Bayramol D.V and Siores E, 2014. Novel "3-D spacer" all fibre piezoelectric textiles for energy harvesting applications. *Energy & Environmental Science*, **7**(5), pp.1670-1679.

[6] Talbourdet A, Raul F, Cayla A, Cochrane C, Devaux E, Gonthier A, Lemort G and Campagne C, 2017. Development of mono-component and tri-component fibres 100% polymer based piezoelectric PVDF to harvest energy. *IOP Conference Series: Materials Science and Engineering* **254**(7,) p. 072026).

[7] Chen X, Taylor L.W and Tsai L.J, 2011. An overview on fabrication of three-dimensional woven textile preforms for composites. *Textile Research Journal*, **81**(9), pp.932-944.
[8] Nilsson E, Lund A, Jonasson C, Johansson C and Hagström B, 2013. Poling and characterization of piezoelectric polymer fibers for use in textile sensors. *Sensors and Actuators A: Physical*, **201**, pp.477-486.
[9] Soin N, Shah T.H, Anand S.C, Geng J, Pornwannachai W, Mandal P, Reid D, Sharma S, Hadimani R.L, Bayramol D.V and Siores E, 2014. Novel "3-D spacer" all fibre piezoelectric textiles for energy harvesting applications. *Energy & Environmental Science*, **7**(5), pp.1670-1679.

Sustainable wool insulation textile products - an opportunity for entrepreneurial initiatives in Romania

Pyerina C Ghițuleasa[1,4], Eftalea Cărpuș[1], Angela Dorogan[1], Cezar Bulacu[2], Ana Enciu[3]

[1] The National R&D Institute for Textiles and Leather - INCDTP, Romania;
[2] S.C. MINET S.A. - SME, Vâlcea County, Romania
[3] The R&D Institute for Sheep and Goats Breeding - ICDCOC Palas, Constanta County, Romania
[4] Author to whom any correspondence should be addressed

ghituleasa@certex.ro

Abstract. The wool processing sector in Romania experienced a regression in terms of fiber quality after 1989, caused by a combination of factors: uncontrolled crossbreeding, low area and low quality pastures and lack of support for sheep breeders. As a result, the spin ability limit of the Romanian wool has decreased, as well as the possibility of using it in the textile industry, in the conditions of increasing demand for fine fabrics and knitwear, leading to the closure of many traditional textile companies and the use of imported wool. Under these circumstances, the use of the Romanian wool for related fields such as construction is much more important, as an efficient and viable alternative for recovery and an alternative for revitalization of several economic sectors. The paper presents aspects regarding the textile valorization of the thick Romanian wool varieties in order to produce materials with the role of insulation and sealing, in the field of eco-friendly constructions.

1. Introduction

Romania's rural development strategy [1] for the coming years is in line with the EU's reform and development context with the Europe 2020 strategy. Following the objectives of the Europe 2020 strategy for a smart, sustainable and inclusive economy, the strategy sets ambitious targets for Member States in the areas of education, innovation, energy/environment, employment and social inclusion and improving competitiveness in general [2].

The National Rural Development Program (NRDP) 2014-2020 [3] contributes to *smart growth* by supporting forms of cooperation between research institutions and farmers and other actors in the rural economy, but also by supporting training, skill acquisition and dissemination of information. The NRDP also envisages a *sustained growth* that focuses on lowering carbon emissions and supporting environment-friendly farming practices.

Last but not least, support for investment in infrastructure and the rural economy leads to poverty reduction and job creation in rural areas, thus contributing to inclusive growth. All these objectives will be possible to materialize only under the conditions of efficient utilization of indigenous raw materials, among which wool fibers, a valuable source both for the textile industry and for related sectors, as is the field of ecological constructions.

The wool processing sector in Romania experienced a regression in terms of fiber quality after 1989, caused by a combination of factors: uncontrolled crossbreeding, low area, low quality pastures and lack of support for sheep breeders. As a result, the spin ability limit of the Romanian wool has decreased, as well as the possibility of using it in the textile industry, in the conditions of increasing demand for fine fabrics and knitwear, leading to the closure of many traditional textile companies and the use of imported wool.

Under these circumstances, the use of the Romanian wool for related fields, such as construction, is much more important as an efficient and viable alternative for recovery and an alternative for revitalization of several economic sectors.

The paper presents aspects regarding the textile valorization of the thick Romanian wool varieties in order to produce materials with the role of insulation and sealing, in the field of eco-friendly constructions.

2. Material and Methods

Figure 1: Romanian sheep breeds a) (semi)fine wool fiber - assimilated Merino breed (Merino of Palas, Constanța); b) semicoarse fiber – Țigaie breed; c) semicoarse fiber - Țigaie-Țurcană metis (Karakul of Botoșani); d) coarse fiber – Țurcană breed; i =1; 2 – details for male; female. Source: Archive of ICDCOC Palas, Constanța.

In Romania there are a variety of sheep breeds (figure 1). Romania, within the EU, but also in the Balkan area, has owned a large number of sheep heads, with a variable weight between sheep and wool sheep.

Țurcană breed derives from the wild species *Ovis Vinai Arkar*, it was domesticated in prehistoric times in the Carpathian Mountains area where it was formed as a breed by natural selection.

From here it has spread through transhumance, in all the countries of Central Europe, in areas varied in relief and level of precipitation. Average wool production per capita is about 2 kg for female and about 3kg for male.

Țurcană sheep has a woolen shell/fleece made of sharp streams that have an outer appearance of spear and corkscrew, being made of different inhomogeneous fibers. The wool is rare and "the seam line" on the upper line is wide. The streamers are long, in conical shape (characteristic of mixed wools). The internal structure of the shell is unclear, tangled.

The Țurcană breed fleece mainly comprises long and thick fiber, on the outside, but also a layer, covered by the previous one, of shorter and thinner fiber.

There are 3(three) varieties Țurcană breed, in Romania, namely white, black and gray varieties.

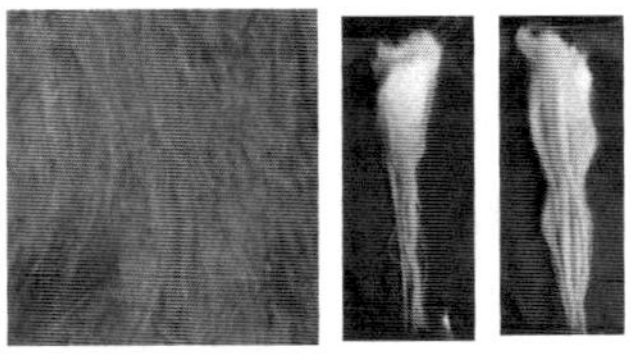

Figure 2. Details for Ţurcană breed wool fleece, exterior layer a) Ţurcană breed unwashed wool fleece, b) (un)washed wool strand.

The thick and long fibers have an average length of 23-25 cm, a fineness of 56-58 µm and a pronounced medullary canal. The short and thin fibers at the base of the fleece have an average length of 9-11 cm, and a fineness of 28-32 µm.

Washing efficiency for Ţurcană breed wool is about 65 -70%. In figure 2, there are details for Ţurcană breed wool fleece exterior layer.

In the research expertise, non –woven infrastructure facilities from SC MINET SA Company; a representative Romanian textile company and the raw materials; washed coarse (Ţurcană wool fibers) and semi-coarse (Tigaie wool fibers) Romanian wool fibers, were utilized.

3. Experimental aspects

The matrix of technological experimentation took into account the following experimental criteria [4]:

- exploiting the technological equipment of SC MINET SA, Vâlcea County (figure 3);
- two distinct technologies were used in processing: i) strengthening the fibrous material by the thermo-chemical process (for Ţurcană wool), using no more than 35% of man-made adhesive fibers and ii) mechanical strengthening (for Ţigaie wool);
- adaptation of technological parameters and processing stages to the characteristics of the processed fiber material;
- the coverage, according to the adjustment parameters, of the entire range of non-woven structures, possible to be obtained, based on the processing technologies used;
- design and fabrication of nonwoven fabrics with different density and thickness, covering a wide range, possible to be used by the constructors in different ways and locations of a building: floor, roof, walls;
- providing improved properties to the materials by applying functional treatments specific to insect/moth protection, flame maintenance and propagation.

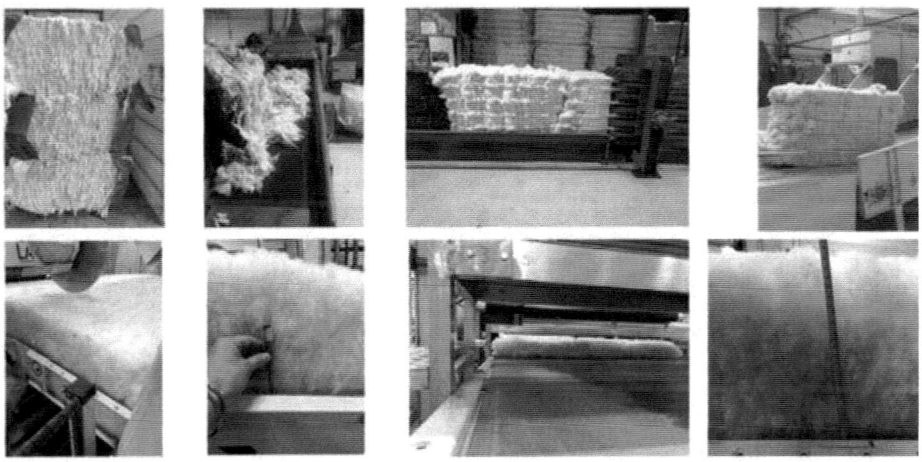

Figure 3. Details for experimental stages, at SC MINET, SA.

The use of thick and semi-thick wools for the production of non-woven textile materials for the sound and thermal insulation of buildings is a high-potential entrepreneurial area, in the context of rising electricity, thermal and energy prices and in the context of national and European trends in green building.

4. Aspects regarding the sustainable development and natural fibers usage as raw material

There was performed also the analysis of the export capacity of Romanian eco-innovative wool products to the European countries and the directions to action for achieving this objective.
Eco Textiles are present not only at clothing and garment products, but also at technical domains. It could say that this is the future, the label as Ecotech, Okeotech and Eco friendly being well known all over the world.

Civil engineering and building engineering are a significant contribution to the human society development, because these industrial fields meant planning, design, construction, manufacturing and maintenance of the infrastructures.

The connexion of the technical textiles and especially high performance technical textiles with these sectors has conducted to an increased impulse to the eco-buildings and green building development.

These textile materials are used for buildings, bridges, tunnels construction and they are generically called "Buildtech".

They offer specific physical-mechanical properties such as: reduce weight, very good mechanical resistance and resilience, as well as resistance to many factors: creep, chemical degradation and pollutants from air and rain, sunlight and acids action. These textile materials play an important role in the modernization of the construction infrastructure.

Natural textile products offer a range of environmentally responsible alternatives to other materials that are resource deficient.

They use both post-consumer and post-industrial waste streams and reuse them for the manufacture of extremely durable and environmentally friendly textile materials.

They not only generate a reduce waste, but more importantly, they save rapidly depleting natural resources.

Some of the characteristics of these textiles are: they use natural fibers, they are processed with less harmful inputs, the processing units are equipped with wastewater treatment plants, the fabrics are of good quality and long lasting use.

They can be used for: soil sealing, textile drainage systems, erosion prevention systems, textiles for protection against dangerous substances, mobile containers, sound barrier systems, filtration systems (air / water), textiles for landfills etc.

Different geo-synthetic products are used as: Geotextiles, geo-synthetic linings, geo- networks, geo-membranes and various geo-specialized composites.

Other potential areas of application of geo-synthetic products: spatial planning (golf courses, ponds, etc.), rehabilitation of mines and tunnel linings, etc.

In Romania, the problem is more complex and difficult, because of the mentality and culture changing need.

The valorization of course wool fibers Turcana for the construction field is a valuable instrument for solving the problem of wool industry, on the one hand, and for the eco-building promoting, on the other hand.

5. Aspects regarding the economic efficiency

These are the arguments for which we analyzed also the economic efficiency of capitalizing Romanian wools by simulating a family entrepreneurial business, that is, a minimal investment, possible to be achieved at the level of young Romanian entrepreneurship.

It should be highlighted that not only the low price aspects must be considered, but first of all the following should be taken into account:

- the technology allows the recovery of Romanian thick and semi-thick wool fibers, a raw material which does not have suitable spin ability characteristics for processing high fineness yarns and which is currently either burned or exported as raw material or collected in peasant farm conditions;
- the technology may be applied to companies producing non-woven textiles with technical use;
- newly created companies will contribute to the use of native resources of thick and semi-thick wools, implicitly in creating new jobs and attracting young people, especially in rural areas, where a depopulation phenomenon is currently occurring;
- the new products will contribute to the development and implementation of the concept of green houses with low construction and operating costs;
- sheep breeding is a basic occupation of most peasant farms, especially (semi)coarse wool fibers;
- the sheep wool is sheared once a year and the raw (greasy) wool obtained is collected and sent to a collecting center, to be forwarded to the wool laundry.

As a consequence, a business plan simulation, based on the technical data obtained in the technological experiments, was elaborated by SC MINET SA, Râmnicu Vâlcea and INCDTP.

The potential users of the research results are:

- sheep breeders,
- breeders' associations,
- economic units with activity in the field of wool manufacturing,
- decision-makers in the field of industry and agriculture,
- young entrepreneurs.

6. Conclusion

The analysis of the construction materials market shows an increased interest in the use of wool as a thermal insulation material, leading to an economically significant impact, as the construction sector is a major energy consumer within the European Union, accounting for 40% of the total energy consumption and 36% of greenhouse gas emissions [5].

In the conditions of increasing Romanian thick and semi-thick wool fibers, their exploitation by producing nonwoven fabrics for efficient constructions, both in terms of construction costs and energy maintenance of buildings, is a good opportunity.

The valorization of course wool fibers Turcana for the construction field is a valuable instrument for:

- Solving the problem of wool industry,
- eco-building promoting,
- strengthening the Romanian technical textile sector by using (semi) coarse indigene wool fibers,
- increasing the entrepreneurial spirit and the capacity to attract and use the European non-reimbursable funds, especially for productive investments in rural areas.

The results will be used by the sheep breeders, wool fibers collecting centers, with the involvement of the Regional Development Agencies in the country and actors in the construction sector, in order to stimulate the regional entrepreneurial initiatives, both in non-woven textiles, and in the green construction fields.

Acknowledgment:
This work was carried out in the SECTORIAL Program implemented with the support of Romanian Ministry of Research and Innovation, project no. 5PS /02.11.2017. The publication of the paper is funded by the Ministry of Research and Innovation within Program 1- Development of the national RD system, Subprogram 1.2 - Institutional Performance - RDI excellence funding projects, Contract no. 6PFE/2018.

References:

[1] NSDS, (2008), National Sustainable Development Strategy of Romania 2013-2020-2030, *Government of Romania*, , Online at: http://strategia.ncsd.ro.

[2] European Commission 2010, A European Strategy for Smart, Green and Inclusive Growth, *Communication from the commission Europe 2020* Brussels.

[3] NRDP 2014, Romanian National Programme for Rural Development 2014-2020 Bucharest, www.fonduri-ue.ro/pndr-2014

[4] Ghituleasa, C, Carpus E, Dorogan A, Bulacu C and Enciu A 2018, Insulation materials for buidings – a successful research&development collaboration for the Romanian wool fibres manufacturing, *Industria Textila* **69**(5), pp. 419-421

[5] Korjenic A, Klaric S and Hadžić A 2015, Sheep Wool as a Construction Material for Energy Efficiency Improvement, *Energies* **8**(6), 5765-5781.

IOP Conf. Series: Materials Science and Engineering **827** (2020) 012017 doi:10.1088/1757-899X/827/1/012017

Mitigation of environmental impact caused by flame retardant textile finishing chemicals

Boisserée Nora[1], Félix de Castro Paula[1,4], De Vilder Ine[2], Detrell Ariadna[3], Casamada Josep[3]

[1] Surface Technologies Unit, LEITAT Technological Center, Spain
[2] Centexbel, Ghent, Belgium.
[3] AEI Tèxtils – The Catalan Technical Textiles' Cluster, Terrassa, Spain.
[4] Author to whom any correspondence should be addressed

pfelix@leitat.org

Abstract. LIFE FLAREX project focuses on the mitigation of environmental and health impacts caused by toxic compounds that are contained in commonly used flame retardants (FRs) in home textiles which are based on halogenated-, antimony trioxide-, formaldehyde release related products. Conventionally used FRs in the home textile sector are decabromodiphenyl ethane, its combination with antimony trioxide and dialkyl phosphono carboxylic acid amides, respectively. The project evaluates alternative FRs and their feasibility at the industrial scale level. Among all the home textile applications, curtains, upholstery, mattress ticking and bedding sheets were selected as the most interesting products in terms of need for substitution. Within these applications, cotton, polyester and polyester/cotton blends were selected as most representatives. Technical performance, flame retardant resistance as well as washing durability was evaluated for up to eight different alternatives and compared to the conventional FRs. The alternative FRs included mixtures of phosphates, ammonium polyphosphate, expandable graphite, guanidine phosphate, ammonium sulphamate and other new patented technologies in the market. A toxicological assessment through skin-FRs contact measurements is being conducted for assessing their absorption and toxicological properties. Additionally, an environmental and health assessment is also being conducted as well as a life cycle assessment comprising environmental, economic and social considerations. The project has at the moment demonstrated the feasible substitution of the aforementioned toxic FRs for greener and safer FR alternatives.

1. Introduction

Flame retardants play an important role in reducing the incidence of fire on a global scale. Statistics that at least 94% of fire deaths occur in homes and buildings [1,2]. Flame retardants are used in home textiles for ensuring protection against the risk of fire by retarding or inhibiting propagation of flame. They are used in European countries in home textile contracting by legislation (i.e. theatres, hotels and public accommodations, public buildings and offices, hospitals and care homes, among others) while in more stricter countries like in the UK their use extends to houses.

Nowadays, current technologies used for obtaining flame retarded textiles are based on halogen, antimony trioxide and formaldehyde releasing compounds [3] which present health concerns and environmental impact [4].

Commercial and greener alternatives to current flame retardant compounds are based on phosphorous compounds, a combination of phosphorous and nitrogen compounds and inorganic materials among others. Phosphorous and their combination with nitrogen compounds have proven to be efficient. However, they are more expensive due to their lower volume production. In some cases, it is necessary to add a higher amount to the flame-retardant formulation in order to achieve good results/flame retardancy. In terms of environmental impact inorganic materials are the best alternatives for flame retardants, however higher amounts of inorganic material needed for achieving good efficiency. In addition to this, the lack of legislation and promotion of greener alternatives [5] lack of user awareness are also other key factors making more difficult.

With this regard, this project aims to demonstrate and evaluate the technical and industrial feasibility of the current flame retardant materials halogen and antimony-trioxide based compounds and formaldehyde releasing compounds greener alternatives. Additionally, an environmental and health assessment as well as a life cycle assessment will be also carried out.

2. Materials and methods

2.1. Selection of materials

Based on a survey and a workshop conducted with the main stakeholders (textile manufacturers, textile finishers, f among other), tree fabric compositions were selected, 100%PES, 100%CO and their blends thereof, PES/CO, as well as four home textile applications as most common and potentially interesting regarding their presence in home textile contracting but also regarding their substitution. The applications selected were curtains, upholstery, mattress ticking and bedding sheets. A screening and selection of the most promising alternatives was carried out.

Table 1 collects the conventional flame retardant compounds and the most promising and commercial alternatives selected. Among the conventional flame retardants are: 1) decabromodiphenyl ethane which has been the drop-in solution in industry of the recently restricted decabromodiphenyl ether 2) decabromodiphenyl ethane used with antimony trioxide used as synergist and 3) dialkyl phosphono carboxylic amide. Selected alternatives include cyclic phosphonate, expandable graphite, ammonium polyphosphate, ammonium, guanidine phosphate, and a new patented technology based on a combination of ammonium, urea and a phosphorus compound. A brominated polymer has also been chosen as a transitional alternative due to its industrial interest, being a bigger halogenated compound therefore presenting lower risk of migration from the fabric.

Table 1. Home textile applications, fabrics selected for each application, conventional, intermediate and alternative flame retardants evaluated.

Applications/ Fabrics	Conventional FR	Transitional FR	Alternative FR
Curtains • 100%PES () • 100%PES ()	• Decabromodiphenyl ethane + melamine cyanurate • Decabromodiphenyl ethane+ antimony trioxide (ATO)	• Brominated polymer	• Cyclic phosphonate
Upholstery • 100PES	• Decabromodiphenyl ethane • Decabromodiphenyl ethane+ antimony trioxide (ATO)	• Brominated polymer	• Expandable graphite

Mattress ticking • PES/CO (50/50) • 100%PES	• Decabromodiphenyl ethane • Decabromodiphenyl ethane+ antimony trioxide (ATO)	• Brominated Polymer	• Ammonium polyphosphate • Guanidine phosphate • Ammonium sulphamate
Bedding sheets • PES/CO (50/50) • 100%CO	• Dialkyl phosphono carboxylic acid amide • Decabromodiphenyl ethane + melamine cyanurate • Decabromodiphenyl ethane+antimony trioxide (ATO)	• Brominated Polymer	• Ammonium sulphamate + Urea + $PO(OH)_2$-R-$PO(OH)_2$ • Phosphorous based

2.2. Experimental part

The experimental part contains two steps. Firstly, the flame retardants were applied and tested at laboratory scale and based on these results the flame retardants were applied at industrial scale. This paper presents exclusively the results obtained at laboratory scale.

The fire-tests are carried out in part B1 of the laboratory-scale project for different textiles used in definite final applications (curtains, upholstery, mattress ticking and bed sheets).

After the pre-industrial tests, the most promising applications are executed at the industrial level in the companies that showed interest in participating in the project. The procedure of the application at the industrial level is similar but with at least 50 m of fabric.

The selected flame retardants were applied to the materials depending on the application. Afterwards the flame retardancy of the treated textiles was tested according to the adequate for the application sector. Another part is to test the toxicity of the flame retardants in-vitro. Also, an LCA (Life Cycle Assessment) is conducted to evaluate different aspects such as raw materials, energy consumed during their application, cost of the products among others important parameters that have to be assessed when promoting substitution of the conventional FR products.

Foulard process is used for impregnation of textile substrates. In this process, after submersion in the aqueous finish solution the fabric is squeezed between 2 rolls with a pressure of 1.5 bar and a speed of 5 rpm to remove the excess of finish solution (Figure 1 Left). This is followed by a drying and curing process that occurs in the stenter machine at the specific drying and curing temperatures for each finish product (Figure 1 Right).

The wet pick-up (amount of wet finish solution absorbed on the textile) is determined after squeezing while the dry pick-up (amount of dry finish product absorb on the textile) is determined after drying and curing.

Figure 1. Left) Foulard; Right) Stenter machine (Mathis coater).

Coating of A4-samples is performed on stenter machine (see Figure 1). The fabric is mounted in the stenter frame and the coating thickness is adjusted by varying the distance between the blade and the roll (knife-over-roll coating). Afterwards the applied coating is thermally dried/cured in the oven; temperature and time depends on the used system. Here the add-on of the fabric is determined by weighting the sample before and after the coating.

The fire behavior of the applied products were conducted according to the EU for the application sectors. The efficiency of the flame retardants was detected by comparing and analyzing the results of the treated samples to the untreated samples.

The flammability of curtain fabrics was determined according to the standard UNE EN 13773:03 which consists of two parts, UNE EN 1101 and UNE EN 13772. It was decided to first conduct the part one (UNE EN 1101) and follow with the second part with the most promising alternatives in case of successful results. For upholstery the flame retardant treated textiles were tested according to the UNE-EN 1021-1:2015. The flammability of the mattress ticking is determined according to EN 597 (1994): Furniture – Assessment of the ignitability of mattresses and upholstered bed bases. Furthermore, the Crib 5-test as described in the BS6807 (methods of test for assessment of the ignitability of mattresses, upholstered divans and upholstered bed bases with flaming types of primary and secondary sources of ignition) was used as a screening method. To determine the flame retardancy of the products applied on bed linen it is performed the EN ISO 12952 Textiles – Assessment of the ignitability of bedding items.

3. Results and discussion

As there were two different PES curtain textiles screened according to UNE EN 1101, it can be said that textile A has better fire results than textile Because less dense material. Also, textile A has a higher absorption potential which means a better wet/dry pick up. Textile B has clearly a handicap for the warp direction. It is quite usual for woven textiles that the warp direction burns faster and more because the yarn is thicker than in weft direction.

Regarding conventional treatments for curtain the products are not meant to be washed. The treatment of DBDPE + MC was not successful for both curtain types at any concentration. Besides, the treatment of 50% DBDPE + ATO worked for the less dense textile A but for textile B it only worked in weft direction.

The results of the alternative cyclic phosphonate are that none of the studied concentrations passed the standard. However, it was noted that the samples passed the fire test if they were not washed. In this regard, a different cyclic phosphonate had to be screened and applied which worked successfully due to its washing resistance.

The samples treated with brominated polymer showed positive results as they were treated by impregnation. On the other hand, the samples treated with brominated polymer showed a high rigidity which is not suitable for curtain applications.

The results for the upholstery treatments are shown in figure 2 and 3. Figure 2 displays the results of UNE-EN 1021-2:2005 of the PES untreated textile (Figure 2 Left) compared to the halogenated products decabromodiphenyl ethane (Figure2 Middle) and the application of brominated polymer (Figure 2 Right). It is clear to see that the coating worked successfully, as the burn marks of both samples are quite small and regular compared to the untreated sample (Figure 2 Left).

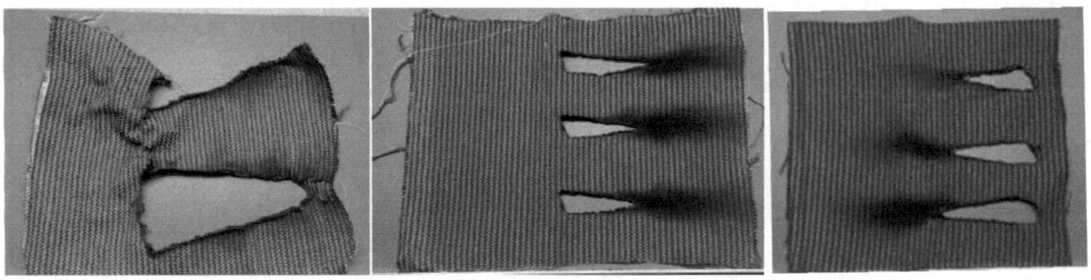

Figure 2 UNE-EN 1021-2:2005, Left) PES untreated, Middle) DBDPE + ATO, Right) brominated polymer.

As it is shown in Figure 3 the burn marks of the sample coated with expandable graphite are quite small, similar to the marks obtained when using halogenated compounds indicating similarity during the burning behavior (Figure 2 Middle and Right).

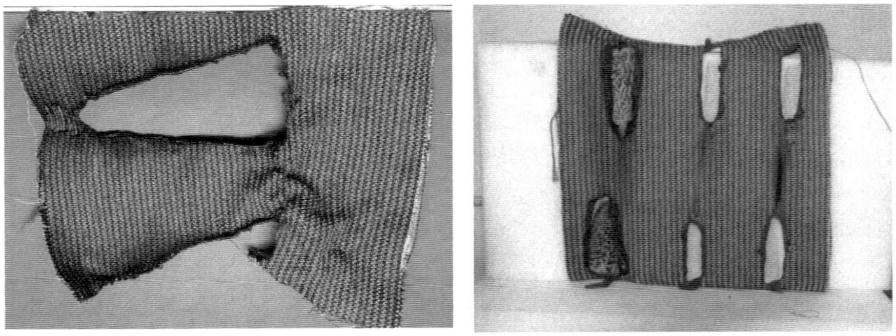

Figure 3 UNE-EN 1021-2:2005, Left) PES untreated, Right) PES treated application with 50% Expandable Graphite.

For mattress ticking the ignition of the untreated CO/PES mattress ticking results in an escalating fire. All FR treatments make the fabric pass the crib 5 test, nevertheless differences between the different treatments are noticeable. Lower concentrations of guanidine phosphate (11 wt%) and ammonium (5 wt%) are sufficient to make the CO/PES FR. The ammonium polyphosphate in higher concentrations (16 wt%) is more efficient than the lower concentration (7 wt%).

The PES 3D spacer fabric passes the crib 5 test without flame retardant treatment, but its performance can be improved by applying a flame retardant treatment. Also, in this case the lower concentrated guanidine phosphate (12 wt%) is more efficient. The higher concentration of ammonium polyphosphate (27 wt%) is significantly better than the concentration of 16 wt%. The difference between higher and lower concentration of ammonium is not decisive

Conventional flame decabromodiphenyl ethane with antimony trioxide is not usually used in the industry for bed linen; this also reflected in the results. As the treatment is not wash resistant it cannot be used for this sector.

The treatment based on Pyrovatex, based on dialkyl phosphono carboxylic acid amide, on the other hand is durable and maintains its flame retardant effect after the washing cycles. Despite the flame retardant is only bound to cotton, the CO/PES 50/50 bed linen still passes the fire test.

As the dry pick-up of the tested cotton bed linen was very low, only 0.4 wt%, the flame retardant effect was not great. The conventional Pyrovatex-like treatment has affinity for cotton which should result in a flame proof fabric.

As the alternatives to Pyrovatex only act on the cotton fraction of a CO/PES blend, non-halogenated alternative identified able to make a wash durable flame retardant CO/PES 50/50 fabriA treatment with the polymeric FR resulted into a wash durable flame retardant blend. The treatment did have an influence on the touch of the fabric: it is stiffened.

The cotton bed linen can be made flame retardant with both the polymeric FR as the ammonium + urea + $PO(OH)_2$-R-$PO(OH)_2$ treatmentP-based FR. The polymeric FR had the most impact on both the color as the touch of the fabric.

Among all the alternatives tested at laboratory scale, the transitional alternative, brominated polymer, showed good efficiency conferred to the fabric for both curtains and upholstery achieving a passed fire test under the standard "UNE EN 13773, part 1: UNE EN 1101" and "UNE EN 1021-part2match flame", respectively. However, for curtains applications the treated textiles showed loss of flexibility. Fire tests are still being carried out with textile treated with the alternative based on a cyclic phosphonate for its application in curtains. Expandable graphite showed promising results being tested under the "UNE EN 1021-part2-match flame", for upholstery applications compared to current alternatives based only on halogenated compounds. The new patented solution consisting of ammonium combined with urea and a phosphorus compound resulted in very promising results being tested under the "UNE EN ISO 12952-1for bedding sheets applications Guanidine phosphate, ammonium and ammonium polyphosphate resulted to pass the fire test for mattress ticking under the standard "UNE EN 597-part1 cigarette and part2-match flame" as well as the crib 5 test.

The technical feasibility of the selected flame retardant alternatives has been demonstrated at laboratory scale. Current work is being performed on running industrial trials for their evaluation. Moreover, work is being conducted on the health and environmental assessment and life cycle assessment of the selected alternatives.

References

[1] Center of Fire Statistics of CTIF 2006 *World Fire Statistics, Report No.10*, https://ec.europa.eu/info/departments/justice-and-consumers_en

[2] World Health Organization 2018 *Burns* https://www.who.int/en/news-room/fact-sheets/detail/burns.

[3] Flameretardants-online 2014 *The flame retardant market* https://www.flameretardants-online.com/flame-retardants/market [accessed on April 5, **2019**].

[4] Lyche J L, Rosseland C, Berge G and Polder A 2015 *Human health risk associated with brominated flame-retardants (BFRs)*. Environment International **74** pp.170-180.

[5] Organisation for Economic Co-operation and Development. Environment Directorate. Joint meeting of the chemicals committee and the working party on chemicals, pesticides and biotechnology 2016 *Synthesis report from the OECD workshop on alternatives assessment and substitution of harmful chemicals. Series on Risk Management*. No.31.

Use Case of Smart Textiles Bio-Sensors Development and Integration for the Automotive Industry

J Decaens[1,2], O Vermeersch[1], D Lachapelle[1] and P Forcier[1]

[1] CTT Group, St-Hyacinthe, Quebec, J2S1H9, Canada
[2] Author to whom any correspondence should be addressed

Jdecaens@gcttg.com

Abstract. As smart clothing is reaching a commercial level of maturity, the technology beyond bio-monitoring is being explored for other fields such as the automotive industry. In this case, the textile sensors integrated into a seat must remain contactless to allow for layer of clothing, seat cover, and such. The use of conductive textile materials is a key in creating an electromagnetic field which allow for contactless sensing. The data gathered by the bio-sensors can be used as prevention tool for detecting driver's over-tiredness and stress or to monitor discomfort among passengers.

1. Introduction

Smart textiles may have been around for only a few decades but it is one of the fastest growing markets in the world; reaching 2.03 billion dollars in 2018 and expected to double these figures by 2025 [1]. Offering various active functions such as lighting, heating and sensing, smart textiles are mostly present in the garment industry and more specifically for workwear and sportswear applications. However, few products are actually commercially available due to important technical challenges related to the wear of the garment such as stretchability and washability. On the opposite end, while the automotive industry is usually perceived has highly innovative and cutting-edge in terms of technology, there are very few applications of smart textiles in cars; despite the fact that an average car is composed of approximately 30 kg of textiles [2].

However, the recent development trend showed a clear emphasis on the driver's safety and comfort. The autonomous cars, which are to be the next generation vehicle, would not only need to perform their main function of driving by assessing its surrounding environment but will also be required to maintain the current entertainment functions encountered in many vehicle : music, lighting, and so on [3]. The same way, no human action is needed to control an autonomous vehicle, it is expected no more actions will be necessary to control accessories in the vehicle cabin. Being able to adjust its surrounding such as ambient temperature, volume of the music, or even engine functions (slow down or stop) based on the monitoring of the driver's physiological state is tomorrow's main innovation for the automotive industry.

2. Context and State-Of-Art Review

Long-distance travels have been known to often be a source of discomfort for both passengers and drivers, independently of the mode of transportation: plane, bus or train. Moreover, tiredness and psychological stress can affect the reflexes of drivers, putting at risk all passengers using public transport system. Professional drivers are considered a vulnerable occupational group due their exposition to environmental stressors and adverse work conditions. Furthermore, studies have been able to create direct correlation between work-related stress and health outcomes such as cardiovascular disease, aggressive driving and fatigue leading to unsafe vehicle operation.

Reports of bus accident are fairly common, though they do not often get as much attention at plane crash much rarer but also involving a higher number of casualties. However, the statistics displayed by the National Transportation Safety Board (NTSB) reached 250 fatal casualty and 20 000 injuries sustain during a bus accident [4]. In over 20% of the cases, the driver was found to be at least partially responsible. These numbers does not take into account the percentage of accidents which may have been averted by the driver, if he would have been in a better state of focus.

Therefore, for safety reasons as well as for insurance purposes, many public transport companies require a check-up of their drivers before letting them on board [5,6]. The regular check- up relies mostly on questions given to the drivers and is solely based on his perceptions and self- evaluation of his condition. Therefore these measures are fairly ineffective as any driver who would want to take on his shift can do so, regardless of his physical and psychological state.

Several projects, such as Harken in Europe or EPIC [7] have looked into integrating sensors in cars; more specifically in the seat, seatbelt or even the steering wheel to monitor variation in heart rate and respiration rate. Indeed, these parameters are good indicators of stress or onset drowsiness and fatigue. However, the constraints related to the integration of such sensors into a vehicle are complex as it requires the car to be built consequently to be able to exploit the data gathered by the sensors [8].

Moreover, most sensing technologies, especially the ones used in smart textiles application, relay on direct electrical resistance variation, and require a very good contact with the human body in order to get a viable signal.

Contactless technologies are a more suitable option for use in transportation system where clothes are a factor to consider [9]. However, because these technologies are still under development, their maturity level is significantly lower and there are no commercial or ready-to- use examples available for direct integration at this stage.

Some major car manufacturers such as Ford and Toyota have contributed to developing DSM systems (Driver Status Monitoring) based on commercial medical technology such as electrocardiogram (ECG) and photoplethysmogram (PPG) [10]. These systems are usually fairly cumbersome, rigid and not compatible with a flexible textile structure. By miniaturizing these electronic devices, applications were found on the steering wheel.

However, it requires the driver to have both hands on the steering wheels at all time. Moreover, not only the degree of accuracy may be reduced but it also limits the type of signal as, for instance, the respiration is out of range.

Figure 1. Example of Toyota prototype vehicle with integrated bio-sensors.

Lately, attention was brought to optical system linked to a camera integrated in the middle rearview and able to capture the blinking rate of the driver [11]. This system is fairly dependent of the quality of the recognition software coupled with the hardware to be effective. Practically, as many drivers wear sunglasses, the technology becomes somewhat irrelevant. After investigation, it appears the more efficient surface to be used at all time by the drivers, independently of any external circumstances is the seat and therefore the seat covers.

In order to bridge this gap and bring forward the opportunities offered by smart textiles, CTT Group has developed contactless sensors using a combination of conductive and electromagnetic materials combined with piezoresistive membranes.

3. Context and State-Of-Art Review

3.1. Heart rate sensing

Two independent technologies were used to measure respectively the heart rate and the respiration rate. The electrical impulse emitted at each heart beat is initiated in the brain and travels through the nervous system to the muscle, in this case the heart. The displacement of such current and its alternation (on/off) lead to the creation of an electrical field and therefore an electromagnetic field. [12-13]

At first, a conductive nonwoven structure, composed of 90% polyester fibers and 10% silver plated polyamide fibers, entangled in a random dispersion to create a 3D effect, was evaluated, as seen in Figure 2a.

However this prototype required to have electrodes in order to connect it to an electronic acquisition system; which involved additional processing steps and a longer production time. Moreover, a nonwoven structure is, by nature somewhat random and therefore the repeatability of the electrical surface resistivity between several samples of a same production roll can vary between 10 to 20%. Lastly the 3D shape of a nonwoven also creates a variable as the electrical conductivity will also vary based on the level of compression.

Research also showed the shape of the textile could influence the sensing sensitivity [10]. Since antenna are usually made in circular spiral shape, a new prototype was designed using a conductive core-spun yarn, made of aramid spun core and cover with very thin copper flat lamella, embroidered according to a spiral pattern, as seen in Figure 2b. The results obtained were similar to the ones with the 3D-conductive nonwoven.

Though, distinct peaks were visible on the oscilloscope signal, an important work was done by

Thales Systems Inc in developing suitable algorithms and filters in order to obtain a clear and exploitable signal. Indeed, any flexion or friction resulting from person's motion on a seat can lead to background noise and drown out the signal emitted by the heart.

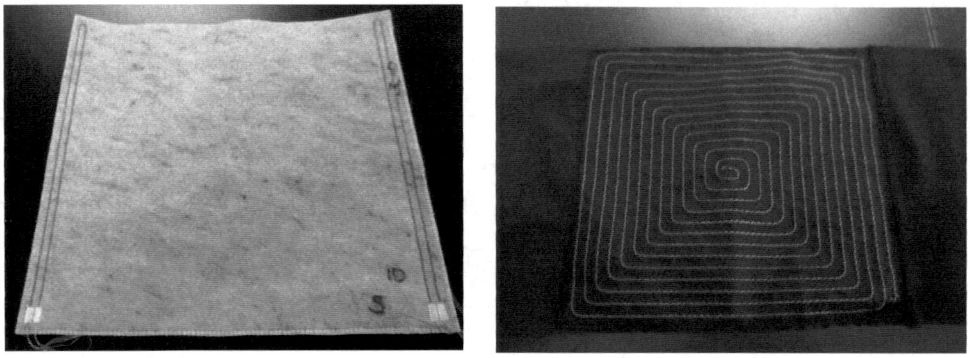

Figure 2. a. on the left: conductive nonwoven with embroidered electrodes; **b.** –on the right: conductive yarn embroidered antenna structure.

3.2. Respiration rate sensing

The respiration rate was measured by using a flexible piezoresistive membrane, placed in-between two highly conductive fabrics. During a respiration, the volume changes of the thorax can be translated in an increased pressure against the backrest of the seat. By using pressure sensing technologies such as piezoresistive materials, it is possible to identify the respiration signal. However, the system must be suitable for various morphologies, meaning both different level of pressure and different positioning of the sensors. The number of sensors, their shape and their position within the seats were tested under multiple configurations.

The sensitivity of the piezo-membrane had to be tailored to the range of pressure applied solely by body contact. The choice of conductive fabric: silver-plated conductive woven fabric, sandwiching the membrane was key to controlling the variation of electrical resistance to the appropriate level; as shown in Figure3. Results for the measurement of the respiration rate were very good and were used directly, without software transformation.

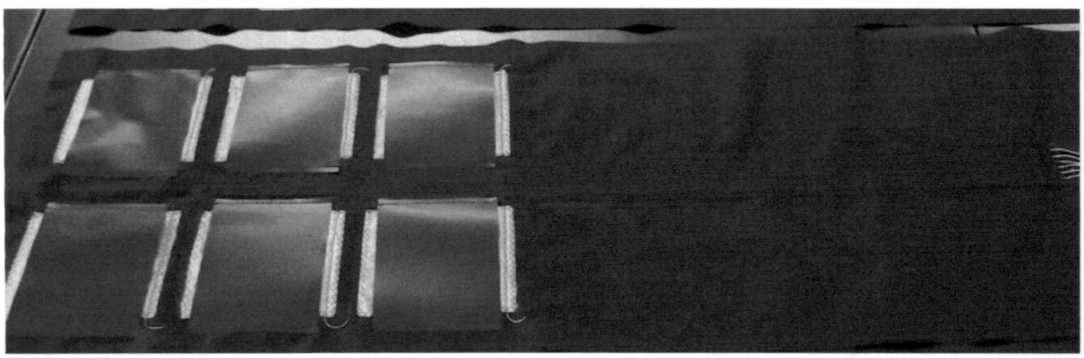

Figure 3. Semi-assembled seat cover with six respiration sensors located at the level of the backrest.

4. Integration, connexion and validation

Each respiration sensor is connected to the others using conductive yarns acting as electrodes. Conductive yarns were embroidered on the surface of the sensor, all the way down towards the connector, using a technique called tailored yarn placement. It allowed for optimal positioning of the conductive yarns, all parallels and close to each other without risk of short-circuit. All sensors were integrated in the backing of a seat cover where the conductive yarns were soldered to a connector allowing plugging the system onto its own circuit board. The connection was made underneath the junction between the backrest and the seat in order to minimize any mechanical strain on the cables. The validation of the system was made by six persons, having various morphologies and wearing different type of clothes, from tee-shirt to winter jacket. The data analysis confirmed that different sensors were triggered for the respiration dependently on the height of the person.

5. Conclusion

The emergence of autonomous cars, already involving numerous on-board technologies, will help launch forward and democratize the integration of bio-sensors in seats. Moreover, it will also provide a structure in which such data can be put to use.

References

[1] Grand View Research, Inc. . (2019, March). Smart Textile Market Size Worth $5.55 Billion By 2025 | CAGR: 30.4%. Consulté le March 2019, 21, sur Grand View Research: https://www.grandviewresearch.com/press-release/global-smart-textiles-industry

[2] Statista. (2018). Global smart, intelligent, digital & interactive fabrics/textile market revenue from 2012 to 2018 . Consulté le March 21, 2019, sur The Statistics Portal : https://www.statista.com/statistics/302526/smart-fabrics-market-revenue-worldwide/

[3] Pettersson I and Karlsson I M, 2015. Setting the stage for autonomous cars: a pilot study of future autonomous driving experiences. *IET intelligent transport systems*, **9**(7), pp.694-701.

[4] Prato C G and Kaplan S, 2013. Promoting safe transit: analyzing bus accident patterns. *Journal of Risk and Governance,* **4**(1), p.13.

[5] Fiser H L and Hopkins P D, 2017. Getting inside the Employee's Head: Neuroscience, Negligent Employment Liability, and the Push and Pull for the New Technology. *BUJ Sci. & Tech. L.*, **23**, p.44.

[6] Vandervort D, 2016, April. Medical Device Data Goes to Court. In Proc. *The 6th International Conference on Digital Health Conference* ACM (pp. 23-27).

[7] Subramaniyam M, Singh D, Park S J, Kim SE, Kim DJ, Lee K S and Min S N, 2018, August. Recent developments on driver's health monitoring and comfort enhancement through IoT. *IOP Conference Series: Materials Science and Engineering* **402**(1) p. 012064. IOP Publishing.

[8] Jung S J, Shin H S and Chung WY, 2014. Driver fatigue and drowsiness monitoring system with embedded electrocardiogram sensor on steering wheel. *IET Intelligent Transport Systems*, **8**(1), pp.43-50.

[9] Scilingo E P, Lanata A and Tognetti A, 2011. Sensors for wearable systems. *Wearable Monitoring Systems* Springer, Boston, MA. pp. 3-25.

[10] Choi Y, Han S I, Kong S H and Ko H, 2016. Driver status monitoring systems forsmart vehicles using physiological sensors: A safety enhancement system from automobile manufacturers. *IEEE Signal Processing Magazine*, **33**(6), pp.22-34.

[11] Dong Y, Hu Z, Uchimura K and Murayama N, 2010. Driver inattention monitoring system for intelligent vehicles: A review. *IEEE transactions on intelligent transportation systems*, **12**(2), pp.596-614.

[12] Mohammad-Zadeh F, Taghibakhsh F and Kaminska B, 2007. *Contactless heart monitoring (CHM)*.2007 Canadian Conference on Electrical and Computer Engineering IEEE , pp. 583-585.

[13] Bortkiewicz A, Gadzicka E and Zmyślony M, 1996. Heart rate variability in workers

exposed to medium-frequency electromagnetic fields. *Journal of the autonomic nervous system*, **59**(3), pp.91-97.

[14] Scalise, L., 2012. *Non Contact Heart Monitoring, Advances in Electrocardiograms-Methods and Analysis*, PhD. Richard Millis (Ed.), InTech.

Design and development of a piezo-resistive sensor based on PEDOT: PSS applied to Sisal's natural fiber for monitoring of 3D warp Interlock fabric

A Abed[1, 2, 3,4], C Cochrane[2], F Boussu[2], O Cherkaoui[3] and R Elmoznine[1]
[1] Laboratory LPMC, Faculty of Science El Jadida, Chouaib Doukkali University, Morocco.
[2] ENSAIT, GEMTEX - Laboratoire de Génie et Matériaux Textiles, F-59000 Lille, France.
[3] Laboratory REMTEX, ESITH (Higher School of Textile and clothing Industries), Casablanca, Morocco.
[4] Author to whom any correspondence should be addressed

ahmed10abed@gmail.com

Abstract. The aim of the current study is to develop and optimize a flexible sensor based on a natural sisal fiber. Based on this sensor yarn, different fibrous reinforcements of 3D warp interlock fabrics used in composite materials could be monitored. This natural yarn (Sisal) has been pre-coated with a layer of PVA solution which makes it possible to homogenize the surface of the yarn and serves as a substrate for the coating phase. A copper metal wire was used as a connecting device to the measurement acquisition system. Some Conductive Polymers Composite (CPC) based on PEDOT: PSS are compared and used as a coating layer in order to get the conductive appearance of the sensor yarn. A sensor calibration step was carried out which consists in testing the sensor yarn on a MTS traction bench while measuring their resistance variation simultaneously. The electromechanical behavior of the different types of sensors will be compared and discussed. Insertion of these developed sensors within 3D woven structures will be helpful for local monitoring and check the inner local elongation of the 3D warp interlock fabric during tensile tests.
Keywords: Mechanical properties; Piezo-resistive sensor; PEDOT: PSS; Sensor calibration; Weaving process; 3D warp Interlock fabric

1. Introduction

Today, metal and petrochemical materials have been decreased in the industrial sectors, especially those related to transportation applications. However, scientists begun looking for new materials, such as composite materials. This will allow concerned companies to reduce mainly the total cost and weight of their mobile structures parts. In addition, this will enable manufacturers to meet current climate and environmental challenges.

Nonetheless, there are some still barriers that restrict the use of these composite materials due to the ageing, fatigue behavior and their failure mode. This lack of knowledge can reduce and limit their use in some areas, where high performance is required. Then, to get out of these limits, composite materials must be more investigated with additional research works, specifically on the fibrous reinforcement. This type of fibrous reinforcement of such composites is often a woven structure. However, few studies have been conducted to investigate the influence of the weaving process (2D or 3D) for the final architecture of textile composite. Since 2003, *Kuo et al* reported a study on failure behavior of 3D woven composites under transverse shear, by measuring the influence of the weaving process (3D) on the properties of the final fabric [1].

Recently, some researchers have highlighted defects caused by the degradation of yarns during the weaving process [2-4]. It was found that the warp yarns suffered a series of damage occurring at most stages of the 3D weaving process, due to the abrasion between the warp yarns and the dynamically moving parts of the loom machinery.

Furthermore, various researches have studied local and in-situ mechanical behavior of fibrous reinforcements used in composite materials [5]. However, few researches have used sensor yarns in textile structures as fibrous reinforcement of composite materials to understand and control their behavior during tensile stresses. The integration of these sensor yarns into the 3D woven structures have allowed also to measure the local and in-situ behavior of one of these yarns submitted to mechanical stresses [6]. This approach could make it possible to identify the various parameters influencing on the mechanical properties of the fibrous reinforcement during static and dynamic stresses.

Physical and mechanical characteristics of natural fibers are very important. Particularly, comparative studies have been done on composites based on natural and synthetic fibers [7-9]. In addition, some natural fibers such as sisal provide better features and performances than other natural fibers. Sisal yarns showed good adhesion and good physical and mechanical properties when is used in reinforcements for composite applications [10].

Therefore, the goal of this study is to develop sisal yarns collected from Morocco and to investigate how to use them both as a fibrous reinforcement of a textile structure and also as a sensor yarn within these structures, and consequently to control their behavior in-situ during mechanical and tensile stresses.

2. Material and Methods

2.1. Preparation of sisal yarn

The sisal yarns used have an average diameter of 2.6 mm, a linear density of 3200 Tex and twisted at 80 t/m. The natural yarn was put into the Soxhlet glassworks with petroleum ether as a solvent for 3 hours. This cleaning would lead to a better contact of the interface between the conductive layer particles and the sisal yarn.

2.2. Mechanical measurement of sisal yarn

After the cleaning of sisal yarns with the soxhlet extractor, the mechanical properties of sisal yarns were investigated using uniaxial tensile tests with five sisal coupon yarns. This investigation of the mechanical behavior of the sisal yarns was important as a substrate of our sensor. It was performed by using a servo-electric MTS 2/M [11-12] corresponding to the standard: ISO 2062:2009. The results of the load–strain curve are shown in Table 1 and Figure 1. The standard test parameters were:

- initial distance between clamps (Li): 200 mm
- preload: 2 N
- speed of the tensile test: 200 mm/min
- number of cycle: 1

Table 1. Load (N) and strain (%) values for sisal yarns.

Yarn	Load at maximum (N)	Strain at break (%)
Sisal 1	645.312	9.148
Sisal 2	551.838	7.739
Sisal 3	700.245	9.239
Sisal 4	702.419	8.753
Sisal 5	564.209	7.721

AVERAGE	**632.805**	**8.52**

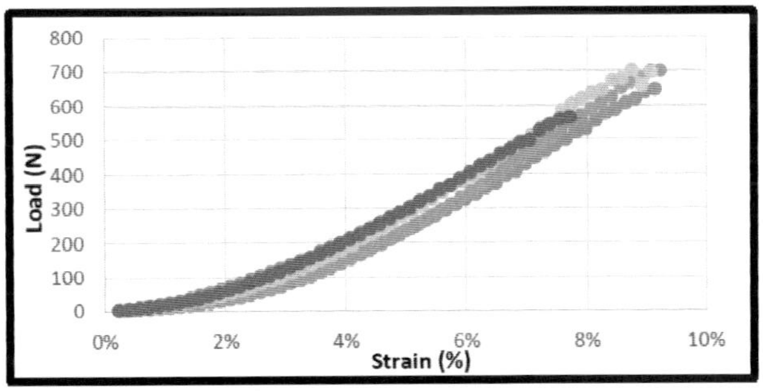

Figure 1. Tensile strength curves of sisal yarns.

2.3. Design and Production of sensor yarns

The main elements of the developed sensor yarns are:

- Sisal yarn
- Piezo-resistive coating.
- Connecting wires

Figure 2 shows a general structure of the sensor wire based on sisal yarn (1). This later was pre-coated with a PVA solution (2) to protect the sensor from abrasion. Then, it was ligated by copper connector wires (3), which was used to measure the electrical resistance of the sensor during its elongation on the tensile bench. The distance between the connecting wires has been chosen to guarantee accurate and precise sensibility of measurement. At the end, layers of piezo-resistive coatings (4) have been deposited on the surface of the sensor yarns to increase its conductivity.

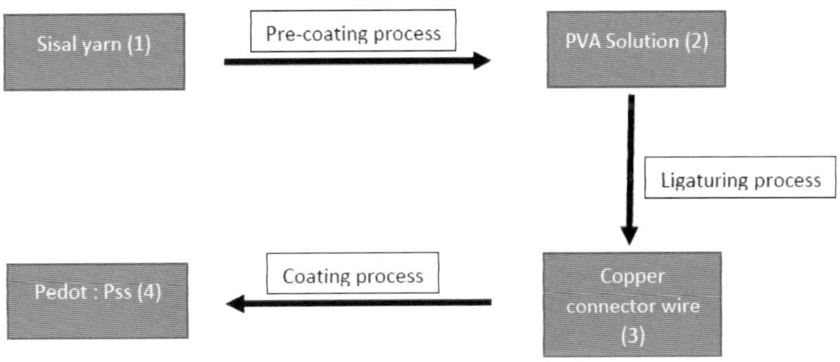

Figure 2. Schematic of the sensor yarn.

3. Results and discussions

3.1. Characterizations of sensor yarns

3.1.1. Mechanical test of sensor yarns. Mechanical behavior of sisal yarns (five yarns in an average) was carried out on the MTS bench, as it was reported in section 2.2, in order to measure their mechanical properties and to show their elastic behavior. The sisal yarn tends to behave as a quasi-elastic material. This mechanical behavior was needed to ease the sensor characterization. Yarns sensor were prepared (refer to section 2.3) and tested on their tensile strength at breaking referring the same test procedure as it was reported in section 2.2. It is based on determine the mechanical behavior of three sensors using sisal yarns. The first was designed without PVA solution (0% PVA). The other two sensors were coated with 3% and 6% PVA layers respectively.

The obtain results from these tests show that the strain at break of the sensor yarns was increased to 10.531%, 10.450% and 10.455% respectively for sensor without PVA, 3% PVA and 6% PVA. The elongation at break was significantly higher on the coating sensors, compared to the sisal yarns (8.520% at max) (see Table 2). Therefore, all the prepared sensor yarns have a solid elastic increased behavior, thanks to the increases rates of PVA. In terms of load, all the sensor yarns has an average value between 753.020 to 836.366 N, whereas the sisal yarns have an average load of 632.805 N (see Table 3). Therefore, the PVA solution could enhance the mechanical properties of the coating. These properties were expected as they had been previously cited in the work of *Chen et al.* [13].

Table 2. Strain at break (%) values for sensor yarns.

	Strain at break (%)	Standard deviation (%)
Sisal	8.520	0.743
Sensor yarns without PVA	10.531	0.772
Sensor yarns with 3% PVA	10.450	0.638
Sensor yarns with 6% PVA	10.455	0.521

Table 3. Maximum load (N) values for sensor yarns.

	Load at maximum (N)	Standard deviation (N)
Sisal	632.805	72.131
Sensor yarns without PVA	753.020	53.538
Sensor yarns with 3% PVA	836.366	127.038
Sensor yarns with 6% PVA	779.631	65.908

3.1.2. Calibration of sensor yarns. The sensor calibration consists in measuring the electrical response by the sensors simultaneously with the tensile bench measurements. In this case, the MTS bench was linked together with a data acquisition system (National Instruments USB-6003) and DAQExpress Processing Software (see Figure 3).

Figure 3. Calibration test of sensor wires on traction bench.

On the other hand, the electrical resistance of the yarns sensors has also been recorded during tensile breaking tests. The electrical measurements were carried out by means of a divider bridge type electric assembly to determine the voltage at the sensor wire terminals which is considered as resistance R2 according to the following relationship (see Figure 4).

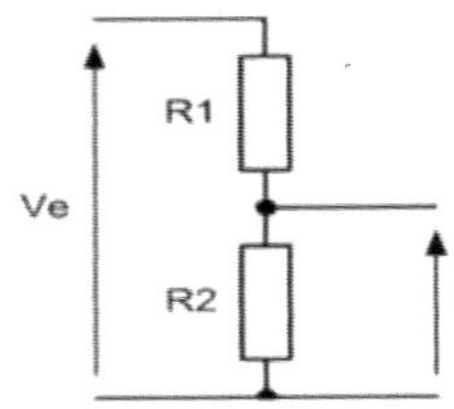

Figure 4. Electrical scheme for measuring the resistance of sensor yarns.

$$R_2 = R_1 \times \frac{V_S}{V_e - V_S} \quad (1)$$

Where the resistance R1 is a decade box serves to balance the divider bridge to choose a value close to R2 when the wire is not stressed. An input voltage Ve has been applied to this arrangement. Subsequently, the determination of sensor's resistance R2, from the output voltage Vs delivered by the acquisition system, is performed by applying the Ohm's law across the resistance R2 using the equation [Equation1].

The evaluation of the sensor's sensibility can be determined by the gauge factor (k). This factor corresponds to the ratio of the relative variation of resistance compared to the initial resistance R0 divided by the measured elongation (ε) using the [Equation 2]:

$$k = \frac{\frac{\Delta R}{R_0}}{\varepsilon} \quad (2)$$

Where:

$$\frac{\Delta R}{R_0} = \frac{R_i - R_0}{R_0} \text{ and } \varepsilon = \frac{l_i - l_0}{l_0} \quad (3)$$

R0 and Ri respectively, represent the initial resistance of the sensor yarn and the electrical resistance value measured at the instance of measurement. l_0 is the initial length of the sensor yarns and li is the length value measured of the sensor yarns at the instance of measurement [Equation 3].

The relative difference of resistance measurements was synchronized with the tensile force measurements applied to the three sensor yarns (without PVA, 3% and 6% PVA), carried out by a MTS tensile bench at fixed speed (200 mm/min).

The gauge factor increases and changes greatly with the elongation of the sensitive coating. However, for our application, strain variation does not exceed 8%. For these series of prepared and tested sensor yarns, the gauge factor (k) was defined and determined. It is observed that a sensor yarn with 6% PVA has a better factor gauge k ≃ 1.31 compared to the sensor yarns without PVA and with 3% PVA, respectively 0.65 and 0.79. In addition, this factor increased with the ratio of 6% PVA solution. The mixture of the PVA solution and the Pedot:Pss could give a better uniformity of the thickness, and then could provide a better compromise between elongation detection sensitivity and electromagnetic noise resistance. This behavior could indicate that resistivity of the coating was changed under tough strain and not only due to a change of geometry.

4. Conclusion

This work consists in the development of a new type of mechanical sensor based on natural fiber. These sensor yarns will be able to monitor and record in situ strains and stresses on sisal yarns, commonly used as natural fibrous for composite application. The work has furthered previous studies undertaken at our institute and has been combined with new technologies to meet the specific needs. A new design to prepare these sensor yarns has been achieved. As a first step, the sisal yarns were cleaned with a

Soxhlet extractor and characterized. Thus, the sisal yarns were pre-coated with a PVA solution used as protection against abrasion. Then, sisal yarns were ligatured by copper connector wires, in order to connect the electrical measurement devices and piezo-resistive coating of the sensor yarns. After that, sisal yarns, pre-coated with a PVA and ligatured by copper wires, were also coated by an aqueous solution of Pedot:Pss. In the second step, three sensors yarns were prepared; the first one without PVA and the two others respectively with 3% and 6% of PVA. These series of sensor yarns have been characterized on a MTS bench linked together with a data acquisition system (National Instruments USB-6003) and DAQExpress Processing Software. The mechanical and electrical behavior were investigated. The sensor yarns with 6% PVA revealed a good factor gauges k greater than 1. In summary, a successful sensor formulation has been established and now it could be ready for forthcoming in-situ measurements during the weaving process.

Acknowledgements
We would like to thank to PHC Toubkal/19/83 - Campus France: 41511ZF and CNRST Maroc agencies for financial support to conduct this research work.

References:

[1] Kuo W, Fang J, Lin H 2003 Failure behavior of 3D woven composites under transverse shear. *Compos. Part A Appl. Sci. Manuf* **34**(7) pp.561–575.
[2] Islam M 1987 *The Damage Suffered by the Warp Yarn during Weaving* (Manchester, UK: s.n).
[3] Lee B, Leong K, Herszberg, I. 2001 The effect of weaving on the tensile properties of carbon fiber tows and woven composites *J. Reinf. Plast. Compos* **20**(8)pp.652-670.
[4] Lee L, Rudov-Clark S, Mouritz A, Bannister, M and Herszberg, I 2002 Effect of weaving damage on the tensile properties of three-dimensional woven composites *Compos. Struct* **57** (1-4) pp .405-13
[5] Rausch J, Mader E 2010 Health monitoring in continuous glass fiber reinforced thermos-plastics: tailored sensitivity and cyclic loading of CNT-based interphase sensors *Compos. Sci. Technol.* **70**(13) pp. 2023-30.
[6] Nauman S, Cristian I, Koncar V 2011 Simultaneous application of fibrous piezoresistive sensors for compression and traction detection in glass laminate composites *Sensors* **11**(10) pp.9478-98.
[7] Joshia S V, Drzalb L T , Mohantyb A K, Arorac S 2004 Are natural fiber composites environmentally superior to glass fiber reinforced composites *Compos. Part A Appl. Sci. Manuf.* **35**(3) pp. 371- 376.
[8] Wu Y, Xia C, Cai L, Garcia A C, Shi S Q 2018 Development of natural fiber- reinforced composite with comparable mechanical properties and reduced energy consumption and environmental impacts for replacing automotive glass- fiber sheet molding compound *J Clean Prod.* **184** pp.92- 100.
[9] Biswas S, Satapathy A 2010 A comparative study on erosion characteristics of red mud filled bamboo- epoxy and glass- epoxy composites *Mater. Des.* **31**(4) 1752- 1767.
[10] Samouh Z, Molnar K, Boussu F, Cherkaoui O, El Moznine R 2018 Mechanical and thermal characterization of sisal fiber reinforced polylactic acid composites *Polym. Adv. Technol.* **30**(3), pp.529-537.
[11] MTS Insight® Electromechanical Testing Systems. MTS, Eden Prairie, MN, USA. http://mts.com (accessed on 12 November 2012).
[12] SE 1200 Single End Roving for Knitting, Weaving, and Filament Winding. Owens Corning:Laudun-l'Ardoise, France 2011.
[13] Chen C H, Torrents A, Kulinsky L, Nelson R D, Madou M J, Valdevit L, LaRue JC 2011 Mechanical characterizations of cast Poly(3,4-ethylenedioxythiophene):Poly(styrenesulfonate)/ Polyvinyl Alcohol thin films *Synth. Met.* **161**(21-22) pp.2259–67.

IOP Conf. Series: Materials Science and Engineering **827** (2020) 012020 doi:10.1088/1757-899X/827/1/012020

Valorisation of waste feathers: treatments and characterization

Ouahiba Mrajji [1,2,3], Mohamed El Wazna [1,2], Abdeslam El Bouari [1] and Omar Cherkaoui[2]

[1] Laboratory of physic-chemical of applied materials, Sciences Faculty of Ben M'sik, University Hassan II Casablanca Morocco.
[2] Laboratory REMTEX, Higher School of Textile and clothing Industries, Km 8, Route d'El Jadida, Casablanca Morocco

[3] Author to whom any correspondence should be addressed

Ouahiba.mrajji@gmail.com

Abstract. Environmental concerns are focused on the search for environmentally friendly, biodegradable, sustainable, renewable and widely available materials. Feather wastes have very special structures and important physiological functions, their valorization offers solutions to these problems. More than 30 million tons of chicken feathers are generated each year by the Moroccan poultry industry. This waste is unfit for valorisation due to the contaminant; debris, skin and blood from the poultry industry. These contaminants should be removed before their valorisation. The aim of this research is to study different treatments (sodium dodecyl sulfate, ethanol, Hydrogen peroxide and Acetone) in order to evaluate their effectiveness in removing the contaminants from the feather's surface obtained from the slaughter, storage or transport of these poultry wastes. Furthermore, the effects of treatments on the physicochemical and mechanical properties are studied. The results of the treatments showed efficiency at higher concentrations and the best results were attributed to treatment with the sodium dodecyl sulfate with no significant differences in structure and the chemical composition. These results were confirmed with Scanning electron microscopy, infrared spectroscopy (FTIR) and EDX analysis.
Keywords: Bacteria, Environmental concerns, Feathers, Physicochemical properties.

1. Introduction

Increasing consumption of poultry meat results in large amounts of feather wastes that need to be disposed of properly. The feather's fiber possesses unique properties such as: low density , acoustic and thermal insulating properties, flame resistance and low moisture absorption which could be used in a number of applications such as cosmetics, pharmaceuticals, composites, textiles and geotextiles. More than 30 million tons of feathers fibers is produced annually in Morocco which is rich in insoluble keratin protein 91%. This protein is responsible for high stability and low degradation rate and the impermeability of the feather fiber [1]. The feathers constitute about 5–10% of the chicken mass and thus they comprise a significant portion of the poultry wastes, which are treated or roasted to form a powder used as a dietary supplement for livestock [2] or incinerated without any use [3]. However, this waste disposal method led to the environmental damage and generation of greenhouse gases [4]. In this perspective, the treatment of feathers waste should be the first step before the valorization. The treatment of fibers waste has been the subject of different researches studies: *Tesfaye et al*, studied the optimization of surfactant decontamination and pre-treatment of waste feathers by using response surface methodology. Treatment of feathers with surfactants was effective in removing contaminants from the fiber, but the decontamination using POE (classic detergent) was more cost effective than using sodium dodecyl sulphate and dimethyl dioctadecyl ammonium chloride [5]. *Tamrat Tesfaye et al*, studied the optimization of decontamination and pre-treatment of feather fiber with bleaching agents using response surface methodology. The results show that the bleaching treatment was effective in removing the impurities and improving the physicochemical properties of the feathers. The untreated feather fibers exhibited the highest microbial counts (1.48E + 07 ± 6.72E05 CFU/g), while the decontaminated samples showed a thousandfold reduction [6]. In this paper, the study will consist on the treatment and characterization of chicken feathers fiber in order to determine the effective treatment, its efficiency and effects on properties of the feathers.

2. Material and Methods:

Chicken feathers were supplied by a slaughterhouse in the province of Casablanca, Morocco. The central axis of the feather fiber (the rachis) presents two lateral series of barb that have a series of small structures, the barbules are held together by a myriad of small hooks (figure 1). The barb fiber has a higher aspect ratio than the quill with a maximum diameter of 50 µm .The feathers samples were treated using Soxhlet mounting in different treatment parameters, as listed in Table 1. Impurity removal is the measure of mass lost from the fiber waste during treatment, represented as a fraction of the weight loss and calculated using equation1:

$$Impurity\ removal(\%) = \frac{M1-M2}{M1} * 100 \qquad (1)$$

The influence of different treatments on the morphology and elemental composition of the feathers was investigated using scanning electron microscopy SH-5000P- EDX integrated system. FTIR spectroscopy (Nicolet iS10 FTIR-ATR spectrophotometer) was used to ascertain if the treatment procedures affected the functional groups of the feathers. The thermal properties of feathers fibers was determined using a Scanning Calorimeter Type 131 SETARAM. Tensile tests were carried out using a universal testing machine "Instron MPK universal" .The force cell is 5 kN and the dimensions of the specimens are 80×20 mm^2.

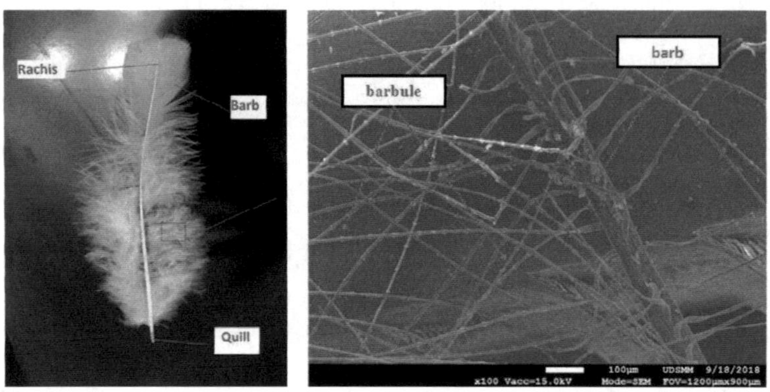

Figure1. Structure of feather fiber.

3. Results and discussion

Untreated feathers appeared dark brown, with a greasy texture, bloody appearance and had an obnoxious odour. The barbs fibers were stuck to the rachis due to bacterial decomposition. The growth of the bacteria on feathers fiber will consume keratin protein, decompose it, and eventually degrade the properties of feathers making them mechanically very weak and dark in colour. The treated feathers showed weight losses compared to the untreated feathers samples. The figure 3(a-b) indicates that there are a large amount of impurities in the initial sample. The treatment processes imparted a sickly-sweet bleach odour to the feathers fibers, changed them from brown to white and remove natural dust particles and waxy substances on the fibers surfaces figure (2-3).

Table 1. Treatment factors and impurity removal of feather fiber.

Treatment	**Treatment factors**				**% Impurity removal**
	Concentration %	Temperature, °C	Time, min	Number of stage	
SDS	1	50	30	1	12.75
SDS	2	50	40	1	16
Ethanol	0.1	25	60	1	6.78
Ethanol	0.5	25	60	1	8.1
Acetone	0.6	25	45	1	9.75
Acetone	1	35	60	1	11.67
Hydrogen peroxide	0.5	52	32	1	5.89
Hydrogen peroxide	1	80	45	1	11.50

Figure 2. Untreated chicken feathers (A), after pretreatment (B) and after treatment with cleaning agent (C).

The removal of impurities blood, skin and other wastes from the poultry industry in feathers fibers depends on different parameters. The impurity removal after washing was about 6–16%. This impurity removal is due to the removal of short fibres, blood, vegetation, dust, suints and some other contaminants from the poultry process. The impurity removal has direct dependency on the treatment agent used. The results revealed that the temperature and concentration have significant effects on impurity removal compared to treatment time. FTIR spectra of untreated and treated feathers fibers are shown in Figure 4. There were no significant differences in the chemical composition and structure of the treated fiber. The treatment with different agents improved the properties of the keratin functions, the keratin peaks are more exposed to the surface of feather fiber, more responsive and reactive for future valorization. Spectrum shows numerous absorption bands that characterize the keratin chains. The transmission band is related to the stretching vibrations of N–H at 3276 cm^{-1} which show the presence of the Amide A. The transmission band at 2939 cm^{-1} represents the symmetrical C-H stretching vibration. The amide I is presented by the C=O stretching vibrations at 1630 cm^{-1}.The transmission band at 1531 cm^{-1} is assigned to amide II with C–H stretching and N–H bending vibrations. The transmission band at 1260 cm^{-1} is attributed to amide III with C–N stretching vibrations and N–H inplane bending vibrations. The treatments effect on the thermal behaviour of waste chicken feathers were studied by DSC in the temperature range of 25–600 °C (figure 5).The DSC indicate an endothermic peak at 125 °C, which is related to the removal of moisture in the keratin structure, whereas the exothermic peak between 230 and 340 °C was attributed to protein denaturation. The complete degradation temperature increases after treatment from 310 to 350°C. The treatments of feathers fiber improve the thermal stability of the feather structure and shift the degradation temperatures higher by increasing the ionic interactions. According to the DSC result we can conclude that the treatments used improves the thermal properties of the feathers waste in order to increase the field of their valorization. Figure 6 shows EDX elemental profiles of the samples. The relative proportions of the sulphur element decreased, Contrariwise the carbon content increased in all the treated feather fiber and the Spectrum of the untreated fibers shows numerous elements have been removed such as phosphorus, chlorine, potassium and magnesium using treatment processing. This result is explained with the removal of dirt and contamination (blood, skin and other waste from the poultry process) from the feathers wastes surface.

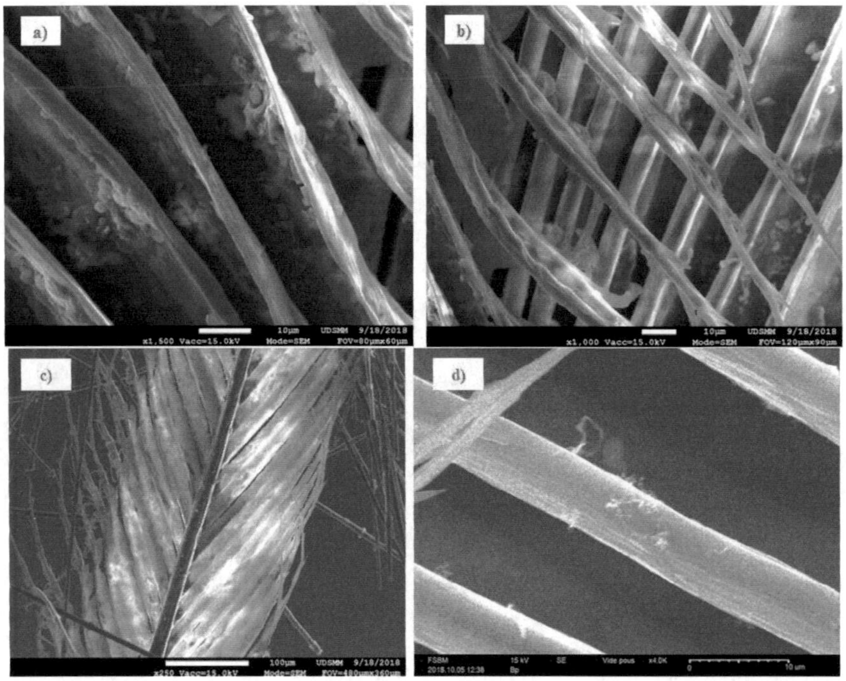

Figure 3. The SEM of untreated feather fibres (a-b), treated with H_2O_2 (c) and treated with SDS (d).

The performance of natural fibers depends on several factors. In order to expand the use of feather fibers and improve their performance, it is essential to know their characteristics. The mechanical properties of treated and untreated feather fibers are summarized in (Figure 7). From figure 7 we clearly see a difference in the mechanical properties of treated and untreated feathers. This may be due to the nature of the fiber "the rate of different constituents (keratin, water and lipid), the level of impurities, treatment and the moisture content in each variety of feathers". The results show that all the treated samples have changed the properties of feather fiber from elastic to plastic.

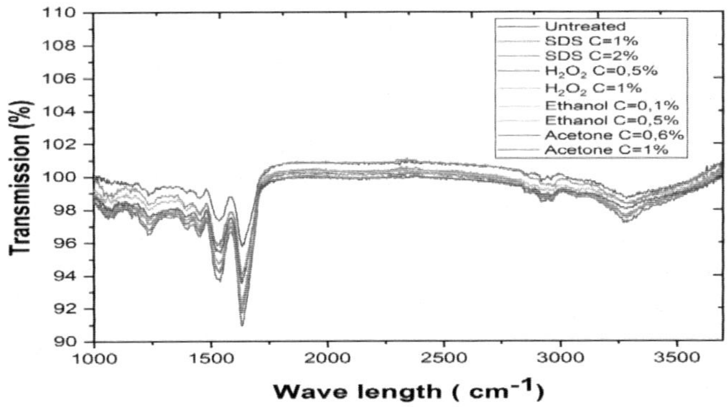

Figure 4. FTIR spectra of treated and untreated chicken feathers fiber.

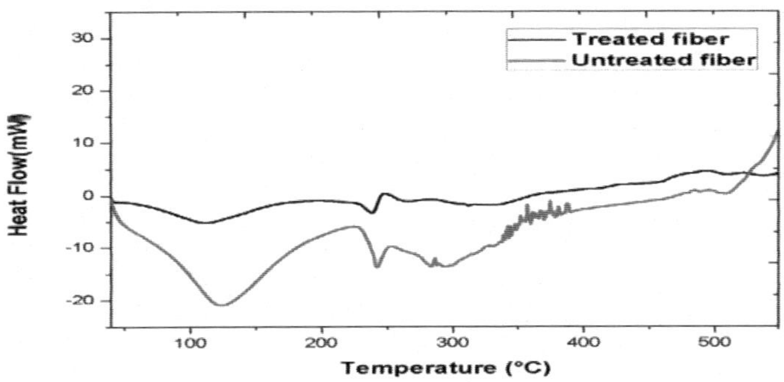

Figure 5. DSC curves of untreated and treated feather fiber.

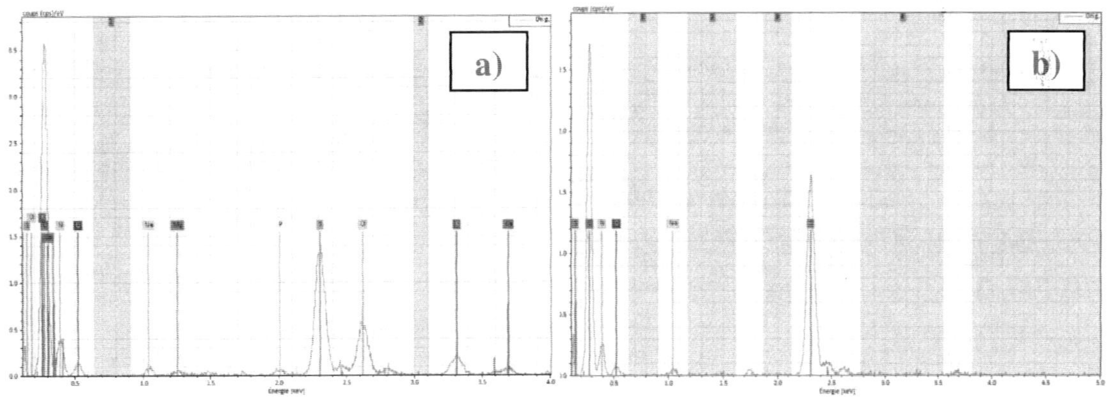

Figure 6. EDX elemental profiles of untreated fiber (a) and treated fibre with SDS (b).

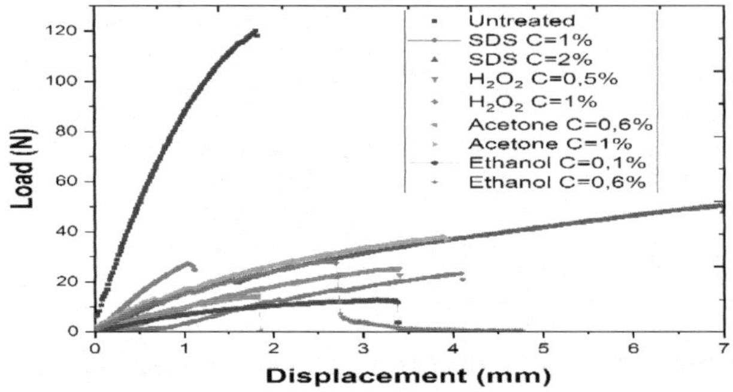

Figure 7. Tensile strength of untreated and treated feather fibers.

4. Conclusion

The feather fibers are an abundant by-product of poultry industry and its valorization offers possible solutions to environmental problems (incineration). These fibers are a hazardous wastes due to contamination by microbiological organisms. Recovery requires feather treatments to eliminate microbial contamination. Morphological analysis showed that feather treatments with sodium dodecyl sulfate were effective in removing lipid matter from the feathers fibers. The chemical properties were not adversely affected by the treatments. The treatments have promoted thermal stability of the fibre structure and shifted the denaturation temperatures at higher levels from 310 to 350°C by increasing the ionic interactions. According to the results, we can conclude that the treatments carried out improve the properties of the feathers in order to increase the field of their application. The unique properties of feather barbs make them suitable for the use in textile applications and offer an environmentally sustainable way to benefit from this organic waste material.

References

[1] Jagadeeshgouda KB, Ravinder Reddy P and Ishwaraprasad K,2014.Experimental study of behaviour of poultry feather fiber – a reinforcing material for composites. *Int J Res EngTechnol*, **3** pp. 362–366.

[2] El Boushy A H and Van der Poel A F, 2013. *Handbook of poultry feed from waste: processing and use.* Springer Science & Business Media

[3] Tesfaye T, Sithole B, Ramjugernath D and Mokhothu T, 2018. Valorisation of chicken feathers: Characterisation of thermal, mechanical and electrical properties. *Sustainable Chemistry and Pharmacy*, **9**, pp.27-34.

[4] Porter M, and Van der Linde C, 2000. Green and competitive: ending the stalemate. *The Dynamics of the eco-efficient economy: environmental regulation and competitive advantage,* pp.33,55

[5] Tesfaye T, Sithole B, Ramjugernath D, et al. 2018, Optimisation of surfactant decontamination and pre-treatment of waste chicken feathers by using response surface methodology. *Waste Manag*. **72** pp. 371–388.

[6] Tesfaye T, Sithole B and Ramjugernath D. 2018,Valorisation of waste chicken feathers: optimisation of decontamination and pre-treatment with bleaching agents using response surface methodology .*Sustain Chem Pharm* **8** pp 21–37.

IOP Conf. Series: Materials Science and Engineering **827** (2020) 012021 doi:10.1088/1757-899X/827/1/012021

Application of Fe_3O_4 nanoparticles on cotton fabrics by the Pad-Dry-Cure process for the elaboration of magnetic and conductive textiles

Ayoub Nadi[1], Sara Jamoudi Sbai[1,2], Aziz Bentiss[1,2] , Mohammed Belaiche[3], Samir Briche[4],Said Gmouh[2,5] .

[1] Laboratoire REMTEX, ESITH, route El jadida, km 8, BP 7731 - Oulfa, Casablanca, Morocco.

[2] Laboratoire LIMAT, Université Hassan II de Casablanca, BP : 9167 Casablanca, Morocco.

[3] Semiconductor and Environmental Sensor Technology Team, Energy Research Center. Faculty of Sciences, Mohammed V University, B.P. 1014 Rabat, Morocco.

[4] Moroccan Foundation for Advanced Science, Innovation and Research (MASCIR), Department of Energy Storage and Multifunctional Coatings, B.P. 10100 Rabat, Morocco.

[5] Author to whom any correspondence should be addressed

s.gmouh1@gmail.com

Abstract. In the present study, a magnetic textile was developed by the application of Iron nanoparticles on cotton fabrics. For this, Fe_3O_4 nanoparticles were synthetized by reverse co-precipitation. Then, the obtained magnetic nanoparticles were applied on cotton fabrics using the Pad-Dry-Cure process. Magnetic behavior of iron oxide nanoparticles was investigated to study magnetic properties by the VSM analysis. Moreover, the effect of iron oxide nanoparticles on the cotton fabrics noticed on the thermal behavior has been studied by thermogravimetric analysis. The thermal stability of cotton fabrics is positively affected after the treatment using Fe_3O_4 nanoparticles. Finally, electrical properties were studied to measure the fabrics conductivity according to the AATCC.
Keywords: Fe_3O_4 nanoparticles, Cotton fabrics, magnetic textile, Pad-dry-cure.

1. Introduction

The nanotechnology, information technologies and biomaterials have continued to evolve towards new stages of maturity[1]. They can be used for discovering new materials and developing industries with new properties such as improving the tensile strength of fabrics, flame retardant, water repellency, antibacterial and self-cleaning properties, etc.[2]. Magnetic nanoparticles continue to attract scientific researchers in order to use them in industrial applications. This interest is due to their controllable dimensions of up to 100 nm[3] and their chemical, magnetic, electrical and mechanical properties[4]. Also, they make the textile fabric recyclable through excellent separation efficiency, good recyclability, durability and better thermal resistance[5].

In the case of textile as a substrate, nanomaterials are incorporated by methods such as the Pad-Dry-Cure process. This latter consists on impregnating a textile matrix in a solution, followed by drying the fabric and curing it under appropriate conditions[6].

We have chosen reverse co-precipitation synthesis as a method for the preparation of Magnetite Fe_3O_4 nanoparticles because of its simplicity (availability of precursors, ease of control of synthesis conditions and recovery of the product) as well as its relatively high yield compared to other methods[7].

Cotton is a widely used fiber type that exhibits high absorbency, softness, and breathability. However, the use of cotton in non-classical applications is limited since its fibers have relatively low strength, low durability, easy creasing and soiling and flammability[8].

Our aim in this work is the functionalization of textiles by the surface treatment of cotton using magnetic iron oxides nanoparticles in order to design a multifunctional material with interesting magnetic, thermal and electrical properties characterized by VSM, TGA and Resistivity measurements, respectively. The principle is based initially on the deposition of the Fe_3O_4 nanoparticles on the surface of the cotton fabric by the Pad-Dry-Cure method.

2. Experimental procedure

2.1. Chemicals

Cotton (CO) woven fabric weighing 168 g.m^{-2} was used. Iron oxide nanoparticles Fe_3O_4 of around 16nm (synthesized as described in our previous study [9]) ; Ethanol (EtOH, 99%), were purchased from Sigma-Aldrich. Co.

2.2. Application of iron oxide Nps onto cotton fabrics

The application of synthesized Iron oxide nanoparticles on the surface of cotton fabric is carried out by the Pad-Dry-Cure method. First, 0.35g of Fe_3O_4 nanoparticles are taken into a solution containing 9ml of Ethanol and 1.5ml on deionized water that we dispersed by sonication for 30min. Then, the fabric is emerged in the prepared solution and padded until reaching a "pick-up" of 80% and dried at 80 °C then at 120 °C for 1 hour.

3. Characterization

The characteristic bonds of cotton fabric and iron oxide nanoparticles has been analyzed by Fourier Transform Infrared Spectroscopy with attenuated total reflection (FT-IR) by Thermo Scientific BRUCKER HYPERION 1000. Surface morphology of the untreated and treated cotton fabrics was observed using Scanning Electron Microscope (SEM) by ESEM FEI QUANTA 200 effect Scanning Electron Microscope (SEM). Thermal properties of the cotton fabrics were estimated by thermogravimetric analysis (TGA) from 25°C to 600°C with a heating rate of 10 °C/min under a nitrogen atmosphere using a thermogravimetric analyzer (Mettler Toledo, TGA2). The magnetic characteristics and behaviors of iron oxide and before and after their application on cotton fabrics were studied using a vibrating sample magnetometer (SQUID Quantum Design PPMS-9) at room temperature (300 K) and at low temperature (10 K). Electrical conductivity was measured according to the AATCC test method of the American Association of Textile Chemists and Colourists. The fabric (10 x 13 cm^2) was placed between two metal electrodes separated by 3 cm under standard conditions (65% RH and 24 °C). The electrical resistance was recorded with a Multimetrix DMM 120.3.

4. Results and discussion

4.1. X-ray diffraction

These results showed that Fe_3O_4 nanoparticles on the surface of cotton fabrics preserve their crystalline nature and size after their application on the cotton fabric surface. The crystallite size that was determined by Scherrer equation on the samples of the treated textiles gave average values around 15 nm[10], which is in agreement with the measurements on the nanoparticles powder according to our previous works [9].

4.2. Morphology analysis

After coating cotton fabrics with iron oxide nanoparticles using the Pad-Dry-Cure technique, we noted that the treated fabric has a smooth surface with a very uniform color. This result is confirmed by the SEM as shown in (Figure.3b, 3b'). However, at the level of the modified fabric (Figure. 3b), it is noted that the nanoparticles covered the surface of the cotton fabric homogeneously.

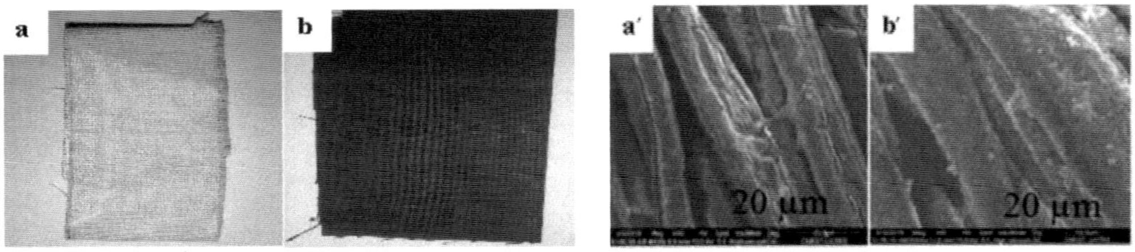

Figure 1. Pictures of untreated (a) and treated cotton fabric (b); SEM images of untreated (a') and treated cotton fabric (b').

4.3. *FT-IR analysis*

The chemical interactions of Fe_3O_4Nps on the surface of cotton fabrics were investigated by FT-IR analysis. Figure 2a represents FT-IR spectra of UCF and TCF-CN and figure 2.b represents spectrum of iron oxide Nps powder. For the uncoated cotton surface (UCF): the absorption peak at 3467.47 cm^{-1} is related to hydroxyl groups -OH vibration and the band at 2896 cm^{-1} is assigned to the C-H stretching. On the other spectrum of Iron oxide coating Cotton fabric (TCF-CN) the band corresponding to the group Fe-O appeared at 526.88 $cm^{-1}$and 434.60 $cm-^{1}$ which is the characteristic band of Iron oxide nanoparticles, as seen in the Fe_3O_4 Nps powder spectrum (figure 2.b)[11].

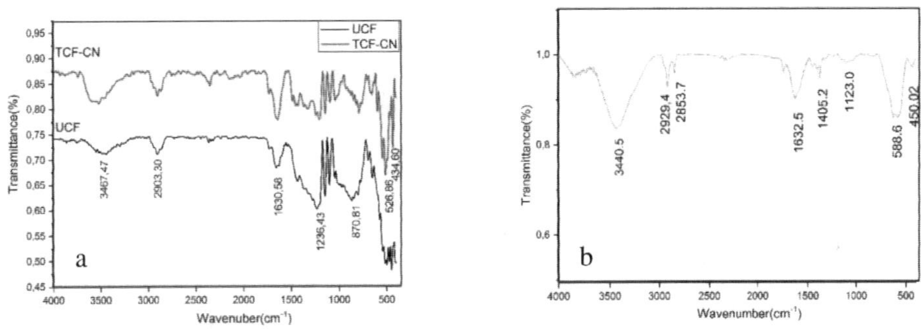

Figure 2. (a) FT-IR spectra of uncoated cotton (UCF) and treated cotton fabrics (TCF-CN) and (b) Fe_3O_4 Nps powder.

4.4. *Thermal properties*

The thermal stability of uncoated cotton fabric (UCF) and coated cotton with magnetic iron oxide nanoparticles (TCF-CN) was studied by thermogravimetry under a nitrogen atmosphere (Figure 3). In the case of cotton treated with nanoparticles (TCF-CN), the sample showed the same stages of degradation of the untreated one until the last stage. In this latter, we observed an improvement in the thermal properties of the sample by retaining about 25% of its mass at 600°C, while the residue of untreated cotton(UCF) is about 3.46% at 600°C[12]. This is due to the good thermal properties of iron oxide nanoparticles as described in the literature [13]. A similar result has been reported by *Harifi*, indicating that the thermal stability of polyester fabric was enhanced after the coating with Fe_3O_4 nanoparticles [14].

IOP Conf. Series: Materials Science and Engineering **827** (2020) 012021 doi:10.1088/1757-899X/827/1/012021

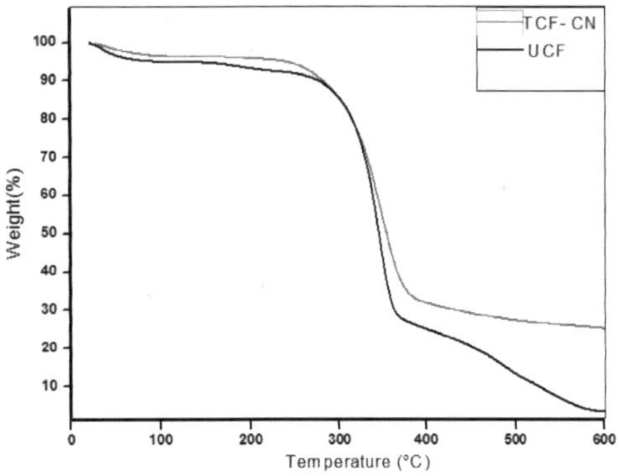

Figure 3.TGA curves of Untreated-CF and functionalized with Fe_3O_4 nanoparticles and TCF- CN.

4.5. Magnetic properties

The measurement of magnetic properties (figure 4) showed that the cotton fabric covered by iron oxide nanoparticles has an identical magnetic behavior as the powder nanoparticles with a difference of Ms=94.1 emu/g in favor of the fabric. This may be due to the agglomeration caused by the pressure of the rollers in the Pad-Dry-Cure process.

Moreover, we can see the absence of coercivity and remanence at 300K, which indicates a superparamagnetic behavior for both samples. However, at low temperature (10K) both samples have a hysteresis cycle, which reflects the transition to ferrimagnetic behavior.

It is established that the size and shape of nanoparticles influence the physical properties of the material, due to dipole-magnetic interactions at long distances, between agglomerations and intra-agglomeration that impact the magnetic response of the sample[15].

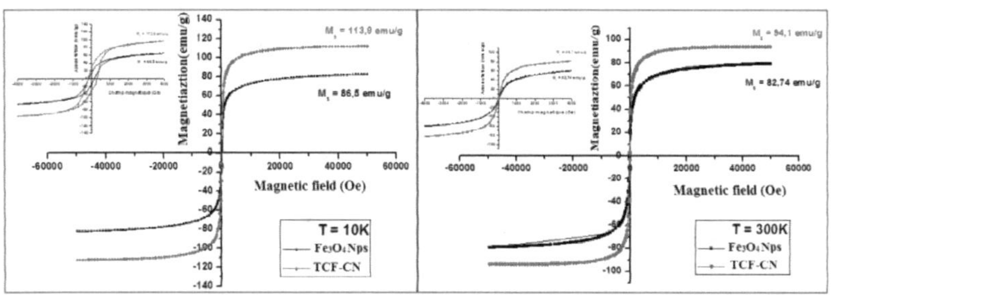

Figure 4. Magnetic measurements of the Fe_3O_4 powder and their application on the cotton fabric. (Left) at 10K; (Right) at 300K.

4.6. Electrical properties

The effect of the coatings on the surface resistivity of the cotton fabrics is shown in Table 1. The samples after padding have similar thicknesses, around 0.32 mm. First, we measured the surface resistivity of the uncoated fabric (UCF) and the fabric treated by the iron oxide Nps (TCF-CN). After the treatment with iron oxide nanoparticles, we noted that the surface resistivity of the fabrics becomes measurable and is of 7.07 $M\Omega/m^2$ (Ohm). These results showed that the coating of cotton fabric by Iron oxide nanoparticles gives an electrical conductivity to the fabric which makes it a semi-conductor material[16].

Table1. Electrical properties measurements of the elaborated cotton fabric.

Samples / Measures	UCF	TCF-CN
Thickness (mm)	0.32	0.30
Resistivity	-	2.83MΩ
Surface Resistivity	-	7.07MΩ/m^2

5. Conclusions

The physicochemical properties of cotton fabrics treated with iron oxide nanoparticles by the Pad-Dry-Cure method were discussed in this article. One of the main characteristics of this work is that the magnetic properties of the Fe_3O_4 nanoparticles on the treated fabrics are more interesting than those of the Iron oxide nanoparticles powder by itself. This is due to the padding process that favors a uniform agglomeration of the nanoparticles on the surface of the fabric. The conductivity of cotton fabrics was significantly increased by iron oxide Nps. The thermal properties of cotton fabrics are positively affected after the treatment by magnetic nanoparticles. In summary, Fe_3O_4 nanoparticles treated cotton fabrics can be more effective through improvements by new functionalization in order to enhance the mechanical and thermal resistance.

References

[1] Afzali A and S Maghsoodlou, 2016, *Modern application of nanotechnology in textile.* Nanostructured Polymer Blends and Composites in Textiles, p. 41-85.
[2] Boukhriss A, 2015, Elaboration de revêtements hybrides par le procédé sol-gel pour conférer de nouvelles fonctions à des matériaux textiles, *Clermont-Ferrand* **2**.
[3] Xu P et al.,2012, Use of iron oxide nanomaterials in wastewater treatment: a review. *Science of the Total Environment,* **424**,p. 1-10.
[4] Qian T et al.,2019, Fabrication of magnetic phase change n-eicosane@ Fe3O4/SiO2 microcapsules on wood surface via sol-gel method. *Journal of Alloys and Compounds,* **772** p. 871-76.
[5] Yan T et al.,2018, A magnetic pH-induced textile fabric with switchable wettability for intelligent oil/water separation. *Chemical Engineering Journal*, **347**: p. 52-63.
[6] Nadi A et al., 2018,Evolution in the surface modification of textiles: a review. *Textile Progress,* **50**(2): p. 67-108.
[7] Chen Y et al., 2016, Preparation of size-controlled magnetite nanoparticles with a graphene and polymeric ionic liquid coating for the quick, easy, cheap, effective, rugged and safe extraction of preservatives from vegetables. *Journal of Chromatography A*, **1448**: p. 9-19.
[8] Yetisen A K et al.,2016, Nanotechnology in textiles. *ACS nano*, **10**(3): p. 3042-3068.
[9] Nadi A et al., 2018 Immobilisation of bacteria onto magnetic nanoparticles for the decolorisation and degradation of azo dyes. *IET nanobiotechnology*, **13**(2): p. 144-149.
[10] Anbarasu M et al., 2015,Synthesis and characterization of polyethylene glycol (PEG) coated Fe3O4 nanoparticles by chemical co-precipitation method for biomedical applications. *Spectrochimica Acta Part A: Molecular and Biomolecular Spectroscopy.* **135**: p. 536-539.
[11] Soares S F et al.,2019, Trimethyl Chitosan/Siloxane-Hybrid Coated Fe3O4 Nanoparticles for the Uptake of Sulfamethoxazole from Water. *Molecules,***24**(10): p. 1958.
[12] Thi V H T and Lee B K,2017 Development of multifunctional self-cleaning and UV blocking cotton fabric with modification of photoactive ZnO coating via microwave method. *Journal of Photochemistry and Photobiology A: Chemistry*, **338**: p. 13-22.
[13] Wang Q et al., 2014, Fe3O4 nanoparticles grown on graphene as advanced electrode materials for supercapacitors. *Journal of Power Sources,***245**: p. 101-106.
[14] Harifi T and M Montazer,2014, In situ synthesis of iron oxide nanoparticles on polyester fabric

utilizing color, magnetic, antibacterial and sono-Fenton catalytic properties. *Journal of Materials Chemistry B,* **2**(3): p. 272-282.

[15] Unni M et al., 2017,Thermal decomposition synthesis of iron oxide nanoparticles with diminished magnetic dead layer by controlled addition of oxygen. *ACS nano,***11**(2): p. 2284-2303.

[16] Liu X et al., 2010,Polyelectrolyte-bridged metal/cotton hierarchical structures for highly durable conductive yarns. *ACS applied materials & interfaces,***2**(2): p. 529-535.

Ultrasound welding device for reinforcement of the invasive textile mesh destined to the reconstruction of the thoracic wall

A G Ene[1,2], C Mihai[1] and C Jipa[1]

[1]IT Research Department in Industrial Engineering, The Research-Development National Institute for Textile and Leather, 16 Lucretiu Patrascanu Street, Sector 3, 030508, Bucharest, Romania.
[2] Author to whom any correspondence should be addressed

alexandra.ene@incdtp.ro

Abstract. The assimilation of synthetic macromolecular compounds in the interdisciplinary fields represented by the medicine and therapeutics, must address the complexity of the problems imposed by the application field, mainly resulting from the temporary or long-term contact of the polymeric materials with the tissues and biological substances. In this context, polymeric biomaterials are defined as polymers or polymeric composite types certified as biocompatible in contact with the biostructures. Comparing to conventional procedures, ultrasonic welding is a domain with a lower overall weight, but with exceptional development and dissemination perspectives in different industrial branches (automotive, electrotechnical, electronics, microelectronics, medical devices), which involves techniques of combining new materials, biocompatible, intelligent composites, with shape memory, etc. Through their geometric, technical, bio functional and biomedical characteristics, the implantable medical devices must meet the imposed requirements, mainly represented by: chemical inertia, resistance to infections, adequate ratio between flexibility and dimensional stability, adaptability to the defect geometry and high bursting resistance etc.

1. Introduction

The process of thermal degradation of biomaterials in the stage of joining the constituent parts of the medical device is very important and can have a negative effect on its biomedical and bio functional properties [1].

In order to avoid this inconvenience, an innovative device was designed and accomplished in order to develop, through ultrasound welding, a reinforced knitted structure for reconstruction of the thoracic wall, in case of rejection of malignant/benign tumors in the sternum (figure 1).

The implantable medical device is intended for parietal stabilization in the case of anterior lateral defects, but also in cases in which a smaller defect (resection of two costal arches) resulted, which was considered as obligatory to ensure the stabilization especially due to the age of the patients or the alteration of the respiratory function (figure 2).

At the designing of the knitted textile structures, the following essential elements were considered:

- usage of raw materials suitable to the demands imposed by the destination, but also to the requirements for mechanical processing at a higher quality level;
- design of structures capable of providing bio functional and biomedical performance products suitable for the field of clinical application.

For the accomplishment of the knitted structure the PES 76 dtex/f34x1 yarns with tenacity of 3,5 $cN \cdot dtex^{-1}$ and elongation at maximum break of 30% was selected.

In order to realize the invasive medical devices intended for the reconstruction of the semirigid wall of the thoracic cavity, two variants of knitting type were designed. These latters were realized in the chain structure with batting threads and executed with two bars with pastes. The difference between the variants is given by the height of the knitting ratio and the chain command.

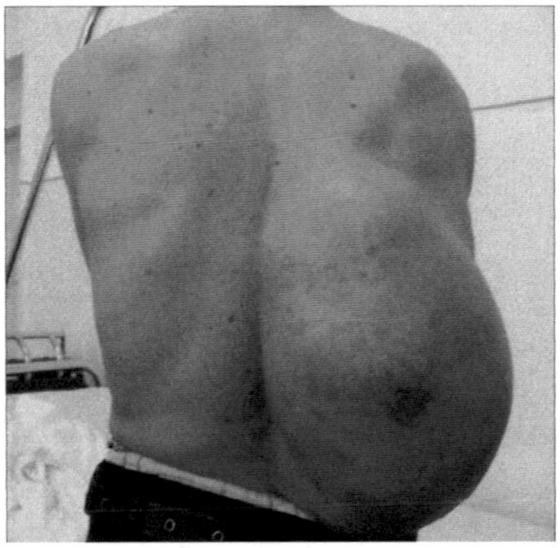

Figure 1. Clinical case study solved at Timisoara County Hospital, Romania, by using the invasive product for the reconstruction of the semi-rigid tissue.

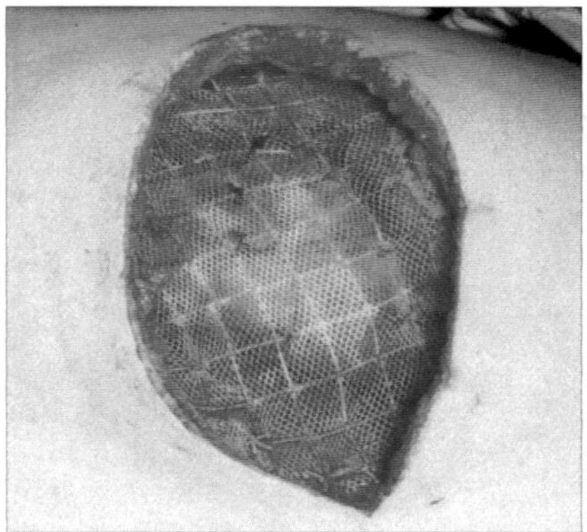

Figure 2. Medical device with textile structure implanted at Timisoara County Hospital, Romania.

The experiments of the reinforcement of the warp knit support was performed by using polyamide monofilaments.

2. Material and methods

The device was designed so that through its components and mechanisms the following operations would be ensured (figure 3):

- the desired displacements, characterized by certain sequences and durations, generated by the task programming system, in accordance with the requirements or commands transmitted. This operation is done by the microcontroller.
- comparing the current motion parameters with those required and making the necessary corrections through the sequence /motion controller.
- amplification of the control signal, usually of low power, in accordance with the requirements of the actuator, by using a power amplifier. Frequently, the power amplifier also performs galvanic separation functions between the low power and the power control circuitry, effective, high power drive.
- the actuator transforms the corrected signal into the input signal (moment, force, speed) in accordance with the requirements of the process.
- adaptation of the actuator parameters to the requirements imposed by the technological process, mechanical mechanisms and transmissions.
- sensor processing of the process parameter information and transmission of the corresponding signals to the motion controller.

The mechanical execution elements in this complex mechatronic system for moving in horizontal and vertical planes, rotating, etc. requires actuation systems powered by different types and forms of energy, that were placed on all three Cartesian coordinates, as follows:

The X-axis actuator is located on the metal frame (chassis) of the machine. The guide axes are located inside the metal frame, being supported at three points in order to avoid the loads that appeared during the ultrasound welding process. The axes are positioned in the same plane, being parallel both with each other and with the frames that form the chassis of the whole system (figure 4).

IOP Conf. Series: Materials Science and Engineering **827** (2020) 012022 doi:10.1088/1757-899X/827/1/012022

The Y-axis actuator has the role of ensuring the sliding and alignment of the ultrasonic welding electrode support (sonotrode). It is based on sliding elements on guide axes and linear bearings with mobile balls (figure 5).

Figure 3. General assembly of the ultrasonic welding device for the accomplishment of the reinforced mesh destined to the thoracic wall reconstruction.

Figure 4. The X-axis actuator.

The Z axis actuator consists of: mechanical support; mobile frame guidance system; transmission system; drive element; race limiting elements (figure 6).

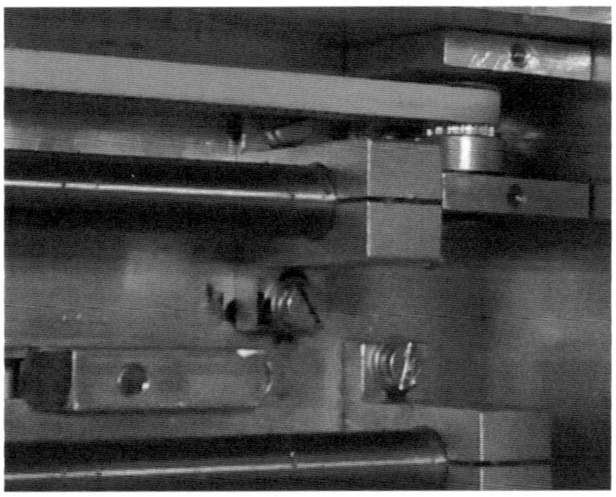

Figure 5. Drive motor, sliding axes, end brackets, Y-belt tensioner.

Figure 6. Mobile frame of Z- axis

All the functional modules of the electrical installation are interconnected and placed in an electrical control panel.

The electrical installation is composed of the following functional modules;

- Electricity protection and distribution module: ensures the distribution of electricity to all other functional modules. The distribution and protection of the modules supplied from the single-phase electric power network is realized with the help of specialized devices of circuit breaker type.

- Automation module: for the control of servomotors and the sequences tracking step by step, an application based on a state-of-the-art microcontroller has been developed.

- Ultrasound generator module having the following components: - oscillator that creates the oscillation frequency itself; - amplifier for the signal level from the oscillator element; - piezoelectric element that receives the electrical signal from the amplifier and transforms it into a mechanical oscillation; - booster with role in the mechanical amplification of the amplitude of the mechanical oscillation coming from the piezo element; - sonotrode - the final element that comes into contact with the welding surface and has the shape of the contact surface depending on the welding material and the shape of the splice and the power supply that distributes electricity for all component modules of the generator.

To drive the execution elements were used servomotor motors powered by DC power. DC motors are intended for the conversion of the electrical signal, in the form of an amplified voltage coming from a transducer, in a rotational motion of a shaft.

The advantages of such a motor consist in the processing of the input signals and the transmission of the initial power from the actuator, reducing the idle time, increasing the resolution and generating a higher maintenance torque.

Considering the intended purpose of the device, meaning the fabrication of the implantable medical, the following functional tests have been performed:

A. Drive control of the component mechanisms of the device and, implicitly, of the logic of the motor's functionality ordered via a microprocessor dedicated to the application.

B. Control of the sequences and durations degree of precision, generated by the tasks programming system, in order to ensure the compliance with the requirements imposed by the designed commands.

Under static and dynamic testing conditions, the ultrasonic welding device has responded to the technical and functional requirements imposed by the final product destination area.

The technological tests was conducted under the conditions presented in the table 1: time of ultrasound welding: [4,5 s; 8,0 s]; power: 800 W; energy: [800 J; 1300 J]; ultrasound initialization: at the contact with the surface ± 0,30 mm, before the contact with the surface up to : [1,0 mm; 5,0 mm]; amplitude: constant, decreasing, increasing.

Tabel 1. Conditions of technological tests.

No.of the functional test	Overall thickness (mm)	Time (s)	Power (W)	Energy (J)	Distance for ultrasound starting (mm)	Amplitude
1.	1,5	8,00	800	1200	at the contact with surface ± 0,30 mm	Constant
2.	1,5	6,50	800	1300	at the contact with surface ± 0,30 mm	Constant
3.	1,5	5,50	800	1100	before the surface contact with 1,0 mm	Incresing
4.	1,5	4,50	800	900	before the surface contact with 1,50 mm	Increasing
5.	1,5	4,50	800	900	before the surface contact with 5,0 mm	Decreasing

The distribution of the pressures when assembling the two components was evaluated using specialized CFD software.

During the experimental program, the structural analysis [2, 3] evidenced the following:

- for an exposure time of 8 s and increasing amplitude have been observed deformations of the base layer (knitted structure) caused by the thermal degradation of the material which can lead to the tears of the knitted fabric at the intersection points of the warp stitches;
- decreasing of the exposure time to ultrasound, leads to the removal of the accidental destruction effect of the assembled elements (monofilament and knitted textile backing);
- if the ultrasonic vibration is switched off at the moment of the contact with the assembled components and these are initialized five millimeters before contact with the polyamide monofilament, a better positioning of the assembled parts and consequently an appropriate stability of the assembly are achieved (figure 7);

- decreasing the exposure time to ultrasound up to 4.50 s, while changing the amplitude from constant to rising, leads to an increase quality of the welding;

The disadvantage, however, is that the assembled parts move from the original position.

This inconvenience can be removed by changing the direction of the amplitude evolution (figure 8).

Based on the results obtained following the structural analysis, several ultrasonic reinforcement tests of the knitted textile support and monofilament yarns made of polyamide were performed and comparative physical-mechanical analyses for two variants of final medical devices were conducted in certified laboratories.

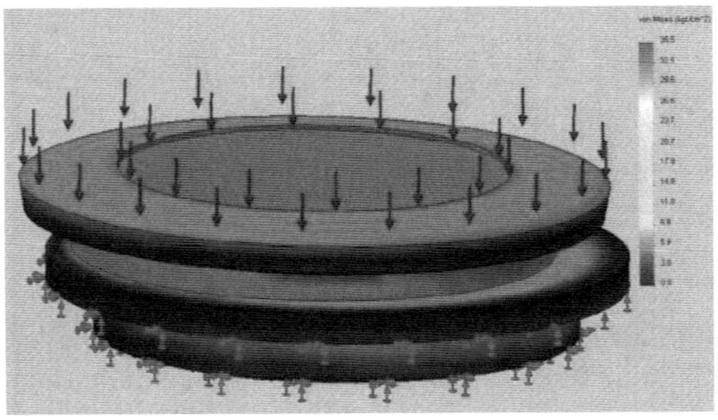

Figure 7. Positioning of the elements to be assembled when the ultrasonic vibration is switched off.

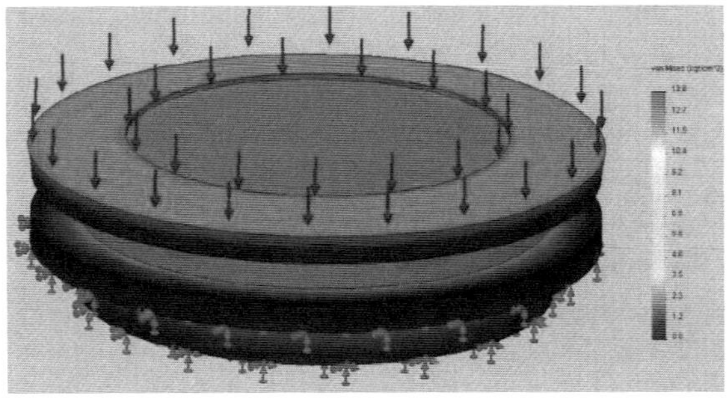

Figure 8. Structural analyses of the welded textile components at increasing amplitude.

The physical - mechanical determinations aimed at: mass $[g(m^2)^{-1}]$; thickness (mm); tear resistance (daN); breaking resistance (daN); elongation (%); impact resistance (kPa) and the results are presented in the table 2.

Tabel 2. Physical-mechanical characteristics for medical device destined for thoracic surgery.

Sample	Mass $[g(m^2)^{-1}]$	Thikness (mm)	Tear resistance (Kgf)	Braking resistenace (Kgf)	Elongation (%)	Bursting resistance, (KPa)
P1	98,1	0,78	2,2	7,60	77,5	220,2
P2	101,20	0,8	1,8	12,9	29	462,5

From the data analysis the following conclusions could be presented:

- There are no significant differences in mass and thickness for the two samples, which ensures that the functional tests were performed under the same conditions.

-Breaking resistance for sample 2 is on average 55% higher in both systems compared to the values obtained for sample 1 and burst resistance is 100% higher than in case of sample 1.

This highlights the stability of the structure and implicitly the balance of the two systems of the warp knit reinforced with polyamide yarns, the perfect positioning of the two ultrasonic assembled elements and the maintenance of the stability of polymers by avoiding degradative phenomena.

3. Results and discussion

The structural analysis of the phenomena that occur during the ultrasound welding process has evidenced that the increase of the assembly resistance and the imposed quality of the surface texture are ensured in the following situations: ultrasound exposure time: 4.50 s; power: 800 W; ultrasonic start: before contact with the surface; decreasing amplitude.

This conclusion was validated both by the results of the physical-mechanical analyses conducted in certified laboratories and clinical applications.

4. Conclusion

The functional, technological and experimental tests demonstrated the feasibility, functionality and suitability of the developed ultrasonic welding device for the medical devices with textile structures accomplishment.

The new innovative technological solution for the reinforcement of the knit textile support made up of polyester yarns with a polyamide monofilament using ultrasound welding equipment ensure the required level of dimensional stability of the product and adequate architecture that enable the semi-rigid neo tissue development.

The components of the implantable medical devices are perfectly ensembled, avoiding the undesirable specific polymeric degradation.

Acknowledgment

This scientific paper is funded by the Ministry of Research and Innovation within Program 1 - Development of the national RD system, Subprogram 1.2 - Institutional Performance - RDI excellence funding projects, Contract no. 6PFE/2018.

References

[1] Naghibi B, Janssen D, Khoshgoftar M and Sprengers A 2016 A comparison between dynamic implicit and explicit finite element simulations of the native knee joint *Medical Engineering & Physics* **38**(10) 1123–30

[2] Arregui M and Margetts L 2014 Practical Application of the Stochastic Finite Element Method *Archives of Computational Methods in Engineering* **23**(1) pp.171–90

[3] Murea C M and Hentschel G 2005 Finite element methods for investigating the moving boundary problem in biological development. In *Nonlinear Elliptic and Parabolic Problems* pp. 357-371 Birkhäuser Basel.

Innovative solution for increasing the performance of the cutting equipment for textile subassembly having variable geometry

C Mihai[1], A G Ene[1, 2] and C Jipa[1]

[1]IT Research Department in Industrial Engineering, The Research-Development National Institute for Textile and Leather, 16 Lucretiu Patrascanu Street, Sector 3, 030508, Bucharest, Romania.
[2]Author to whom any correspondence should be addressed

alexandra.ene@incdtp.ro

Abstract. Although the textile industry is in worldwide recession, the field of technical textile articles is in a continuous development, proving that the manufacturers of these products are concerned about finding solutions and creating high performance articles. Considering the need to increase productivity and product quality while reducing the consumption of materials, in the last decades, processing technologies specific to textile industry had a quick dynamic, that has led to the development of new, very efficient technological methods and processes, characterized by an advanced degree of automation. The proposed technological solution was developed based on the correlational analysis of the technological parameters of the variable geometry subassemblies, targeting the following technological aspects: monitoring of the functional parameters during the technological process development and real-time warning of the occurring dysfunctions; display of functional parameters and controls on a HMI touch panel display; adjusting the settings of the parameters of each inverter according to the dynamic parameters of the ordered devices (couple, speed, acceleration-deceleration, brake moment, etc.); PLC circuit logic programming via a software program dedicated to the application.

1. Introduction

The main result of the automation of the installation of textile subassemblies with variable geometry for technical articles consists in the production of a quasi-continuous production, associated with the increase of the number of subassemblies realized in a reduced time with the following performances: the speed of the cutting device 100-500 RPM; roller speed 150-500 rpm; feed rate of cutting device 1-100 mm/sec; positioning accuracy of the cutting device 0.025 mm; sharpening time of the cutting device 1-10 sec; the deburring time of the cutting device 1-5 sec; average number of cuts per minute 2-10 pcs/min. (depending on the type of material and its geometry).

At the same time, the automation of the installation ensures the reduction of non-periodic losses due to working bodies, control and control devices, scrap production, improper organization, etc. Thus, the considerable increase of the main indicator of the economic efficiency represented by the labor productivity is obtained [1].

Also, of particular importance is the improvement of exploitation indices by: early detection of the non-productive sequences of the cutting devices functionality; detection of accidental stops; quickly adoption of the imposed measures for a correct functioning regime restoring [2].

In addition, the automation will allow to obtain a higher quality of the textile subassemblies, generated by the technological results represented by: - precise control of the movement of the cutting devices; - the correction of the cutting parameters even during the operation of the installation; - easy visualization of possible errors and damages by using the programmable terminal; - reduction of material losses by about 35 - 40%; - increasing the cutting efficiency and the sectioning accuracy (<0.1 mm) with a positive impact on the quality of the subassembly and implicitly of the final application; - 80% reduction of material deformation, so as to ensure a very good dimensional reproduction, with high precision markings and cuts; - reduction of waste by about 40%; - intuitive, friendly and easy usage.

Thus, it should be noted that between competitiveness, quality and productivity there is a direct link, both quality and productivity being determinants of a company's competitiveness.

In this context, it is obvious that the research results had a very good contribution to the SME's competitiveness development by increasing the: quality of the products through the automatic and efficient control of the factors that influence it (the technological process, the quality control, the personnel qualification); diversification of production through production renewal, modernization and improvement of existing products in accordance with consumer demand; the profitability of the production process; labor productivity; innovative character of the automation system; celerity of the innovation process implementation, design, production, delivery to the consumer; level of flexibility and adaptability as well as the ability to adapt to the markets 'changes and requirements; the professionalism of the personnel and the activity in the group; strategies adopted etc.

2. Material and methods

As it results from the operating algorithm of the variable-geometry sub-assembling textile cutting equipment, under conditions of increased productivity and efficiency, the main parameter influencing the performance of the device is represented by the cutting advance rate.

On the basis of an experimental program implemented in the industrial environment, a multilinear regression analysis, that allowed to predict the performances of the device, both technologically (e.g. speed, productivity, product quality) and - implicitly – financially (i.e. costs) has been carried out [3].

Experimental data obtained as a result of running the testing program allowed to process through a software package dedicated to the application more than 50 results that have been processed using a specialized software package, based on the following working hypotheses:

- independent variables are represented by: X_1 – ballot weight, X_2 – length of the material, X_3 – speed of the ballot positioning shaft, X_4 – speed of the cutting device;
- dependent variable: Y- advanced cutting speed.

The sequential commands for database initialization are shown in figure 1.

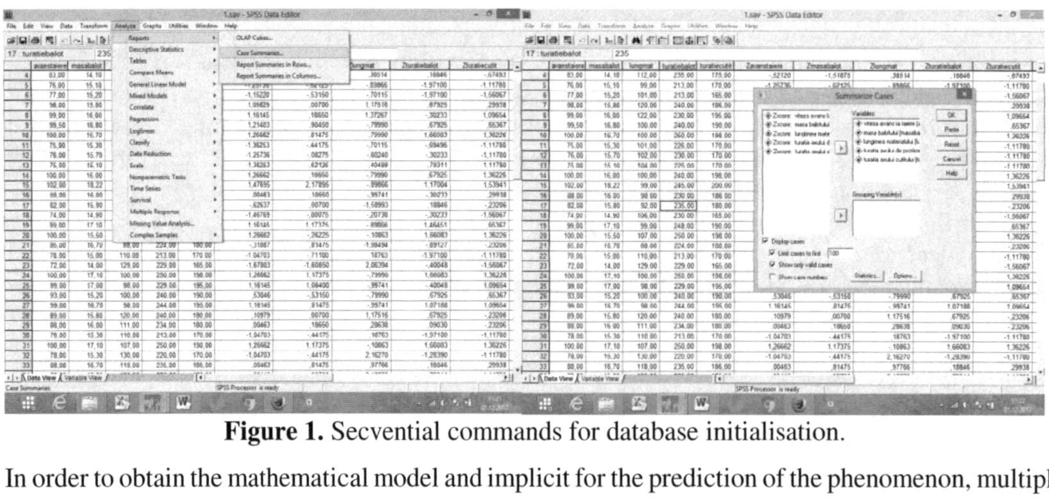

Figure 1. Secvential commands for database initialisation.

In order to obtain the mathematical model and implicit for the prediction of the phenomenon, multiple regression analysis was used, starting from the hypothesis that the regression equation is of the form:

$$Y = A_0 + A_1 X_1 + A_2 X_2 + A_3 X_3 + A_4 X_4 \qquad (1)$$

where: $A_0, A_{1,} A_2, A_3, A_4$ – represent the coefficient of the regression equation and Y, X_1, X_2, X_3, X_4 are the dependent, respectively the independent variables defined above.

The initialization commands are presented in the figure 2 and the histogram of the standard notes of the model residues in figure 3.

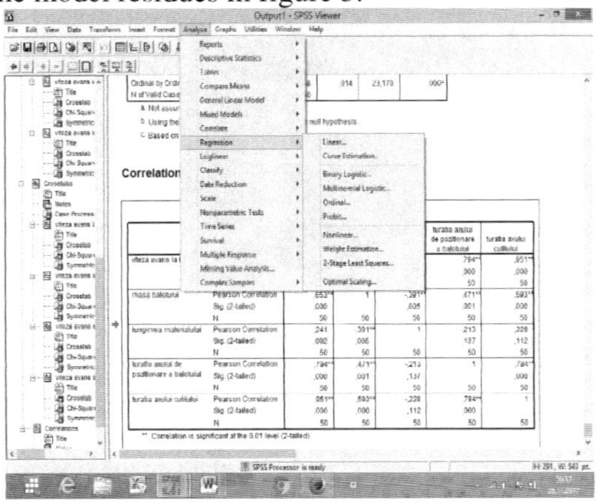

Figure 2. Specific commands for regression analyses initialisation.

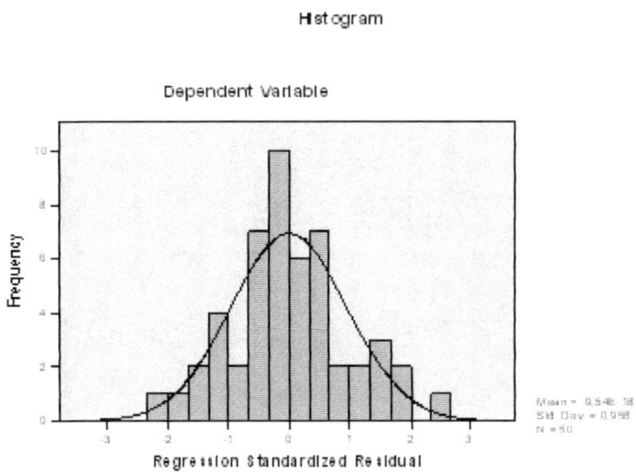

Figure 3. Histogram of standard residue notes (deviations from the model from reality)

The regression equation coefficients are presented in the table 1 and the correlation between these coefficients is summarized in table 2.

Table 1. Coefficients of the model.

Independent variables	Non-standardised coefficients		Standardised coefficients	t	Sig.
	B	Std.Error	Beta		
Constant.	-79,069	11,785		-6,709	0,000
Ballot weight.	1,224	0,462	0,143	2,652	0,011
Length of the material.	0,017	0,042	0,018	0,397	0,694
Speed of the ballot positioning shaft.	0,115	0,062	0,124	1,861	0,069
Speed of the cutting device.	0,652	0,061	0,774	10,644	0,000

The main output of regression equation is highlighted in the figure 4, the dependence between the analyzed variables is presented in the figure 5.

Table 2. Coefficients correlations.

Independent variables	Non-standardised coefficients		Standardised coefficients	t	Sig.
	B	Std.Error	Beta		
Constant.	-79,069	11,785		-6,709	0,000
Ballot weight.	1,224	0,462	0,143	2,652	0,011
Length of the material.	0,017	0,042	0,018	0,397	0,694
Speed of the ballot positioning shaft.	0,115	0,062	0,124	1,861	0,069
Speed of the cutting device.	0,652	0,061	0,774	10,644	0,000

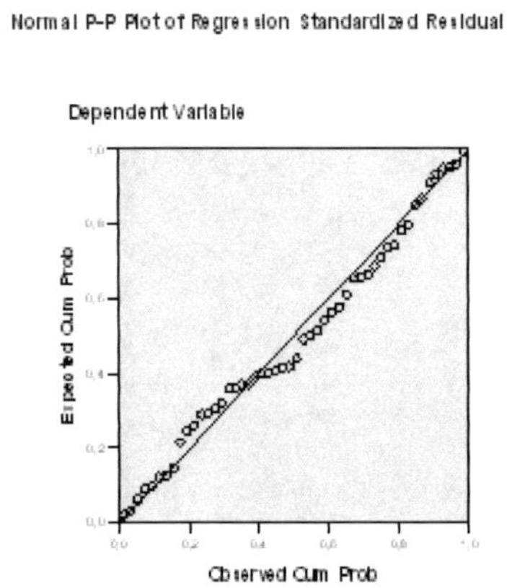

Figure 4. Output of the regresion equation.

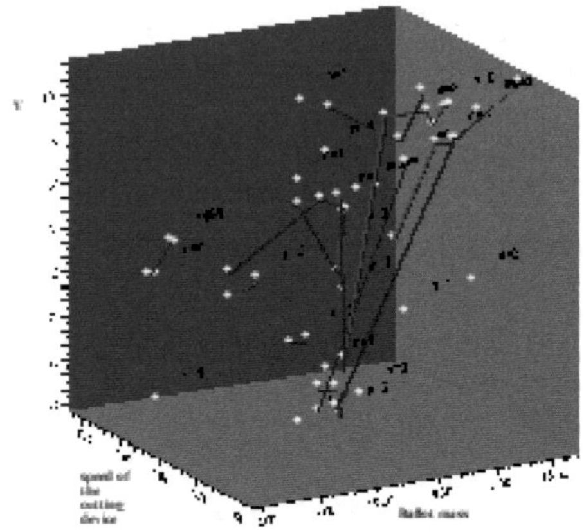

Figure 5. The interdependence between dependent variable and mass of the ballot and speed of the cutting device.

In order to verify the regression model, three experiments were performed and the obtained values are centralized in table 3.

Table 3. Experimentally recorded values versus those calculated according to the multivariate regression model.

Variables	U.M	Experimentally recorded values			Dependent variable values calculated according to the model (regression equation)		
		Exp. 1	Exp.2	Exp.3	Exp. 1	Exp. 2	Exp. 3
X_1	kg	15,50	16,60	16,70			
X_2	m	107	122	88			
X_3	rot/min	250	230	224			
X_4	rot/min	198	195	180			
Y	m/min	100	99	85	100.65	97.94	86.91

The differences between the model predicted value and those determined experimentally are: for experiment 1 with 0,65 m /min higher; for experiment 2 with 1.06 m /min smaller, and for experiment 3 with 1.91 m /min higher, but all values are within the standard deviation of the dependent variable, of 2.74 m /min. The residual statistics for the developed model are presented in the table 4.

Table 4. Residual statistics of the model for the dependent variable: advanced cutting speed.

	Minimum	Maximum	Mean	Standard deviation	N
Predicted value	73,3144	103,4762	87,9560	9,13857	50
Residual value	-5,56661	6,83497	0,00000	2,62749	50
Standard predicted value	-1,602	1,698	0,000	1,000	50
Standard residual value	-2,030	2,493	0,000	0,958	50

The same algorithm can be used for the prediction of other important parameters of the device, respectively: time of rolls, the density of wrapping of the material cut, the speed of variation of the type-dimensions. This will contribute to the elimination of the stationary time for sizing and implicitly to the increase of the labor productivity.

3. Results and discussion

The standard scores histogram of the residues follows the normal curve, so the model accurately predicts the cutting advance speed, depending on the considered predictors.

The graph of the cumulative probabilities of the residues standard scores reveals a normal distribution because it follows the line of the cumulative percent described by the normal curve (the right line on the diagonal of the graph, from left to bottom to right - up), so the prediction model demonstrated to be accurate.

All the predictors used for the dependency variable – advanced cutting speed- helps very well in the prediction phase, because neither for low values of the dependent variable (on the lower part of the graph – left side) neither for high values (superior part of the graph – right side), the model does not have the tendency of underestimate or overestimate the reality.

The obtained regression equation is on the form:

$$advanced\ cutting\ speed = -79.10 + 1{,}22\ "x\ ballot\ weight\ +\ 0.02\ x\ length\ of\ material\ +\ 0{,}12\ x\ speed\ of\ the\ ballot\ positioning\ shaft\ +\ 0{,}65\ x\ speed\ of\ the\ cutting\ device" \quad (2)$$

The value of the square multiple correlation coefficient R^2 reveals that 92.4% of the variance of the dependent (predicted) variable represented by the advanced cutting speed is explained by the model, with the prediction security of 91.7%.

The standard error of the predicted dependent variable (cut-off speed) demonstrates that the standard deviation of the cutting advance speed is of 2.74 m/min when the value of the independent variables in the model is known, respectively: the mass of the ballot, the length of the material, the ballot and the knife shaft speed.

The value of 0.924 demonstrates that R^2 does not change when additional variables are considered to the model.

Analysis of the variance highlights, with a probability of error of 5%, that the model significantly explains the variance of the dependent variable. In addition, the F-values demonstrate that the variance explained by the model is significantly higher than the residual one, so the created model is effective in prediction.

4. Conclusion

The created mathematical model allowed the prediction of the advanced cutting speed, knowing the ballot weight, the length of the material, the speed of the ballot positioning shaft, and the knife axis speed.

Experimental tests of the regression equation revealed that the differences between the predicted values of the model and those determined experimentally fall within the standard deviation of the dependent variable.

Acknowledgment

This scientific paper is funded by the Ministry of Research and Innovation within Program 1 - Development of the national RD system, Subprogram 1.2 - Institutional Performance - RDI excellence funding projects, Contract no. 6PFE/2018.

References

[1] Wenchang L, Feng L 2017 On single-machine scheduling with workload-dependent maintenance duration *Omega* **68** 119-122

[2] Zhu H, Miu L, Zhangjing Z 2015 Machine scheduling with deteriorating and resource-dependent maintenance activity *Computers & Industrial Engineering* **88** pp.479-486

[3] Atkinson A, Donev A, Tobias R Optimum Experimental Designs, with SAS 2007 *International Statistical Review* **3***(75) pp.* 409-413

Fit evaluation of sportswear for Belgian elite male rowers in static and dynamic rowing postures

S Vasile[1,4], J Cools[1], A De Raeve[1] , B Malengier[2] and F Deruyck[3]

[1] Department of Fashion, Textiles and Wood Technology, University College Ghent, Buchtenstraat 9-11, 9051 Gent, Belgium
[2] Centre for Textile Science and Engineering, Department of Materials, Textiles and Chemical Engineering, Ghent University, Technologiepark-Zwijnaarde 70A, 9052 Zwijnaarde, Belgium
[3] University College Ghent, Faculty of Science and Technology, Department of Exact Sciences, Valentin Vaerwyckweg 1, 9000 Gent, Belgium
[4] Author to whom any correspondence should be addressed

Simona.vasile@hogent.be

Abstract. The main objective of the SHAPE study was to assess the anthropometry of elite male rowers and develop customized size charts and sportswear with adequate fit in static and dynamic rowing postures. This paper briefly discusses the anthropometry of Belgian elite male rowers and focuses on the fit features of a related unisuit called SHAPE which is compared against a reference (SMARTFIT) with similar materials and design, whereof the garment construction is based on body charts of average Belgian males. Four elite male rowers evaluated both unisuits in garment size 52 and 58. Most of the ten fit features investigated for SHAPE unisuit were allocated average scores between 3 (adequate fit) and 4 (very good fit). In static posture, the overall fit of the SHAPE unisuit was found slightly better than of SMARTFIT unisuit (average score 3.8 and 3.1 respectively). Unlike SMARTFIT, SHAPE unisuit based on rowers body dimensions scored equally well in static and dynamic rowing postures catch and finish and will likely successfully accommodate repetitive sport movements.

1. Introduction

To ensure great freedom of movement, perfect fit and comfort, regardless of any sport-specific movement, sportwear shall consider body proportions and shape of elite athletes who categorically differ from the average population. Several international studies [1-7] reported deviation of rowers anthropometry from average population but they did not link anthropometry to garment patterning and fit of the sportswear in sport- specific posture was rarely investigated.

The main objective of the SHAPE study [8] was to develop sportswear with adequate fit in both static and dynamic rowing postures. Therefore the anthropometry of Belgian elite male rowers (age 18-35 years) was first assessed using a 3D body scanning technique and eventually customized size charts were developed. Significant differences were found between the body measurements of heavy weight male rowers and average Belgian males of same age [8]. This study quantified moreover the influence of the catch and finish rowing postures on selected body measurements and investigated the fit of the corresponding sportswear [9]. This paper briefly discusses the anthropometry of Belgian elite male rowers assessed by 3D body scanning and zooms in on fit features of related SHAPE unisuit which were evaluated in static and dynamic rowing postures. Fit of SHAPE unisuit was also compared against a reference unisuit (SMARTFIT) whereof the construction is based on average body measurements of Belgian males [10].

2. Materials and methods

2.1. Rowers anthropometry assessment and body size charts development

The target group was scanned using 3D body scanners with structured light (i.e., Symcad and TC^2). A number of n=20 body measurements were extracted [11] for a total number of n=54 elite male rowers (of which 32 were Belgian), age 18-35 (mean age 21.0 ± 4.0 years). A number of 41 subjects belonged to the heavy weight category (>72.5 kg) and 13 subjects to the lightweight category (< 72.5 kg). The rowers had the following demographics (Mean± SD): height 183.6±6.8 cm, body mass 80.5±9.2 kg and Body Mass Index (BMI) 23.8±1.9 kg/m^2. Thirteen significant differences ($p<0.05$) in the body measurements were identified, especially the chest girth and stature (i.e., heavy weight rowers > 4 cm larger than average Belgian male of same age). Length of chest, back and legs was also significantly larger (up to max 4 cm) for the heavy weight rowers.

A dataset of n=52 rowers was used to develop body sizing charts. According to ISO 8559-2:2017 [12] chest girth was defined as the primary dimension. The subjects were categorized in 8 different size groups (sizes 44-58) according to the recommended size ranges [13]. For the whole size range a proportional interval of 4 cm between two adjoining body measurements was chosen. For each chest girth range, the average of the other sizes was calculated. The standard size was set on size 52 and except chest, all girths and shoulder length were graded proportionally. Other length measurements are based on an average height of 183,7 cm and are equal for the whole size range. Table 1 shows comparatively the average body measurement of Belgian males SMARTFIT [10] and elite rowers SHAPE [8] based on a dataset of 307 Belgian males and 52 rowers respectively.

Table1. Sizing charts size 52 and 58 according to average Belgian males (SMARTFIT) and elite rowers (SHAPE) showing the body differences between the two groups.

Body measurements (cm)	Size 52 SMARTFIT	Size 52 SHAPE	Size 52 Difference SHAPE-SMARTFIT	Size 58 SMARTFIT	Size 58 SHAPE	Size 58 Difference SHAPE-SMARTFIT
Body height	180	186.1	+ 6.1	180	184.9	+ 4.9
Chest girth	104.0	104.0	0.0	116.0	116.0	0.0
Waist girth	88.0	84.0	- 4.0	100.0	91.5	- 8.5
Top hip girth	92.5	86.0	- 6.5	105.0	95.0	- 10.0
Hip girth	101.5	100.5	- 1.5	110.5	107.0	- 3.5
Waist height	111.7	115.0	+ 3.3	111.7	115.0	+ 3.3
Across back width	40.0	42.5	+ 2.5	41.8	44.0	+ 2.2
Across front width	40.5	41.0	+ 0.5	43.5	45.5	+ 2.0
Back neck point to waist	43.0	45.5	+ 2.5	43.0	45.5	+ 2.5
Neck girth	40.0	39.0	- 1.0	43.0	42.0	- 1.0
Neck base girth	45.5	42.0	- 3.5	48.5	45.0	- 3.5
Shoulder length	15.8	16.3	+ 0.5	16.7	16.9	+ 0.2
Shoulder slope	6.0	5.9	- 0.1	6.0	5.9	- 0.1
Arm length	60.1	62.5	+ 2.4	60.1	62.5	+ 2.4
Upper-arm girth	33.0	32.5	- 0.5	36.0	35.5	- 0.5
Wrist girth	17.6	17.0	- 0.6	18.8	17.6	- 1.2

Inside leg length	81.9	85.0	+ 3.1	81.9	85.0	+ 3.1
Thigh girth	59.0	59.0	0.0	63.5	63.5	0.0
Knee girth	39.5	40.0	+ 0.5	42.0	43.0	+ 1.0
Calf girth	39.0	38.5	- 0.5	42.0	41.5	- 0.5
Ankle girth	25.5	26.0	+ 0.5	27.0	27.5	+ 0.5
Knee height	49.9	52.0	+ 2.5	49.9	52.0	+ 2.1

2.2. Materials and prototypes development

Three types of fabric for sportswear exhibiting various composition, elasticity and moisture management properties were selected for the development of the prototypes. To envisage elevated comfort level, fabric 2 with enhanced moisture management, air permeability and short drying time (Table 2) was placed on the back side, which is potentially prone to a large amount of sweat during rowing. Fabric 1 was used in the front panel and fabric 3 on the buttock respectively (Figure 1a).

Table 2. Physical and comfort properties of the fabrics 1-3 used for the rowing unisuit.

		Fabric 1 knitted	Fabric 2 knitted	Fabric 3 woven
Composition (%)		81 PES/ 19 EA	76 PA/13 PES/ 11 EA	71 PA/ 29 EA
Mass (g/m^2) [14]		156±2	156±5	147±4
Thickness (mm) [15]		0.5±0.00	0.6±0.01	0.42±0.00
Elongation (%) [16]	Wales/warp	155±2	128 ±4	89±1
	Courses/weft	223±3	99±2	95±1
Drying time (min) [17]		46±0.4	28±0.3	-
OMMC [18]		0.4±0.03	0.55±0.01	0
Air permeability (mm/s) [19]		258.5±3	1394±45	105±11

Two unisuits called SHAPE were developed in garment size 52 and 58 (Table 1). Design was decided considering the preferences of the rowers and aiming at improving shortcomings of existing rowing suits. [20] Equally two unisuits called SMARTFIT with similar design and materials (Figure 1a) were produced starting from average body measurements of Belgian males.

(a) (b)

Figure 1. Front (left) and back view (right) of the unisuit indicating the fabrics 1-3 used (a) and virtual fit of SHAPE unisuit size 52 in front, back and lateral view (b) [20].

2.3 Garment fit assessment in rowing postures

During a rowing stroke, four positions can be distinguished, among which flexed posture "catch" and extended posture "finish" potentially lead to significant body changes, both in lengths and girths. During the catch (figure 2a) the legs are compressed, the shins are vertical, arms are extended, triceps work to extend the arms, and the flexor muscles of the fingers and thumbs grip the handle. The back muscles are relaxed, and abdominals are flexing the torso forward. [21]At the finish (Figure 2c), the abdominals stabilize the body, and the glutes and quads are contracting. The biceps and many of the back muscles are also contracting to help keep the torso in the finish position and to internally rotate the upper-arms. [21] Changes of both upper and lower body upon posture are relevant for the purpose of garment development in general and they likely change most upon catch and finish posture.

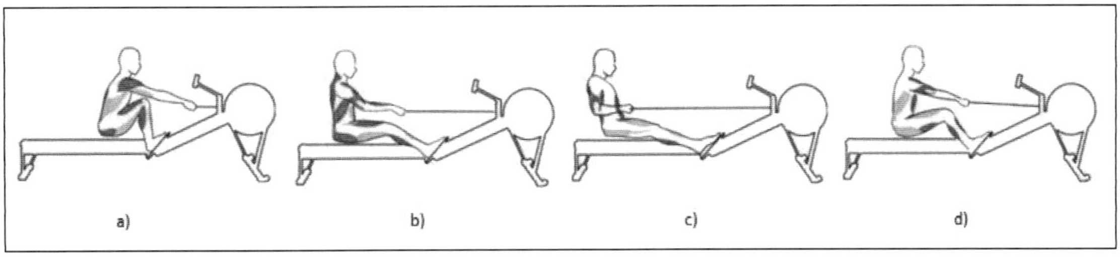

Figure 2. Rowing stroke and muscles involved in a) catch, b) drive, c) finish, d) recovery (permission from Concept2, Inc.).

Four elite male rowers having previous experience with fit assessment, evaluated the fit of the two outfits standing as well as in catch and finish rowing postures on an ergometer. The overall fit and ten other fit features of outfit SHAPE and reference unisuit SMARTFIT were assessed on a 5-point scale (1= extremely poor fit, 5= excellent fit).

3. Results and discussions

Most of the average scores assigned to the fit features of SHAPE unisuit (figure 3) were between 3 (adequate fit) and 4 (very good fit). Fit was found adequate at level of chest girth CG and thigh girth TG regardless the posture. Similarly neck line NL, pants length LL and arm cut AC_B (of back panel) were equally good (average score 4 in static and dynamic postures). A decrease of the average fit on waist WG and hip HG was noticed upon dynamic postures as well as a slight decrease of fit for back length BL and arm cut front AC_F. Three test persons were slightly smaller than the size of the suit they evaluated, which likely explains the large variability (standard deviation) of the scores assigned to most of the fit features.

The overall fit of the SHAPE unisuit (figure 4) in static posture was found slightly better than of SMARTFIT unisuit (average score 3.8 and 3.1 respectively). Moreover SHAPE protype scored equally well both in static and dynamic postures. On the contrary SMARTFIT unisuit based on body size charts of average Belgian males had a poorer fit (average score 2.8) in catch and finish posture and will likely impede repetitive rowing movements. SHAPE unisuit was found statistically significant better (t-test, $p<0.05$) than SMARTFIT, in static and dynamic postures by three of four test persons with respect of all fit features considered. Figure 1b displays the virtual fit of unisuit SHAPE on an avatar with body dimensions of a rower size 52.

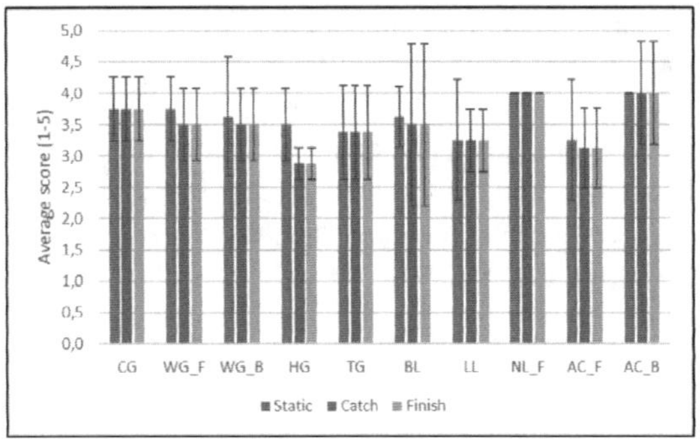

Figure 3. Average fit features of SHAPE outfit in static and two dynamic rowing postures.

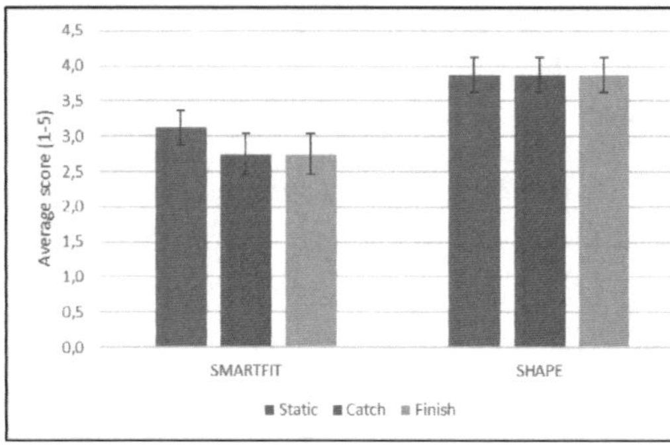

Figure 4. Overall fit evaluation of SMARTFIT and SHAPE outfits in static and dynamic posture.

4. Conclusions

The overall fit of unisuits SHAPE was positively evaluated by the rowers regardless of the posture which suggests that the body size charts were correct, despite a rather limited number of subjects considered (n=52 rowers). Fabrics 1 and 2 used for the garment's body had an elongation between 99-223 % depending on testing direction, which seems to be sufficient to accommodate body changes due to dynamic postures. Nevertheless, in case of fabrics with lower elasticity, extra ease may be necessary to ensure adequate fit regardless of the posture. These aspects along with the feedback of the test persons will be considered in development of future prototypes. Special attention will be payed to selection of test persons whose body sizes perfectly match the size of the garment tested.

Finally, unisuit SHAPE developed based on anthropometry of elite male rowers exhibited better fit than outfit SMARTFIT based starting from average body measurements of Belgian males, especially in dynamic rowing postures.

Acknowledgments

The authors would like to express appreciation for the financial support of the sponsors VLAIO [IWT TETRA HBC.2016.0078] and industrial partners of project SHAPE [8].

References

[1] Bourgois J, Claessens A L, Vrijens J, Philippaerts R, Van Renterghem B, Thomis M, Janssens M, Loos R and Lefevre J 2000 Anthropometric characteristics of elite male junior rowers *British Journal of Sports Medicine* **34** pp 213-216

[2] Bourgois J, Claessens A L, Janssens M, Van Renterghem B, Loos R, Thomis M, Philippaerts R, Lefevre J and Vrijens J 2001 Anthropometric characteristics of elite female junior rowers *Journal of Sports Sciences* **19** pp 195-20

[3] Claessens A, Bourjois J, Van Aken K, Van der Auwera R, Philippaerts R, Thomis M, Vrijens J, Loos R and Lefevre J 2005 Body proportions of elite male junior rowers in relation to competition level, rowing style and boat type *Kinesiology* **37**(2), pp 123-132

[4] Forjasz J 2011 Anthropometric Typology of Male and Female Rowers Using K-Means Clustering *Journal of Human Kinetics* **28** pp 155-164

[5] Schranz N, Tomkinson G, Olds T and Daniell N 2010 Three-dimensional anthropometric analysis: Differences between elite Australian rowers and the general population *Journal of Sports Sciences*, **28,** pp 459-469

[6] Mikulic P 2008 Anthropometric and physiological profiles of rowers of varying ages and ranks *Kinesiology* **40** pp. 80-88

[7] Mikulic P, Ruzic L and Oreb G 2007 What distinguishes the Olympic level heavyweight rowers from other internationally successful rowers? *Collegium Antropologicum* **31** pp. 811-816

[8] Malengier B,Vasile S, De Raeve A, Cools J, Deruyck F,Teyeme Y and Langenhove L 2019 *Results of SHAPE project Adapted performance wear* IWT TETRA HBC.2016.0078

[9] Vasile S, Cools J, De Raeve A, Mallengier B and Deruyck F,2019, Effect of rowing posture on body measurements and skin-sportswear interface pressure and implications on garment fit, *Journal of Industrial Textiles p.*1528083719877005

[10] SMARTFIT Adapted performance wear [project IWT 110139]

[11] ISO 8559-1:2017 *Size designation of clothes - Part 1: Anthropometric definitions for body measurement*

[12] ISO 8559-2:2017 *Size designation of clothes - Part 2: Primary and secondary dimension indicators*

[13] EN 13402-3:2017 *Size designation of clothes – Part 3: Size labelling based on body measurements and interval*

[14] ISO 3801:1977 *Textiles -- Woven fabrics -- Determination of mass per unit length and mass per unit area*

[15] ISO 5084:1996 *Textiles -- Determination of thickness of textiles and textile products*

[16] EN 14704-1:2005 *Determination of the elasticity of fabrics - part 1: Strip tests*

[17] ISO 17617:2014 *Textiles -- Determination of moisture drying rate*

[18] AATCC 195:2011 *Liquid Moisture Management Properties of Textile Fabrics*

[19] ISO 9237:1995 *Textiles -- Determination of the permeability of fabrics to air*

[20] De Bisschop C 2019 *A study on the optimisation of garment fit and comfort of roing suits for male Belgian rowers,* bachelor thesis, University College Gent

[21] https://www.concept2.com/indoor-rowers/training/muscles-used Accessed on 10 April 2019

IOP Conf. Series: Materials Science and Engineering **827** (2020) 012025 doi:10.1088/1757-899X/827/1/012025

Development of E-textile electrodes: washability and mechanical stresses

S U Zaman[1,2,3], X Tao[1,2], C Cochrane[1,2] and V Koncar[1,2]

[1,2] ENSAIT/GEMTEX, 2 Allée Louis et Victor Champier, Roubaix, F-59100, France
[1,2] Université of Lille, Cité Scientifique, Villeneuve d'Ascq, F-59650, France
[3] Author to whom any correspondence should be addressed

shahood-uz.zaman@ensait.fr

Abstract. E-textile, as a growing field in the current era, gained a lot of attention and currently, some tremendous outcomes of these fields can be seen. They are being used in specific functional requirements depending on application types in various fields including medicals, military, safety and leisure, etc. E-textile structures can contain various elements including wearable motherboards, sensors, electrodes, antennas, actuators, and power sources. These components are integrated with available textile materials according to their usage requirements. For the use of e-textiles in daily life and to make them popular with consumers, these products must be reliable and reusable for a prolonged time. The e-textile market has developed rapidly in the past few years especially in the applications of wireless communication, monitoring, gaming, and data collection. Among these applications, products related to health monitoring gained attraction due to their importance in diagnosing the real- time health problems of the customers. The current study is focused on wearable ECG electrodes to make them reliable and washable. Embroidered electrodes are prepared and washed multiple times to check the washing impact on ECG signals performance. Available mechanical tests are performed to predict equivalent damage as during washing. Hence these predictions, in the coming future, can be used to check, analyse and predict the washability of e-textile prototypes at different stages of manufacturing without actually washing of samples for a multiple numbers of wash cycles.

1. Introduction

In the current era of the world, we are using a large number of smart devices in our daily life. These devices can be of different types ranging from mobile phones to smart wearing garments. The conventional textile can be converted into smart textile by integrating some state-of-the-art user- defined product on it. These smart textile systems can be used for performing various tasks including detection and monitoring from outside environment and giving responses or information to a centralized database. These e-textile systems can also be used for personal human data monitoring based on medical, sports or any other sort of purposes. These products are penetrating in our daily life now days depending on their usage and output we received [1].

For the use of e-textiles in daily life and to make them popular with consumers, these products have to be reliable and reusable for a longer period time. The E-textile market has developed rapidly in past few years especially in the applications of wireless communication, monitoring, gaming, and data collection. Among these applications, products related to health monitoring gained attraction due to their importance in diagnosing the real-time health problems of the customers. One common example is real-time ECG monitoring for the intensive care customers and recording their data in the central data system [2][4]. A wide range of products covering these applications is available in developed markets. But these products lacks in terms of reliability and washability for their multiple usages during a given specific lifetime. To make these applications useful and adoptable by the general public, they should be reliable and reusable with acceptable results in general household conditions [3].

Different studies have been published based on the washability of electronic textile components. *Dabby et al* [5] examined the washability of their developed wearable electronic sensors system for ECG and heart rate measurements. They claimed that all samples' results were still acceptable after 10 washing cycles and they were damaged after completing 25 washing cycles. They concluded in their research that severe stretching and twisting forces spoiled the samples during the wash process. *Chow et al* [6] embedded textile pressure sensors into smart fitness socks with the help of conductive threads. Then these socks were washed for 15 wash cycles. Wash results revealed that conductive threads increased their resistance significantly which reduced the performance of sensors in socks. *Molla et al* [7] washed these samples containing solder joints, LEDs and traces in a household washing machine. Change of their resistance and delamination at different points after 10 washing cycles were discussed. They concluded that the failure prospect was higher on solder joints near LED samples.

The current study is focused on wearable ECG electrodes to make them reliable and washable. Different types of embroidered electrodes are prepared and washed multiple times to check the washing impact on ECG signals. In our previous studies, the wash cycle in a normal household washing machine was analysed. These analyses were used to find possible available mechanical actions undergone during the washing process. Afterwards, these mechanical actions were performed separately on these embroidered ECG electrodes. The results of these mechanical tests were then used to predict the wash damage without washing of these samples. Hence these predictions, in the coming future, can be used to check, analyse and predict the washability of e-textile prototypes at different stages of manufacturing without actually washing the samples for a multiple number of wash cycles. These steps can be helpful for manufacturing organizations especially electronic companies that are working on parts manufacturing of e-textile products but don't have strong experience in textile fields.

2. Material and Methods

Embroidered electrodes, made by embroidery machines using conductive silver-coated polyamide yarn, were used in this experiment (Figure 1) [8]. Connection yarns are used to connect these electrodes with pressor snap buttons. These buttons were made of pure conductive metals and do not impact the output results. Connection yarns have been tested separately with different types of protection on them. This practice was performed to check connection yarn's damage and to reduce their destruction as much as possible during the wash process. For embroidered electrodes, Shieldtex 117/17-2 ply HCB yarn was used. The same yarn was used as connection threads.

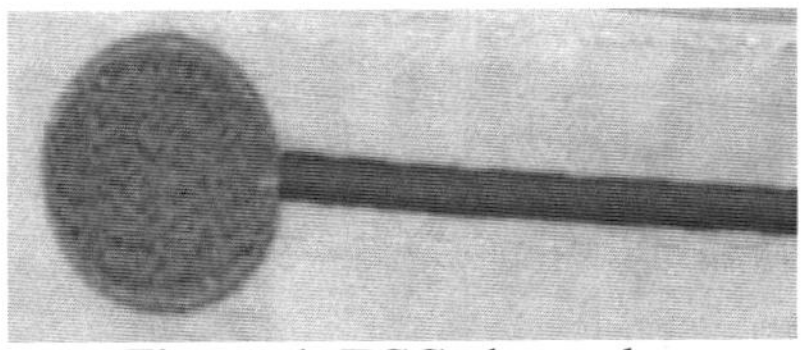

Figure 1. ECG electrodes.

Produced skin electrodes output was recorded in terms of ECG signals. ECG signals were recorded with the help of SHIELD-EKG-EMG (OLIMEX) processing board and ARDUINO software [4].

These signals were then processed on MATLAB. These signals were filtered on the Butterworth passband and notch filter. A notch filter was used to remove 50 Hz power line noises.

Linear resistance of connection threads was recorded in terms of an ohm per unit length and change in resistance was calculated. Agilent 34401A digital multi meter (Figure 2) was used to measure the linear resistance of these samples. The surface resistance of conductive electrodes was measured with the help of four probe surface resistance measurement devices. Ossila four probe surface resistance device was used for this experiment.

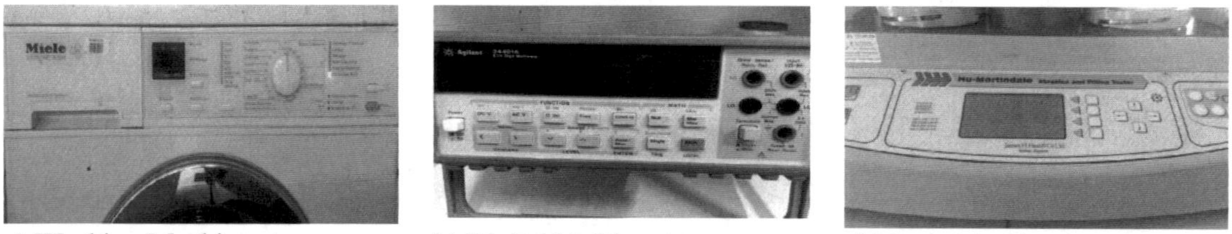

a) Washing Machine b) Digital Multi meter c) Martindale Abrasion tester

Figure 2. Testing equipment.

Front loading washing machine MIELE W3240 (Figure 2) was used for the washing of these samples. This is one of the most commonly used household washing machines. The commercial washing machine was preferred over laboratory washing machines because ultimately these e-textile products will be used in normal household machines by customers. Similarly, commercial detergent was preferred on standard detergent for better simulation with normal usage of these products. All washing conditions were maintained as described in ISO 6330 washing procedure standards, to obtain standard repetition of these results in the future. Mild washing process (silk) with 400 rpm tumbling speed was used in all experiments.

Martindale abrasion test was performed on a separate set of each sample. Change in resistance was recoded after abrasion cycles. A total of 10000 abrasion cycles were performed for each set of samples. Martindale abrasion tester, by "James H. Heals & Co Ltd" with 9 K.Pa load on samples, was used in these experiments (Figure 2).

3. Results and discussion

3.1. Washing

All the electrodes and connection yarns were washed up to 50 washing cycles. Changes in linear resistance from original resistance for connection threads were calculated (Figure 3). Connection yarn was washed in three different ways including connection yarns without any protection on it, connection yarns with TPU protection placed on it and connection yarn with simple yarn stitched above connection yarn forming a bridge to protect the inner material from a different type of damages during washing.

Figure 3 shows the results of each type of connection yarns after washing. Connection yarns protected with simple yarn showed the least damage as compared to the other two types. The ratio of change in resistance with original resistance remains near 1 after 40 number of wash cycles in case of simple yarn protection. Whereas, it goes higher in the other two types of connection yarns experimented here. Hence embroidered ECG electrodes were prepared to protect connection yarns with simple yarns on it. These electrodes were then washed up to 50 washing cycles and change in linear resistance was measured.

Figure 4 shows the results of ECG electrodes after 50 washing cycles. The ratio of change in resistance to original resistance was under two for 15 numbers of washing cycles and goes above 3 after 50 numbers of washing cycles.

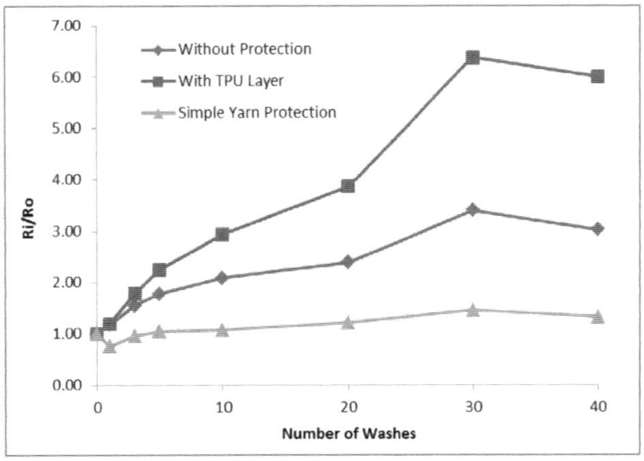

Figure 3. Change of resistance for connection yarns.

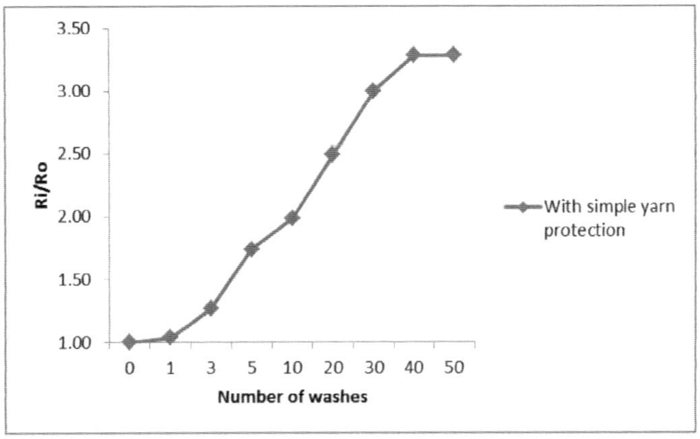

Figure 4. Change of resistance for embroidered ECG electrodes.

3.2. ECG Measurement

ECG signals were recorded before and after fifty washing cycles with the help of SHIELD- EKG-EMG (OLIMEX) device. Figure 5 shows the ECG plots before and after washing cycles. These curves are plotted with the help of MATLAB. The sampling frequency was 250 Hz. These signals were filtered on the Butterworth passband and notch filter. A notch filter was used to remove 50 Hz power line noises.

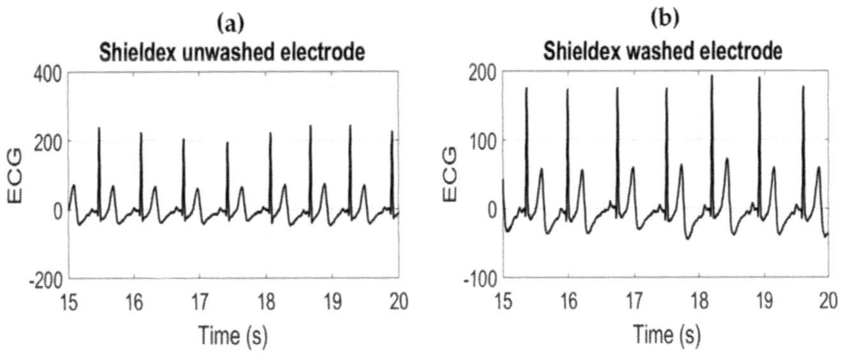

Figure 5. ECG of embroidered electrodes.

ECG plots are still under acceptable conditions after fifty wash cycles. Signal to noise ratio before washing was 39.41 dB and after 50 washes it was 21.27 dB. As a silver coating on conductive yarn surface was damaged during wash cycles, this removal increases the electrical resistance of conductive yarn. This phenomenon is also clear from linear electrical resistance measurement where the ratio of change in resistance to the original resistance is increased almost three times after fifty number of wash cycles.

3.3. Martindale Abrasion Test

A separate set of samples for ECG electrodes was used to perform Martindale abrasion tests. 10000 abrasion cycles were performed on four sets of samples and change in resistance was noted. Surface resistance was increased about two times after exercising 10000 abrasion cycles (Figure 6). Change of resistance is observed to be linear for these abrasion cycles experiment. Figure 7 explains the relationship between damages done by washing and abrasion cycles. We can co-relate these damages with each other and in future, this can be used to predict the damages that washing can create in e-textile prototypes without actually washing samples. In the current experiment, we can predict that damage occurred by 10 wash cycles is equivalent to damage that was observed by 10000 abrasion cycles. This is an initial prediction and it can be validated by further experiments and also by increasing the number of samples.

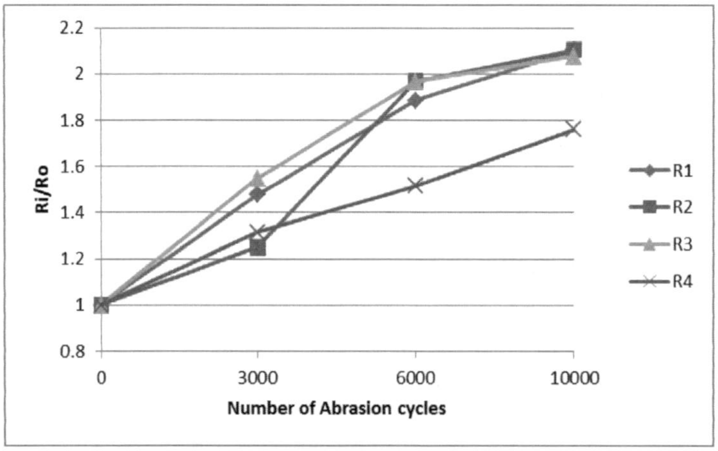

Figure 6. Martindale Abrasion test.

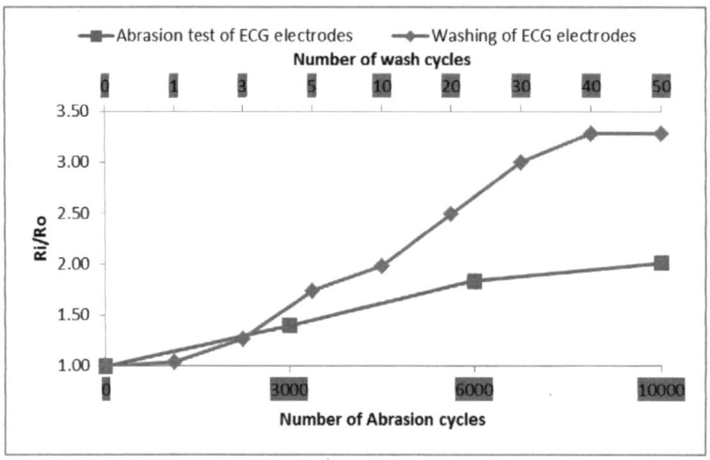

Figure 7. Comparison of washing and abrasion cycles.

4. Conclusion

This research was performed to enhance the use of flexible, easy to use and user-friendly skin electrodes for real time ECG measurement. But these products should be reliable and washable. These electrodes were washed up to 50 washing cycles to check the impact of washing on these samples. Equivalent mechanical stresses were performed on these samples to predict the resistance to wash damage without washing the samples in actual wash conditions. After these experiments, we can predict the relationship between damages by washing and abrasion tests. These mechanical tests can be used to prepare the e-textile prototypes by electronics experts without the actual washing samples to predict the lifespan and quality of outcomes from these samples.

In current consideration, we can predict that damage by 10 wash cycles is equivalent to damage occurred by approximately 10000 abrasion cycles. Although mechanical damage is not the only one that occurs during the wash process. There can be other damages such as chemical, temperature, and water itself that can affect the e-textile performance during the wash process, however the mechanical damage remains one of the most impacting forces in the wash process. All other forces are being investigated on electrodes and other e-textile prototypes and will be published separately.

References

[1] Tao X, Huang T H, Shen C L, Ko Y C, Jou G T and Koncar V 2018. Bluetooth Low Energy-Based Washable Wearable Activity Motion and Electrocardiogram Textronic Monitoring and Communicating System. *Advanced Materials Technologies*, **3**(10), p.1700309.

[2] Ankhili, A, Tao X, Cochrane C, Coulon D and Koncar V 2018. Washable and reliable textile electrodes embedded into underwear fabric for electrocardiography (ECG) monitoring. *Materials*, **11**(2), p.256.

[3] Zaman S U, Tao X, Cochrane C and Koncar V 2018, December. Market readiness of smart textile structures-reliability and washability. In *IOP Conference Series: Materials Science and Engineering* **459**(1) p. 012071 IOP Publishing.

[4] Ankhili A, Tao X, Cochrane C, Koncar V, Coulon D and Tarlet J M 2019. Ambulatory Evaluation of ECG Signals Obtained Using Washable Textile-Based Electrodes Made with Chemically Modified PEDOT: PSS. *Sensors*, **19**(2), p.416.

[5] Dabby N, Aleksov A, Lewallen E, Oster S, Fygenson R, Lathrop B, Bynum M, Samady M, Klein S and Girouard S 2017, September. A scalable process for manufacturing integrated, washable smart garments applied to heart rate monitoring. In *Proceedings of the 2017 ACM International Symposium on Wearable Computers* (pp. 38-41). ACM.

[6] Chow J H, Sitaraman S K, May C and May J 2018 Study of Wearables with Embedded Electronics Through Experiments and Simulations. In *2018 IEEE 68th Electronic Components and Technology Conference (ECTC)* (pp. 814-821) IEEE.

[7] Molla M T I, Compton C and Dunne L E 2018 October. Launderability of surface- insulated cut and sew E-textiles. *In Proceedings of the 2018 ACM International Symposium on Wearable Computers* (pp. 104-111) ACM.

[8] Ankhili A, Tao X, Cochrane C, Koncar V and Coulon D 2019. Washable embroidered textile electrodes for long-term electrocardiography monitoring. *In Textile Leather Revision* 2019

Theoretical model of green supply chain management: Support elements

Sana Elhidaoui[1,2], Khalid Benhida[1]
[1] Sciences d'ingénieur, LAPSSII, EST de Safi, Université Cadi Ayyad, Maroc
[2] Author to whom any correspondence should be addressed

sanaelhidaoui@gmail.com

Abstract. Nowadays, works on supply chain management continue to expand, which is associated with the speedy growth of technology on one hand, and the global issue of climate change on the other hand. All these factors have contributed to the emergence of a new field of research, namely the green supply chain management especially in the early 2000s. Through this paper, a brief literature review was conducted in order to summarize the published research work on green supply chain management, followed by a theoretical model that combines a set of tools and practices as support elements for the green supply chain management.

1. Introduction

Supply chain management (SCM) is among the most attractive topics for industrials and researchers, in the logistic field. However, new constraints are appearing, which incite to improve the existing methods, in addition to adapting them to traditional management, especially the case of the environment, where the company is requested to balance between respecting regulations in this sense, and the competitiveness, in terms of environment respect, and ensuring green products.

The paper is intended to provide management researchers with a general framework in green supply chain management (GSCM) research, and also a conceptual model for future research directions. Through this paper, we tried to draw up a general framework of green supply chain management (GSCM), we review current research works in green supply chain management (GSCM), and then provide a brief discussion of research opportunities and future research.

This article has been divided into five parts, the first section devoted to the introduction. The second section provides the state of the art of green supply chain management (GSCM), section 3 is devoted for material and methods, and section 4 to discuss results. And finally section 5 is devoted to the conclusion and some perspectives.

2. Green supply chain management

Supply chain is a network, linking several parties like suppliers, distributors, retailers, and manufacturing site[1], another definition: "a supply chain is defined as all the companies involved in the processes of component procurement, manufacturing, distribution and sale of the product, from the first supplier to the end customer"[2]. While green supply chain is considered as a traditional supply chain that involves environmental practices during all processes, such 'life cycle assessment of product' and 'environmental analysis' etc., it is also based on sustainable development. Moreover, the green supply chain aims to minimize and ideally eliminate the negative effects of the supply chain on the environment [3].

SCM is to manage the inputs of goods or services as well as a variety of activities, not only in a single department of an organization but in the different departments as well as outside the organization; for final users from procurement of raw materials to the end of the products' useful life [4].

GSCM can be regarded as one of the fastest expanding management areas. Several relevant reviews have been written over times that demonstrate various aspects of GSCM related research works. Today a set of tools and methods of GSCM are employed across all industrial companies, due to many factors, such competitiveness, technological development, and the appearance of new constraints as the climate change which involve companies in setting up an action plan to protect the environment. We define (GSCM) as a set of managerial practices including the following:

-Environmental axis (impacts, issues, regulation)

-Chanel relationship (attributes, suppliers, retailers, customer…)

-Coordination between each process

All these elements are further seen critical for the GSCM, including their interaction between the main activities of the supply chain. It is necessary to distinguish the difference between the two dimensions, sustainable and green, where sustainable SCM is an extension to the conventional concept of SCM by encompassing environmental, social, and ethical aspects [5].

3. Material and methods

As previously mentioned, this paper aims to give a general framework of GSCM, based on the collected research works. To achieve this purpose, initially, a comprehensive literature review is carried out; hence a literature review has proved to be a crucial step in structuring a field of research [6], followed by a qualitative approach where the theoretical model was limited to some practices and tools, as support element of the GSCM.

The selection of publications for this study was evaluated using the Web of Science and Scopus database, includes peer-reviewed journal articles and conference paper with managerial impact on the topic. For the period of search, it is chosen from 2000 to 2019. A set of search keywords is planned to effectively search for GSCM. Our methodology is summarized in figure 1 below:

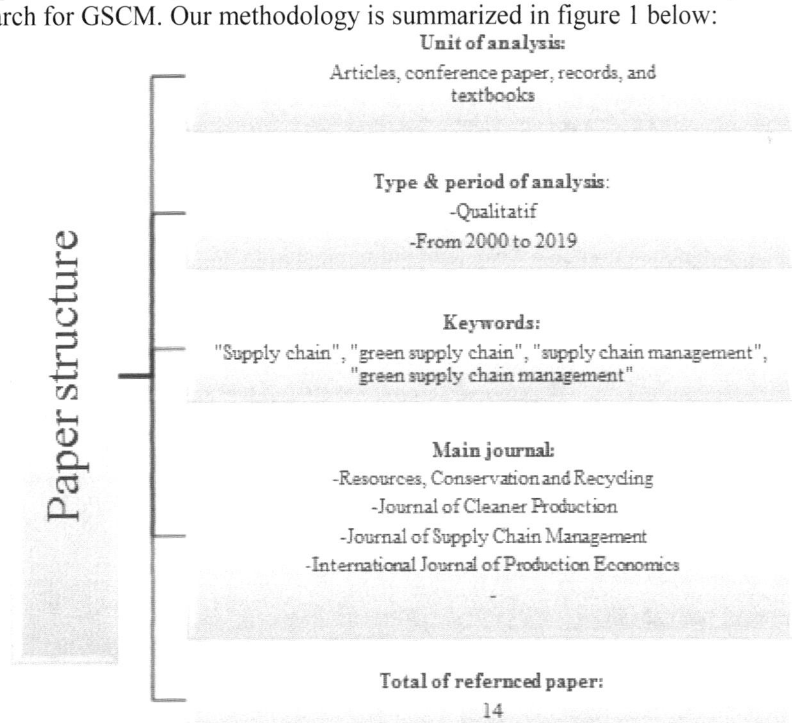

Figure 1. Structure of the research methodology.

Furthermore, we proposed, in the next section, a conceptual model that includes a set of actions and tools to be implemented, which combines both the SCM pillars and the support elements of a green supply chain.

4. Results and discussion

Several GSCM practices are proposed in the literature covering the majority of the functions of the supply chain [7] [8]. An efficient management of supply chains is essential for the smooth running of their activities and directly affects the global performance of the supply chain [9] [10]. Efficiency poses real challenges at the level of supply chain components on the one hand, and supply chain partners on the other hand. Supply chain members are requested to move to a unified system and coordinate with each other to meet these challenges [11]

Most of the environmental practices of companies mainly characterized by initiatives focused on vehicle use, on the reduction of CO2 emissions throughout the supply chain [12] [13], and also waste recovery or waste management [14]. To build a sustainable model of supply chain management, it stays primary to focus on the sustainability of the supply chain, and introduce the new tools into the management system. It appears that practicing continuous improvement and lean management are among the best methods that allows integrating easily the environmental practices into the SCM.

The proposed conceptual model of GSCM below covers a set of practices and tools, as support elements, to be integrated into the traditional SCM. Reverse logistics is among the primary keys of the management of the green supply chain, it allows reducing, managing and eliminating waste from products and inputs. Thanks to Life Cycle Assessment (LCA), a company can measure, improve and report on the environmental performance of a product throughout its life cycle, and the supply chain at all levels (design, transport production, etc.). Also, from now on, more companies are forced to adopt the principles of sustainable development in their activities in accordance with the regulations .

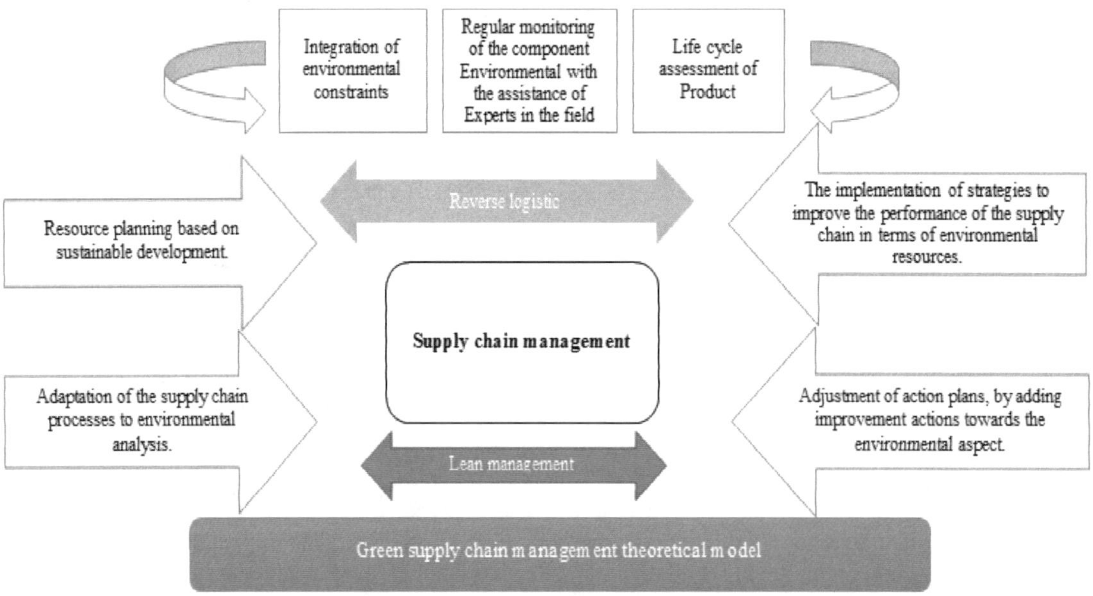

Figure 2. Theoretical model for the GSCM.

5. Conclusion and perspectives

This research work presented a theoretical model of GSCM. First of all, a brief literature review is proposed summarizing a body of research on SCM and GSCM. We also highlighted the difference between the two terms: sustainable and green.

Then we proposed through a general framework, the support elements of GSCM under the shape of a theoretical model. It is about a set of tools and practices which fall within the framework of environmental respect, as well as key elements that aim to enhance the competitiveness and performance of the supply chain. Despite the enormous amount of research work in this field, it is clear that research opportunities remain to be developed in the future:

- There is limited knowledge of the primordial practices that should be taken into account in GSCM.
- As research on the GSCM further expands, it will be increasingly important to deal with inconsistencies, in terms of the proposed models.
- It is becoming important to establish a global and standard model of GSCM that integrates all main elements that relate both greening and management tools.

References

[1] Afolayan A, White GR and Mason-Jones R 2016 Why knowledge acquisition is important to effective supply chain management: the role of supply chain managers 'as knowledge acquisitors' British Academy of Management 2016.

[2] Rota-Franz, C. Thierry and G. Bel 2001'Gestion des Flux dans les chaînes logistiques'. In Performances industrielles et gestion des flux (P. Burlat, J.P. Campagne) Hermès Traité IC2, 2001, pp 153-186.

[3] Andic, E., Yurt, O. and Baltacıoglu, T. 2015 'Green supply chains: efforts and potential applications for the Turkish market'. Resources, Conservation and Recycling, 2012, 58, 50e68.

[4] Eng, T.-Y., 2005 'The influence of a firm's cross-functional orientation on supply chain performance'. J of Supply Chain Management, 2005, **41** (4), pp.4-16.

[5] Wittstruck, D and Teuteberg F 2011 'Understanding the success factors of sustainable supply chain management: empirical evidence from the electrics and electronics industry *Corporate Social Responsibility and Environmental Management* **19** (3) pp.141-158.

[6] Easterby-Smith M, Thorpe R and Lowe A 2002 Management Research: an Introduction *Sage Publications*, London, 2002.

[7] Statler H and Kilger C 2000 *Supply Chain Management and Advanced Planning: concepts, models, software and case studies* Editions Springer Verlag.

[8] Tseng M L, Islam M S, Karia N, Fauzi F A and Afrin S 2019.A literature review on green supply chain management: Trends and future challenges. *Resources, Conservation and Recycling* **141** pp.145-162.

[9] Das D 2017 Development and validation of a scale for measuring Sustainable Supply Chain Management practices and performance *J of Cleaner Production* **164**, pp.1344-1362.

[10] Foo P Y, Lee V H, Tan G W H and Ooi K B 2018 A gateway to realising sustainability performance via green supply chain management practices: A PLS–ANN approach. *Expert Systems with Applications* **107** pp.1-14.

[11] Kanda A and Deshmukh S G 2008 Supply chain coordination:perspectives, empirical studies and research directions *Inter J of Production Economics* **115**(2) pp. 316- 35.

[12] Ahmed W and Sarkar B 2018 Impact of carbon emissions in a sustainable supply chain management for a second generation biofuel *J of Cleaner Production* **186** pp.807-820

[13] Shih-Chang Tseng and Shiu-Wan Hung 2014 A strategic decision-making model considering the social costs of carbon dioxide emissions for sustainable supply chain management *J of Environmental Management* **133** pp.315-22.

[14] Kucukvar M, Egilmez G and Tatari O 2016 Life cycle assessment and optimization-based decision analysis of construction waste recycling for a LEED-certified university Building *Sustainability* **8**(1) p. 89.

Integration of Lean management for the growth of Green manufacturing

K Jbira[1,4], M Hlyal[2] and J El Alami[3]

[1], LASTIMI Laboratory. University Med V, Rabat, Morocco

[2], ESITH Casablanca, Morocco

[3], LASTIMI Laboratory. University Med V. Graduate school of technology of sale, Morocco.

[4] Author to whom any correspondence should be addressed

jbira.kaoutar@gmail.com

Abstract. The transition to green products or operations has forced companies to find solutions to balance their activities, services and products. This has increased business challenges due to rising costs. The quest for a way to reduce costs has become a necessity for companies. In this article, we will shed light on several definitions given by different researches, and we will focus on the relationship between lean and green, see if we can combine them and discuss the relationship between them. We will also investigate the benefits for the industry, especially the manufacturing sector. Furthermore, we will answer how can the integration of lean management contribute to the growth of green manufacturing.

1. Introduction

Today, companies are confronted with many challenges that can impact their efficiency; poor environmental performance can negatively impact economic performance.

These latters have a significant negative effect on economic growth, human health and business performance. Among the main business challenges, we find that costs, mainly the cost of energy and resources are constantly increasing due to increased demand and it is practically impossible to predict supply price trends. succeed in producing with a wide range of energy and resource prices. We also find the costs spent to reduce environmental damage caused by pollution by increasing investments in pollution control, many of which sacrifice efficiency and increase operational costs.

2. Terminology

In order to prepare an understandable literature review, a terminology sequence will be used to facilitate the understanding of the article:

2-1. Green management:

Green management, or eco-management, refers to the management methods of an entity (company, department, etc.) aimed at considering the environmental impact of its activities, assessing this impact and reducing it. Environmental management is part of a sustainable development perspective, i.e. a sustainable development perspective. It strengthens product stewardship capacity by integrating stakeholders into the procurement process and considering environmental concerns, which improves environmental performance, integrates environmental protection techniques into the product life cycle. However, it requires specific investments in green practices, which are time-consuming and cost-effective.

2-2. Green manufacturing:

Green manufacturing has been created to describe any manufacturing that uses different green strategies and techniques, the objective is to become more eco-efficient by creating systems or products that consume less equipment and energy, or by changing old resources and equipment to reduce unwanted outputs or converting outputs into inputs (recycling)[1].

In general, green technology is the application of one or more of the environmental sciences, to monitor, model and conserve the environment and natural resources, and to reduce the negative effects of human participation. Researchers and managers are debating the importance of investing in green technologies and green transformation. *Heinkel R et al.* [2] argue that environmental and green testing should change an environmental strategy. This will lead to a win-win situation. Manufacturers could improve performance while achieving environmental objectives.

Below is a figure showing the losses most detected by green manufacturing:

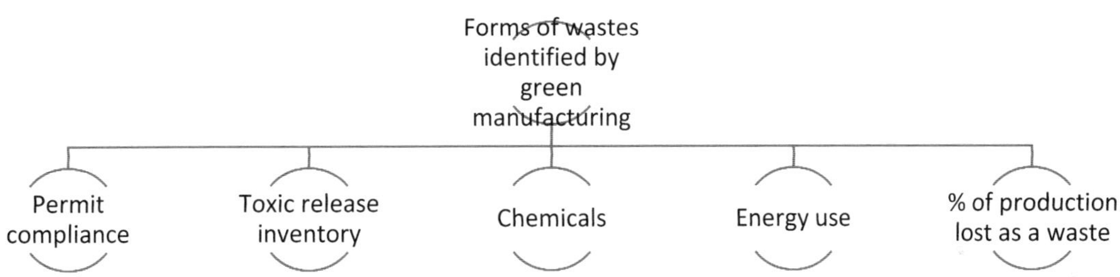

Figure 1. Some types of wastes identified by green manufacturing.

The addition of green to anything certainly reflects an ecological conscience; similarly, when it comes to manufacturing, we are talking about production that respects the environment. Moving towards green production is the result of the market and technological drivers.

2-3.Lean manufacturing:

Lean manufacturing has its origins within the Toyota Production System where it was integrated with just-in-time practices in order to improve quality and delivery time (*Krafcik1988*). And because of the company's high production rate, Toyota has adopted a production policy, as well as a new management system, which aims to reduce losses in all these aspects [3], Since then, lean manufacturing has gained notoriety in many industries around the world [4].

Today, lean manufacturing is seen as the most influential new paradigm in production industry. It leads to a competitive organization [5] by reducing wastes and time production, and improving productivity and quality at all stages of the supply chain [6]. Lean is also an operational management approach that aims to eliminate waste.

The five main elements of lean manufacturing are: Flow Manufacturing - Organization - Process Control-Logistics and Metrics.

These components represent the various facets necessary to support the activities of a strong lean manufacturing program, and it is the full deployment of these components that will propel a company on the path to becoming a world-class manufacturer.

The following is a basic definition of each of the five primary elements:

- Manufacturing flow: Volume produced per unit of time. All the steps taken by the product up to the final stage of its production.
- Organization: This allows the identification of the identity of people, their functions, roles, as well as training in new working and communication methods.
- Process control: The aspect oriented towards monitoring, control, stabilization and finding ways to improve the process.
- Logistics: The aspect that provides the definition of operating rules and organization and mechanisms for planning and controlling the flow of materials.
- Parameters: The visible, results-oriented performance aspect of measures, targeted improvements and team rewards/recognitions.

These main elements make it possible to cover all the challenges that can be encountered in setting up lean production. Each element focuses on an area of interest and compartmentalizes activities.

While each element is important for the deployment of a successful lean manufacturing program, energy comes from the integration of the following elements. For example, the manufacturing flow provides the basis for redesigning. In this sense, the Lean concept has not only been consistent with organizational objectives such as profitability and efficiency, but also with customer satisfaction, quality and responsiveness.

2-4.Lean management:

Lean management is a method that aims to improve the company's performance by developing all employees. The method makes it possible to find the ideal operating conditions by working together with personnel, equipment and sites in such a way as to add value with as little waste as possible[7].

Lean management (LM) aims to improve productivity and waste reduction, (LM) has several other impacts on the company.

- Demonstrate strong leadership and governance.
- Engage staff in daily improvement.
- Coaching.
- Make the measures more effective and show the results on a permanent basis.
- Integrate standard work into the company's culture.

The dual objective of Lean management is the complete satisfaction of the company's customers translated into turnover and the success of each employee translated into motivation and commitment.

3. Research method

The purpose of the present study is to identify and evaluate the current states of the art of research to see the points of convergence and divergence of lean and green, and how can lean integration influence on the growth of green manufacturing, then identify the different gaps in order to see if the result of integrating lean is always the in terms of the growth of a green policy.

4. Finding and discussion

In this paragraph we will mainly examine what the literature says about the subjects mentioned above and then discuss the data found.

4-1. Points of convergence and divergence of lean and green management:

During the last 10 years, many articles cited several times have dealt with the divergence and convergence between lean and green. Organizations were forced to meet several regulations for the control of environmental impacts [8], their environmental initiatives became a key factor in market differentiation. Lean management (LM) was adopted by firms to reduce waste and to achieve such improvements. Its practices are more likely to be adopted by large enterprises than small and medium-sized companies

LM focuses on eliminating waste within organizational processes [9], its implementation requires a customer–supplier relationship capable of encouraging a continuous learning and a high level of trust.

Green management was also adopted for the same reason, it aims to reducing environmental risks and negative impacts, while improving the eco efficiency of companies and theirs partners [10]. Both approaches aim to reduce waste, through the adoption of preventive initiatives [11]. The integration of both Lean management and green management results in an innovative approach called lean green.

In both approaches, suppliers are no longer treated only as entities that must comply strictly with customers' requirements; they are essential to increase value-added to goods and services through a proactive relationship [12]. Both concepts provide approaches to manage organizations and their supply chains with the aim of improving organisational performance. The literature confirms the convergence of both approaches, lean & green, waste reduction techniques.

The two approaches do not always go in the same direction, sometimes they diverge. according to the literature, there is still a gap regarding the assessment of the level of integration between both approaches. The researchers found that some successful Lean methods should be evaluated because they are not "eco-friendly" [13] and that only judoka and Frap had positive environmental effect.

4-2. The influence of lean integration on the growth of green, or in general on environmental performance:

The Green Lean research concludes that the use of Lean methods seems to have a positive influence on environmental performance[13], These methods include 5S, Cellular Manufacturing, Lean Supply Chain Management, Total Productive Maintenance, and Value Stream Mapping. The integration of lean and green concepts is promising for addressing the sustainability performance or triple bottom line: economic, environmental and social dimensions. Lean manufacturing practices (LMPs) have been a subject of interest for organisational and operational completeness for decades [12]. Key principles of LMPs include the identification and elimination of all non-value-added activities, or waste, and involve employees in efforts toward continuous improvement [14].

Research on lean manufacturing suggests that while LMPs do not necessarily incorporate environmental responsibility, such practices can contribute to mitigate some of these environmental impacts because of their intrinsic focus on waste elimination [15].

LMPs may not have a direct intent to reduce environmental impact, their implementation had improved energy efficiency, reduced waste and emissions, and reduced inventory waste [16]. Several researchers have investigated the link between LMPs, operational and environmental performance [11]. The literature has reported mixed findings for the green and lean linkage; LMPs are positively associated with environmental performance.

King and Lenox found it evident that lean manufacturing, as measured by ISO 9000 adoption and low chemical inventories, is complementary to waste reduction and pollution reduction [17].

Other researchers postulate that LMPs show similar positive effects on both operational and environmental performance [18]. It has been argued that LMPs lead to environmental efficiency due to the core principle of zero waste [19]

4-3.Different gaps in the relation between lean and green:
As already mentioned, lean and green do not always converge towards the same results, in many cases, the objectives of lean management are not in line with those of green, nor with respect for the environment. Some Lean Management Practices do not positively relate to environmental performance [11]; *Rothenberg, Pill, and Maxwell* found that lean management and reduction of air emissions of volatile organic compounds (VOCs) are negatively associated [18]. For example; the objectives of just in time (JIT) are to produce only the products that are needed, when needed and in the amount needed. It aims to reduce cycle times and eliminate unnecessary delays in processing time [20], [21].

When these objectives are achievable, JIT does not appear to respect any environmental principles. The JIT makes it possible to ensure a fluidity of flows, but also a significant emission of gases, especially CO_2, caused by the increase in the frequency of transport [22]. This will consequently reduce environmental performance.

This means that firms may adopt Lean management practices that are valuable for improving operational results without thinking too much about respecting the environment, so automatically without having to be green [15].

5. Conclusion

Lean management goes in most cases in the same context of green, i.e. in the respect of the environment. But there are cases when being green becomes an option that does not go with the company's lean objectives. Several questions arise, which may be the subject of future studies:
- What are the lean unfriendly environment approaches?
- How can they be replaced by green approaches?

Reference

[1] Deif A M 2011 A system model for green manufacturing *Journal of Cleaner Production* **19**(14) p. 1553-9.
[2] Heinkel R, Kraus A and Zechner J 2001 The Effect of Green Investment on Corporate Behavior *The Journal of Financial and Quantitative Analysis* **36**(4) p. 431.
[3] Herron C and Hicks C 2008 The transfer of selected lean manufacturing techniques from Japanese automotive manufacturing into general manufacturing (UK) through change agents *Robotics and Computer-Integrated Manufacturing* **24**(4) p. 524-31.
[4] Garza-Reyes J A 2015 Lean and green – a systematic review of the state of the art literature *Journal of Cleaner Production* **102** p. 18-29.
[5] Hines P Holweg M and Rich N 2004 Learning to evolve: A review of contemporary lean thinking *Int Jrnl of Op & Prod Mnagemnt* **24**(10) p. 994-1011.
[6] Wahab A N A, Mukhtar M and Sulaiman R 2013 A Conceptual Model of Lean Manufacturing Dimensions *Procedia Technology* **11** p. 1292-8.
[7] Kiran D R 2017 Lean Management in *Total Quality Management*, Elsevier p. 363-372.
[8] Melnyk S A Sroufe R P and Calantone R 2003 Assessing the impact of environmental management systems on corporate and environmental performance *Journal of Operations Management* **21**(3) p. 329-51.
[9] Droste A 2007 Lean thinking, banish waste and create wealth in your corporation *Action Learning: Research and Practice* **4**(1) p. 105-106.
[10] Campos L M S 2012 Environmental management systems (EMS) for small companies: a study in Southern Brazil *Journal of Cleaner Production* **32** p. 141–8.

[11] Martínez-Jurado P J and Moyano-Fuentes J 2014 Lean Management, Supply Chain Management and Sustainability: A Literature Review *Journal of Cleaner Production* **85** p. 134-150

[12] Dües C M,Tan K H and Lim M 2013 Green as the new Lean: how to use Lean practices as a catalyst to greening your supply chain *Journal of Cleaner Production* **40** p. 93-100.

[13] Barth H and Melin M A 2018 Green Lean approach to global competition and climate change in the agricultural sector – A Swedish case study *Journal of Cleaner Production* **204** p. 183-192.

[14] Anand G and Kodali R 2008 Selection of lean manufacturing systems using the PROMETHEE *Jnl of Modelling in Management* **3**(1) p. 40-70.

[15] Bai C, Satir A and Sarkis J 2019 Investing in lean manufacturing practices: an environmental and operational perspective *International Journal of Production Research* **57**(4), p. 1037-51

[16] Zhu Q, Johnson S and Sarkis J 2018 Lean six sigma and environmental sustainability: a hospital perspective *Supply Chain Forum: An International Journal* **19**(1) p. 25-41

[17] King A A and Lenox M J 2001 Does It Really Pay to Be Green? An Empirical Study of Firm Environmental and Financial Performance: An Empirical Study of Firm Environmental and Financial Performance *Journal of Industrial Ecology* **5**(1) p. 105-16.

[18] Larson T and Greenwood R 2004 Perfect complements: Synergies between lean production and eco-sustainability initiatives *Environ. Qual. Manage.* **13**(4) p. 27-36.

[19] Rothenberg S, Pil F K and Maxwell J 2009 Lean, green, and the quest for superior environmental performance *Production and Operations Management* **10**(3 p. 228-43.

[20] Sugimori Y, Kusunoki K, Cho F and Uchikawa S 1977 Toyota production system and Kanban system Materialization of just-in-time and respect-for-human system *International Journal of Production Research* **15**(6) p. 553-564.

[21] Dowlatshahi S and Taham F 2009 The development of a conceptual framework for Just-In-Time implementation in SMEs *Production Planning & Control* **20**(7) p. 611-21.

[22] Wu P and Low S P 2012 Lean Management and Low Carbon Emissions in Precast Concrete Factories in Singapore *J. Archit. Eng.* **18**(2) p. 176-86.

IOP Conf. Series: Materials Science and Engineering **827** (2020) 012028 doi:10.1088/1757-899X/827/1/012028

Effect of water immersion, laundering, and abrasion on the conductivity of reduced graphene oxide coatings on aramid fabrics

Chungyeon Cho[1], Anastasia L Elias[1], Jane Batcheller[2], Hyun-Joong Chung[1] and Patricia I Dolez[2,3]

[1] Department of Chemical and Materials Engineering, University of Alberta, Edmonton, AB, T6G 1H9, Canada
[2] Department of Human Ecology, University of Alberta, Edmonton, AB, T6G 2N1, Canada
[3] Author to whom any correspondence should be addressed

pdolez@ualberta.ca

Abstract. Opportunities for developing end-of-life sensors for fire resistant fabrics are explored using reduced graphene oxide coatings on textiles. Fire resistant fabrics are known to experience significant losses in performance over time. Large reductions in mechanical properties have also been recorded when these fabrics were subjected to accelerated aging conditions simulating the use in service. In addition, the fabric loss in performance may exceed the safety requirement threshold before any sign of damage is visible to the naked eye. Electrically conductive coatings and tracks were prepared on an m-aramid woven fabric using graphene oxide that was further reduced. The preparation technique allowed wrapping the individual aramid fibers with rGO sheets. No significant change in sheet resistance was recorded after up to 120h of immersion of the rGO-coated fabric specimens in water. An increase in resistance after 10 accelerated washing cycles was measured on the rGO-coated specimens prepared with 5 coating cycles while no significant effect was detected for specimens prepared with 10 and 15 coating cycles. Under abrasion exposure, the electrical resistance of rGO tracks increased gradually until 150 cycles, after which the conductivity dropped abruptly. These results show the potential of reduced graphene oxide applied as a coating on m-aramid fabrics to prepare end-of-life sensors for fire resistant fabrics.

1. Introduction

Graphene is well-suited to flexible electronic applications. The oxygen-containing functional groups of graphene oxide have allowed good bonding to be achieved for coatings prepared on fabric materials of natural (cotton, wool) and synthetic origin (nylon, polyester, para-aramid) [1]. After reduction of the graphene oxide, a reasonable electrical conductivity can be obtained. The main application of these coatings is strain gauges and heaters.

Opportunities for developing end-of-life sensors for fire resistant fabrics are explored using reduced graphene oxide coatings on textiles. Fire resistant fabrics, which are made with high performance fibers such as para- and meta-aramids, polybenzimidazole, and polybenzoxazole [2], are known to experience significant losses in performance over time, for example when used in firefighter protective garments [3]. Large reductions in mechanical properties have also been recorded when these fabrics were subjected to accelerated aging conditions simulating the use in service, for instance thermal [4], UV [5], moisture [5], and exposure to a combination of abrasion and laundering cycles [6]. In addition, the fabric loss in performance may exceed the safety requirement threshold before any sign of damage is visible to the naked eye [7].

This study investigates the production of conductive coating and tracks on aramid fabrics using graphene oxide and their stability upon exposure to water immersion, laundering and abrasion.

2. Material and Methods

Reduced graphene oxide (rGO) coatings were prepared on a meta-aramid woven fabric (2/1 twill, 222 g/m^2) using a technique described in [8]. Fabric specimens were first dipped in a graphene oxide solution (4 wt%) for 15 min and dried at 65°C for 30 min. Then, the graphene oxide coating was reduced by immersion in L-ascorbic acid at 90°C for 4 hours, rinsed in DI water and 2-propanol, and dried at 80°C for 30 min. This dip-dry-reduce process was repeated for a number of times (up to 15 times). Between each coating cycle, the rGO flakes not firmly attached to the fabric substrate were removed with adhesive tape.

Rectangular rGO tracks were also prepared on the fabric surface by covering the backside of the fabric with a wax film (figure 1). The specimen was then sandwiched between two acrylic molds, a solid one located on the specimen face covered with the wax film and the other one with cutout tracks. The GO solution was deposited on the patterned specimen areas using a pipette. The rest of the process was the same as when coating the entire fabric.

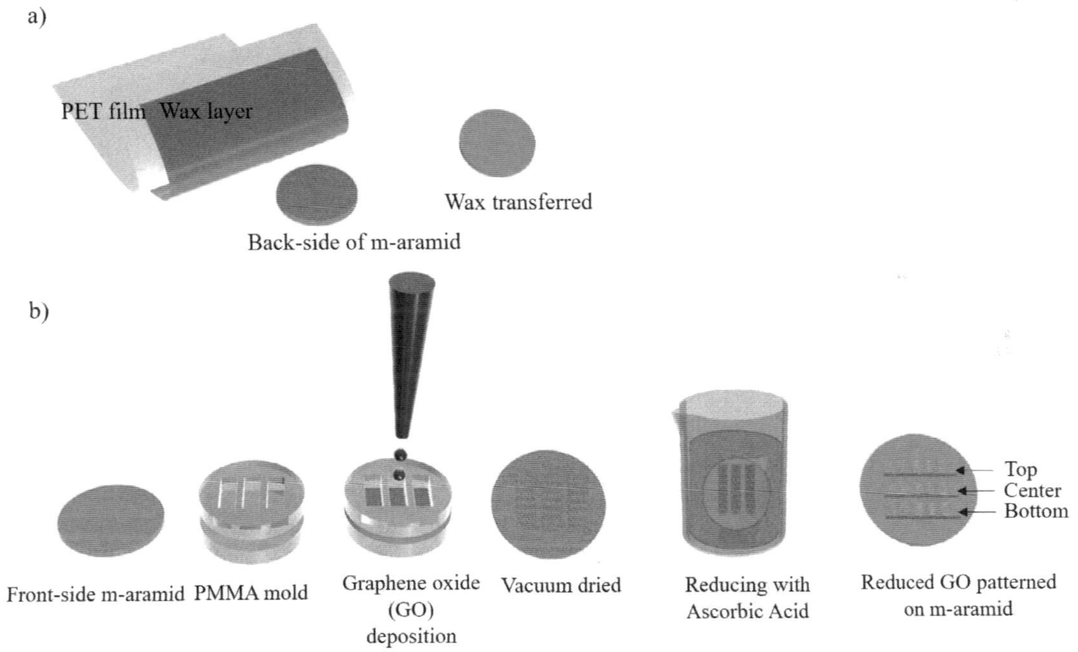

Figure 1. Preparation of rectangular rGO tracks on fabric samples a) wax layer deposition, b) patterning and reduction process [8].

The durability of the rGO coating on the m-aramid fabric was assessed by immersing specimens in DI water at room temperature for 5 successive 24h periods, for a total of 120h. Between each immersion, the specimens were dried at 65°C for 30 min and their sheet resistance measured with a 4-point probe. The washing fastness was assessed using a Launder-Ometer according to the ISO 105-C06 test method (test procedure A1M). After being subjected to a defined number of accelerated washing cycles (up to 10), the specimens were rinsed with DI water and dried at 65°C for 15 min before their sheet resistance was measured. For the determination of the abrasion resistance, fabric specimens with three 1.5cm wide rGO tracks were abraded for a defined number of cycles with a Martindale Abrasion Tester using the standard wool abradant fabric at 47.5 rpm according to ASTM D4966. The residual electrical resistance of the conductive tracks was monitored. All measurements were performed in triplicate. Complementary imaging was carried out by field emission scanning electron microscopy (FE-SEM) and helium ion microscopy (HIM).

3. Results and discussion

3.1. Initial condition

In an unaged condition, the rGO coating on the m-aramid fabric exhibited a sheet resistance of about 860 kΩ/□ after 5 dip-dry-reduce cycles with an add-on ratio of 6.8% [8]. After 10 cycles, the sheet resistance decreased to 310 kΩ/□ and the add-on value increased to 8% (figure 2). They reached about 225 kΩ/□ and 9.2% after 15 cycles. Figure 3 illustrates that each individual aramid fiber is wrapped by rGO sheets. This is attributed to the removal of the loose rGO flakes between each dip-dry-reduce cycle. It is essential to ensure that the bending of the fabric during normal use does not lead to a premature decrease in the electrical conductivity of the coating.

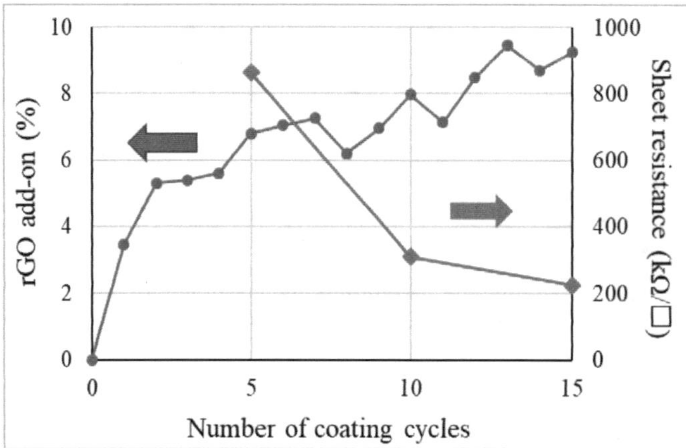

Figure 2. rGO add-on (left axis) and sheet resistance (right axis) as a function of the number of coating cycles.

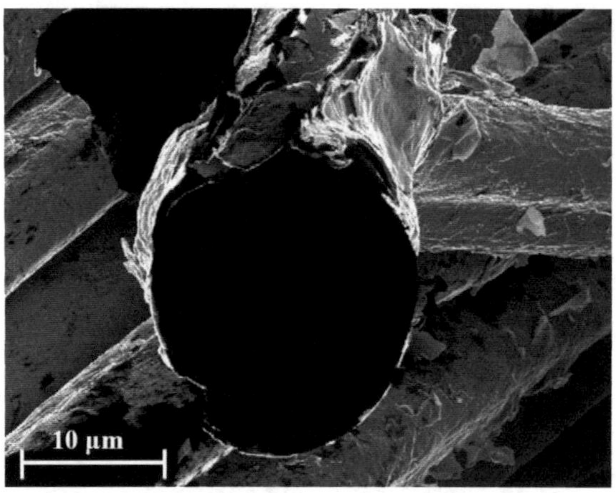

Figure 3. HIM image of an rGO-coated m-aramid fiber [8].

3.2. Effect of water immersion

The effect of water immersion was studied with specimens prepared with 15 coating cycles. No statistically significant difference in sheet resistance (p-value of 0.73) was obtained between the 5 successive 24h periods of water immersion. This could be attributed to the scaly nature of the rGO coating around each fiber, with free spaces being able to accommodate the potential slight increase in the aramid fiber volume due to moisture absorption. In addition, the individual graphene sheets may have some ability to slide relative to each other.

3.3. Resistance to laundering

When exposed to 10 successive washing cycles (corresponding to 50 domestic launderings), rGO-coated fabric specimens prepared with 5 coating cycles showed an increase of 50% in the sheet resistance (figure 4). This was attributed to damages suffered by the rGO coating as a result of the washing cycles; FE-SEM inspection of the surface revealed cracks in the rGO coating and partially delamination of the rGO sheets. On the other hand, no statistically significant change in the sheet resistance with the number of washing cycles was obtained when the specimens were prepared with 10 (p-value of 0.16) and 15 (p-value of 0.29) coating cycles. The coatings did not show any signs of cracking or delamination either.

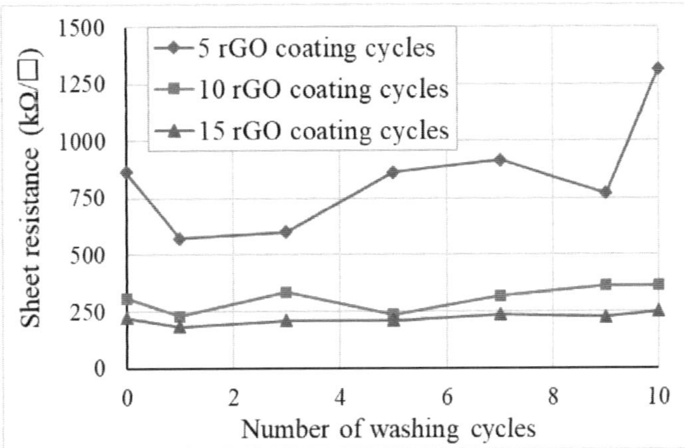

Figure 4. Sheet resistance of the rGO-coated fabric as a function of the washing cycles.

3.4. Effect of abrasion

The rGO coating behavior under abrasion is illustrated in figure 5. The three series of data correspond to the three tracks formed on each fabric specimen (prepared with 15 coating cycles). The electrical resistance of the tracks increased gradually with the number of abrasion cycles. Above 50 abrasion cycles, the track situated in the center of the specimen experienced a larger decrease in conductivity, which was attributed to an imperfectly distributed stress applied across the specimen width by the foam padding inserted between the fabric specimen and the sample holder. Figure 5 also shows a dramatic increase in resistance for all three tracks above 150 abrasion cycles. At that point, the ability of the rGO conductive tracks to conduct electricity was completely lost.

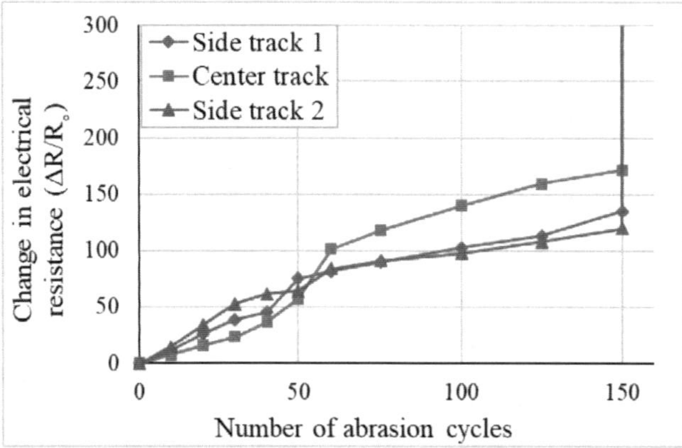

Figure 5. Electrical resistance of the rGO tracks as a function of the number of abrasion cycles.

4. Conclusion

Electrically conductive coatings and tracks were prepared on an m-aramid woven fabric using graphene oxide that was further reduced. The preparation technique allowed wrapping the individual aramid fibers with rGO sheets. No significant change in sheet resistance was recorded after up to 120h of immersion of the rGO-coated fabric specimens in water. An increase in resistance after 10 accelerated washing cycles was measured on the rGO-coated specimens prepared with 5 coating cycles while no significant effect was detected for specimens prepared with 10 and 15 coating cycles. Under abrasion exposure, the electrical resistance of rGO tracks increased gradually until 150 cycles, after which the conductivity dropped abruptly. These results show the potential of reduced graphene oxide applied as a coating on m-aramid fabrics to prepare end-of-life sensors for fire resistant fabrics.

Acknowledgments

The authors would like to thank S. A. Manzoor Bukhari for his help with the HIM imaging. This work has benefitted from the financial support of NSERC [Strategic Project Grant STPGP 521866].

References

[1] Molina J 2016 Graphene-based fabrics and their applications: a review *RSC Adv.* **6**, pp. 68261-91

[2] Bourbigot S and Flambard X 2002 Heat resistance and flammability of high performance fibres: A review *Fire Mater.* **26**, 155-68

[3] McQuerry M, Klausing S, Cotterill D and Easter E 2015 A Post-use Evaluation of Turnout Gear Using NFPA 1971 Standard on Protective Ensembles for Structural Fire Fighting and NFPA 1851 on Selection, Care and Maintenance *Fire Technol.* **51**, 1149-66

[4] Dolez P I, Tomer N S and Malajati Y 2019 A quantitative method to compare the effect of thermal aging on the mechanical performance of fire protective fabrics *J. Appl. Polym. Sci.* **135**, 47045 (15p)

[5] Arrieta C, David E, Dolez P and Vu-Khanh T 2011 Hydrolytic and photochemical aging studies of a Kevlar-PBI blend *Polym. Degrad. Stab.* **96** (8), 1411-1419

[6] Vanderschaaf C J, Batcheller J C and Torvi D A 2015 Combined Effects of Laundering and Abrasion on the Protective Performance of Flame Resistant Fabrics *Proc. Spring Technical Meeting - Combustion Institute – Canadian Section (Saskatoon, Canada)* 6p

[7] Rossi R M, Bolli W and Stampfli R 2008 Performance of firefighters' protective clothing after heat exposure *Int. J. Occup. Saf. Ergon.* **14** (1) 55-60

[8] Cho C, Elias A L, Batcheller J, Dolez P and Chung H.-J 2019 Electrical conduction of reduced graphene oxide coated meta-aramid textile and its evolution under aging conditions *J. Ind. Text.* DOI: 10.1177/1528083719869387 18p

IOP Conf. Series: Materials Science and Engineering **827** (2020) 012029 doi:10.1088/1757-899X/827/1/012029

The investigation of crimp and spin-finish effects on the processing behavior of polyaramide fibers during the spinning process

I Boughamoura[1,2,3], **A Sinoimeri**[1,2]

[1]Laboratory of Physics and Mechanics of Textiles (LPMT), University of Upper Alsace (UHA), Mulhouse, France
[2] Université de Strasbourg, France
[3] Author to whom any correspondence should be addressed

imen.boughamoura@uha.fr

Abstract. The spinning process optimization has been the subject of many articles and publications. They mainly study cohesion and friction forces presented as the most influent mechanical phenomena on fiber workability. However, they were limited to natural fibers studies and more specifically cotton. Therefore, and to enlarge these studies, this article focuses on the processing behavior of fire-resistant fibers of polyaramide type. Besides, inner fiber characteristics have a significant effect on their mechanical behavior. Consequently, the object of this article is to investigate the effect of fiber material and its characteristics, mainly spin-finish and crimp, on mechanical processing behavior, which includes fiber cohesion, and frictional resistance.

1. Introduction

Frictional and cohesive forces have been identified as the most influencing forces during the spinning process. They are the main mechanical phenomena affecting not only the homogeneity of the yarn but also its mechanical properties [1]. Add to that, the crimp is a parameter that affects both the opening of the floss and its workability. It is therefore important to determine the optimal crimp area for the workability of the polyaramide fiber. To that end, it is necessary to be sure about the measurement of the crimp's degree.

This article focuses on two different batches of tests. The first one concerns the crimp measurement, and the second one the inter-fiber friction and cohesion analysis. The main aim of this study is to determine the optimum levels of crimp, cohesion and friction forces for better processing of polyaramide fibers.

Cohesion is defined as the force that is opposed to the relative movement between two distinct bodies in contact when no external normal forces are applied. On the other hand, external normal force, when applied between two bodies in contact, gives rise to some additional forces, called frictional ones, which equally are opposed to the relative motion of these bodies [2]. Generally, two types of friction forces have to be considered: static frictional force, just before the movement takes place, and dynamic one, which is generated during the movement.

When studying inter-fiber friction and cohesion forces, several parameters need to be considered, namely:

- The intrinsic characteristics of the fiber, including the nature and extent of the contact surfaces,

- The treatments applied to the fiber,
- The fibers arrangement in the tested sample. This parameter is not treated in the study, as the fibers are considered to be mostly aligned (sliver samples).

Friction is directly related to the surface properties and characteristics of the fiber. *Gupta and El Mogahzy* [3] divided the characteristics of the fiber into two categories: parameters related to the morphology of the surface of contact (surface nature, the presence of convolution, crimp ...) and those related to the surface junction's properties depending on the chemical and physical structures of the fiber. It has been proved that the irregularity of the yarn increases with the friction force [4]. The main role of the latter is to limit the slippage between fibers by ensuring uniform linear distribution. This study [4] shows a correlation between the friction force and the tenacity of the yarn (between -0.56 and -0.74) which leads to thinking that the resistance of the yarn increases by decreasing the friction force. This can be explained by the possibility of mechanically damaging the fiber with high frictional force.

Unlike frictional forces, the cohesive ones increase the yarn homogeneity thus enhances its mechanical properties [4]. Cohesion is beneficial to fiber processing as it prevents the fiber flow's breakage. As far as friction is concerned, very minimal values will not ensure a sufficient tenacity, whereas higher values give rise to fiber package slippery and irregularities.

2. Material and Methods

Different types of polyaramide fibers have been analyzed which include Conex, Kermel, Nomex, Yantai, etc... The fiber fineness is 1.7 or 2.2 dtex. The fibers were selected having a different level of crimp and spin-finish percentage, in the range of 10-26 and 0.32-1% respectively.

2.1. Crimp test

The measurement method is based on image analysis using laboratory software: Visiolog 6.9. For each sample 3 parameters have been measured: the segment length, the range and the frequency (the number of patterns/cm), as shown in Figure 1.

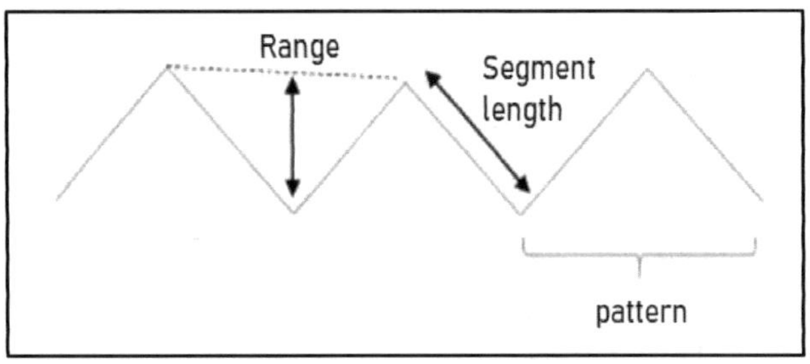

Figure 1. Crimp's parameters.

The theoretical frequency of the pattern was calculated using the following geometrical considerations made by segment length and range (Pythagoras theorem):

$$f = \frac{1}{2 * \sqrt{s^2 - r^2}} * 10 \quad (1)$$

Where:
f is the theoretical frequency (pattern/cm),
s is the segment length (mm),
r is the range (mm).

2.2. Friction and cohesion tests

17 samples were prepared using a laboratory sized carding machine model MDTA developed by Uster Technologies. It provides a sliver of 90 cm of length for 5 g of floss. The sliver was afterward divided

into smaller samples of 15 to 20 cm of length each. Every sample had a linear density of approximately 5.56 ktex.

The device used for this study (Figure2) is a static friction tester (SFT). According to *Nowrouzieh* [5], it is composed of "two identical carriages. One of them (left) was fixed, whereas the second (right) was moving through a linear guide. A piece of sliver was put down in the channel of the two carriages, which were initially in zero displacement position i.e. the two carriages were in contact. The sliver was compressed with the upper carriage sides, where two identical weights were loaded. The moving one was tracked at a constant speed, whereas the fixed one was attached, by the intermediate of a force sensor, to the frame. The distance between the two carriages was measured by the displacement sensor". The sensor 100N has been used with a speed of 0.350 mm/s [5].

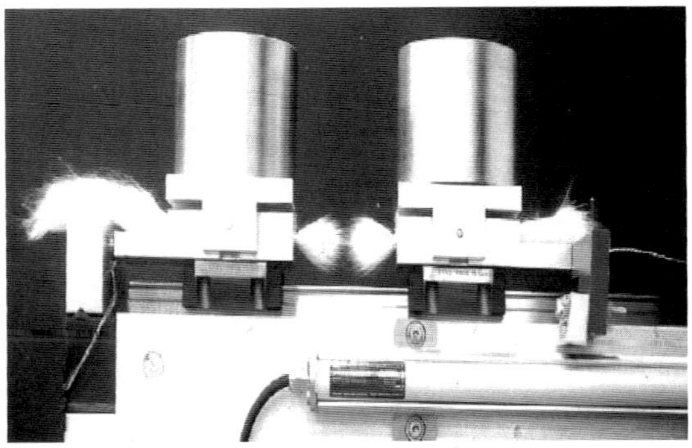

Figure 2. SFT device.

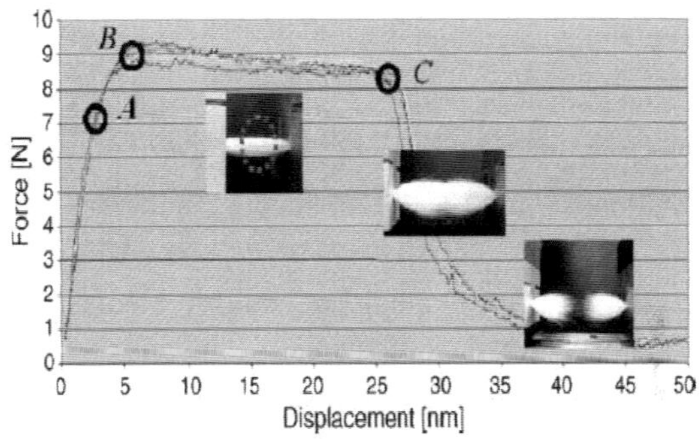

Figure 3. The usual friction curve [2].

Preliminary tests have shown that this speed, initially intended for cotton fibers, was also adequate for polyaramide fibers. The test finishes at a 90% break rate. The normal loads applied on the samples vary from 1 to 5 kg. Six tests have been carried for each load constituting 30 tests per sample.

Figure 3 shows the usual friction curve of a tested sample: In the first section AB the fibers are parallelized (elimination of the crimp) and elongated. The point B represents the friction force: it is the maximum force the sliver withstands at the beginning of the slip/slide of fibers with each other inside sliver and is a function of the total number of fibers. BC represents a stage during which the force slightly decreases. However, a small increase in strength is often noticed. This may be due to the overlapping of other forces: cohesion (for the fibers between the clamps and held by only one side), friction (for the fibers between the clamps and held only by the neighboring fibers) and the fiber sliding. Beyond the C point, sliver breakage takes place rapidly.

The results obtained from the SFT/ MTS device give rise to the following equation:

$$\frac{F}{Nf} = K\,\frac{W}{Nf} + C \tag{2}$$

Where:

F is the frictional resistance force (cN) at point C (Figure 3),
W is the normal load applied to the fibers (cN),
Nf is the number of fiber in the sliver cross-section,
K is the friction coefficient,
C is the cohesion strength (cN).
PS: The friction strength for a normal load of 0 Kg is equivalent to a cohesion strength. This criterion allows deducing the cohesion strength from the friction test.

3. Results and discussion

3.1. Crimp test

The following table (Table 1) summarizes the results for all fiber types.

IOP Conf. Series: Materials Science and Engineering **827** (2020) 012029 doi:10.1088/1757-899X/827/1/012029

Table 1. Crimp test validation results.

Sample	Measured crimp			Theoretical crimp		
	Average	*Standard deviation*	*CI*	*Average*	*Standard deviation*	*CI*
1 to 10	17.4	16.17	4.6	16.3	21.07	6.0
11 to 13	26	2.72	0.8	24.8	6.84	1.9
14 to 17	10.2	22.61	6.4	12.4	8.98	2.6

A clear difference was observed between the crimp of the tested samples. One represents a higher crimp of the order of 26 while the other one does not exceed 10 (rounded). Samples with satisfying spinning behaviour show less crimp.

The statistic studies showed a high difference between the tested samples. This may be due to the number of used samples.

3.2. Friction and cohesion test

Figure 4 describes the curve frictional behavior of polyaramide fibers:

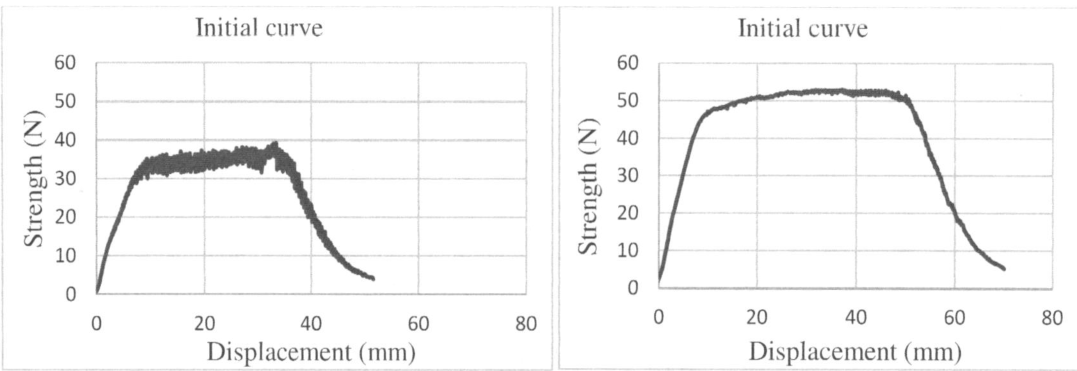

Figure 4. Friction curve: With and without spin-finish.

The stick-slip phenomenon (left) was observed during tests. It is particularly harmful to the spinning process. It affects the proper functioning of the production line (the speed of the machines) and the homogeneity of the produced material, and therefore the quality of the yarn.

As a first hypothesis, this phenomenon may be due to the spin-finish. Thus, it was necessary to test spin-finish-less fibers. No stick-slip was observed on this curve. Moreover, the friction parameters and the cohesive forces were calculated with and without spin-finish. All samples are of the same fiber and using the same spin-finish. The results are shown in Figure 5.

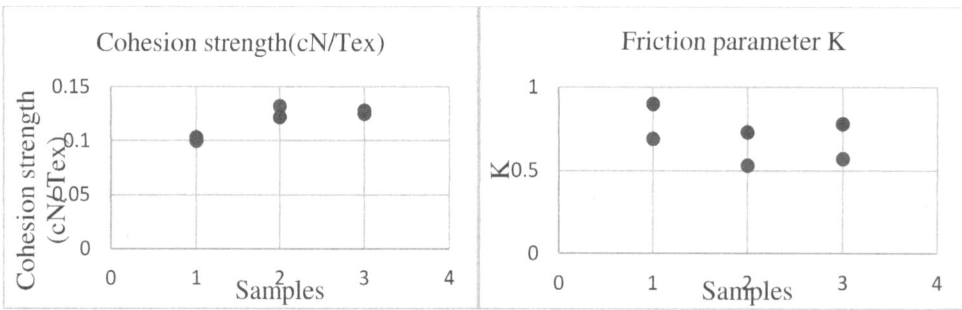

Figure 5. Spin-finish effect on cohesion and friction strength (blue for fibers without spin-finish, orange for fibers with spin-finish).

The spin-finish affects more friction force compared to the cohesive ones and decreases it by approximately 10 N. This result is observed for all sample types.
The following table summarizes the found results:

Table 2. Cohesion and friction test results.

Sample code	Crimp (1/cm)	Spinfinish rate (%)	Cohesion strength (cN/Tex)	K
1	14	0.68	0.517	0.75
2	14	0.76	0.425	0.82
3	16	0.32	1	0.74
4	16	0.63	0.448	0.81
5	16	0.72	0.5	0.81
6	18	0.71	0.425	0.82
7	18	0.71	0.402	0.74
8	20	0.85	0.420	0.86
9	20	0.85	0.736	0.78
10	22	0.60	0.460	0.81
11	18	0.71	0.414	0.82
12	26	0.43	0.184	1
13	26	0.61	0.701	0.73
14	10	0.81	0.529	0.99
15	12	1	0.598	0.85
16	14	0.74	0.592	0.95
17	16	0.56	0.626	0.84

For samples 1 to 10, presenting a good spinning behavior, the cohesive force varies inversely with the spin-finish rate and crimp. Also, the spin-finish parameter seems more influential than the crimp on the inter-fiber cohesion. The friction parameter K varies in the same way with crimp and spin-finish rate. The latter parameter was found to be the more influencing one.

Add to that, the crimp seems to act more on the friction than on the inter-fiber cohesion.

4. Conclusion

Different polyaramide fibers presenting 17 variant samples were analyzed for this study. As far as the friction test is concerned, an unexpected phenomenon of stick-slip was highlighted for a particular commercial class of polyaramide fibers. The results show that it is due to the spin-finish rate. It is probably one of the origins of the fact that the tenacity of the fiber does not effectively affect one of the yarn.

For the second study, starting from the crimp 18, spinning problems start to appear. It seems that reducing the crimp up to less than 16 could help avoid these problems.
Crimp seems to dominate the behavior of the polyaramide fiber for samples 14 to 17, while the spin-finish governs the ones from 1 to 11.

For a spin-finish rate of 0.3% the cohesion increases and reaches about 0.18 for a crimp value of 16. This corresponds to the "optimal" crimp level for the tested polyaramide (found above in the section dedicated to the study of crimp). Therefore, the reduction in the spin-finish rate may optimize the response parameters. Therefore, a separated study of spin-finish (chemical nature and quantity) seems needed to complete these results. Add to that, it seems that the spin-finish limits the cohesion. This is clear with samples of a crimp 16 and a spin-finish rates higher than 0.56%.

References

[1] Sinoimeri A 2009 Friction in textile fibres and its role in fibre processing *Wear Elsevier* **267** p 1619-24.
[2] Sheraz A 2012 *Etude du frottement inter-fibres pour les fibres du coton* (Mulhouse : Université de Haute-Alsace (ENSISA)).
[3] El moghazy Y and Gupta 1993 Friction in fibrous materials: Experimental study of the effect of structural and morphological factors *Textile Research Journal* **63** p 219-230.
[4] Guo H, El Mogahzy Y and Broughton R 1998 The contribution of fiber friction to yarn quality *Beltwide Cotton Conferences Textile Research Journal* p 719-722.
[5] Nowrouzieh S 2007 *Etude des phénomènes de cohésion et friction inter-fibres : Cas du coton.* (Mulhouse : Université de Haute-Alsace (ENSISA)).
[6] Olsen J 1974 Measurement of sliver drafting forces *Textile Research Journal* **44** p 852-855.

IOP Conf. Series: Materials Science and Engineering **827** (2020) 012030 doi:10.1088/1757-899X/827/1/012030

Analysis of the power transfer and electrical performances of an embroidered textile loop antenna for near field communication (NFC) application

Baptiste Garnier[1,4], François Rault[1], Philippe Mariage[2], Savvas Vassiliadis[3], Nikos Stathopoulos[3], Stelios Mitilineos[3], Cédric Cochrane[1] and Vladan Konkar[1]

[1] ENSAIT, GEMTEX - Laboratoire de Génie et Matériaux Textiles, F-59000 Lille, France

[2] Univ. Lille, CNRS, Centrale Lille, ISEN, Univ. Polytech. H-d-F, UMR 8520 - IEMN, France

[3] Department of Electrical and Electronic Engineering, University of West Attica, Greece

[4] Author to whom any correspondence should be addressed

baptiste.garnier@ensait.fr

Abstract. We present a new class of wearable textile antenna and analyse their performances through an electrical study. The developed prototypes have been realized by an embroidery process and are made of conductive twisted copper/polyester yarns on a cotton substrate. These antennas aim to transfer power and data thanks to the NFC technology, so the operating frequency is 13.56MHz and the maximal reading distance is around 10cm. The strong resistance of the yarn (4.20Ω/m) make us overlap three yarns to reach satisfactory electrical characteristics. Antennas are composed of a textile plan circular coil to which we add a usual capacitor. The study of coils electrical characteristics highlights a linear evolution of coil resistance and inductance in function of embroidery geometrical parameters. Also, these data show the feasibility of such antennas, especially through quality factor values reaching 45.

1. Introduction

The use of textile materials for wireless communication and energy harvesting is supposed to improve communications around human body and to interconnect users within new concept of Internet of Clothing (IoC). Nowadays, there are different ways to develop wearable antennas, as patches or prints [1], [2]. This concept is also under development throughout French ANR CONTEXT project funded by National Ministry of Research. The main objective focuses on body-centric communication and energy and data transfer. It aims also at development of power supply for textile structures embedded electronic components such as low consumption sensors, actuators and processing and communication electronic modules. Therefore, NFC 13.56MHz wireless communication will be used for energy transfer from smartphones to connected clothing and for data transfer from connected clothing to a smartphone [3]. Data collected could be sent to a cloud for further processing, analysis and if necessary decision making.

ANR CONTEXT project aims also to solve the problem of unreliable connections among sensors, actuators and processing and communication electronic modules, by using textile metamaterials as surface wave guides at ISM band (2.4GHz) for communications purposes [4]. This project should have repercussions in the telemedicine field, security, sports and leisure, etc. [5].

The NFC technology is a short-range radiofrequency identification system operating at 13.56MHz. It uses the magnetic induction principle to transfer power and data between two antennas, i.e. to transform an electric current into an electromagnetic field and vice versa.

In the field of usual radio frequency identification antennas are composed by very conductive materials as copper or silver, and their electromagnetic characteristics, such as inductance, capacity or emitted magnetic field, are well known by the literature [6]. It is also possible to make accurate simulation to forecast antennas behavior. Nevertheless, the transition from usual materials to textile materials reconsiders theories mentioning previously. Indeed, the complex structure of textile materials, in particular yarns, makes previous equations very inaccurate. So, it is necessary to understand evolution of electromagnetic characteristics under several parameters, in particular geometrical, for textile antennas.

An experimental study of this subject could enables to make textile antennas resonating at the desired frequency repeatedly. Also, the electromagnetic characteristics knowledge is necessary to make efficient and optimize the power and data transfer.

2. Materials and method

2.1. Materials

This study is dedicated to energy and data transfer using NFC protocol. First step of this study aims to design an RLC resonant circuit at 13.56MHz compatible with textile structures composed of textile flat loop coil on a fabric support made by embroidery and an adapted capacitor. The use of textile conductive yarn, instead of classic conductive yarn, involves an increasing of the circuit resistance. In order to prevent power and transfer issues, the coil has been made by overlapping three textile conductive yarns "DataTrans" from TibTech Innovation to reduce the total circuit resistance. This yarn is composed of twisted polyester, polyamide and four copper micro filaments and its electric characteristic is presented in table 1.

Table 1. Electric characteristic of TibTech Datatrans yarn

Yarn	Resistivity [Ohms/m]	Tex [g/m]	Outer wall [m]	Max Voltage [V]
TibTech - Datatrans	4.20	0.18	2.8e-7	48

Different designs of textile flat loop coil are made by a ZSK industrial machine on a cotton fabric support to study the impact of geometrical parameters, such as the radius, the number of turn and the distance between two current lines, on electrical parameters. The figure 1 shows the ZSK embroidery machine (a), and a zoom on needle during an embroidery process (b).

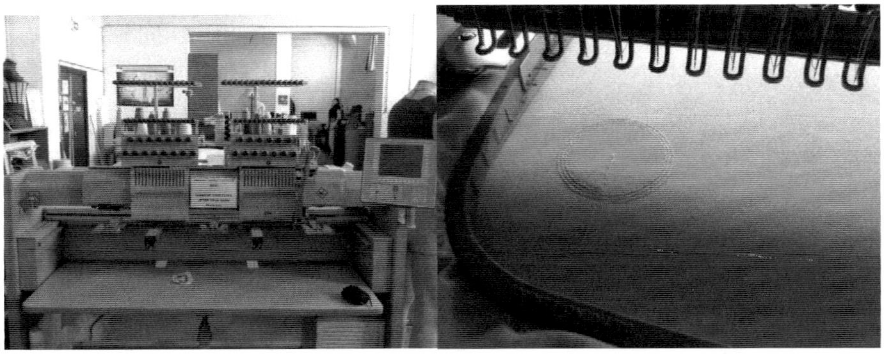

Figure 1. Photography of the ZSK embroidery machine (a) and a zoom on needles during an embroidery process (b).

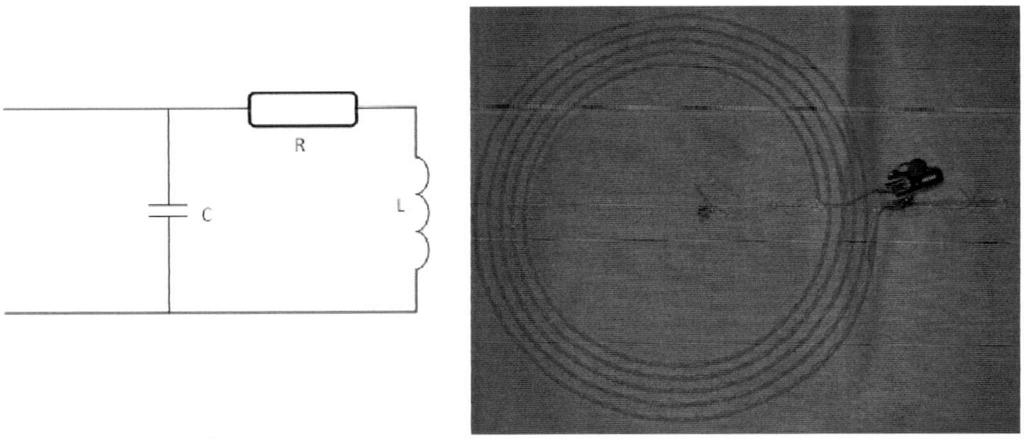

(a) **(b)**

Figure 2. Schematic of a RL circuit in series with the capacity C in parallel (a) and photography of the antenna (b).

The embroidery process requires the use of two yarns: the sewing yarn, which is delivered by the needle, and the bobbin yarn, which is delivered by the bobbin under the fabric support. Prototypes have been realized by using the conductive yarn as a bobbin yarn and a thin cotton thread as sewing yarn. This setup enables to get a flat positioning of the conductive yarn, so a better linearity of the coil. Indeed, during the embroidery process, the sewing yarn go through the fabric support, moreover the bobbin can deliver larger and stiffer yarn.

Then, once flat coil prototypes are made, it is necessary to provide a way to connect them to measurement devices. The presence of copper filaments in the conductive yarns enables to solder them on a SMA connector.

2.2. Methods

In order to evaluate performances of our antennas, it is necessary to measure coils electrical characteristics first. The resistance, the inductance and the quality factor are measured by a calibrated impedance analyzer Agilent 6945A. Coils are simply connected to the impedance analyzer with a coaxial cable via the SMA connector. The influence of the coaxial cable is included into the preliminary calibration. So, only the SMA connector will lightly influence the measurement.

From these data, the inductance is calculated to adapt the capacity value in order to create 13.56MHz resonant antennas thanks to the equation 1, with f_0 is the resonant frequency, L the inductance and C the capacity.

$$f_0 = \frac{1}{2\pi\sqrt{LC}} \tag{1}$$

Then, antennas' electrical parameters measurements have been realized through the calibrated impedance analyzer Agilent 6945A in order to evaluate its resonant frequency and quality factor. The antenna is built as in the figure 2: a schematic of a RL circuit in series with the capacity C in parallel (a) and photography of the antenna (b).

3. Results

First electrical parameters results obtained show that the coil resistance is directly dependent on the length of the embroidery, so coil radius and number of turn are two influent parameters. Indeed, intrinsic yarn linear resistance contribute to a resistance evolution. Figure 3 highlights the resistance evolution in function of the number of turn, at constant radius (a) and the radius, at constant number of turn (b)

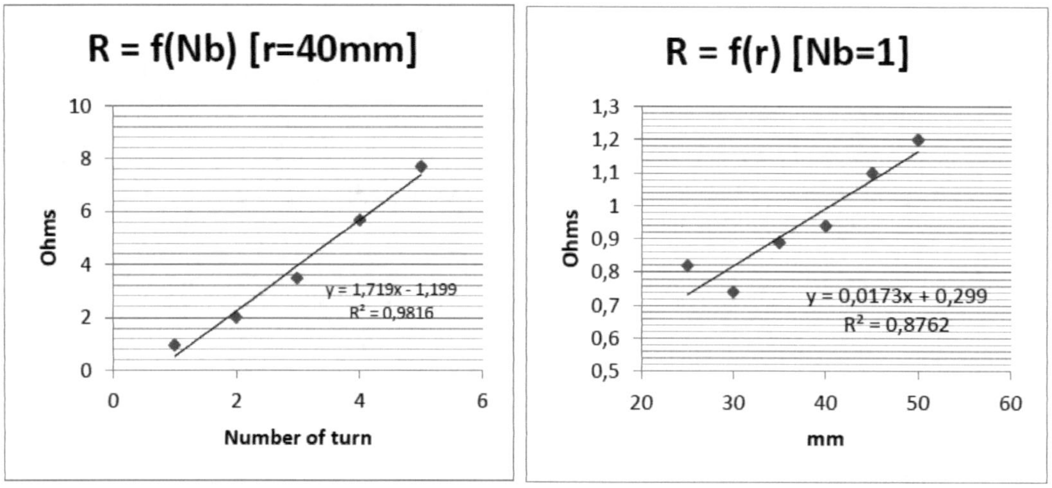

(a) (b)

Figure 3. Evolution chart of the resistance in function of the number of turn, at constant radius (a) and the radius, at constant number of turn (b)

Then, the coil inductance values are a slight increase from 1.97e-7H to 3.7e-7H when the radius of the coil increases from 25mm to 50mm. The most influent parameter is the number of turns, indeed the inductance values increase from 2.85e-7H to 3.18e-6H with number of turn of the coil increase from 1 to 5. The figure 4 represents charts of the inductance evolution in function of the radius, at constant number of turn (a) and in function of number of turn at constant radius (b).

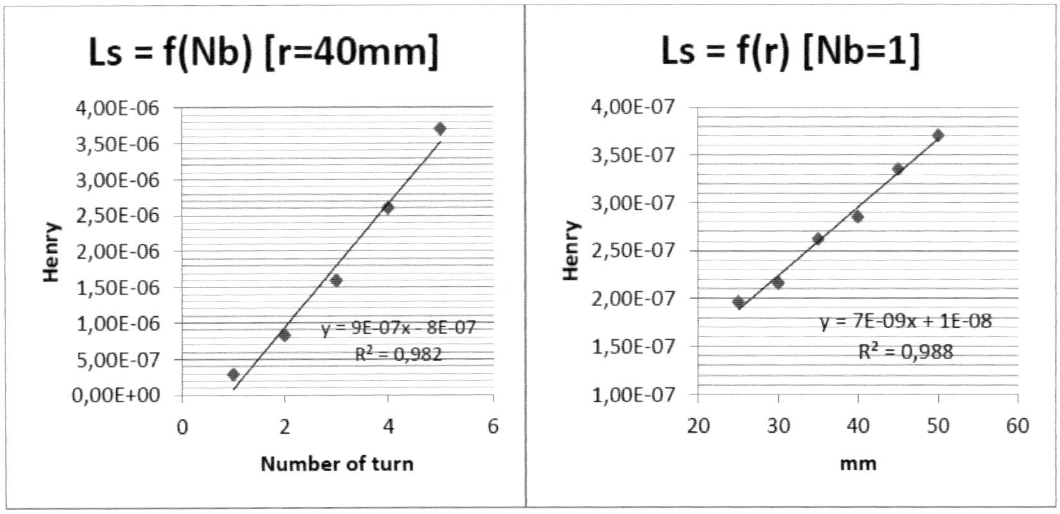

(a) **(b)**

Figure 4. Evolution charts of the inductance in function of the number of turn, at constant radius (a) and the radius, at constant number of turn (b).

In this circuit, the quality factor can be calculated by the equation (2), where Q is the quality factor, Ls the inductance, ω0 the resonant pulse and R the resistance.

$$Q = \frac{Ls\omega 0}{R} \qquad (2)$$

The equation (2) highlights the influence of the resistance and the inductance on the quality factor. Therefore, in order to obtain an antenna that can carry enough electric power in the desired frequency range, it is necessary to establish a compromise between these two values.

According to previous graphs, evolution of the inductance and the resistance can be considered as linear. Moreover, the number of turn appears as the most influent parameters, so it is natural to study the quality factor through this. Diagrams from figure 3 (a) and figure 4 (a) enable to make an equation system, which brings the equation (3) representing quality factor evolution in function of number of turn, with x = number of turns.

$$\begin{cases} R = 1{,}72(x) - 1{,}2 \\ L = 9.\,10^{-7}(x) - 8.\,10^{-7} \end{cases}$$

$$Q = \frac{76.64x - 68.13}{1.72x - 1.2} \qquad (3)$$

According to this model, the figure 5 represents the quality factor evolution in function of the number of turn.

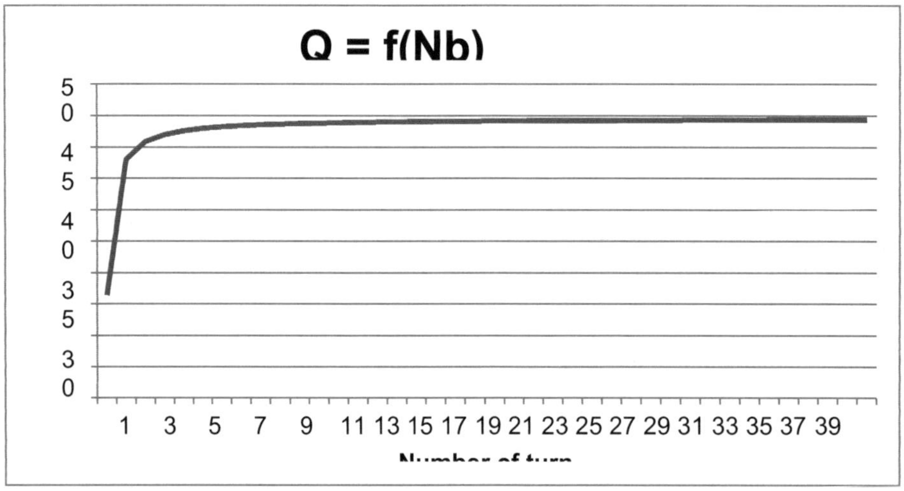

Figure 5. Evolution charts of the quality factor in function of the number of turn, at constant radius r=40mm

Even if this model is only an approximation, it makes it possible to highlight a maximum value of 45 for the quality factor. This maximum is mainly due to the significant resistance of the wire used. Indeed, when increasing the number of turn, increases both the inductance but also the resistance.

4. Conclusion

The fabrication of a new class of textile antennas was made possible by using the embroidery process as printing and overlapping three yarns. Also, the study of electrical characteristics in function of embroidery machine geometrical parameters enables to evaluate antennas performances at the desired frequency, i.e. 13.56MHz. Then, despite the structural complexity of conductive yarn and coil design, the evolution of electrical characteristic is approximately linear, which allows to create a model. So, quality factor and resonant frequency values can be predicted to certain extend by simulation.

References

[1] Ortego I., Sanchez N., Garcia J., Casado F., Valderas D. and Sancho, J.I., 2012. Inkjet printed planar coil antenna analysis for NFC technology applications. *International Journal of Antennas and Propagation,* **2012**.

[2] Locher I, Klemm M, Kirstein T and Tröster G, Design of Purely Textile Patch Antennas, 2006 *IEEE*, **29**(4), 1521-3323

[3] Kishore Kumar Reddy N G, Rajeshwari K, 2017 Interactive Clothes based on IOT using NFC and Mobile Application, *IEEE 7th Annual Computing and Communication Workshop and Conference (CCWC)* Las Vegas, Nevada, USA (pp. 1-4).

[4] Burgnies L, Cochrane C, Rault F, Sadaune V, Leurette E, Koncar V and Lippens D, 2015 Experimental phase-advanced in woven textile metasurface, *Applied physics letters*, **107**.

[5] Rao S and Llombart N 2013 Body-Worn Antennas Making a Splash: Lifejacket-Integrated Antennas for Global Search and Rescue Satellite System, *IEEE Antennas and propagtion Magazine*, **55**(2)

[6] Mohan S S, del Mar Hershenson M, Boyd S P and Lee TH 1999. Simple accurate expressions for planar spiral inductances. *IEEE Journal of solid-state circuits* **34**(10) pp.1419-24.

IOP Conf. Series: Materials Science and Engineering **827** (2020) 012031 doi:10.1088/1757-899X/827/1/012031

A method to reduce the number of sizes in apparel industry

B Hamad[1,2,3], S Thomassey[1] and P Bruniaux[1]

[1] ENSAIT, GEMTEX, F-59100 Roubaix, France
[2] Univ. Lille Nord de France, F-59000 Lille, France
[3] Author to whom any correspondence should be addressed

balkiss.hamad@gmail.com

Abstract. The aim of this research is to define a methodology for reducing the number of garments sizes after building a sizing system for a given population. Thus, we introduce the process of development of sizing system in accordance with the literature. Despite the importance of developing appropriate sizing systems in the textile industry, there has been very little research focused on the task of reducing the huge number of sizes generated. The intention to keep the number of garment sizes to a minimum is to reduce confusion among customers, in addition to the logistics costs associated with maintaining a large number of sizes. Therefore, we propose a method to reduce the number of sizes by removing the sizes that cover fewer people. The challenge is to find the most convenient order of removal, how to maintain the linearity of the sizing system as well as to accommodate a larger number of people in the population.

1. Introduction

A sizing system can be defined as a table of numbers that represents the value of key measurements used to classify variety bodies in the target population into homogenous subgroups [1]. People of the same subgroup have the same body shape characteristics, and share the same garment size.

The main aim of a body sizing system is to provide the best fit for most customers with the optimum number of sizes that benefit both consumers and manufacturers. Since too many sizes to choose from will lead to confusion in finding the right size and economically unfeasible, in contrast too few sizes will yield dissatisfaction because of loosing of fit. Therefore, the most important criteria for evaluating sizing system are the percentage of population covered, number of sizes proposed and goodness of fit [2]. Some research has been done and improved throughout the years to find a compromise between these conflict criteria. Consequently, different methods were used: statistical techniques such as principal component analysis (PCA) and factor analysis [1,3]; data mining techniques such as cluster and decision tree analysis[2,4];and artificial intelligence techniques like self-organization methods (SOM)[5]. Some of these techniques are briefly outlined in the following paragraph to highlight the range of methods available for the development of a sizing system.

Salusso et al. [6] pointed out that a well-designed sizing system should be based on a classification of data covering the largest part of the population with a minimum number of sizes. In this perspective, some researchers start with a segmentation of the population into groups. Afterwards, they establish a sizing system.

Gupta [7] propose a simple statistical approach for developing a sizing system from anthropometric data for Indian women. Principal component analysis was carried out to identify the key body measurements: bust girth for the upper body and hip girth for the lower body were identified as the critical dimensions affecting garment fit. The population was divided into three height categories were determined based on the standard deviation values of the height of the subjects and six bust categories based on the bust to waist girth ratio. Validation of sizing system was done by calculating the aggregate loss of fit, which means determining the fit of clothing by measuring the distance between actual body measurement and assigned sizes introduced by *McCulloch et al.* [8]. Euclidean distance is one fundamental and the most popular way used to determine this distance. The aggregate loss should be as low as possible for conformance to better fit. According to him, an efficient and effective sizing system must satisfy the following criteria:

- The sizing system must be in accordance with body measurements to ensure the best fit.
- The coverage of the target population should be at the maximum.
- The number of sizes proposed should be at a minimum.

Doustaneh et al. [5] use a two- stage self -organizing maps to generate a sizing system. They start by using SOM on the height attribute to generate three sizes, and then for each size, they use second SOM on the bust girth, also with three sizes.

Chung and et al. [3] established systems for Taiwanese by performing a two-stage cluster analysis. The number of sizing groups for all the developed sizing systems was between 20 and 36. The same clustering method was used for girls aged 13–20 years in Croatia[9]. As a result, three body types were obtained named T1,T2,T3 and 56 sizes for a girl's lower body with a coverage rate of 77.7% were determined. Following the practice as adopted by *Petrova and Zakaria*[1,10] the size categories which cover less than 2% of the subjects in the class has been excluded from the sizing system as an attempt to reduce the number of sizes. The problem of this method resides in finding many sizes with less than 2% coverage rate and if the size to delete is in the center of the sizing system, it will lead a non-linear system (holes in system). This has the disadvantage of difficult labelling and size grading.

A variety of methods have been experimented to develop effective sizing system. However, research on reducing the number of sizes is lacking. Hence, this paper proposes a neighbor sum matrix method to delete the size with minimum persons inside.

2. Methodology

The process of sizing system development has been the subject of much work exposed in the literature [1,11]. The general principal is divided into three steps:

i) Sizing system development**:** define key measurements or control dimensions that are required to be measured on the person intended to be fitted. These are usually body girth (bust, waist, hip, neck) or height measurements according to ISO 8559-2:2017 [12-14]. This standard specifies primary and secondary dimensions for specified types of garments to be used in combination with ISO 8559 1 (anthropometric definitions for body measurement). Deciding the size range and size intervals for each dimension. For example, if the size ranges of height for a sample group is 145–185cm that makes a range of 40 cm. The size interval is set at 8 cm, thus, the number of sizes is estimated to be five sizes. Therefore, the development of a good sizing system depends heavily on the size range and size intervals.

ii) Sizing system validation: validate the system by calculating the coverage rate, the aggregate loss, and the number of sizes. A size system with a coverage rate between 65% and 80% is considered satisfactory by *Petrova* [10]. The aggregate loss represents the distance between the actual measurement of the human body and the assigned sizes.

iii) Sizing system designation: labelling the system. Size labelling can be numeric, alphabetical or graphic. One of the methods is the use of pictograms introduced in standard NF EN 13402-3 with key measurements in cm.

Once the sizing system is developed, the number of sizes founded is extensive. The idea of this work is to reduce the number of sizes one-by-one and to recalculate the aggregate loss and the coverage rate each time. The question was which size should be deleted first and how to prevent the possibility of a gap in the sizing system.

Figure 1 shows the flowchart for illustrating the overall of our method which ensures that the sizes with fewer number of person inside and fewer person surrounded by it are removed in forward order. This method called Erosion technique. The advantage of this method is to keep the linearity of the system. For this purpose, a matrix is constructed from a computation of the number of persons in each size (each person is assigned to the nearest size according to his measurements). The initial matrix is therefore a matrix representing in each element the number of person by size. The next step is to calculate for each element the sum of the numbers of people in each element from his neighbors. Then, it is necessary to sort all the elements in ascending order to find the element with the minimum number of surrounds and we save the index of this element in an array. We repeat the process until the condition of stop is reached: there are no more sizes to delete. Indeed the algorithm must take into account the update and it should not give back an already recorded index.

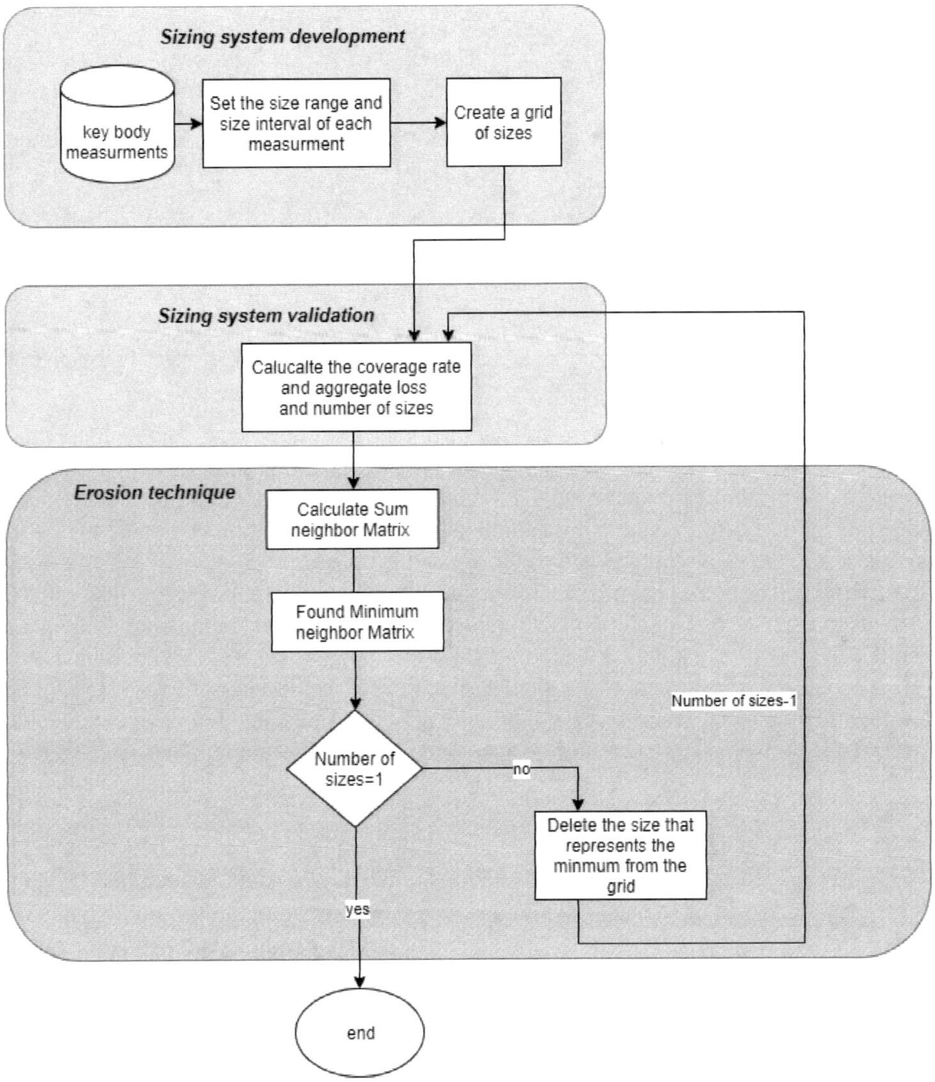

Figure 1. Flowchart of the proposed method.

The algorithm to calculate the sum of neighboring elements in matrix has been summarized in **Erreur ! Référence non valide pour un signet.**.

Table 1. Neighbor sum matrix algorithm.

Algorithm : Neighbor Sum Matrix

Input : m, n //m: number of height sizes, n: number of bust sizes
M : Matrix of m*n reflecting the number of people at each size

output T: table containing the index order of the sizes to be deleted

while number of sizes deleted < Total number of sizes **do**

For each size i from M do

Neighbor[i]= calculate the sum of the neighbors for each size

Fin

NeighborS=sort(Neighbor) // sorting the given matrix in increasing order

[Index, Value]= min(NeighborS) // find the index of the size with minimum number of person size

Save index in T;

M[i]=0// reinitialize the size to remove

end

3. Results and discussion

A representative sample of 452 French women taken from the anthropometric survey conducted in 2006 by IFTH was used.

To illustrate the algorithm for reducing the number of sizing system for an upper body garment, the height and bust girth are taken as the control dimensions as mentioned in section 2, the population is first divided according to height and then further divided according to bust girth. After examining the range of the control dimensions, size interval for height is 8 cm and 6 cm for bust girth as seen in Figure 2 which depicts the scatter plot of sample size according to the control dimensions of height and bust girth. Overall, 49 sizes to accommodate the entire sample population with an aggregate loss for the assigned sizes equal to 2.82 cm. Hence, the algorithm to reduce the number of sizes was applied. Figure 3 indicates the outcome of the matrix of neighbors by showing the size number to be deleted first and so on. The colorbars display the current colormap and indicate the mapping of data values into the colormap.

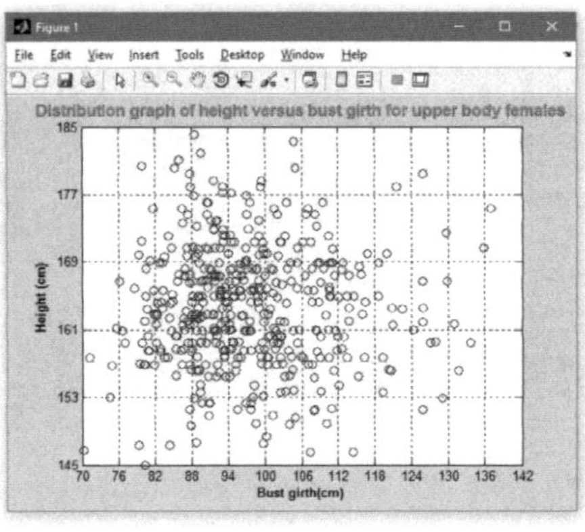

Figure 2. Distribution graph of height versus bust girth for upper body.

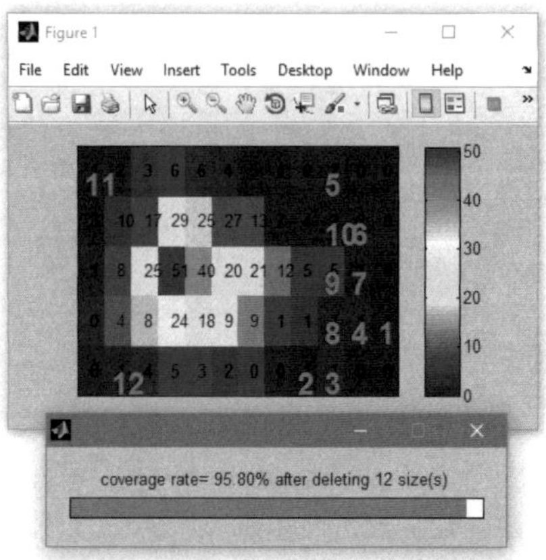

Figure 3. Reducing the number of sizes step-by-step.

Figure 4 illustrate the distribution of the coverage rate of the population by the number of sizes. It is possible to see that the percentage has a tendency to increase as the number of sizes increases. From this distribution, the textile expert can decide how many sizes to keep according to the desired percentage of coverage he wants, for example if he wants to ensure that his system keeps 80% of the population he can delete 28 sizes.

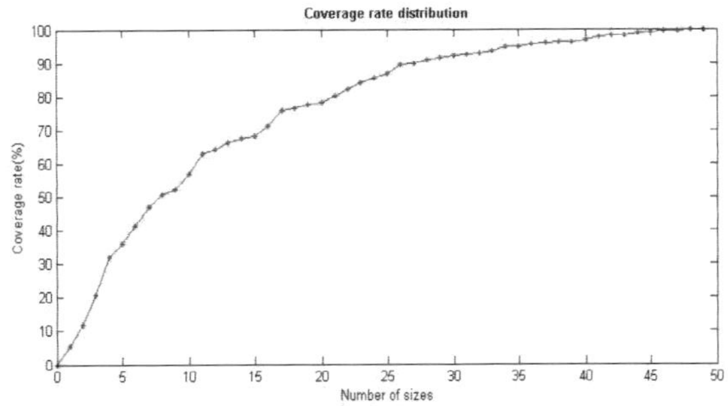

Figure 4. Coverage rate distribution of the population.

Figure 5 illustrates the distribution of the average Euclidean distances between morphological measurements and their closest sizes, by the number of closest sizes. It is possible to see that these distances tend to decrease as the number of sizes increase. This is the expected behavior as the more sizes there are, the bigger the probability that a customer finds a suitable size.

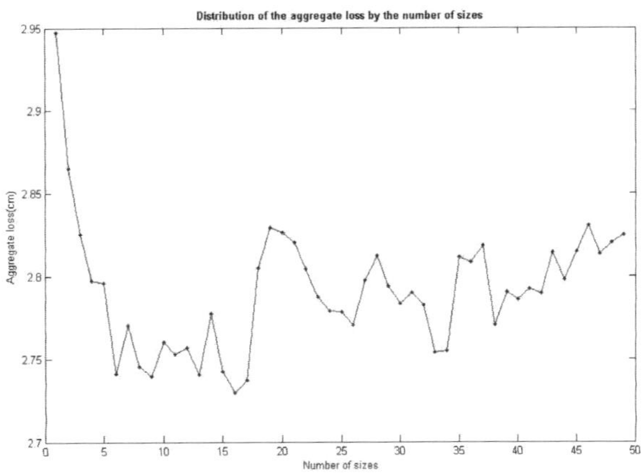

Figure 5. Distribution of the aggregate loss by the number of sizes.

4. Conclusion and future work

A new method of reducing number of sizes was designed and then tested on the French data set. The method can also be used on any other anthropometric data sets and help in creating sizing systems with less number of sizes economically feasible for apparel industry. The method involves a process of selecting key measurements, size interval based on published standards. Once the sizing system is created, the algorithm for calculating neighbor matrix was applied. A very precise order of the sizes to be deleted was implemented.

In a future study, a combination of different size intervals could be applied and compared to see which combination works best. An additional aspect to study would be to determine the best way to apply the sizing system not on the whole population but on the groups representing the result of the clustering.

Acknowledgements
This study is funded by the Single Inter-Ministry Fund (FUI).

References

[1] Zakaria N 2016 *Clothing for Children and Teenagers: Anthropometry, Sizing and Fit* Woodhead Publishing.

[2] Chung M J, Lin H F and Wang M J J 2007 The development of sizing systems for Taiwanese elementary-and high-school students *Int. Journal of Industrial Ergonomics* **37**(8) pp.707-16.

[3] Xia S and Istook C 2017 A Method to Create Body Sizing Systems *Clothing and Textiles Research Journal* **35**(4) pp. 235-248

[4] Hrženjak R, Doležal K and Ujević D 2015 Sizing system for girls aged 13–20 years based on body types *Textile Research Journal* **85**(12) pp.1293-1304.

[5] Doustaneh A H, Gorji M and Varsei M 2010 Using Self Organization Method to Establish Nonlinear Sizing System *World Applied Sciences Journal* **9**(12) pp.1359-64

[6] Salusso C J, Borkowski J J, Reich N and Goldsberry E 2006 An alternative approach to sizing apparel for women 55 and older *Cloth. Text. Res. J.* **24**(2) pp. 96–111.

[7] Gupta D and Gangadhar B R 2004 A statistical model for developing body size charts for garments *Int. Journal of Clothing Science and Technology* **16**(5) pp.58-469.

[8] McCulloch C E, Paal B and Ashdown S P 1998 An optimisation approach to apparel sizing. ', Journal of the Operational Research Society **49**(5) pp.492-499.

[9] Hrženjak R, Doležal K and Ujević D 2015 Sizing system for girls aged 13–20 years based on body types *Textile Research Journal* **85**(12) pp.1293-304.

[10] Petrova A 2007 *Creating sizing systems.' Sizing in clothing: Developing effective sizing systems for ready-to-wear clothing* ed. S.P. Ashdown Woodhead Publishing pp.57-87

[11] Pei J, Park H, Ashdown S P and Vuruskan,A 2017 A sizing improvement methodology based on adjustment of interior accommodation rates across measurement categories within a size chart *Int. Journal of Clothing Science and Technology* **29**(5), pp.716-31.

[12] ISO: ISO 8559-1, 2017 *Part 1: Anthropometric definitions for body measurement.*

[13] ISO: ISO 8559-2, 2017 *Part 2: Primary and secondary dimension indicators.*

[14] ISO: ISO 8559-3, 2018 *Part 3: Methodology for the elaboration of the body measurement tables and intervals.*

IOP Conf. Series: Materials Science and Engineering **827** (2020) 012032 doi:10.1088/1757-899X/827/1/012032

Collection of requirements for teaching in the area of Smart Textiles

Manuela Bräuning[1,2]

[1]Faculty of Engineering, Textile and Clothing Technology and Management, Albstadt-Sigmaringen University, Germany

[2] Author to whom any correspondence should be addressed

braeuning@hs-albsig.de

Abstract. The textile and clothing industries are facing new challenges. Smart Textiles are a huge topic with a high market potential, however there are only few approaches in teaching this interdisciplinary field. Moreover, individuals are expected to continuously adapt to the new requirements by following the concept of lifelong learning. We propose a new concept that will integrate a smart textiles kit. This application-oriented construction kit should help to extend the frontiers of knowledge, stimulate creativity, and give students the ability to solve real-world problems. As added benefit it will accelerate the understanding of innovation and interdisciplinary challenges. The blended learning concept aims to prepare students to think interdisciplinary, broadly, deeply, and last but not least critically. Students will be ready to contribute to the development of new products and face the challenges of structural change along their working life.

1. Introduction

1.1. Lifelong learning for personal development and innovation

Globalization contributes to the fact that our societies are increasingly changing, especially in the world of education and work. Each individual is expected to continuously adapt to new circumstances. This only works through further education in the form of lifelong learning. Lifelong learning is understood as the personal willingness to actively participate in shaping the future. It helps each individual to be better prepared for the challenges of the job market and to face changed conditions in all areas of life. Furthermore, lifelong learning counts as one of the success factors for personal development and can decisively improve the quality of life. Lifelong learning is therefore becoming increasingly important and is regarded as a key issue in education policy in innovative regions. The state of Baden-Württemberg in Germany, where Albstadt-Sigmaringen University is located, is particularly worth mentioning here as the state occupied the top position in the European Union in the Innovation Index 2018 and continues to be the region with the highest innovative capacity [1-2].

1.2. The special position of the textile and clothing industries

The textile and clothing industries in particular, a traditional branch of industry that is constantly reinventing itself and is thus regarded as very innovative, must face the challenge of systematically developing new knowledge. Especially so in Baden-Württemberg, which is Germany's strongest textile location according to the Southwest Textile Industry Association [3]. In addition, many companies benefit from the increasing demand for high-tech materials that are subsequently processed in other industries, however this requires up-to-date know-how. With smart materials, today's companies are already developing products that will, for example, bring even more safety, sustainability, energy efficiency, and comfort into everyday life in the future.

1.3. The interdisciplinary field of Smart Textiles

One of the most important fields of activity in the Textile and Clothing Sector will be Smart Textiles – products with extended functionalities. These will contribute to the next revolution in textile and clothing technology. Although the development of smart textiles has been going on for decades, everyday use is still in its infancy and there are numerous definitions with different emphasis. In the following, Smart Textiles are defined as textile products that interact with their environment and can thus actively support users. Often there is also the distinction between I-textiles, so-called intelligent textiles, and E-textiles, which are textiles with integrated electronic components. The following Figure 1 illustrates the different levels of Smart Textile products.

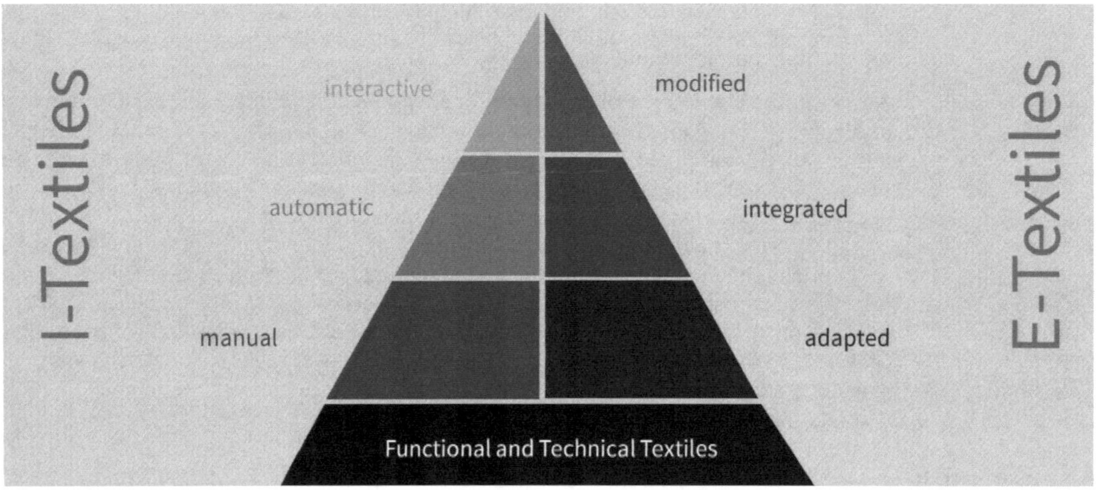

Figure 1. Smart Textiles definition

1.4. Market development of Smart Textiles and the need for new educational concepts.

The Smart Textiles segment is a fast-growing market with very good forecasts [4], picking up on current megatrends [5]. It is therefore of immense importance to prepare future technical specialists and management personnel competently for the challenges in this topic area. Overall, one of the biggest challenges for textile companies is the search for skilled employees, especially against the background of some companies wanting to relocate their production back to Europe [6]. In the field of Smart Textiles, interdisciplinary cooperation between specialists from the textile and clothing industry as well as computer science and electrical engineering is required, especially in research and development. Therefore, the transfer of competences must be interdisciplinary for a heterogenic target group, not only for current students, but also for participants from industry. Lifelong learning is a key competence in professional life and significantly contributes to a company's success [7]. To cater for participants from industry, the learning contents will be communicated via a blended learning module that includes both e-learning and classroom teaching suitable for the heterogenic target group. In this way, a tip from the manual "New Media and Mobile Learning" will be realized and a successful introduction into a new topic area will be guaranteed [8]. In order to strengthen the application-oriented competence of participants, a learning kit for hands-on exercises is to be provided.

2. Models behind the didactical approaches

Our didactical approach will include the areas of research-based learning, problem-oriented learning in projects, learning group coaching, and gameful experience in learning and edutainment. The gameful experience or gameful design puts the user's needs first while making education more engaging and fun. This is in contrast to gamification, which stands for the use of game thinking and game mechanics in non-game contexts to engage the users in solving problems, that motivation helps a successful learning process is also proven by the so-called ARCS model of motivational design. ARCS stands for Attention, Relevance, Confidence, and Satisfaction and has the goal to stimulate the curiosity to learn and to focus on the learners' needs [9]. In the following, less known didactical approaches are shortly explained and the last paragraph deals with the fact, that the learning module will be designed for a blended learning approach.

2.1. Learning group coaching

Learning group coaching was developed in the engineering sciences at the Heilbronn University of Applied Sciences and takes place in 3 phases. The aim is to strengthen the ability for independent further education as well as cooperation and organisational competence. During self-learning in phase 1, the students independently work out a learning text created by the lecturer especially for this purpose and note down any uncertainties and questions. When learning in a team in phase 2, the students discuss the questions that were raised in the first phase with each other. Work on the remaining open questions happens in phase 3, the actual learning group coaching, together with the lecturer [10].

2.2. Gameful design

The biggest crowd puller in the world is a game, soccer. So why is a game so fascinating for us humans? It is often the combination of having fun, social interaction, and a possibility to satisfy yourself and your curiosity. The focus on the user's needs leads to increased motivation. The implementation with consideration of playful approaches generates positive emotions, strengthens the social networks, communicates the usefulness, and makes success directly perceptible. According to *Ehrenberg* [11], the following characteristics are important for creating good games:

1. Player autonomy, competence and social inclusion
2. Balanced addressing of intrinsic and extrinsic motivation for user retention
3. Endogenous values in the game, i.e. something happens in the game that makes sense there.
4. the playful elements have to be connected sensibly with the purpose of application, if this is not the case, ludonarrative dissonance arises and with it a lack of meaningfulness and in turn a lack of motivation.

Playful elements should therefore enrich the user experience and thus have a positive effect on the learning experience. Moreover, a game offers the possibility to handle a failure in a positive way. [11]

2.3. Blended learning

As for Smart Textiles interdisciplinary teamwork is a key factor for success, it is mandatory that the educational module considers the social interaction between users with different backgrounds. Moreover, the module must be realised in a way that directly combines further education and training with the job requirements, a blended learning concept. Hands-on work on products and the direct interaction with other players is extremely important and cannot be provided by e-learning or distance learning alone. The concept of blended learning combines the advantages of classic presence learning scenarios with the advantages of online based learning. It describes a flexible teaching-learning setting that can deal with a very heterogenic target group comprising diverse teachers and learners. At the beginning of the course, a face-to-face event is useful so that the lecturers and participants get to know each other personally and the organizational process can be explained. The e-learning phase offers participants the advantage that any content can be learned and repeated individually at their own pace and at any location – therefore it is helpful if the content is multimedia and caters for different types of learners. [12]

3. Experience with different didactical approaches

Digitalisation and numerous changes in societies brought new challenges to lecturers and students. This sometimes leads to a fight for attention and a challenge to focus on learning. Some students have developed a so called "consumer" attitude and hope that they don´t have to engage in the learning process. This problem seems to be rooted in the traditional educational system that encourages extrinsic motivation. But how do we get intrinsically motivated students? Teaching students in the field of textile and clothing technology and management with different didactical approaches has led to a bunch of helpful experiences concerning this topic. We combined research-based learning with edutainment in the "Smart Textiles" lecture pursued with Master students. The combination of developing a new product in an interdisciplinary team and at the same time creating a crowdfunding campaign did lead to more fun, higher motivation and a more in-depth learning. Moreover, the students had to face the challenge of organizing themselves and handling the freedom they got for this task. In the lecture "production and processing of technical textiles" the students were given the task to develop their own learning game and writing a specification for this game in a group setting. This was a dual use task and the students afterwards were very proud of their achievements. We also observed this with another group of undergraduate students who got tasked with developing a cuddly toy instead of a standard part of clothing in a smaller scale. As soon as the motivation and the fun in a lecture are higher than normal the learning experience was rated better than in a classical setting.

4. The challenge of heterogenic groups and interdisciplinary work within the Smart Textiles educational module

Especially in the field of Smart Textiles, the challenge exists that specialists with different backgrounds have to work closely together in order to create innovative products that are fit for use. Moreover, there might be a mixture of novice students straight from school, and professionals doing further education, but having already experience e.g. with product development or project management. The module has to be generated in a way that different initial positions can be considered.

4.1. The difference between idea, prototype and reality

Whilst teaching smart textiles we used the WEARIC Smart Textiles Kit to bring different heterogenic groups of students in contact with this innovative new field of combining textiles and electronics. The WEARIC Kit is promoted as "the world's first do-it-yourself package with textiles sensors" and is intended to help people to discover the potential of Smart Textiles. Therefore, the set includes textile-based sensors for pressure and wetness, heating, push-buttons, sewable LEDs, and as core piece an expansion board with an Arduino nano-controller. The board and actors are easily connected by conductive snap fasteners. [13] The following Figure 2 shows the WEARIC Smart Textiles Kit with attached sewed on LEDs, a pressure sensor and a push-button.

Figure 2. Wearic Smart Textiles Kit.

Students rated working with this kit as a good entry point to the topic of smart textiles, but as soon as they wanted to realize their own products based on their new experiences, they found the limits of the system. Due to its size and construction the kit cannot be integrated in a prototype and is not washable.

In order to support students in developing new products and to realize a prototype the whole system must be miniaturized and become more flexible. This is the idea behind the application-oriented construction kit which is in development right now. An application-oriented construction kit gives users the chance to experiment and to experience learning by doing and trial and error. This is the key factor for learning according to *Hüther and Quarch* in "Save the game: Because life is more than just functioning" [14]. While this will be a big step towards enabling innovations this kit will at the moment not be suitable for the industrial manufacturing of the products. Following serial industrial production this will then be the next step which could be realized either in a virtual manner or directly with a specialised company.

4.2. Realisation

The bespoke educational module with the new concept will be developed within the framework of the SEKT research project. The SEKT project systematically investigates the integration and IT security of electronic communication systems in smart textile products and develops practical security concepts as well as smart demonstrators. The main objectives of the research project are the scientific investigation of the area of IT security of smart textile products, the transfer of existing security concepts as well as the development of secure and innovative prototypes. In the last module, "Knowledge and Technology Transfer", the newly collected knowledge is prepared in an application-oriented way using guidelines and a teaching module in blended learning format. This allows direct incorporation into the development and improvement of new Smart Textiles. At the same time, the publication of guidelines can raise awareness the general public. Whilst self-learning phases are supported by didactically prepared study materials available to participants, a practical part utilizes a learning kit for exercises aimed at strengthening the competence of participants. The application-oriented construction kit together with the blended learning module should help to enable cross innovations instead of linear innovations. In order to develop the new module in line with the target group regular surveys and revisions are planned. Finally we will evaluate if the new concept met the outlined specific requirements for this important and innovative new field and helped to successfully prepare students for the upcoming challenges.

5. Conclusions

In the field of smart textiles, interdisciplinary cooperation between specialists from the textile and clothing industry, as well as information technology and electrical engineering, is one of the most important key factors, specifically in the area of research and development. This requires that knowledge be prepared in a suitable way for a heterogenic and interdisciplinary target group. The focus of this work is therefore the development of an innovative educational concept for the teaching of Smart Textiles competences via a blended learning module. This learning module is suitable for the bespoke target groups because it uses appropriate didactic concepts and is interlinked with an application-oriented construction kit for smart textiles applications.

Acknowledgement

The Federal Ministry of Education and Research (BMBF) funds interdisciplinary research and development projects in the disciplines of engineering, natural sciences and economics, in which universities of applied sciences cooperate with partners from industry and science. The current research project SEKT (funding code FKZ: 13FH180PX6) is the first to focus on evaluating the cybersecurity of electronic communication systems in smart textile products. The project started in December 2018 and will last until November 2021. The project team works with well-known brands with backgrounds in textiles from different areas such as fashion, sports, military or medicine and cooperates with various partners from the IT sector. In addition, the group cooperates with various associations and the universities of Erlangen-Nuremberg and Hamburg – a consortium of 12 partners is involved. The SEKT project is funded by the FHprofUnt2016 initiative with approx. 500,000 Euros.

References

[1] Ministerium für Kultus, *Jugend and Sport des Landes Baden-Württemberg*, Weiterbildung: Lebenslanges Lernen [Online]. Available: https://www.km-bw.de/,Len/Startseite/Kultur_Weiterbildung/Lebenslanges_lernen. [Accessed 14 04 2019].

[2] Statistisches Landesamt Baden-Württemberg, 2018, *Innovationsindex 2018: Baden-Württemberg belegt in der Europäischen Union den Spitzenplatz*, 14 12 2018. [Online]. Available: https://www.statistik-bw.de/Presse/Pressemitteilungen/2018291. [Accessed 14 04 2019].

[3] Südwesttextil e V 2016, *Südwesttext*, **107**.

[4] Ohnemus D J and Rasel D F 2018, FashionTech – Smart Textiles: Kurzexpertise im Auftrag des BMWi, *ZEW - Zentrum für Europäische Wirtschaftsforschung GmbH*, Mannheim,.

[5] Zukunftsinstitut GmbH 2018, *Megatrends.* [Online]. Available: https://www.zukunftsinstitut.de/dossier/megatrends/. [Accessed 14 04 2019].

[6] Ruchay K 2018 *Intelligente Textilien aus Baden-Württemberg sind gefragt," Stuttgarter Nachrichten*. [Online]. Available: https://www.stuttgarter- nachrichten.de/inhalt.textilindustrie-intelligente-textilien-aus-dem-suedwesten-sind- gefragt.b3fdc5c4-f0c9-4b7f-9cd3-e2aa8e674477.html. [Accessed 14 04 2019].

[7] Europäische Kommission 2007, SCHLÜSSELKOMPETENZEN FÜR LEBENSBEGLEITENDES LERNEN: EIN EUROPÄISCHER REFERENZRAHMEN, *Amt für amtliche Veröffentlichungen der Europäischen Gemeinschaften*, Luxemburg.

[8] Bündnis für Lebenslanges Lernen 2016, Neue Medien und Mobiles Lernen, *Ministerium für Kultus, Jugend und Sport Baden-Württemberg*, Stuttgart,.

[9] Dichev C, Dicheva D, Angelova G and Agre G 2014, From Gamification to Gameful Design, *Cybernetics and information technologies* **14**(4), pp. 80-100

[10] Fortbildungszentrum Hochschullehre, Lernteamcoaching. Eine Methode zur Förderung des eigenverantwortlichen und kooperativen Lernens, 30 06 2014. [Online]. Available: https://www.fbzhl.fau.de/2014/06/30/lernteamcoaching/. [Accessed 06 09 2019].

[11] Stephanie Ehrenberg D E 2019, "Gamification im UX-Design: Behaltet eure blöden Badges," *t3n Magazin* ***57*** Q4, pp. 168-173,.

[12] Leibniz-Institut für Wissensmedien, "e-teaching.org," 28 03 2017. [Online]. Available: https://www.e-teaching.org/lehrszenarien/blended_learning. [Accessed 08 09 2019].

[13] Wearic / Texible Gmbh,2018, "Do it! Make it! Wear it! - Wearic,". [Online]. Available: https://www.wearic.com/wp-content/uploads/2018/07/WEARIC_Datasheet_Smart-Textile-Kit.pdf [Accessed 06 09 2019].

[14] Hüther G and Quarch C 2016, *Rettet das Spiel!: Weil Leben mehr als Funktionieren ist (engl. Save the game: Because life is more than just functioning),* München: Carl Hanser Verlag GmbH & Co. KG,.

IOP Conf. Series: Materials Science and Engineering **827** (2020) 012033 doi:10.1088/1757-899X/827/1/012033

Quantification of the silver content of a silver-plated nylon electrode according to the nature of the laundering detergent

V Gaubert [1,2,3,4,5,6], H Gidik [1,3], N Bodart [5] and V Koncar [1,2,4]

[1] GEnie et Matériaux TEXtiles (GEMTEX) Laboratory, 59100 Roubaix, France
[2] École Nationale Supérieure des Arts et Industries Textiles (ENSAIT), 59100 Roubaix, France
[3] UCLille, Hautes Etudes Ingénieur (HEI) - YNCREA, 59000 Lille, France
[4] University of Lille, 59650 Villeneuve d'Ascq, France
[5] Petit Bateau, 115, Rue du Lieutenant Pierre Murard, 10000 Troyes, France
[6] Author to whom any correspondence should be addresse

valentin.gaubert@ensait.fr

Abstract. In this study, the loss of the silver content was determined accurately when silver plated nylon electrodes dedicated to a measurement of human physiological parameters are washed with detergents containing bleaching agents. Thermogravimetric analysis and morphological observations based on SEM images have been conducted to determine the quantity of remaining silver on the electrodes. This loss of the conductive silver content is critical for the electrodes as it reduces their electrical conductivity and therefore their capacity to be used to measure either low voltages from the skin generated by ECG (electrocardiograph), EMG (electromyograph) or even EEG (electroencephalograph) or to determine the bioimpedance of human body or specific organs. The reduced surface conductivity leads also to a very complex contact impedance between the electrodes and the skin and therefore to their high sensitivity to any movement. This also results in noisy signals that are not useful to monitoring of physiological parameters and the concept of predictive medicine.

1. Introduction

Although the market of smart textile has been predicted to grow exponentially in the last few years, no connected garment has been a commercial success yet. However, since 2000s many e-textile prototypes have been designed and fabricated by researchers [1] but only a few of them were put on the market yet [2]. The poor reliability and washability of smart garment prototypes can account for that [3] even though the withstanding to washes has been identified as a critical element to focus on, when designing a wearable textronic [4]. Whatever the conductive material used (intrinsically conductive polymers like PEDOT [5], metal plated yarns [6] or fabric printed with conductive ink [7]), the washing process tend to impact the conductivity, more specifically to reduce it. Indeed, this phenomenon has been mostly explained by the rubbing of the conductive material due to the mechanical constraint of the washing drum. However, chemical impact of the washing process has not been much studied although electronic parts are mostly made with metal, which have a great tendency to oxidize into water. This oxidization was reported by *Kellomäki et al.* [8] who studied the color change of an antenna made with silver which has darkened with the washes. Very few similar studies can be found in the literature as most of them used bleach free detergents. One study by *Slade et al.* [6] concluded that the performance of their developed metal-coated fabrics were poor after washing them with a detergent containing bleaching agents. Nonetheless their specific roles were not studied.

In a previous study [9], the impact, on the electrical conductivity, of oxidizing bleaching agents on a silver-plated-nylon fabric used as a sensing skin electrode in a smart garment was investigated. It has been found that using detergents containing these agents reduces dramatically the overall electrical conductivity of the electrode. The oxidation of the silver layer, evidenced by the darkening of the electrode, can account for the loss of conductivity as silver oxides tend to be less conductive. However, a morphological study, using a scanning electron microscope (SEM), shows also that the silver coating has been damaged. As a result, the silver content tends to diminish with the washings, according to the detergents used. In this study, the loss in silver is precisely quantified. To this purpose silver plated nylon yarns were washed with detergents with and without bleaching agents, then their silver content was determined by thermogravimetric analysis (TGA).

2. Materials and Methods

2.1. Design and manufacturing of textile electrodes

Three silver-plated-nylon yarns were bought from three different suppliers. In the following study they will be referred to as yarn A, B, and C. Double jersey fabrics (reverse side in cotton with a polyester bonding, see Figure 1.) were knitted with yarn A, B and C respectively. An industrial circular knitting machine (CMO4A, Orizio) was used.

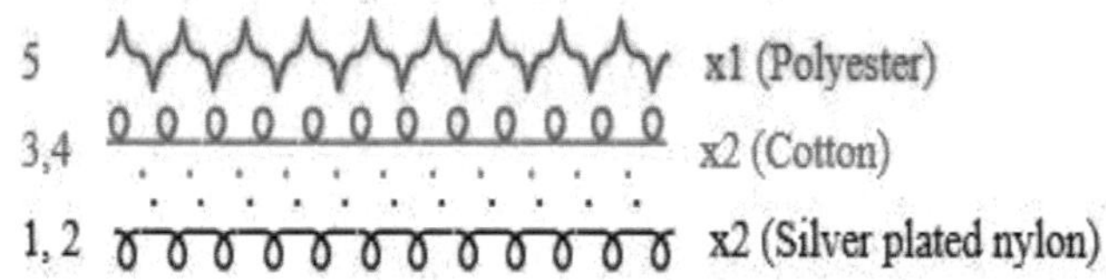

Figure 1. Diagrammatic representation of the fabric.

2.2. Washing Conditions

Domestic washing conditions were tried to be reproduced as this study was funded by a clothing manufacturer. That is why a domestic washing machine (ENF 1486 EHW, Electrolux) was used using the cycle delicate which consists in a one-hour-washing at 30°C with a maximum spinning speed of 400 rpm). Different detergents were used: A powder detergent which contains bleaching agents (Omo Concentrated Professional, Diversey S.A) was used for lot n°1 whereas a liquid one (Eco bulle Bio from Bulle Verte), without any bleaching agents was used for lot n°2. which does not contain any bleaching agents. A large sample (60x50 cm^2) was divided into smaller sampler of 10x10cm^2 and placed inside the washing with 1.8kg of random everyday clothes to simulate a real household washing, in compliance with the AATCC135 standard for laundry. Every cycle one sample was withdrawn and line dried, until 30 cycles. As a result, 90 (3*30) samples were collected in each washing condition (lot n°1 and lot n°2) but fewer samples (samples washed 10, 15, 20, 30) were deeply analyzed in this study.

2.3. Scanning Electron Microscope (SEM) Observation

A scanning electron microscope (Phenom ProX, ThermoFischer Scientific, US) was employed to look closer at the silver layer of the metal-plated-yarn and observe the potential damages.

The different chemical elements present at the surface of the fabric were identified by the Element Identification (EID) software package which included an Energy Dispersive Spectrometer (EDS). It was used to determine the percentage of silver which remains on the washed samples. But also, to investigate the potential formation of a silver oxide. The sample that has not been washed (reference sample) and samples washed respectively 10, 20 and 30 times have been observed and compared.

2.4. Thermogravimetric Analysis (TGA)

To quantify the impact of the laundering on the silver content of the different fabrics, thermogravimetric analyses were carried out by a thermogravimeter (TGA 4000, Perkin Elmer). To improve the readability of the TGA curves and to obtain more accurate silver contents, silver plated nylon yarns were unknitted from the fabric which also contains cotton and polyester yarns (Figure 1). For each measurement, approximately 20 mg of silver yarn were put in the crucible and heated from 30° to 900°C in an oxygen atmosphere, then cooled down to 30°C. As the degradation of nylon starts at around 330 °C and silver is not degraded till 900 °C, the weighed residue is only composed of pure silver at such temperature. It turned out that measurements have a high standard deviation. It can be explained because on the one hand the plating of silver is not strictly identical from one bobbin to another and on the other hand mechanical constraints from the washing drum are not equally applied on the fabric during the washing cycles. That's why a deeper analysis was done on yarn A for investigating the impact of the bleaching agents on the silver content. Seven successive measurements were done for each of the searched silver content (before washing and after being washed 30 times with and without bleaching agents). A comparison between yarns was also done. For each yarn (A, B and C), five measurements were done: before washing (reference), washed 15 and 30 times with and without bleaching agents.

3. Results and Discussion

At first glance, it can be noted that the 30-time-washed sample from lot n°1 has darkened visually compared to sample from lot n°2 which looks like the reference sample. This darkening of the samples washed with bleaching agents has been further characterized in a previous study [9]. At the yarn level, SEM observations (Figure 2) reveal that the silver layer on the filaments is significantly damaged after 30 washing cycles with bleaching agents but much less when washed without.

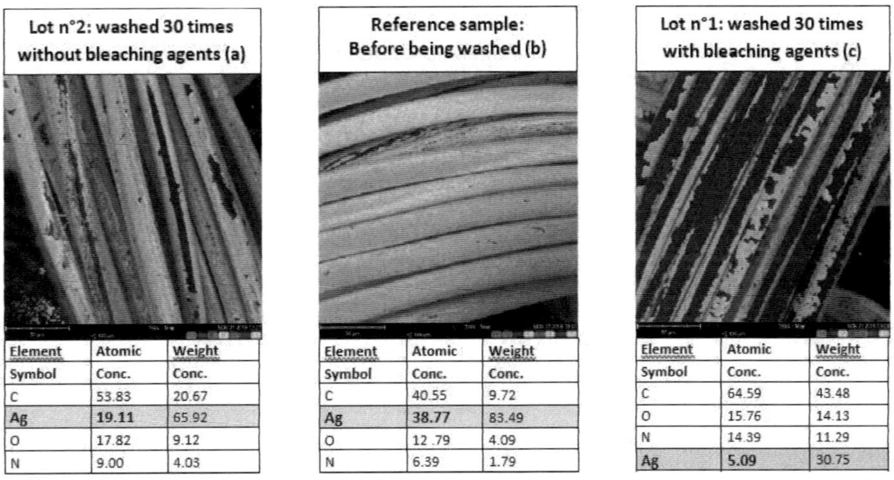

Element Symbol	Atomic Conc.	Weight Conc.
C	53.83	20.67
Ag	19.11	65.92
O	17.82	9.12
N	9.00	4.03

Element Symbol	Atomic Conc.	Weight Conc.
C	40.55	9.72
Ag	38.77	83.49
O	12.79	4.09
N	6.39	1.79

Element Symbol	Atomic Conc.	Weight Conc.
C	64.59	43.48
O	15.76	14.13
N	14.39	11.29
Ag	5.09	30.75

Figure 2. SEM observations and EDS analysis of the yarn A (a) 30 washing cycles without bleaching agents, (b) reference sample without washing and (c) 30 washing cycles with bleaching agents.

As both samples underwent the same mechanical stress, this difference can only be explained by the presence of bleaching agents during the washing process. They have probably oxidized the silver layer which made it more vulnerable to the rubbing from the machine drum.

Regarding the EDS analysis (Figure 2), it seems that half the silver (19.11 compared to 38.77) has been removed by 30 washing cycles without bleaching agents, which is quite high value. However, it is relatively low when compared with the same number of washings with bleaching agents where only an eighth (5.09 compared to 38.77) of the silver layer seems to remain. However, such comparisons can't be drawn directly between yarns as the observed material can't be strictly the same. Indeed, due to the surface inhomogeneity of yarn, it is impossible to observe strictly the same number of filaments in the same spatial configuration, although the magnification is the same. As a result, the analyzed silver

surface varies from one observation to another. Furthermore, the observed zone is only 0.16 mm² which is not significant for determining the average silver content of a sensor electrode which is 8 cm².
That's why a more precise quantification of the silver content was done by thermogravimetric analysis. In the first place, the silver content of the reference sample was measured. As the silver coating is not strictly homogenous, due to the metallization process, seven measurements were done to obtain an average reference content of silver of 18.88% ± 0.23% for the yarn A.

Then seven measurements were done for each 30-time-washed sample from the two lots. Results show that the average silver content of a yarn washed with bleaching agents is 15.44% ± 0.34% compared to 17.12% ± 0.26% when washed without. To put this figure into perspective, a comparison of the effective silver weight on a silver-yarn sensor has been drawn. Indeed, as silver is the conductive material, every single microgram is involved in the final conductivity of a sensor. For this comparison, a 0.2 g silver-plated-nylon electrode (which corresponds approximately to an 8 cm² ECG sensor) was considered. As a result, this electrode would lose almost two times more silver if it is washed with a detergent with bleaching agents compared to without (Table 1).

Table 1. Comparison of the remaining silver content regarding the presence of bleaching agents in the detergent.

	Electrode weight (mg)	Silver weight (mg) 0 wash	Silver weight (mg) 30 washes	Silver Loss in weight (%)	Loss Ratio
With Bleaching agents	200	37,76	30,88	18,21	**1,95**
Without Bleaching agents			34,24	9,33	

The fact of losing silver leads inevitably to a loss in conductivity at the yarn level and so a loss of information at the sensor level. Indeed, the silver removed from the conductive layer will decrease its thickness, so increase its resistance, in the first washing cycles. As the conductive layer is thin, this small removal will rapidly create discontinuities (cf. Figure 2) in that layer that hinder the electrons' move. As a result, a complete loss of conductivity can be observed after less than 30 washing cycles. More details can be found in our previous study [9].

Thermogravimetry (TG) and corresponding derivative thermogravimetry (DTG) curves from the unwashed (reference) yarn A are given in Figure 3.The plots show that yarn A undergoes thermal degradation between 392°C and 764°C with a DTG peak at 493,50°C which almost corresponds to the nylon 6,6 T_{max} [10]. Nevertheless, the degradation of silver-plated nylon is not as sharp as pure nylon which loses more than 90% of its mass between 480 and 520°C. Silver plated nylon tends also to lose sharply 75% of its weight between these temperatures but the remaining 25% decrease slowly and steadily from 520 to 800°C.

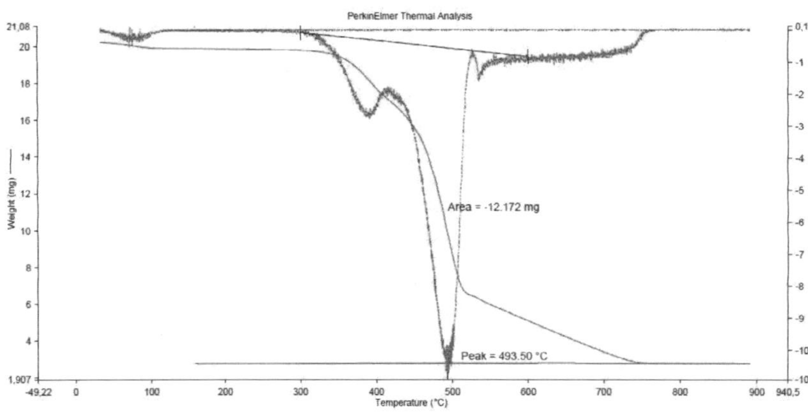

Figure 3. TG and DTG curves of the silver-plated nylon yarn (sample A) before washing cycles.

A very slight decrease in weight can be noticed at the early beginning of the heating process. It corresponds to the small amount of ambient moisture absorbed by the nylon polymer. Figure 4 shows TGA curves of the three silver plated yarns. Not surprisingly their shapes are quite similar, even nearly identical for yarn A and C. However, it should be noted that their silver contents are significantly different. Indeed, yarn C contains only 13.88% whereas yarn B contains 23.42% and yarn A 18.8% as seen before. This difference in the silver content can account for the difference in prices of those yarns, which varies accordingly to their silver content.

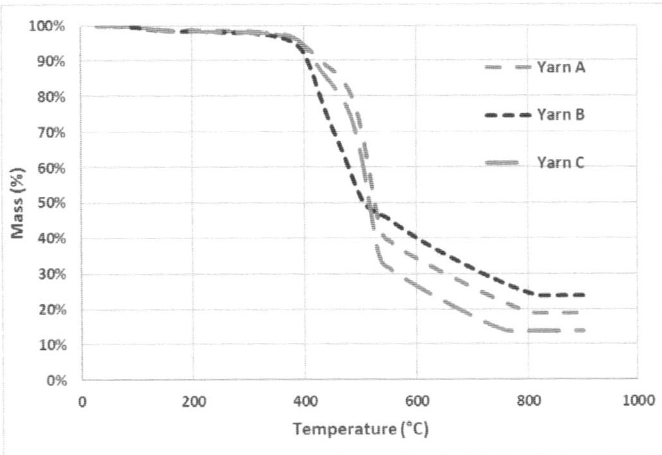

Figure 4. TG curves of the three yarns.

Furthermore, regarding their washing resistance, it is interesting to note that the more expensive yarn, presenting the higher silver content, is also the most resistant one to washing cycles (Table 2). It suggests that the thicker the silver layer, the more resistant it is. Indeed, yarn C is highly degraded by the washing cycles. Curiously the silver loss is even more significant when washed without bleaching agents.

Table 2. Comparison between yarns of their silver loss, according to the number of washings.

	Yarn A				Yarn B				Yarn C			
	Lot M1		Lot M2		Lot M1		Lot M2		Lot M1		Lot M2	
Washes	Silver Content	Silver Loss	Silver Content	Silver Loss	Silver content	Silver Loss	Silver content	Silver Loss	Silver content	Silver Loss	Silver content	Silver Loss
0 (ref)	**18,88%**				**23,42%**				**13,88%**			
15	16,68%	11,64%	17,58%	6,89%	22,18%	5,27%	23,51%	-0,40%	10,22%	26,38%	8,84%	36,31%
30	15,44%	**18,21%**	17,12%	**9,33%**	21,94%	**6,29%**	23,16%	**1,10%**	10,65%	**23,25%**	8,23%	**40,72%**

4. Conclusion

The silver layer appears to be more damaged when the textile is washed with a detergent containing bleaching agents (powder detergents). In fact, on one hand, the sensor tends to be visually darker. On the other hand, at a microscopic level, the silver layer appears rubbed and scratched. Indeed, it turns out that silver layer has been rubbed nearly twice (in weight) in such washing conditions compared to when washed without bleaching agents. Bleaching agents have probably oxidized the silver layer making it more amorphous and so more easily removable under mechanical stress. This loss in silver, which makes the yarn conductive, results in a decrease of conductivity during the first washing cycles (<15 cycles) but drives to a total loss of conductivity after 30 washing cycles. It means that sensors are no longer able to neither record bio-signals nor transmit them. As a result, "smart garment" will no longer be "smart" under those oxidizing laundering conditions unless sensors are replaced. However, in order to have better acceptability, sensors should be preferably integrated seamlessly in the garment which makes them unreplaceable. Moreover, sensors are mostly responsible for the high retail prices of such garments, as conductive materials and their integration are expensive, so replacing them every 15 washing cycles won't be accepted by customers. That's why the focus should be put on making sensors more resistant to washings and finding washing processes that are less aggressive for them. According to this study liquid detergents (without bleaching agents) are highly recommended for the laundering of textiles incorporating silver plated nylon sensors for the sake of their durability. Otherwise, the pure silver content of the sensors will decrease more quickly (it will be oxidized and then torn off more easily), resulting in reducing the overall conductivity of the electrodes so its long-term reliability.

Acknowledgements

This research was supported and carried out at the factory of the French brand Petit Bateau located in Troyes (10), under the supervision of Nicolas Bodart. Funding of this work was obtained from FEDER, region Grand-Est, ANRT (French National Research Agency) and the company Petit Bateau.

References

[1] Fernández-Caramés T and Fraga-Lamas P 2018 Towards The Internet-of-Smart-Clothing: A Review on IoT Wearables and Garments for Creating Intelligent Connected E-Textiles *Electronics* **7**(12) p. 405.

[2] Vagott J and Parachuru R 2018 An overview of recent developments in the field of wearable smart textiles *J. Text. Sci. Eng.* **8** p.364.

[3] Tao X, Koncar V, Huang T H, Shen C L, Ko Y C and Jou G T 2017 How to Make Reliable, Washable, and Wearable Textronic Devices *Sensors* **17**(4) p. 673.

[4] Gniotek K and Krucińska I 2004 The Basic Problems of Textronics *Fibres Text. East. Eur.* **45**(1) pp. 13-16.

[5] Ankhili A, Tao X, Cochrane C, Coulon D and Koncar V 2018 Washable and Reliable Textile Electrodes Embedded into Underwear Fabric for Electrocardiography (ECG) Monitoring *Materials* **11**(2) p. 256.

[6] Slade J *et al.*Washing of electrotextiles 2002 *MRS Proc.***736**.

[7] Karaguzel B*et al.* 2009 Flexible, durable printed electrical circuits', *J. Text. Inst.* **100**(1), pp.1-9.

[8] Kellomäki T, Virkki J, Merilampi S and Ukkonen L 2012 Towards Washable Wearable Antennas: A Comparison of Coating Materials for Screen-Printed Textile-Based UHF RFID Tags *Int. J. Antennas Propag.* **2012** pp. 1–11.

[9] Gaubert V, Gidik H, Bodart N and Koncar V 2019 2A3_0644_ towards washable smart textiles: investigating the impact of oxidizing bleaching agents on silver-plated textile electrodes *Proc. 19th World Text. Conf. - Autex 2019*, pp. 6–6.

[10] Pashaei S, Siddaramaiah S, Avval M and Syed A 2011 Thermal degradation kinetics of nylon6/GF/crysnano nanoclay nanocomposites by TGA *Chem. Ind. Chem. Eng. Q.* **17**(2) pp. 141–151.

IOP Conf. Series: Materials Science and Engineering **827** (2020) 012034 doi:10.1088/1757-899X/827/1/012034

Enhanced piezoelectric properties of PVdF-HFP/PZT nanocomposite for energy harvesting application

K Oumghar [1,3,6], N Chakhchaoui [1,2], R Farhane [3], A Eddiai [3], M Meddad [4], O Cherkaoui [1], Lieva Van Langenhove [5]

[1] REMTEX Laboratory, Higher School of Textile and Clothing Industries (ESITH), Casablanca, Morocco

[2] BGIM Laboratory, Higher Normal School (ENS), Hassan II University, Casablanca, Morocco

[3] Laboratory of Physics of Condensed Matter (LPMC), Faculty of Sciences Ben M'Sik, Hassan II University, Casablanca, Morocco

[4] LAS Laboratory of Setif, Mohamed el Bachir el Ibrahimi BBA University, Algeria

[5] Centre for Textile Science and Engineering, Department of Materials, Textiles and Chemical Engineering, Ghent University, Belgium

[6] Author to whom any correspondence should be addressed

oumghar.khadija11@gmail.com

Abstract. The use of piezoelectric nanocomposite in detection and actuation applications for the development of electromechanical microsystems (MEMS) has become quite common over the last decade. In this paper, we present a flexible piezoelectric nanocomposite films, composed of lead zirconate titanate (PZT) nanoparticles, embedded in poly(vinylidene-difluoride hexafluoro propylene) (PVdF-HFP) matrix. Piezoelectric and ferroelectric properties evolution is proportional to the evolution of the crystalline β-phase. The evaluation of the interactions between PZT and PVdF- HFP, performed by Fourier transform infrared spectroscopy (FTIR), revealed a dramatic improvement in these characteristics over pure PVdF-HFP, and attributed to a better crystallinity of the PVdF-HFP matrix and uniform distribution of nanoparticles. These films nanocomposites were done by solvent casting method, with various concentrations of PZT. Results of these experiments indicate that the investigated thin films nanocomposites are appropriate for various applications in energy storage and energy harvesting application.
Keywords: Energy harvesting; Piezoelectric properties; β-phase; Polyvinylidene difluoride-hexafluoropropylene (PVdF-HFP); Lead zirconate titanate (PZT)

1. Introduction

In recent years, a great deal of research has been done on autonomous energy and in particular on the supply of sensors. Embedded sensors have gained accuracy, reliability, robustness while miniaturizing but they are still limited by the lifespan of their power system [1]. A promising alternative is to recover the ambient mechanical energy to ensure nomadic energy autonomy [2]. At present, these sensors are powered by batteries that require regular maintenance and pose environmental issues (recycling in particular) [3]. If it is possible to replace these batteries by micro-generators, then the use of this type of sensor will become widespread and will contribute in developing more efficient mechatronic systems. The piezoelectric solution is the most promising because it has the highest electromechanical coupling coefficient, which does not require polarization. Piezoelectric systems are well suited to Microsystem Manufacturing Technologies (MEMS) [4].

A piezoelectric nanocomposite is designed to convert daily human activities and acoustic vibrations into usable electrical energy. When a mechanical energy such as an acoustic wave is applied to the piezoelectric polymer film, electric charges are induced. With this property, a piezoelectric material can be capitalized as an electromechanical energy converter [5].

In this paper, a different tippers of lead zirconate titanate ceramic nanoparticle (PZT) has introduced into the polyvinylidene fluoride-hexafluoropropylene matrix in order to nucleate the crystalline β-phase and eventually improve the piezoelectric properties of the films nanocomposite. This procedure was made by solvent casting-evaporation technique. The characterization of nanocomposite for their structural and morphological analyses are observed by Fourier transform infrared studies and Scanning electron microscopy to confirm the evolution of physicomechanical, thermal, and barrier properties of polymeric nanocomposite for energy storage and energy harvesting application.

2. Experimental section

2.1 Materials

Lead Titanate-Zirconate (PZT) nanoparticles have been used as additives to improve the properties of the PVdF-HFP matrix, was obtained from Sigma Aldrich. The PVdF -HFP (pellets) used has a density of ρ = 1.78 g/cm 3, Mn = 130,000 g/mol and Mw = 400,000 g/mol, was obtained from Sigma Aldrich. N, N-dimethylacetamide (DMAc) is used as the organic polar solvent often used to dissolve hexafluoropropylene polyvinylidene fluoride (PVDF-HFP). It has a density of 0.949 g/cm 3 at 20 °C and a boiling temperature of 163 to 165 °C, was obtained from Sigma Aldrich.

2.2 PVdF-HFP / PZT composite film Preparation

Polyvinylidene fluoride-hexafluoropropylene (PVdF-HFP) was dissolved in N, N-Dimethyl Acetamide (DMAc). In the other hand, different percentages of the nanoparticles of titanate lead zirconate (PZT) are prepared by dispersing the nanoparticles of PZT (powder) in N, N-Dimethyl Acetamide (DMAc). Then, the two solutions are mixed. Afterwards, the mixture is poured and scraped onto a glass plate to enhance the uniformity of the film thickness that will be placed by the glass in an oven at 85 °C for 2 hours, in order to drying the nanocomposite film by evaporation of the solvent.

3. Results and discussions

3.1 Scanning electron microscope (SEM)

Figure 1 shows an SEM image of the Neat PVDF-HFP and its composites of weight of 0.5; 0.8; 1 and 1.5 wt.%, respectively. Figure 1.a shows the PVdF-HFP micrograph. The morphology observed for the copolymer used as a matrix for the composite thus being locked and compressed indicates an indefinite morphology. The PVdF-HFP / PZT micrograph shown in Figure 1.b; 1.c; 1.d; 1.e reveals the homogeneity of the PZT grain distribution. As found in a previous study [6], non-porous composites are only obtained if the ceramic volume fraction is adjusted below about 0.5.

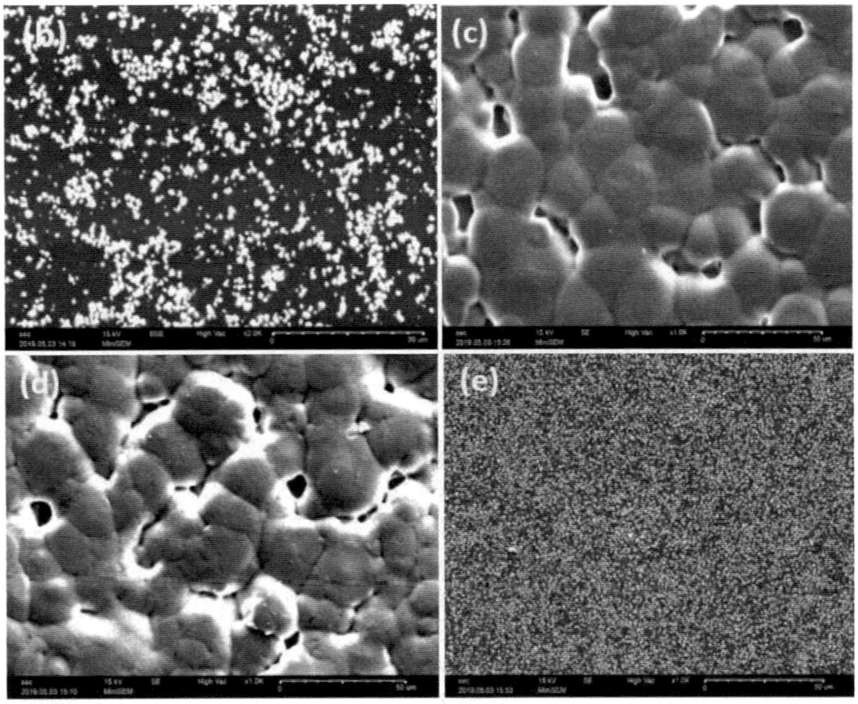

Figure 1. Morphologies of neat PVdF-HFP (a), nanocomposites containing 0.5 wt.% PZT (b), 0.8 wt.% PZT (c), 1 wt.% PZT (d) and 1.5 wt.% PZT under a scanning electron microscope (SEM). PVdF-HFP: polyvinylidene fluoride-co-hexafluropropylene, PZT: Lead zirconate titanate.

Therefore, the composite structures shown in Figure 1 represent almost the maximum PZT loading qualities that can be incorporated into PVdF-HFP copolymers while getting autonomous, supple and non-porous films. With higher ceramic weight fractions, brittleness increases considerably and the nanofillers agglomerate and the ceramic particles are not completely covered by the polymer which leads to a porous structure [7-8].

3.2 Fourier Transform Infrared Spectroscopy (FTIR)

The crystallinity of PVdF-HFP/PZT nanocomposites were further analyzed by FTIR spectra and represented in Figure 2. The vibrational peaks observed at 1071, 1176, 1397 cm^{-1} were due to bending vibration C – C, swinging vibration CH_2, motion vibration CF_2 group in neat PVdF-HFP respectively [9]. The vibrational peaks observed at 796 and 1071 cm^{-1} were due to the non-polar crystalline α-phase of PVdF-HFP [10]. The absence of a band corresponding to the bending of OH confirms that the neat PVdF-HFP and composite PVdF-HFP-PZT films are hydrophobic in nature. This is noticed from the addition of a fraction of 0.1 wt. % of PZT nanofillers up to 1.5 wt.%.

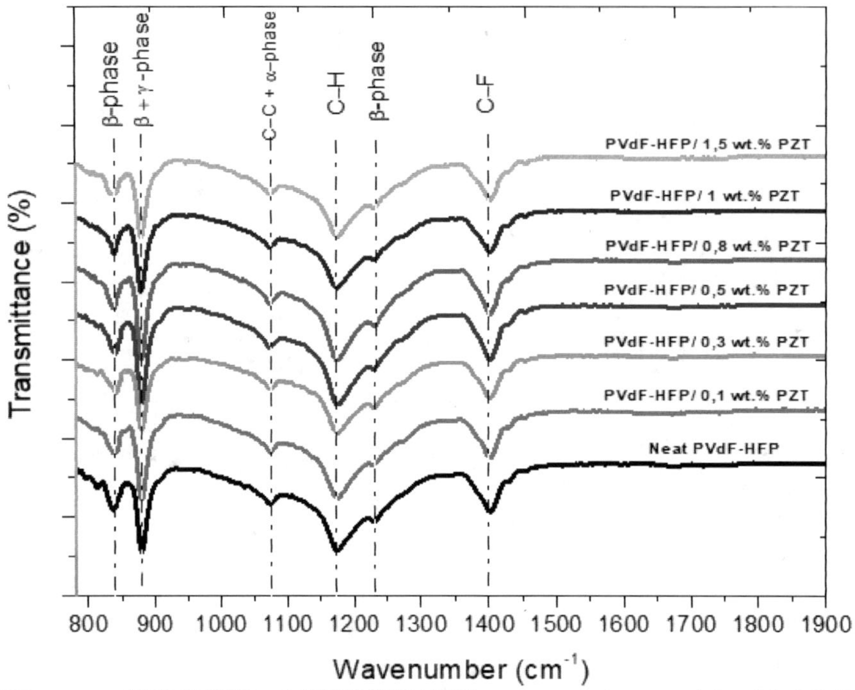

Figure 2. FTIR spectra of PVDF-HFP and PVDF-HFP / PZT nanocomposites containing 0.1, 0.3, 0.5, 0.8, 1, and 1.5 wt. % PZT, respectively. FTIR: Fourier transform infrared spectroscopy; PVdF-HFP: poly (vinylidene fluoride-co-Hexafluoride propylene); PZT: Lead zirconate titanate.

The nucleation of the fully polar β-phase in the PVdF-HFP/PZT composites film were clearly shown from the appearance of the intense vibrational band at 1232 cm^{-1}, which is not present in the Neat PVdF-HFP film (Figure 1). The semi polar γ-phase is present in both the Neat PVdF-HFP and the PVdF-HFP/PZT composite films as observed from the 1176 cm^{-1} peak. However, it gets reduced because of the β-phase nucleation in the composite film. The dual character peaks (β and γ-phases) at 882 cm^{-1} also indicate the β-phase formation in the PVdF-HFP/PZT composites film.

4. Conclusion

In order to evaluate the physicomechanical, morphological, and barrier properties of polymeric nanocomposites for energy harvesting, PZT was mixed with PVDF-HFP in a specific ratio following the simple method of solvent casting-evaporation. The composites obtained showed a good piezoelectric response, which has been demonstrated in terms of crystallinity derived from FTIR analysis. The FTIR analysis shows the interaction between PZT and PVdF-HFP, where the α-crystal content decreases while the β-crystal content increases. Indeed, the formation of the β-phase was well observed in these nanocomposites showing the efficiency of the functionalization of titanate lead zirconate (PZT) nanofillers on the PVdF-HFP matrix.

References

[1] Layek R K, Samanta S, Chatterjee D P, and Nandi A K 2010 Physical and mechanical properties of poly (methyl methacrylate) -functionalized graphene/ poly(vinylidine fluoride) nanocomposites: Piezoelectric β polymorph formation *Polymer* **51**(24) 5846–56.

[2] Eddiai A, Meddad M, Farhan R, Chakhchaoui N, Mazroui M, Rguiti M and Guyomar D 2019 Using PVDF piezoelectric polymers to maximize power harvested by mechanical structure *Superlattices and Microstructures* **127** 20-26

[3] Kim G H, Hong S M and Seo Y 2009 Piezoelectric properties of poly (vinylidene fluoride) and carbon nanotube blends: β-phase development. *Physical chemistry chemical physics* **11** 10506-10512

[4] Chakhchaoui N, Jaouani H, Ennamiri H, Eddiai A, Hajjaji A, Meddad M, Van Langenhove L, Boughalen Y 2018 Modeling and analysis of the effect of substrate on the flexible piezoelectric films for kinetic energy harvesting from textiles. *Jounal of composite materilas* **53** 3349-3361.

[5] Chakhchaoui N, Ennamiri H, Hajjaji A, Eddiai A, Meddad M and Boughaleb Y 2017 Theoretical modeling of piezoelectric energy harvesting in the system using technical textile as a support. *Polymers for Advanced Technologies* **28** 1170-78

[6] Tian G, Deng W, Gao Y and al. 2019 Rich lamellar crystal baklava-structured PZT/PVDF piezoelectric sensor toward individual table tennis training. *Nano Energy* **59** 574-81

[7] Shrout T R and Zhang S J 2007 Lead-free piezoelectric ceramics: Alternatives for PZT *Journal of Electroceramics* **19** 113-126

[8] Zak A K, Gan W C, Majid W A, Darroudi M and Velayutham T. S 2011 Experimental and theoretical dielectric studies of PVDF/PZT nanocomposite thin films. *Ceramics International* **37** 1653-60

[9] Thakur P, Kool A, Bagchi B, Das S and Nandy P 2015 Effect of in situ synthesized Fe2O3 and Co3O4 nanoparticles on electroactive β phase crystallization and dielectric properties of poly (vinylidene fluoride) thin films. *Physical Chemistry Chemical Physics* **17** 1368-1378

[10] Malmonge L F, Malmonge J A and Sakamoto W K 2003 Study of pyroelectric activity of PZT/PVDF-HFP composite. *Materials Research* **6** 469-73

IOP Conf. Series: Materials Science and Engineering **827** (2020) 012035 doi:10.1088/1757-899X/827/1/012035

Interactive Textile Vestimentary Systems for Wellbeing

Eftalea Cărpuş[1,3], Angela Dorogan[1], Ioana Doretty Cărpuş[2]

[1] The National R&D Institute for Textiles and Leather - INCDTP, Romania;
[2] Titu Maiorescu University, Medicine Faculty, Romania
[3] Author to whom any correspondence should be addressed

eftalea.carpus@certex.ro / eftalea.carpus@incdtp.ro

Abstract. In response to the changes, international research and innovation has as a priority, the development of a sustainable health system with tools to ensure both the personalization of the medical act and the implementation of prevention activities. The motivation for the use of textile elements, in this context, addresses functional, aesthetic, social, communicative, psychic and moral aspects. The balance between functionality, durability, economy, ergonomics and beauty in the case of textile systems, that change the quality of life and add value, is the result of modern design focused on meeting the requirements of use. The paper presents the results of an experimental study for the realization and testing under static conditions of textile structures used as a support for attachment sensors or electronic functions.

1. Introduction:

The textile sector has, as a driving force, the development of the technical articles and the implementation of the results in the field ensuring the sustainable development of the economic agents and the social inclusion of the labour force. [1]

In EU-28, in accordance with EUROSTAT data, technical textiles represent 15% of total textile employment, 8% of the company's active in the textile sector and 23% of the turnover of the sector. The dispersion is large inside EU countries. People's health, being a precondition for economic prosperity, influences the productivity of the economy, human capital and expenditure levels. The Union's health strategy sets out measures to prevent the emergence of diseases and to work on lifestyle- related health determinants such as working conditions, nutrition and physical activity. Promoting health is particularly important in the context of an aging population, a phenomenon stemming from increased life expectancy, and birth reduction. The projections show that by 2060 the percentage of people over 65 will increase in Europe from 17% (in 2010) to 30% [2].

In this context, textile systems used as instruments that ensure both the medical presentation and the implementation of prevention activities, define the clothing that goes beyond the type of body that approaches the spectrum of body types and ensures comfort, accessibility and style. Textile elements used as a flexible interface for attaching sensors or as a surface with textiles and electronics are part of the body-clothing-environment system. It is an open system that is permanently in dynamic interaction with the environment; physical, sensory, psychological and informational.

The human being is a complex bio-psycho-social entity; therefore, regardless of the nature of an external factor, whether harmful, aggressive or beneficial, the human individual's organism reacts as a whole on all its manifestations: physical, physiological, neuroendocrine, psycho-behavioural, social.

The trend in the textile-garment sector to include consumer requirements in the technical aspects of the product is the key to the sustainable development of the sector. In theory, satisfying consumer requirements and technical aspects (functionality, ergonomic apses, comfort, etc.) are just as important as ensuring the success of the product [3-5].

Companies have quickly learned that clothing needs to be functional, versatile and durable. There are two product-making philosophies: a so-called "product-out" where the manufacturer's decision takes into account the technological and product design specifications, and a "market-in" that embraces the consumer's requirements in the product making process.

According to this philosophy, the demands and needs of the consumer, who want attractive and functional products in correlation with his personality, are transformed into technological and design requirements.

The motivation for the use of textile elements, in this context, addresses functional, aesthetic, social, communicative, psychic and moral aspects.

The balance between functionality, durability, economy, ergonomics and beauty in the case of textile systems that change the quality of life and add value is the result of modern design focused on meeting the requirements of use. The paper presents the results of an experimental study for the realization and testing under static conditions of textile structures used as a support for attachment sensors or electronic functions.

2. Material and Methods

The base textile yarns are spun type: polyester with special cross section, which improve the psychosensorial and thermal comfort, at the skin surface, respectively the cellulose type.

The interactive yarns is double component type, with safety torsion. A spun cotton yarn and a monofilament copper wire, resin covered. The yarns are knitted and woven on conventional textile machine. No additional sets, or accessories, or development are necessary. All the experiments realized in the Experimental Station of INCDTP.

The structure characteristics of the yarns are conventional, usually used for textile clothes, garments.

3. Results and discussion

Generally, the requirements for interactive textile systems, are the same from macro to nano levels, and from component parts to whole textile systems, like a fractal. There are two types of requirements, namely, conventional/ classical and specific.

Specific requirements are vectors which define the structure of the system. A textile system allows the textile component to support and embed nontextile behaviours which cumulate functionalities, includes interactive functionalities.

Conventional requirements are associated with a textile assembly with well-known attributes, such as:

- to allow to human body to move, which means to take into considerations the dimensional modification/ variations, in dynamic way,
- to be a changing media for heat, humidity, air or different fluids, between human body and a defined environment, which means to take into consideration of physiological reaction of the body,
- to ensure protection and safety.

Specific requirements depend from a domain to other, between target groups, and between persons, inside a population of a target group.

The requirements result from temporary or constant needs of a defined population. Specific requirements allow attributes like, monitoring in real time of a status, improve the behaviour of the textile components of the system through nontextile active, passive interactive functionalities

Clothing motivation may vary depending on what is being considered. Thus, we have practical, aesthetic, social, combative and psychological motivations.

Regardless of the motivation, the clothing ensemble has an important role in managing the interactions between the human factor and the action environment in order to optimize the active and passive connections (figure1).

These factors influence each other, being in almost total interdependence, in the body- product-climate system, the only free variable is the clothing ensemble.

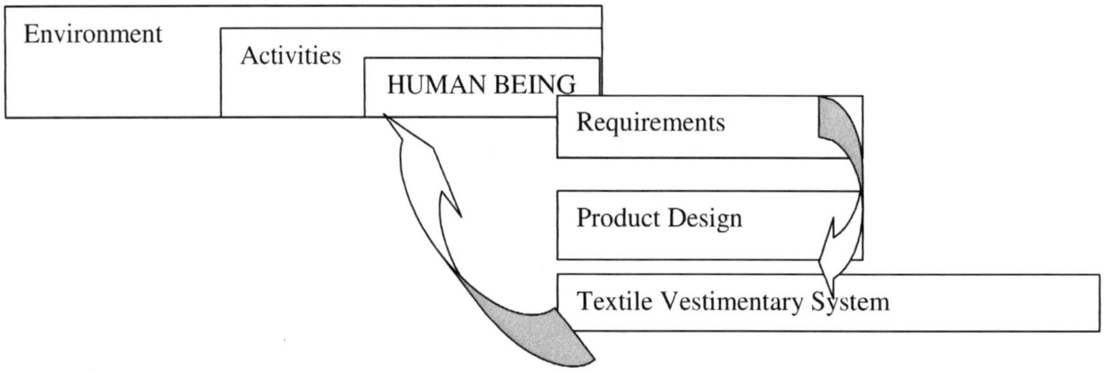

Figure 1. Textile vestimentary system concept.

Considering that the multifunctional textile support, which is used as a flexible interface for attaching sensors, is an important centrepiece, its realization has taken into consideration the following aspects:

- understanding the causes and mechanisms underpinning the maintenance of health and / or well-being;
- definition of new potential tools to be used in disease management and remote monitoring, secondary prophylaxis, drug treatment according to personalized protocols;
- establishing the relationships between the physical and mechanical characteristics and the functional ones;
- the potential sensors attached to the textile support for the manufacture of clothing used for non-invasive monitoring of physiological signals can be: body temperature - both central and average cutaneous; heart rate including heart rate; blood pressure; angular acceleration and tilt, all integrated by specific algorithms in formulas that allow description and quantitative hierarchy of comfort, respectively problematic / including critical

Considering the above-mentioned considerations, the experiments have resulted in the realization of knitted textile structures by processing different composition yarns and different structures, such as basic or additional yarns, to provide psycho- sensorial comfort on knitting machines with a link design system, (seamless) in "anatomical cuts" for on-the-go comfort.

Determination of physical-mechanical characteristics, in static conditions, indicates performance values in comfort,

Experimental models are knitted underwear, with insertion of narrow belts/ stripes. the narrow-woven fabrics are realized with interactive yarns, for monitoring the diaphragm movement. This movement could characterize the respiratory activity, in terms of frequency, of amplitudes.

In figure 2, are presented underwear structures.

a)

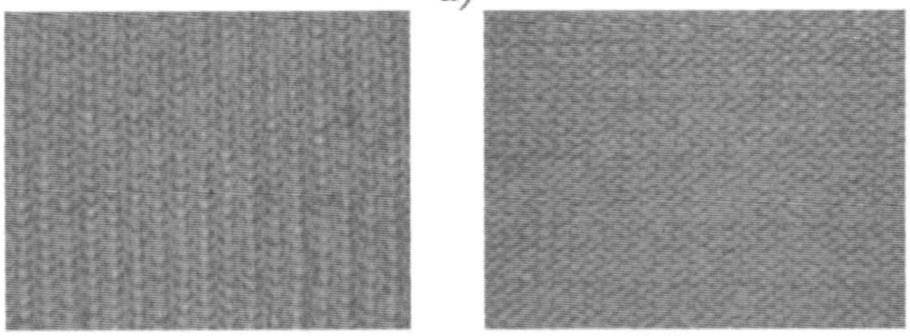

b)

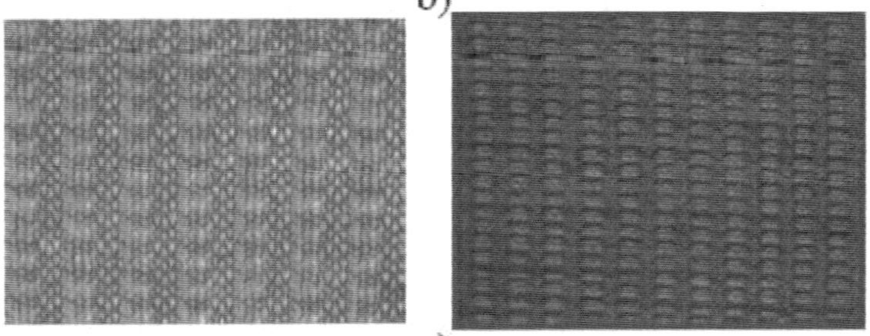

c)

Figure 2. Knitted underwear, seamless type (a), with different patterns, derivate of jersey (b) and (c) support for attaching narrow woven fabrics, ribbon type.

The incorporation/ embedding of the interactive elements was realized in the area of the diaphragm, respectively of the heart.

The textile support, in which the embedding took place, was the woven structure, narrow woven ribbon. This structure is characterized by greater dimensional stability, compared with knitted fabrics. This float length, optimal for fabrics, is complemented by the evolution of a yarn system, on a single coordinate. And are 2 yarn systems, orthogonal, one to each other.

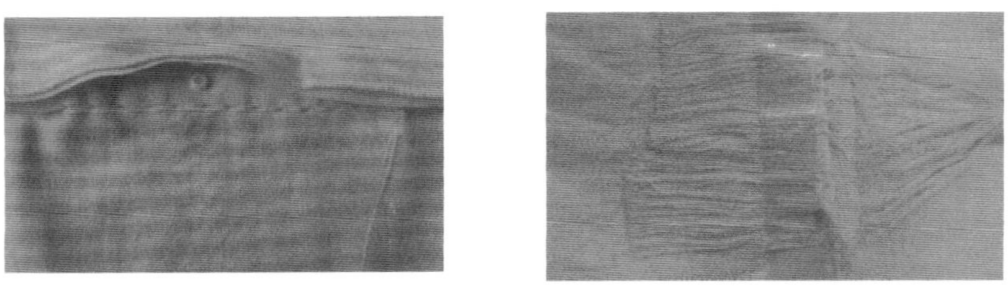

Figure 3. Details for interactive yarns embedding.

The embedding is done at the level of textile thread, in the direction of the warp or warp system or on a predetermined contour by depositing the thread by sewing / embroidery (Figure 3). Another option is by weaving narrow strips, with a width that allows a milling in the monitoring area, by ensuring permanent contact with the extremities of the body in the defined anatomical areas, for monitoring.

The results obtained show that the underwear- can be used as a component of a textile monitoring system for the following reasons:

- The combination of used yarn systems (cotton, elastane and functionalized yarns) ensure stability, flexibility, comfort
- Use the "anatomical cut", ensure the comfort of movement, and the implementation of the medical indications on the sensor location areas.
- Experimentation, through static specific measurements in the context of adaptive behaviour, has highlighted the potential of the product to be used as a support for attachment of sensors or as components of a vital function monitoring system (heart rate, respiration).

4. Conclusion

Textile knitted fabrics by processing combined yarn systems on knitting machines with a link design system can be used as a flexible interface to integrate through specific algorithms, some vital function monitoring sensors.

Acknowledgment:
The authors would like to express appreciation for the support of the sponsors: the work is funded by the Ministry of Research and Innovation within the Romanian R&D "Nucleu" Program, Project PN 19 17 03 03 and the publication of the article is funded by the Ministry of Research and Innovation within Program 1- Development of the national RD system, Subprogram 1.2 - Institutional Performance - RDI excellence funding projects, Contract no. 6PFE/2018

References:

[1] The European apparel and textile confederation EURATEX 2016, *EURATEX Annual report 2016*, Brussels.
[2] European Commission 2012 *Ageing report: Europe needs to prepare for growing older*
[3] Akao Y 1990. Quality Function Deployment: Integrating Customer Requirements into product design, G.H Mazur (trans) Cambridge, M.A: Productivity Press ;
[4] Francis F 2016 Engineering Approach with Quality Function Deployment for an ABET Accredited Program: A Case Study, *American Journal of Mechanical Engineering.* **4**(2), pp. 65-70.
[5] Donald A. Norman, Emotional design, Published by Basic Books, ISBN 0-465-05135-9,
[6] Green P E and Srinivasan V 1990 Conjoint analysis in marketing: new developments with implications for research and practice. *Journal of marketing*, **54**(4) pp.3-19.

The color fastness properties of conductive leather improved by the use of mordants

M Hylli [1,4], A Shabani [2], I Kazani [1], S Drushku [3], G Guxho [1]

[1] Departament of Textile and Fashion, Faculty of Mechanical Engineering, PolytechnicUniversity of Tirana, Albania.
[2] Department of Electrotechnics, Polytechnic University of Tirana, Tirana, Albania
[3] Departament of Industrial Chemistry, Faculty of Natural Science, University of Tirana,
[4] Author to whom any correspondence should be addressed

mhylli@fim.edu.al

Abstract. Leather has limited uses in smart or advanced applications because it doesn't conduct electricity. Applying double in situ–polymerization of pyrrole gives leather conductive properties. The treated leather changes its natural color into black color going through this double in–situ polymerization method and show very good conductivity. The problem we faced was the losing in color while using it in different applications. Therefore, in this research, we have concluded an increasement of the color fastness resistance of the black leather, which was demonstrated by the electro conductive measurements and good resistance to rubbing.
Keywords: color fastness, conductivity, dyeing, leather, mordants, and polymerization.

1. Introduction

Leather is one of the most unique and fashionable material used in different products. It is so desirable and very useful, because it owns a wide range of positive qualities. Leather has good flexibility, tear strength, abrasion resistance and good insulating properties [1-2]. However leather has limited uses in smart or advanced applications because it doesn't conduct electricity. There are some techniques that can be used to produce smart products such as printing, application of a conductive agent such as carbon black, embroidering a conductive thread or knitting a conductive complex [3].

Pyrrole is used a lot in research and applications because of its high conductivity, less toxicity and environmental friendly nature [3-4]. Polypyrrole coated fabrics have shown electrical conductivity, thermal properties and flexibility suitable for a number of applications. In addition, deposition of thin layers of conductive polymers on the fibre surface of fabrics or yarns does not noticeably change and sometimes even improves the mechanical properties of the original material [5 -6].

In our previous research, the double in situ–polymerization of pyrrole was applied to give to the leather conductive properties [4]. After the treatment, the leather changes its white color into black color and shows very good conductivity, eliminating in this way the dyeing process with toxic dyes. Moreover, it was observed that these treated leathers lose color, while touching and operating with them. Consequently, the aim of this research is to increase the color fastness resistance of the black leather in order not to lose color, while using it in different smart applications.

2. Materials and Methods

2.1. Materials and chemical reagents

In this study was chosen white sheep crust leathers of Albanian origin, produced in a tannery in Berat (Albania).

The chemical used for the transformation of leather from nonconductive to conductive were pyrrole, ferric chloride and anthraquinone-2- sulfonic acid sodium salt monohydrate (AQSA). The selected mordants used to improve the color fastness resistance of the black leather samples were sodium sulphate, and iron sulphate.

All the chemical used were of laboratory grade and high purity (97-100%).

2.2. Methods, Chemical technology

The method used to produce conductive leather was double in- situ polymerization of pyrrole. At the beging of the experiment the parameters (concentration of the solutions, time of the treatment and the number of rotations) were optimised to provide maximum conductivity [3].

The dimensions of the leather samples were 10 x 10 cm [7] and a thickness of 0.97 ± 0.2 mm [8]. Leather samples were treated at the beginning with a mixed solution of 0.15 M pyrrole and AQSA (10 wt % based on the weight of pyrrole) for 1 hour at room temperature rotating manually at 10 rpm. Then 0.4 M ferric chloride solution is added to the mixture as an oxidant to initiate the polymerization. The polymerization was carried out for 2 h at 5°C in 10 rpm. Treated leather was washed 4 times with distilled water and dried at 35 °C. Then the leather samples were treated again in a second bath containing half of concentrations of reactants following the same procedure to obtain double in situ PPy coated leather.

This polymerization method produces conductive and black coloured leather sample in one chemical treatment.

After this, the black leather samples are treated with selected mordants (sodium sulphate 2 % and iron sulphate 2 %) in order to improve the resistivity of the color.

The selected mordant solution is added along the PPy treatment following three mordanting techniques: before, along and after treatment.

2.3. Methods used to measure electrical properties

Shabani et. al. [9] mentioned that to measure the electrical properties of the conductive transformed leather throughout a polymerization chemical process, there are a variety of methods which are used before on conductive textiles. One of them is measuring the surface resistance by placing a different number of electrodes over the surface of thin film material [9-10]. The most used measurement methods are two and four probe measurement method. '*Van Der Pauw*' proposed a different four probe system for measuring surface electrical resistance for any arbitrary shape thin material. In our study, we used Van Der Pauw technique for resistance measurement. In order to measure the square resistance with this method, four contacts are required on the boundaries of the conductive simple [9-12].

The electrical resistance was measured three times on the grain side of the coated leather and an average resistance of the sample was calculated.

This method has proved to be very useful for measuring the sheet resistance of conductive coated and printed fabrics.

Figure 1. Scheme of Van Der Pauw measurements of the coated leather.

Furthermore, the samples treated and not treated with mordants were subject of the rubbing test, where their color fastness were assessed with grey scale [13-14].

3. Results and discussion

After the leather samples were treated with double in situ polymerization method and the selected mordants, the electric properties were measured. From the results was found that the selected mordants and the technique of mordanting has an influence to stabilise and increase the color fastness.
As it can be seen from the figure 2 where the mordant used is $FeSO_4$ the leather samples with very good conductivity were from the simultaneous – mordanting where the value of resistance is 142.95 Ω/□ and for the post-mordanting the value of resistance is 89.67Ω/□. Leather samples treated with pre-mordanting technique has given the higher value of resistance 719.17 Ω/□

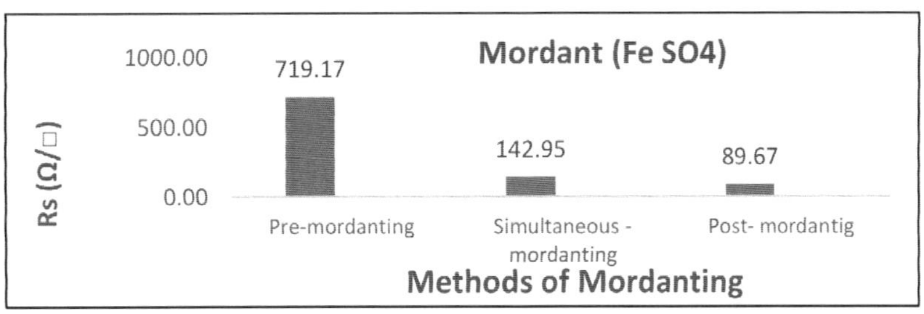

Figure 2. Resistance of coated grain side leather samples treated with double in situ polymerization and $FeSO_4$ as a mordant.

In the figure 3 are shown the values of resistance where the second selected mordant used is Na_2SO_4 X 10 H_2O. The leather samples with very good conductivity were from the simultaneous – mordanting where the value of resistance is 136.55 Ω/□ and for the post-mordanting the value of resistance is 107.15 Ω/□. Leather samples treated with pre-mordanting technique has given the higher value of resistance 557.5 Ω/□ .

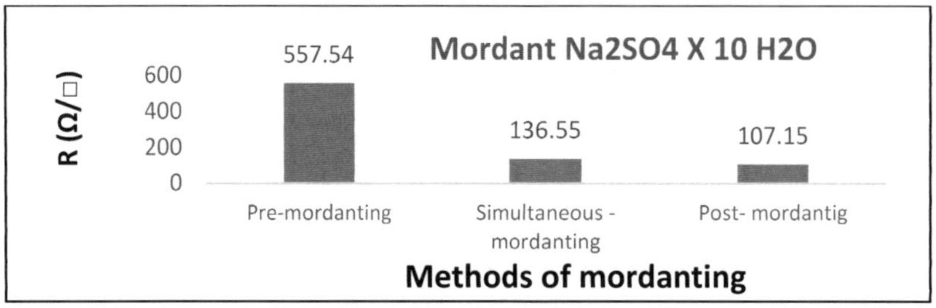

Figure 3. Resistance of coated grain side leather samples treated with double in situ polymerization and Na_2SO_4 X 10 H_2O as a mordant.

The use of pre-mordanting technique, compared to the other two techniques of mordanting has given lower conductivity of the treated leather samples, which may be due to the less absorption of the pyrrole. From the figure 2 and 3 we can conclude that mordant Na_2SO_4 show to have higher conductivity for the pre-mordanting and simultaneous-mordanting methods. When it comes to the post-mordanting method, it is the mordant Fe SO_4 which has the best conductivity, not only for this method but also compared to all methods used.

In order to observe the color fastness resistance of the treated leather, the analyses are done according the dry rubbing method, and the evaluation is done with the gray scale. The results are shown in table 1 and 2 for both selected mordants.

From the color fastness resistance tests (dry rubbing), in comparison with samples not treated with mordants, table 1 and 2 was found that the color fastness to rubbing has shown better results for the leather samples treated with mordants.

Table 1. The color fastness resistance to dry rubbing for leather samples treated with $FeSO_4$ as a mordant.

Samples	Methods of mordanting ($FeSO_4$)	Grade of Color fastness to dry rubbing
1	Pre-mordanting	4
2	Simultaneous -mordanting	4
3	Post- Mordanting	3
4	No mordants	2-3

Table 2. The color fastness resistance to dry rubbing for leather samples treated with (Na_2SO_4 x 10 H_2O) as a mordant.

Samples	Methods of mordanting (Na_2SO_4 x 10 H_2O)	Grade of Color fastness to dry rubbing
1	Pre-mordanting	4
2	Simultaneous -mordanting	4
3	Post- Mordanting	3
4	No mordants	2 -3

Moreover the samples treated in the pre and simultaneous mordanting technique have shown the grade 4, compared to post- mordanting technique which has shown a lower grade of 3. The reason may be that when we use post- mordantig technique the mordant is added after the polymerization process, and in this way the conductive coating that is formed cannot allow the penetration of the mordant.

4. Conclusion

In this research the polypyrrole treated leather became black in color after polymerization process. In this way we not only obtain conductive black leather but also along this treatment we proposed to avoid using toxic dyes. Double in-situ polymerization method of pyrrole with optimized parameters has given the conductive colored leather with a resistivity ranged from 89.67 up to 719.17 Ω/□. All the treated samples could operate successfully, even with different resistivity, on smart devices and in addition, the method used is environmentally friendly. The use of the mordants increases the color fastness of the dyed leather, reaching the aim of this research by not losing color, while using it.

References

[1] Hylli M, Guxho G and Drushku S 2014 The influence of environmental conditions on the mechanical properties in finished leather *6th Int. Conf of Textile 2014* Tirana, ALBANIA.
[2] Covington T 2009 *The Science of leather* The Royal Society of Chemistry pp 2-27
[3] Wegene J D and Thanikaivelan P 2014 Conducting leathers for smart product applications *Industrial & Engineering Chemistry Research* **53**(4), pp. 18209-15
[4] Hylli M, Shabani A, Kazani I, Beqiraj E, Drushku S and Guxho G 2018 Application of double in-situ polimerization for changing the leather properties *8th Intern Conf of Textile 2018* Tirana, Albania pp. 42-47
[5] Varesano A, Dall'Acqua L and Tonin C 2005 A study on the electrical conductivity decay of the polypyrrole coated wool textile Polymer degradation and stability 89 , pp. 125-132
[6] Gasana E, Westbroke P, Hakuzimana J, Clerck K, Priniotakis G, Kiekens P and Tseles D 2006 Electroconductive textile structures through electroless deposition of polypyrrole and copper at polyaramide surface Surface and coatings Technology 201, pp 3547–3551
[7] ISO 2418:2017Leather- Chemical, Physical and Mechanical and Fastness Tests, Sampling Location
[8] ISO 2589:2016 Leather- Physical and Mechanical Tests, Determination of thickness
[9] Shabani A, Hylli M, Kazani I and Berberi P 2019 Resistivity behaviour of leather after electroconductive- treatment Textile and Leather Review Vol 2, Issue 1, pp. 15-22
[10] Shabani A, Hylli M, Kazani I, Berberi P, Zavalani O and Guxho G 2019 The anisotropic structure of electroconductive leather studied by Van Der Pauw method *Textile and Leather Review* **2**(3), pp. 136-144
[11] Banaszczyk J, Mey G D, Schwarz A and Langenhove L V 2010 The Van Der Pauw method for sheet resistance measurements of polypyrrole coated woven fabrics *Journal of Applied Polymer Science* **117**, pp. 2553-8
[12] Kazani I, Mey G D, Hetler C, Banaszczyk J. Schwarz A, Guxho G and Langenhove L V 2011 Van Der Pauw method for measuring resistivities of anisotropic layers printed on textile substrates *Textile Research Journal* **81**(20), pp. 2117-2124
[13] ISO 20433:2012 *Leather- Tests for colour fastness -Colour fastness to crocking*
[14] ISO 105-A03:1993 *Textiles- Tests for colour fastness-Part A03 Grey scale for assessing staining*
[15] Hylli M, Kazani I, Shabani A, Komici D, Guxho G and Drushku S 2017 Transforming leather properties from nonconductive to conductive *Ist Intern. Conf. Eng. and Entrepreneurship* 2017 Tirana Albania . pp 214-219
[16] Zhang H, Zou X, Liang J, Ma X, Tang Zh and Sun J 2011 Development of Electroless Silver Plating on Para-Aramid Fibers and Growth Morphology of Silver Deposits *Journal of Applied Polymer* Science **124**, pp. 3363–3371
[17] Lacerda S N, Gonçalves L. M and Carvalho H 2013 Deposition of conductive materials on textile and polymeric flexible substrates 2 *Journal of Materials Science: Materials in Electronics* **24**, pp. 635–643
[18] Cetiner S 2014 Dielectric and morphological studies of nanostructured polypyrrole-coated cotton fabrics *Textile Research Journal* **84**, pp. 1463–1475
[19] Petersen P, Helmer R, Pate M and Eichhoff J 2011 Electronic textile resistor design and fabric resistivity characterization *Textile Research Journal* **81**(13) pp. 1395–1404

IOP Conf. Series: Materials Science and Engineering **827** (2020) 012037 doi:10.1088/1757-899X/827/1/012037

The predictive development of the electroconductive textile using artificial neural network

R M Aileni[1,2], L Chiriac[1], L Surdu[1]

[1]National Research & Development Institute for Textiles and Leather, Romania
[2] Author to whom any correspondence should be addressed

raluca.maria.aileni@gmail.com

Abstract. This paper presents several aspects concerning the predictive development of the electroconductive textiles based on the artificial neural network. The electroconductive textiles have been obtained using thin film coating with metal micro/nanoparticles and printing. For these textiles, we have made a mathematical model and we optimized dependent parameters, such as pH, dispersion conductivity, mass, thickness, air permeability, and surface resistance, by neural networks based on optimization algorithm Levenberg-Marquardt back propagation. The methodology consists in developing three feed-forward neural networks that are capable of estimating the surface conductivity of the textile materials, based on six parameters obtained in the laboratory. Moreover, this work is significant in the development of surfaces with conductivity directed to obtain the antistatic, electroconductive textiles or insulators, starting from input matrix values.

1. Introduction

The concerns in the use of the artificial neural network (ANN) for predictions in the field of textile was demonstrated through numerous researches such as prediction of the fabric behavior as output by yarn properties and machine settings as inputs [1] or in fiber to yarn prediction [2]. Neuronal networks can be used for the prediction of the parameters of the textile materials based on advanced micro/nanoparticle submitted through various methods of finishing (foulard, thin film deposition, and direct printing).

In scientific literature, the neurons are defined as units of account with a k-output and input (i_1, ..., i_k). The output is neuronal terminal which provides the result of the calculations which were carried out. The inputs are the terminals through which the neurons collect the information from various sources to be processed. Each entry, i_h, is associated with a coefficient of transfer, weighted synaptic (w_h) [3, 4]. Each neuron has a function of the response to the excitation, f (depending on neuronal, depending on the Exit or the activation function) [5, 6, 7].

The mathematical model for a process can be achieved by an analytical or experimental method using dependent and independent variables. The identification of the parameters for the experimental method involves the construction of the mathematical model based on the processing of functional variables (inputs, outputs) associated with the process. The model identified in the process is determined from an experiment conducted in the laboratory (offline) based on representative data sets acquired during the experiment, and does not change because it is assumed that the parameters of the process remain constants during the experiment. Identification of the parametric modeling, experimental, involves the construction of the mathematical model in the field of discreet time, based on the operational processing variables (input, output) associated with the process.

In general, the predictive modeling and advanced solutions for intelligent data analysis require powerful software applications, specially designed for the rolling of complex algorithms on large data sets. For modeling and prediction of the parameters in the complex experiments, we used multiple CPUs such as grid type Systems Network or Cluster Computing. For the prediction, we used neuronal networks. This paper presents the prediction of the parameters for electroconductive textiles, based on input/output variables and the ANN method [3].

2. Material and Methods

To achieve this objective was designed three neuronal networks, feed-forward (figure 1), to estimate the surface conductivity of the textile material based on six parameters obtained in the laboratory (earth, thickness, conductivity, pH, air permeability, surface resistance). Functional units within the neural network consisted in:

- Units of input (input) represented the values of the matrix Inputconductivtex 6x9, which defines the six attributes for nine tests.
- The hidden units (Hidden) are identified by the number of neurons (10, 30 and 40 respectively of neurons).
- The output (output) represented by the array Outputconductivtex values, which defines the conductivity attribute the surface for the nine tests.

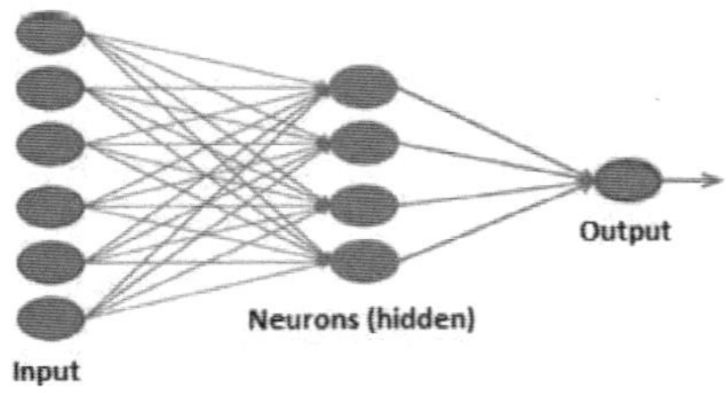

Figure 1. Feedforward neural network.

The input data used in the matrix inputconductivtex (Table 1) correspond to the values obtained by laboratory tests.

Table 1 Input data matrix inputconductivtex

	1	2	3	4	5	6	7	8	9
pH	5.45	5.96	4.86	7.15	7.53	5.41	6.25	6.7	5.76
Conductivity [μS]	211	155	163	268	191	129	90	101	164
Thickness [mm]	1.041	1.148	1.112	0.821	1.041	1.13	1.07	1.079	1.165
Air Permeability [l/m^2/sec]	16.24	8.21	8.73	16.28	23.78	16.36	13.62	8.662	21.48
Mass [g/m^2]	619	565	592	596.5	494	604	509	592	781
Surface Resistance Rs[Ω]	$1.3x10^9$	$1.29x10^{10}$	$1.2x10^9$	$1.1x10^9$	$1.2x10^{10}$	$1.4x10^9$	$1.8x10^{10}$	$1.4x10^{10}$	$1.5x10^{10}$

In the framework of the neuronal network ten layers hidden neurons (figure 2), 30 neurons (figure 3) and 40 of neurons (figure 4) have been used for the approximation of functions between the conductivity of the surface and the six input parameters Inputconductivtex matrix. For the optimization Levenberg-Marquardt algorithm that uses an approximation of the Hessian matrix (H) has been used as followed:

$$H=J^TJ \tag{1}$$

IOP Conf. Series: Materials Science and Engineering **827** (2020) 012037 doi:10.1088/1757-899X/827/1/012037

Where:
-J is a Jacobian matrix which contains the error function derivatives about the weights (w) and bias (b)
-J^T is transposed Jacobian matrix
-e is the vector of the errors.

The Levenberg-Marquardt algorithm uses the following rule for updating the parameters:

$$x_{k+1} = x_k - [J^T J + \mu I]^{-1} J^T e \qquad (2)$$

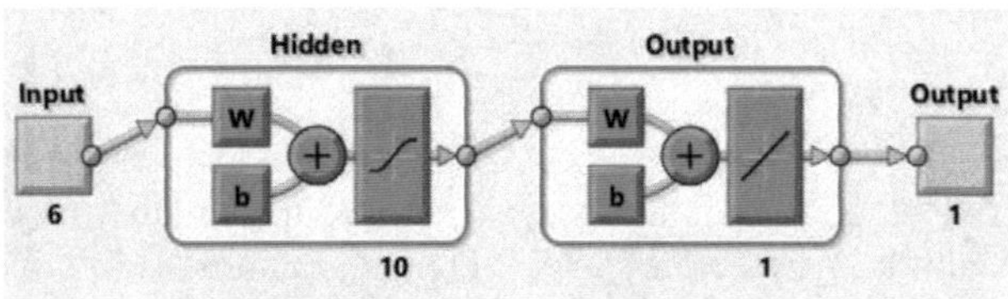

Figure 2. Neural network based on 10 neurons

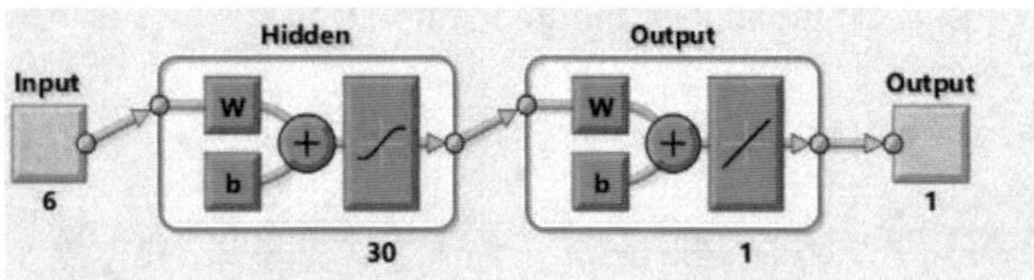

Figure 3. Neural network based on 30 neurons

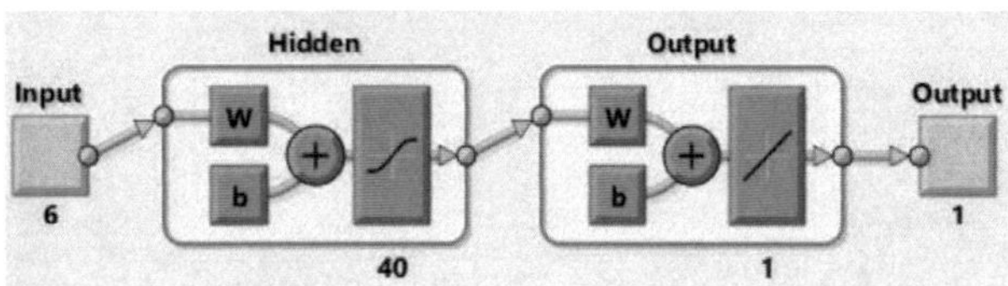

Figure 4. Neural network based on 40 neurons

For the networks based on 10, 30 and 40 neurons, the regression R (figures 5, 6 and 7), which measures the correlation between the outputs and objectives, and histogram errors (figure 8) have been represented. From graphs associated with the validation (figure 5 and figure 6) the value of the regression R= 0.95941 and R=1 shows a good match between outputs and the objectives. From the regression graph (figure 7), it can be observed that the value of the regression to the test, training, and validation is 0.84013, which indicates a good match between inputs, outputs, and objectives.

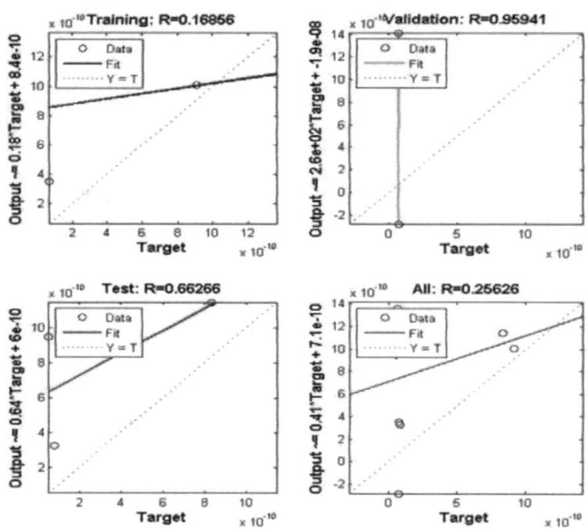

Figure 5. Graphical representation of the regression – neural network based on 10 neurons

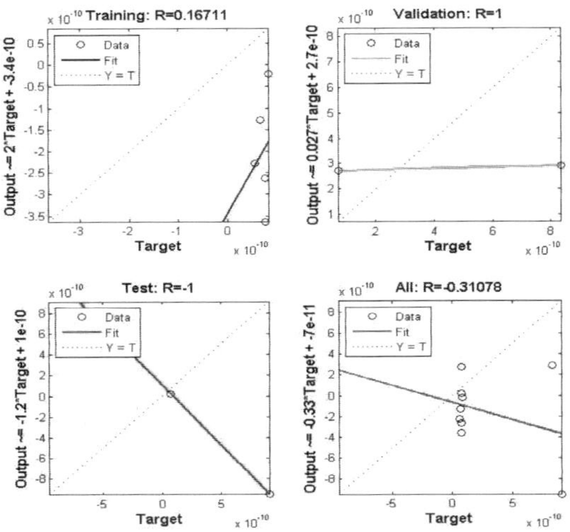

Figure 6. Graphical representation of the regression – neural network based on 30 neurons

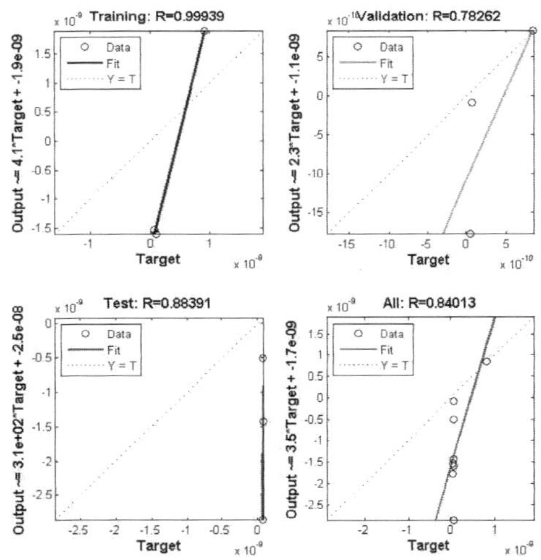

Figure 7. Graphical representation of the regression – neural network based on 40 neurons

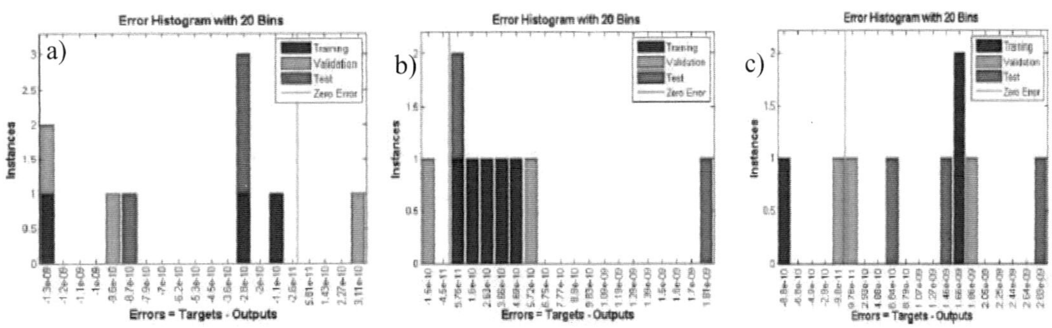

Figure 8. Error Histograms: a) Neural network - 10 neurons; b) Neural network – 30 neurons; c) Neural network –40 neurons.

3. Results and discussion

Using the experimental data obtained, a mathematical model for surface conductivity of the fabric has been developed using the parameters (surface resistance, dispersion conductivity and pH) leading to the obtaining of the antistatic, conductive or insulator materials. The mathematical model obtained for the expression of the surface conductivity based on the dispersion conductivity (x) and pH (y) is given by the math expression (3):

$$z=f(x,y) \Longleftrightarrow z=a+b*x+c*y+d*x^2+e*x*y+f*y^2 \qquad (3)$$

Where:

a=0.00000001582; b=0.0000000000321; c=-0.000000004078; d=0.00000000000006406; e=0.0000000000027; f=0.0000000002843.

In order to analyse the outliers values, the coefficient of determination (correlation between multiple square values of response and the response values predicted R –square), the value of R-square adjusted (Adjusted R-square = 0.386), the amount of the range due to errors (SSE) and the square root of the root mean square error (RMSE = 2.767e-10) have been used.

$$\text{R-square}=\frac{SSR}{SST} \Longleftrightarrow \text{R-square}=0.7698 \quad (4)$$

$$\text{SSE}=\sum_{i-1}^{n} w_i(y_i - \hat{y}_i)^2 <=> SSE = 2.296e - 19 \quad (5)$$

Where:

-SSR is the sum of squared residuals (6);
-SST is the sum of squares due to regression (7);

$$SSR = \sum_{i=1}^{n} w_i\,(\hat{y}_i - \bar{y})^2 \quad (6)$$

$$SST = \sum_{i=1}^{n} w_i\,(y_i - \bar{y})^2 \quad (7)$$

In figures 9 and 10, the surface conductivity according to the pH and conductivity of the dispersion based on metal micro/nanoparticles (C) using the mathematical model (3) are represented in 3D and 2D, respectively.

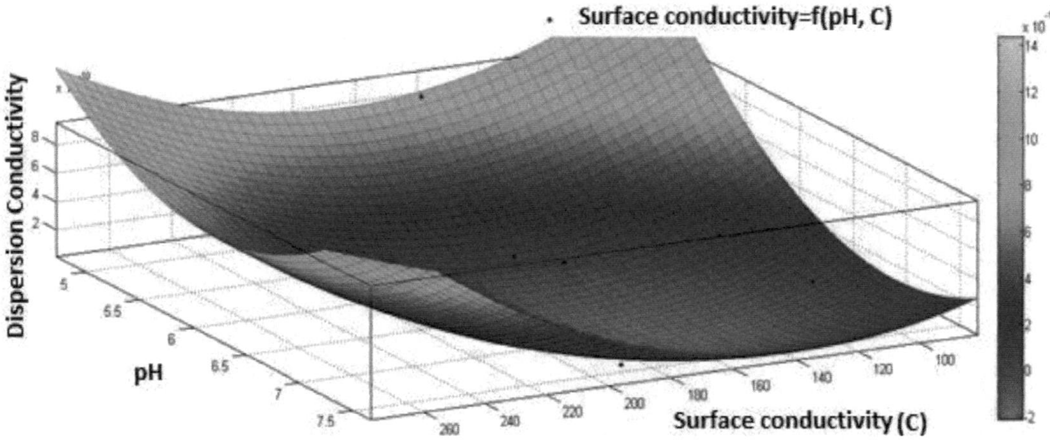

Figure 9. 3D representation- surface conductivity according to the dispersion conductivity and pH based on the mathematical model.

By analyzing the covariation between two variables C and pH (8) containing the numerical values for the parametric model, it can be concluded that between pH and C there is a direct dependence.

$$cov(C, pH) = 1.0e + 03 * \begin{vmatrix} 3.0630 & 0.0119 \\ 0.0119 & 0.0008 \end{vmatrix} \quad (8)$$

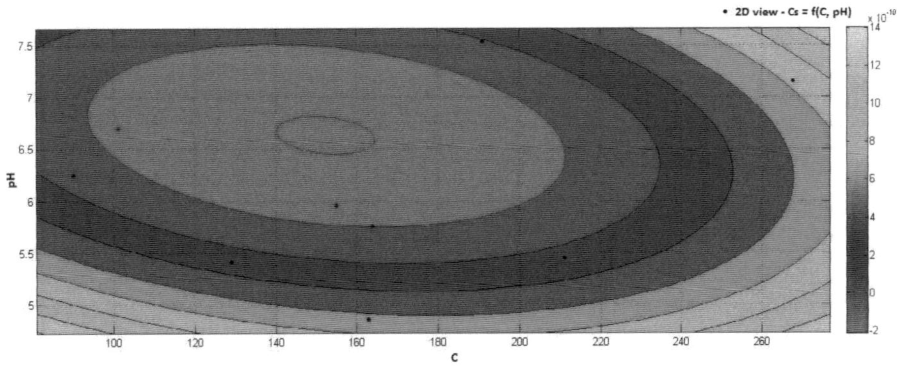

Figure 10. 2D representation of the surface conductivity according to the dispersion conductivity (C) and pH.

4. Conclusion

In conclusion, the increase in the number of neurons in the neuronal networks leads to the values of the appropriate regression that indicates a good match of the mathematical model. By variation of micro/nanoparticles quantity, conductivity obtained for dispersions at pH >7 is 30% higher than the one obtained for dispersions with acid pH (<7). The values of the regression R obtained, close to 1 for testing, learning (training) and validation in neuronal networks show a good match with the values predicted for the conductivity of the surface using the mathematical model and the experimentally obtained values. The use of the neuronal networks based on the optimization algorithms leads to the obtaining of the predictive values for the surface conductivity and the choice of the optimum set of experimental parameters that lead to the obtaining of textile surface with the electrical resistance values appropriate for antistatic and conductive properties.

Acknowledgments

The research presented in this paper was prepared in the INCDTP laboratories. This scientific paper is funded by the Ministry of Research and Innovation within National Project "Optimizing the performance of the functional textile by advanced technologies", Contract PN 18 23 01 05 and Program 1 - Development of the national RD system, Subprogram 1.2 - Institutional Performance - RDI excellence funding projects, Contract no. 6PFE/2018.

References

[1] Yang S, Gordon S 2018 Fiber-to-yarn predictions *Engineering of High-Performance Textiles* ed Woodhead Publishing pp 81-106
[2] Neha C, Nirmal Y and Nisha A 2018 Applications of Artificial Neural Network in Textiles *International Journal of Current Microbiology and Applied Sciences* 7 (4) pp 3134- 3143
[3] Cochocki A and Unbehauen R 1993 *Neural networks for optimization and signal processing* ed John Wiley & Sons
[4] DeGroff D and Neelakanta P S 2018 *Neural network modeling: Statistical mechanics and cybernetic perspectives* ed CRC Press
[5] Krogh A and Vedelsby J 1995 Neural network ensembles, cross-validation, and active learning *Advances in neural information processing systems* pp 231-238
[6] Zhang Z 2018 Artificial neural network *Multivariate Time Series Analysis in Climate and Environmental Research* ed Springer Cham pp 1-35
[7] DeGroff D and Neelakanta P S 2018 *Neural network modeling: Statistical mechanics and cybernetic perspectives* ed CRC Press

IOP Conf. Series: Materials Science and Engineering **827** (2020) 012038 doi:10.1088/1757-899X/827/1/012038

CALM scale – Assessment of the total comfort of the popular female jeans

I Braga[1,2], M J Abreu[2] and M Oliveira[3]

[1]Federal University of Piauí, Course de Bach .in Fashion, Design e Stylism, Campus Universitário Ministro Petrônio Portella - Bairro Ininga - Teresina - PI, Brazil
[2] Department of Textile Engineering, University of Minho, Campus de Azurém, Guimarães, Portugal
[3] Institute of Social Sciences, University of Minho, Campus Gualtar, Braga, Portugal

iarabraga@yahoo.com.br

Abstract. The present paper consists on the experimental stage of a PhD research, on the study of the popular female jeans comfort, referring to the process of subjective evaluation of comfort to wear, specifically the comfort objective evaluation of the jeans to wear, using CALM scale. The popular jeans in study, refers to the jeans created, produced and consumed by the poor classes of Fortaleza, in the state of Ceará, Brazil. The purpose is to investigate the female jeans, with the objective of collecting information about the comfort / discomfort perceptions of the consumers of the Fortaleza popular markets. When the collected responses were correlated, it was found that the perception of comfort/discomfort of the jeans is directly related to the touch sensation.

1. Introduction

The present study consists on the presentation of the experimental phase of the Phd research, the study of the comfort of the popular female's jeans, referring to the process of evaluation of the subjective sensations of the comfort to the use.

The jeans popular in study, refer to jeans created, produced and consumed by the popular poor Brazilian classes. The popular jeans market in Fortaleza, in the state of Ceará, Brazil, has 47.8% of the population of the city as potential consumers. It is the equivalent of approximately 623,790 women and it plays an important role as a source of income generation for the poor populations of the suburbs and metropolitan cities, which is surprising from the point of view of the expansion of production, commercialization and consumption.

The purpose is to investigate female's jeans, a symbol of popular fashion and popular consumption, with the objective of collecting information about the perceptions of comfort / discomfort of consumers in Fortaleza's popular markets when they wear the jeans sold in popular shopping malls in Fortaleza, such as Buraco da Gia, Beco da Poeira and José Avelino Street Fair. To carry out the experiment, five (5) models of jeans were bought at the popular marketing centers of Fortaleza Beco da Poeira, Buraco da Gia and the José Avelino Street Fair.

The group of evaluators consists of 101 women volunteers, aged between 18 and 40 years, consumers of the popular markets of Fortaleza. The tests were applied in real environments of use of jeans by these women. The sensorial analysis to the use of the pieces of jeans used as instrument the application of inquiry. The investigation was constructed from the combination of three different question models [1]:

with the combination of numerical, attributes and faces scales, body maps [2] and with the final question of the CALM (comfort affective labeled Magnitude) scale [3]. The data presented in this article refers specifically to the data obtained from the CALM scale.

The CALM (Comfort Affective Labeled Magnitude) scale was developed through the combination of test instruments with people and equipment developed a methodology consisting of 3 series of experiments with the completion of the third series of tests developed a scale of comfort measurement, using words related to the magnitude of the semantic meaning of sentence construction that indicate different levels of comfort / discomfort [3].

The authors applied the scale in two experiments in order to evaluate the performance and reliability of the scale. With the scale validation tests, it was possible to conclude that the "CALM" scale presents sentences of sensitivity and reliability, identifying these factors as advantages of this scale over previous scales [4].

Cardello and Winterhalter(2003) [3] report that the CALM (Comfort Affective Labeled Magnitude) scale is a non-complex scale, has high levels of sensitivity to comfort and facilitates comparisons of different percentages and consumer perception ratios for tissue evaluation and of clothing.

The realization of the sensorial test to the use of the popular jeans by means of the application of the investigation, from the analysis of the data collected and related among the parameters evaluated was identified by the evaluators that the pants named as model 3 and 5 jeans pointed as the most comfortable.

2. Material and Methods

The sensorial evaluation to use of the jeans, was made according to the following phases:

a) The inquiry was the instrument used for data collection [1]. The inquiry was constructed from the combination of three different steps: with the combination of scales (Figure 1) (numerical, attributes and faces); With two leaves consisting of a front and back body map, with each side (front and back) (Figure 2) being divided in 15 parts, referring to the body zone dressed in jeans, adaptation of the Discomfort Opinion Scales developed by *Corlett; Bishop (1976)*; And the final question concerning the assessment of the total comfort of the jeans through the CALM scale (Comfort Affective Labeled Magnitude) (Figure 3).

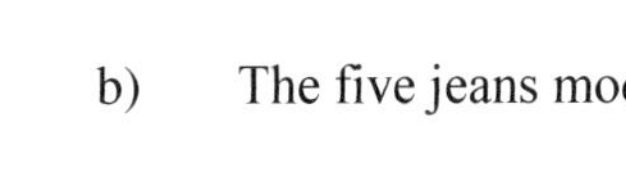

Figure 1. Numerical, attributes and faces scales.

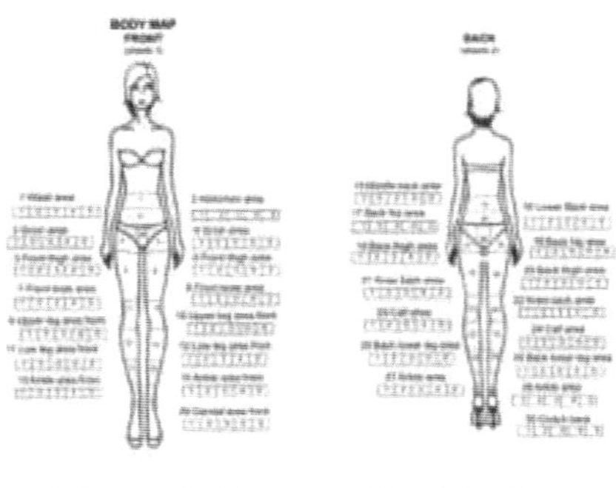

Figure 2. Front and back body map.

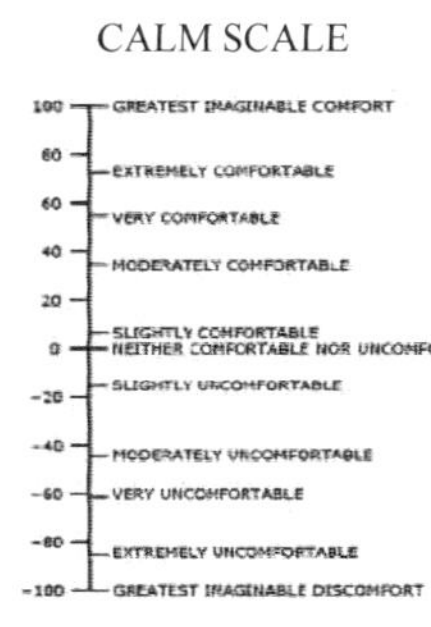

Figure 3. CALM scale.

b) The five jeans models were bought at popular malls and fairs in the city center.

IOP Conf. Series: Materials Science and Engineering **827** (2020) 012038 doi:10.1088/1757-899X/827/1/012038

Table 1. Characterization of jeans material.

Jeans Models	Jeans 1	Jeans 2	Jeans 3	Jeans 4	Jeans 5
Image	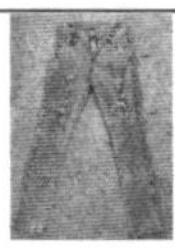	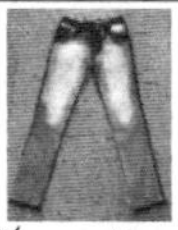			
Composition	98%contton 2%spandex	77%contton 21%polyester 2% spandex	98%contton 2% spandex	96,5%contton 3,5% spandex	77%contton 21%polyester 2% spandex
Structure	Twill (2/1 1)	Twill (2/1 1)	Twill (3/1 1)	Twill (3/1 1)	Twill (3/1 1)
Mass (g/m)^{2}	281	332	322	289	323
Thickness (mm)	0,70	0,83	0,75	0,72	0,75
Count threats (yarns/cm)	29 x 21	37 x 23	39 x 22	40 x 26	38 x

c) The group of evaluators consists of 101 female volunteers, aged between 18 and 40 years, consumers of the Fortaleza popular markets.

d) All the tests were performer in real environments (Figure 4). The means of temperature and humidity were between 28°C with variation of ± 2°C and 65%, with variation of ± 2%.

e) The time of the experiment: 30 minutes for each jeans model.

Figure 4. Tests application

3. Results and discussion

Through the analysis of the data collected, the evaluators pointed out: the jeans 1 and jeans 2 showed very close data in all parameters analysed, and the jeans 1 (Figure 5) and jeans 4 (Figure 6) showed the same data, as the most uncomfortable jeans.

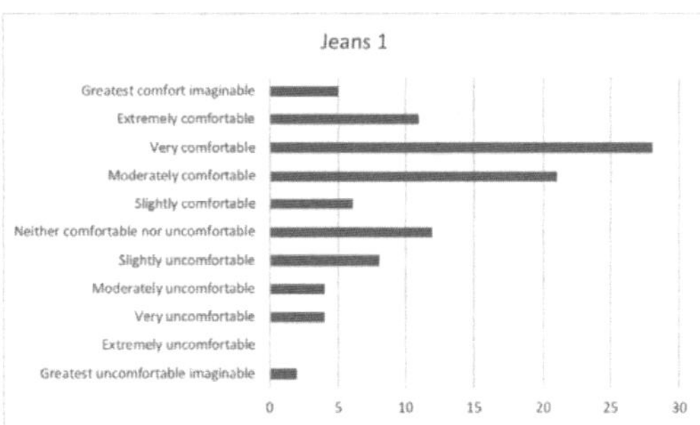

Figure 5. Jeans 1 data.

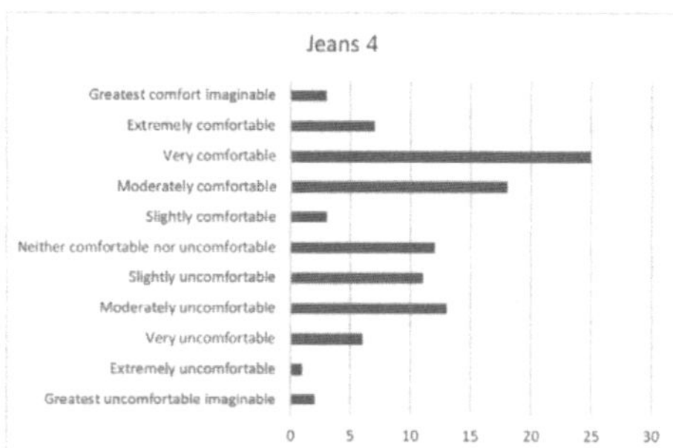

Figure 6. Jeans 4 data.

The jeans 3 and jeans 5 showed the very close data, as the most comfortable imaginable jeans.

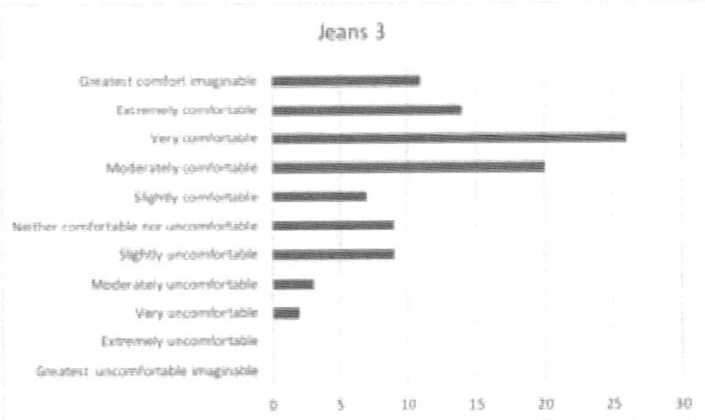

Figure 7. Jeans 3 data.

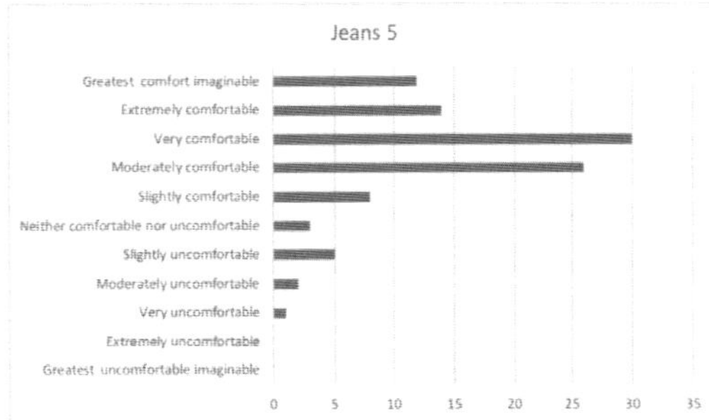

Figure 8. Jeans 5 data.

When comparing the data of all models of the jeans tested it was observed that, in general, the models were evaluated as comfortable as showed in the Figure 9. The jeans 5 stood out by having the greater number of positive evaluations of comfort and therefore it was considered the most comfortable jeans. While the data of jeans 4 demonstrated to be regarded as the most uncomfortable.

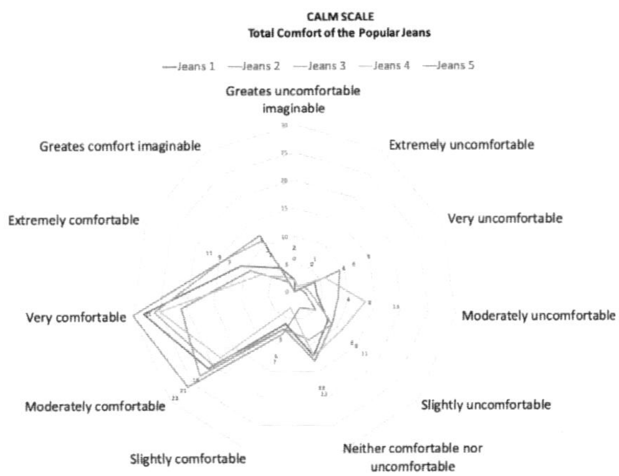

Figure 9. Comparison of jeans evaluation data

During the application of the tests, it was possible to observe the level of satisfaction/dissatisfaction to wearer of popular jeans, relating not only to the aesthetics attributes of the tested jeans, but also to the sensorial aspects brought when the skin is in contact with the jeans. The recognition of the sensory test applied to the popular jeans by means of the application of this investigation, starting from the analysis of the collected data and related among the evaluated parameters identified that the jeans as model 3 and 5 jeans pointed as the most comfortable.

When the collected responses it was found that the perception of comfort/discomfort of the jeans is directly related to the sensation of touch.

Acknowledgment:
This work is financed by FEDER funds through the Competitivity Factors Operational Program - COMPETE and by national funds through FCT – Foundation for Science and Technology within the scope of the project POCI-01-0145-FEDER-007136.

IOP Conf. Series: Materials Science and Engineering **827** (2020) 012038 doi:10.1088/1757-899X/827/1/012038

References

[1] Braga I, Abreu M J and Oliveira M 2017 Subjective evaluation of the comfort of popular denim: Elaboration and validation of the the data, *IOP Conference Series: Materials Science and Engineering*. **254**(18), p. 182001 IOP Publishing.

[2] Corlett E N and Bishop R P 1976 A Technique for Assessing Postural Discomfort, *Ergonomics* **19**(2) pp 175–182. http://www.tandfonline.com/doi/abs/10.1080/00140137608931530

[3] Cardello A V, Winterhalter C and Schutz HG 2003 *Predicting the handle and comfort of military clothing Fabrics from sensory and instrumental data: development and application of new psychophysical methods.* Text. Res. J. **73** pp.221–37.

[4] Gagge A, Stolwijk J and Nishi Y 1972, *An effective temperature scale based on a simple model of human physiological regulatiry response.* Memoirs of the Faculty of Engineering, Hokkaido University, **13** pp 21-36 http://eprints.lib.hokudai.ac.jp/dspace/handle/2115/37901

Polyurethane coated non-woven: A Promising Solution for Building Insulation

Mohamed El Wazna [1,2,6], Ouahiba Mrajji [1,2], Salma Ouhaibi [4], Mohammed Garoum [3], Naoual Belouaggadia [5], Omar Cherkaoui [2], Abdeslam El Bouari [1]

[1] Laboratory of physic-chemical of applied materials, Sciences Faculty of Ben M'sik, University Hassan II Casablanca Morocco
[2] Laboratory REMTEX, Higher School of Textile and clothing Industries, Km 8, Route d'EL JADIDA, Casablanca Morocco
[3] LEME Structure, Superior School of Technology, Salé city, University in Rabat-Morocco
[4]Laboratory of Engineering and Materials (LIMAT), Faculty of Sciences Ben M'Sik, Hassan II University of Casablanca, Morocco.
[5]Laboratory of signals, distributed systems and artificial Intelligence, ENSET, Hassan II University, Mohammedia, Morocco.
[6] Author to whom any correspondence should be addressed

elwazna-mohamed@hotmail.fr

Abstract. The aim of this work is to evaluate the potential application of textile and polyurethane aqueous dispersion in insulation application in the form of a nonwoven polyurethane composite. Based on their properties, nonwoven are promising insulation solution for buildings. However, the major disadvantage is their flammability and their low mechanical property. The main purpose of this research is to improve physical, thermal and acoustical properties of these materials, by using a thin layer of polyurethane. According to the experimental results, it can be concluded that the PU layer enhanced the barrier properties of textile nonwoven. Moreover, it has a great effect on the sound absorption coefficient and thermal conductivity. The results of this study provide an effective solution for insulation application. Indeed, these alternative materials will contribute to the cost advantage as well as the green building initiative.
Keywords: Energy efficient, insulation material, nonwoven, sound absorption, thermo-physical characterizations.

1. Introduction

In Morocco as in Europe, the regulations around the energy performance of building are in strong evolution. The new regulations set ambitious targets for energy efficiency, which include minimizing energy consumption, controlling energy consumption and implementing good practices to reduce energy costs and to achieve a reduction in energy costs [1]. Thermal insulation in buildings is an important factor for the thermal comfort of its occupants. It reduces unwanted heat loss or gain and the energy demand of heating and cooling systems by conduction, convection and radiation. Several thermal insulation materials are used commercially for the insulation of buildings, but few are completely satisfactory for this purpose. The main problems are the lack of mechanical resistance, flammability and moisture absorption. Therefore, it is important to choose easily recyclable, renewable, locally available and environmentally friendly insulation materials.

During last few years, textile materials have found a range of applications in the field of thermal insulation. In the literature, researchers have studied the potential application of nonwovens based on acrylic fibers and they found that the developed sample exhibited excellent insulation performance with a conductivity value of 0.035 to 0.0355 W/(m.K) [2, 3]. Other studies [4] have evaluated the use of textile fibers such as recycled polyester, wool as insulation materials, these materials have been tested in terms of thermal insulation, sound absorption and moisture absorption, and they found that applying these wastes as a possible thermal insulation material seems to be an adequate solution due to their ecological benefits.

Non-woven materials, because of space and weight savings, are one of the most important products in textiles that are being used as thermal insulation materials, due to these the technical advantages, they replaced traditional materials such as foam and panels. There is an increasing number of products made of nonwovens for roofing and construction, such as insulation and membranes. However, the major disadvantage is their flammability and their low mechanical property [5].

The main goal of this paper is to develop cost effective coated insulation materials for building and automotives industries, also to improve physical, thermal and fire behavious properties of nonwoven made form reclaimed fibers, by using an aqueous dispersion polyurethane foam coatings.

2. Materials and methods

2.1. Material

The needling technique was used for the manufacture of the mechanically entangled polyester nonwoven fabric to obtain a consolidated shape with an average porosity of 99% (Table 1).

Table 1. Sample code and its physical properties

Sample	Thickness [mm]	Mass per unit area [g/m^2]	Bulk density [Kg/m^3]	Porosity (%)
Polyester nonwoven [PN]	13	142	10.93	99.2

The PU used in this study were produced by the HUNSMAN Company. A mixture of commercial aqueous dispersion of PU and thickening drop was used, the mixture was subjected for 30 min at a speed of 900 rpm to achieve a viscosity of 3510 cP.

2.2. Methods

After the PU's preparation step, they are applied to the polyester nonwovens through a MATHIS type SV coating machine and dried for 3min at T=160 °C.

The thickness (*e)* of samples was determined based on the procedure cited in ISO 9073-2. The coating thickness C percentage was determined according to equation (1).

$$C\ (\%) = \frac{E - E_0}{E_0} * 100 \qquad (1)$$

With e : thickness of coated nonwoven, e_0 : thickness of uncoated nonwoven.

Experimental observation of coated non-woven materials were carried out using optical microscope (the Leica DME). The objective lens used was a 506226 Hi plan 4x/0.1.

Thermal conductivity was measured using the guarded hot plate apparatus λ-Meter EP500e as per the EN 12667. Sample size used for measurement was (200 mm × 200 mm × samples thickness). In this study, the measuring temperature was 10, 25 and 40°C. Moreover, the temperature difference between the hot plate and the cold plate is set at 15°C in all measurements.

The uniaxial tensile tests of different samples were conducted at room temperature using an MPZ3k universal testing machine, at constant crosshead speed of 5 mm.min^{-1}.

The acoustic performance was evaluated using a vertically disposed impedance tube, two microphones and a digital frequency analyzer (Figure.1). This device known by Kundt's consists of a tube of 1.2m in length and 100 mm in diameter, which was kindly provided by LEME l'EST Structure (University in Rabat-Morocco). The measurements were based on the method described in ISO 10534-2: 2001 Standard. The samples were placed in the middle of the tube, and the coated surface is located in front of loudspeaker. The frequency range used for the measurement was 50–1600 Hz. Five readings were taken randomly from each sample for evaluating acoustic properties.

3. Experimental results and discussion

According to Figure 1 and table 2, a non-linear decrease in the thermal conductivity coefficients with the increase in coating thickness. The good coefficient of thermal conductivity can be explained by the unique structure of nonwoven during the manufacturing process, it consists of dense fibers and pores [2]. The PU coated acts as an additional insulating layer with a conductivity that is constant for a given thickness (Figure.3), which explains the stabilization observed in Figure.2. By developing the thermal resistance equation (2) and measured value, it was possible to determine the thermal conductivity of the coatings. The thermal conductivity of PUF at 10°C was found 22.60 mW/(m.K).

$$R_{th} = R_N + R_c = \frac{E}{\lambda} = \frac{E_0}{\lambda_0} + \frac{E_c}{\lambda_c} \qquad (2)$$

$$\lambda_c = \frac{\lambda * \lambda_0 * E_c}{\lambda_0 * E - \lambda * E_0}$$

Where, R_{th}= thermal resistance of coated sample; R_N = thermal resistance of uncoated sample, R_c=thermal resistance of coating layer, E = thickness of coated sample, E_0 = thickness of uncoated sample, E_c = thickness of coating layer. λ = thermal conductivity of coated sample, λ_0 = thermal conductivity of uncoated sample, λ_c = thermal conductivity of coating layer.

Table 2. Measured thermal conductivity

Coating thickness (%)	Thermal conductivity mW/(m.K)		
	10°C	25°C	40°C
5	34.35	36.77	39.63
15	34.26	36.498	39.75
20	34.22	36.71	39.54
25	33.95	36.49	39.72
30	33.76	36.17	38.97

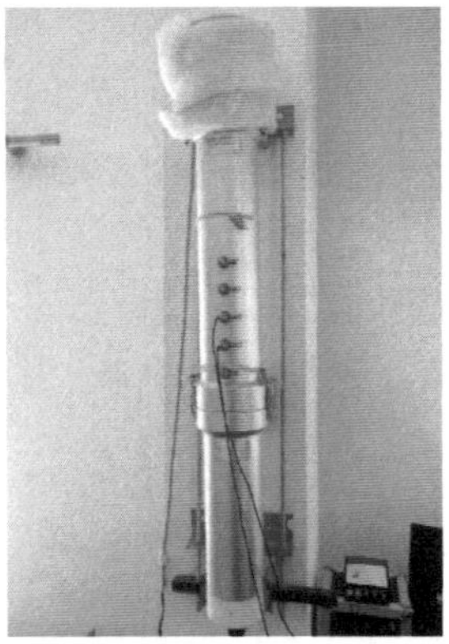

Figure 1. Kundt's tube.

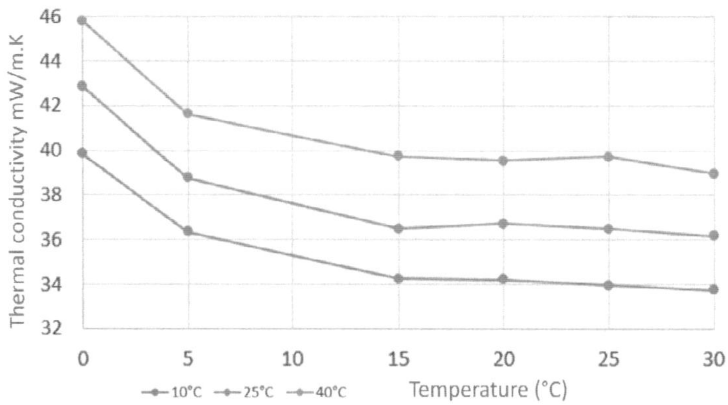

Figure 2. Thermal conductivity in function of temperature.

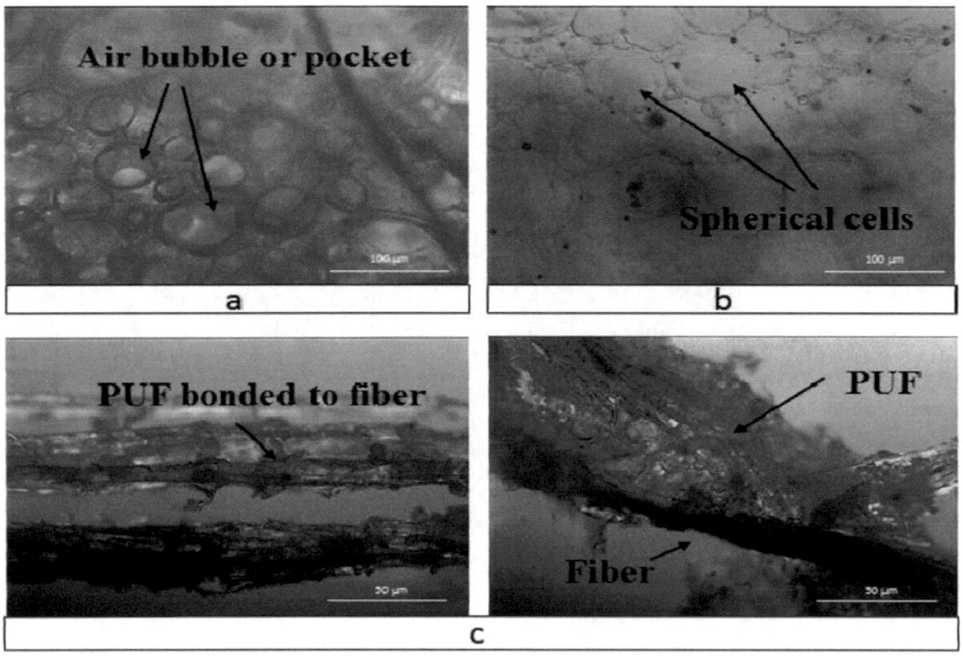

Figure 3. Optical image of coated sample.

The variation in the breaking strength of PN nonwoven coated sample is presented in Figure.4. It is clearly observed that the breaking strength increase with increase in PU coating thickness. At 5% coating thickness (CPUF), the breaking strength of the PN coated samples increases by 77.16 and 30% respectively.

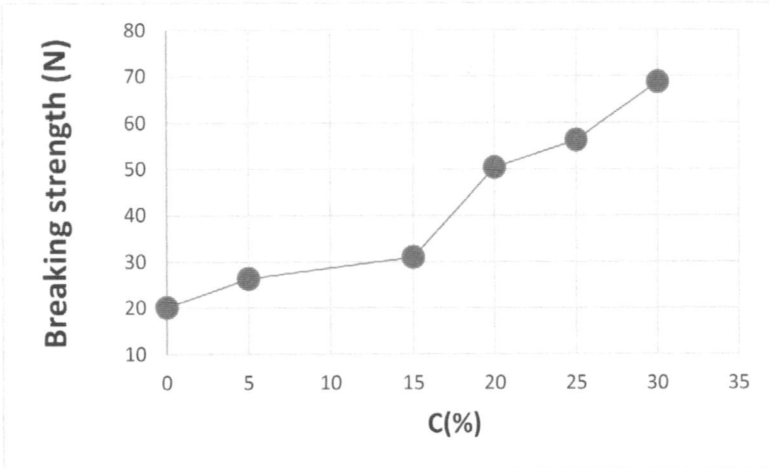

Figure 4. Breaking strength in function of thickness percentage.

The Figure 5 shows the influence of the coating thickness on the sound absorption behavior of PN samples. Is clearly seen that the PUF coating has a great effect sound absorption behavior in the high frequencies (800-1400 Hz). The sound absorption was lower at low frequencies (0-600 Hz) and increased from medium (800-1000 Hz) to high frequency range (1000-1400Hz) for all the samples, and also its also observed that the absorption curve of all samples has been moved to the right (higher frequency), this change is due to the thickness of the sample, because by increasing the thickness of the sample, the acoustic absorption performance increases significantly, this observation is in agreement with the previous study [22, 23].

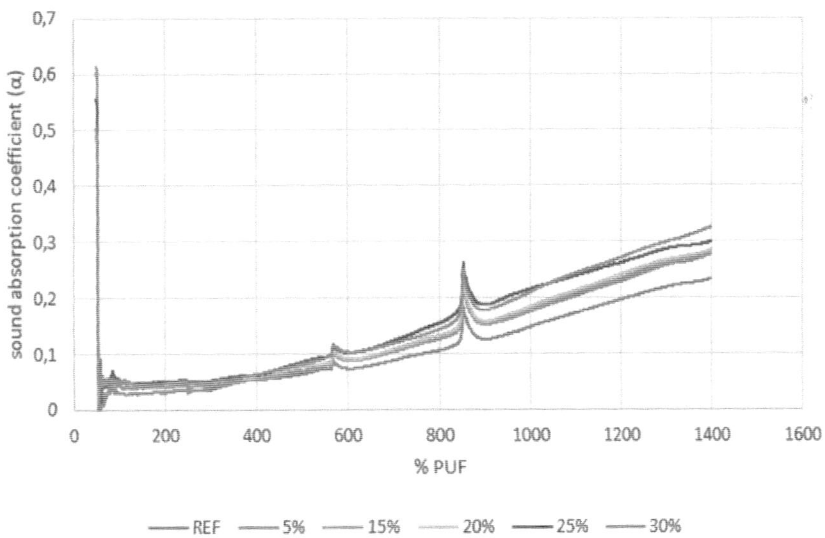

Figure 5. Sound absorption coefficient in function of coating percentage.

4. Conclusion

This article presents a study on the effect polyurethane coating on thermal and physical properties of samples made from polyester fibers for thermal insulation applications. Based on the results, it can be stated that the PU coating enhanced the physical and barriers properties of coated nonwoven. Only 5% in PUF coating thickness improves the conductivity value at 10°C. The breaking strength of PN coated samples increases by 30%. Results also have shown that coating has a great effect on sound absorption behavior in the high frequencies (800-1400 Hz) with a maximum rate of improvement of 39.3%.

References

[1] Kylili A and Fokaides P A 2015 European smart cities: The role of zero energy buildings. Sustainable Cities and Society **15** pp. 86-95.

[2] El Wazna M, El Fatihi M, El Bouari A and Cherkaoui O 2017 Thermo physical characterization of sustainable insulation materials made from textile waste. *Journal of Building Engineering*.

[3] Briga-Sa A, Nascimento D,Teixeira N, Pinto J, Caldeira F, Varum H and Paiva A 2013 Textile waste as an alternative thermal insulation building material solution. *Construction and Building Materials* **38** pp.155-160.

[4] Patnaik A, Mvubu M, Muniyasamy S, Botha A and Anandjiwala R. D 2015 Thermal and sound insulation materials from waste wool and recycled polyester fibers and their biodegradation studies *Energy and Buildings* **92** pp.161-169.

[5] Bhat G S 1995 Nonwovens as three-dimensional textiles for composite *Material and manufacturing process* **10**(4) pp. 667-688.

A DEA-ANN framework based in Improved Grey Wolf Algorithm to evaluate the performance of container terminal.

Mouhsene Fri[1,2,3], Kaoutar Douaioui[1], Samir Tetouani[1,2], Charif Mabrouki[1] and El Alami Semma[1]

[1] LMII- Faculty of Sciences and Technology, Hassan 1st University, PO Box 577, Settat, Morocco
[2] CELOG-ESITH, PO Box 7731 OULFA, Casablanca, Morocco
[3] Author to whom any correspondence should be addressed

frimouhsene@gmail.com

Abstract. Managing the performance of port container terminals is one of the major challenges in the supply chain management. In response to this challenge, we propose a new framework to assess managers in evaluating the global performance of operations in port container terminals. The new framework integrates the Data Envelopment Analysis (DEA) and Artificial Neural Network (ANN). The DEA is used to compute the efficient score of the system. In ANN, we use the improved grey wolf optimizer based on Levy's flights to improve learning. In order to prove the efficiency of our model, we apply the framework in 2 ports container terminal: Tangier Med Port and Casablanca Port. The result is compared to standard algorithms and outperforms the cited algorithm in order to avoid local minima. The new trainer improved grey wolf optimizer is also evaluated using four known classification datasets and on three approximation functions datasets.
Keywords: Port Container Terminal, Performance Measurement System (PMS), Data Envelopment Analysis (DEA), Artificial Neural Network (ANN), Levy Flights.

1. Introduction

Port is the link between two geographical space; the seaside and the landside. One of the main goals of the seaport is to transport cargo from suppliers to customers in excellent conditions. In this regard, the important role of port container terminal has evolved from the first generation to the fourth generation. However, the issue of managing the performance present one of the most challenges in the port terminal container. Port performance has largely been treated by researchers [1], [2]; *Bichou and Gray* highlight the important role of logistics performance in Port containers [3]. On the other hand, *Chul-hwan* demonstrate the impact of the supply chain integration into port performance by proving that the customer integration and cost performance impact the quality performance [4]. European seaports launch a project named "Performance Indicators: Selection and Measurement Project" in order to standardize the performance indicators in all European ports [5]. Despite this, we call attention to the lack of unique performance measurement system able to reflect the global performance in order to evaluate and classify different ports. In this study, we define the performance measurement system, in addition, we developed a new framework to evaluate the global performance based on two steps. In the first step we compute the efficiency score by using DEA oriented input models and in the second step we use ANN based in improved grey wolf optimizer. Grey Wolf Optimizer is enhanced by the so-called

Levy flights.

This study is organized as follows: The first section aims to provide a performance measurement system. The second section presents the methods and algorithms used to elaborate the proposed framework. The third section contains the numerical results as well as the benchmarking of result with the known algorithms. Finally, the fourth section aims to present the conclusion and future works.

2. Performance Measurement System

Globalization and liberalization of business have a direct impact on the number of goods transported through port container terminals. In this regard, the intention to manage performance has increased in recent years. Many researchers purposed performance measurement systems but the majority of those systems focus on the financial performance [6] and neglect the industrial and logistical performance which represents one of the primary goals of the seaport in the supply chain management. However, the unfulfilled gap here is the absence of a method able to evaluate and measure the global performance system. Regarding this, we used the proposed key performance indicators developed by *Fri et al.* [7](see table1).

Table 1. The overall performance measurement system

Key driver [7]	
Administrative Management (AM)	Safety and Security (SS)
Management of Ship Services (MSS)	Port Sizes (PS)
Management of Quay Operations (MQO)	Port Equipment (PE)
Yard Operations (YO)	Exploitation of Technology (ET)
Employee Performance (EP)	Services and Organizations (SO)
Financial Health (FH)	Competitive Position (CP)
Competitive Position (CP)	Safety and Security (SS)

3. Hybrid Framework

The proposed hybrid framework is based on three steps illustrate in figure 1. In the first step, we identify the performance measures objectives and the indicators, the sub-indicators affecting our system. In the second step, we compute the efficiency scores by the CCR model (oriented inputs) [8]. In the last step, the improved grey wolf algorithm employed as a new method for training ANN to determine the efficiency scores. In our case the learning will be enhanced by improving the grey wolf algorithm.

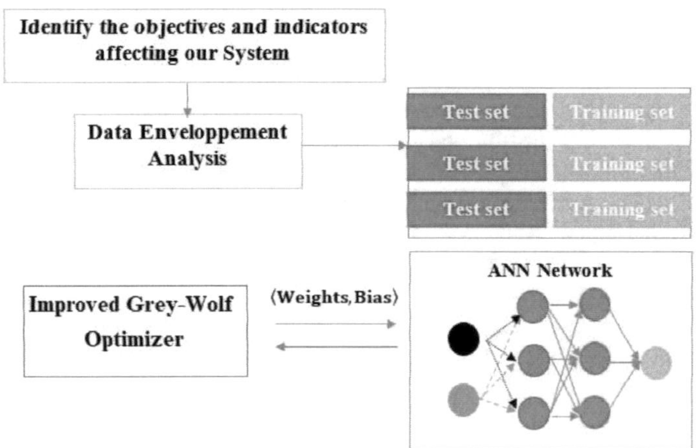

Figure 1. The proposed hybrid framework.

3.1. Improved Grey Wolf Algorithm

Grey wolf optimizer (GWO) algorithm is one of the latest bio-inspired optimization techniques, which mimicking the social leadership and hunting technique of grey wolves in nature. This algorithm was proposed by [9]. Grey wolves live in group, they have very strict rules in the dominant social hierarchy. This hierarchy is simulated by categorizing the population of search agents in four types of individuals [9] , i.e., alpha, beta, omega and delta.

1. The alpha wolf is the leader of the group; he is the maker of all decisions. The alpha's decisions are ordered to the group.
2. The beta wolf is the subordinate of alpha. Beta wolf help alpha wolf in making decisions and the other activities. The beta is most qualified to be the alpha in the future.
3. The omega wolf is the scapegoat. He is all time obliged to submit to the dominant wolves and he is the last one allowed to eat.
4. The delta wolf must obey alpha and beta, but he dominates the omega. This category includes the Scouts, the sentinels, the elderly, the hunters, and the caretakers. The scouts are responsible for observing the frontiers of the territory and alerting the gro0up if there is any danger. The sentinels safeguard and protect the group. The elders have extensive experience who has been in the level of alpha or beta. The hunters support the alpha and beta in hunting and providing food for the other members of the group. Finally, the caretakers are responsible for watching over the sick wolves, the weak and the wounded wolves in the group.

Grey Wolf Optimizer Algorithm

Initialization. Population of grey wolf, NP, the maximum of iterations T_{max} , A, C, a;
Initialize the grey wolf ($X_i = 1,2,\ldots,N$);
For each wolf, calculate the fitness value;
Rank the wolf pack as alpha (X_α), beta (X_β),), delta (X_δ);
for $i = 1\ To\ \frac{N_p}{5}$ **do**
  $X_i^{t+1} = X_{i,worst}^t + \alpha \oplus Levy(\lambda)$
  if $(F(X_i^{t+1}) < F(X_{i,worst}^t))$ **then**
    $X_{i,worst}^t = X_i^{t+1}$
    $F(X_{i,worst}^t) = F(X_i^{t+1})$
  end if
end
while (stop criterion) **do**
  for $i = 1$ = 1 to NP **do**
    Update the position by equations (5)-(8);
  end for
  Obtain the updated wolf pack;
  Update $A, C\ and\ a$;
  Recalculate the fitness values of the wolves;
  Rank the updated wolf pack as alpha (X_α), beta (X_β), delta (X_δ);
end while
Return the position of alpha (X_α) as the final solution.

Figure 2. The improved Grey Wolf Optimizer Algorithm.

In conventional Grey Wolf Algorithm, it may converge prematurely without enough exploration of search space. In order to increase the diversity of the population against premature convergence and accelerate the convergence speed, this paper proposes an improved grey wolf optimizer based on Levy flights. Levy flights, represents a kind of non-Gaussian stochastic process whose step sizes are distributed based on a Levy stable distribution to generate new solutions. When a new solution is produced, the following Levy flight is applied:

$$X_i^{t+1} = X_i^t + \alpha \oplus \mathrm{Levy}(\lambda) \quad (1)$$

Here, α is the step size that is relevant to the scales of the problem. The product ⊕ means entry-wise multiplications. Levy flights essentially provide a random walk while their random steps are drawn from a Levy distribution for large steps:

$$Levy(\lambda) = u = t^{-\lambda};\ 1 \leq \lambda \leq 3) \quad (2)$$

In this paper, we will use the algorithm proposed by [9], which is one of the most efficient algorithms used to implement Levy flights.

4. Numerical Results

In this section, we conduct investigations in the efficiency of the proposed hybrid framework for evaluating the global performance of port container terminal. The dataset is partitioned based on DEA into 66% for training data and 34% for testing data. The results of port datasets are presented in Table 02 based on the average (AVE) and standard deviation (STD). The main scope to employ these two measures is to indicate the ability of algorithms to avoid local minima. By analyzing the Table 2, the first thing that can be observed in the results of the highest performance obtained by the proposed method, this behavior is a result of the great ability to avoid the local optima, significantly better than other algorithms. In addition, a convergence comparative experiment was carried out to confirm that Improved Grey Wolf Algorithm has better convergence performance than Algorithm Genetic and Particle Swarm Optimization. Figure 3 –(a) shows the convergence of Improved Grey Wolf Algorithm, Biogeography-Based Optimization, Particle Swarm Optimization and Algorithm Genetic. To validate our model, we conducted a case study based on two ports container terminals namely: Tangier Med and Casablanca port. Table 03 shows the obtained results. Additionally, the show the efficacy of the proposed method, we are benchmarked it on four selected standard classification data sets from the University of California at Irvine (UCI) Machine Learning Repository: XOR, balloon, Iris, breast cancer. Table 4 – 8. The results of Improved Grey Wolf Algorithm-ANN follow by those of Biogeography-Based Optimization-ANN, Algorithm Genetic-ANN and Particle Swarm Optimization-ANN for all datasets. Since the difficulty of this dataset and MLP structure is high for this dataset, these results are strong evidence for the efficiency of Improved Grey Wolf Algorithm in training MLPs. The results testify that this algorithm has superior local optima avoidance and accuracy simultaneously. In addition, a convergence comparative experiment was carried out to confirm that Improved Grey Wolf Algorithm has better convergence performance than the benchmark for data classification and approximation function datasets. Figure 3 (b-e) shows the convergence of Improved Grey Wolf Algorithm for XOR, Baloon, Iris and Cancer data sets respectively.

(a) Ports dataset (b) XOR dataset (c) Balloon dataset

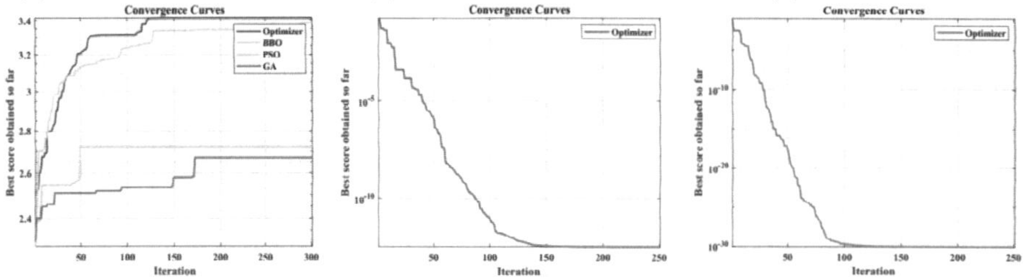

(d) Iris dataset (e) Breast Cancer dataset

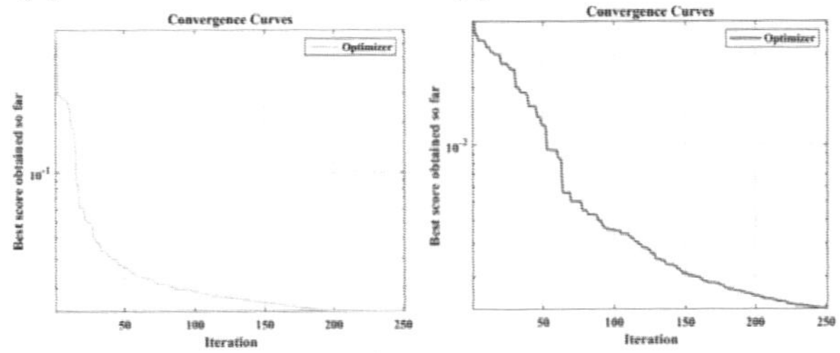

Figure 3. Overall Convergence Curves.

Table 2. Experimental results for Ports datasets.

Algorithm	AVG ± STD	Performance – Error
Improved Grey Wolf Algorithm	**3.3002 ± 1.549e – 07**	**0.5725**
Biogeography-based optimization [10]	3.1126 ± 0.04898	1.2210
Particle Swarm Optimization [10]	2.0106 ± 0.23041	2.8324
Algorithm Genetic [10]	2.0096 ± 0.18295	3.8081

Table 3. Moroccan Case Study

Ports	Target	Obtained-Performance				Rank
		BBO	PSO	GA	IGWO	
Tangier Med	200%	**157%**	108%	120%	**164.22%**	**1**
Casablanca Port	200%	**122%**	92.7%	95%	**127.35%**	2

Table *4*. Classification datasets

Classification datasets	Number of attributes	Number of training samples	Number of tests	Number of classes
3-bits XOR	3	8	8 as training samples	2
Balloon	4	16	16 as training samples	2
Iris	4	150	150 as training samples	3
Breast Cancer	9	599	100	2
Heart	12	80	187	2

Table 5. *Experimental results for XOR datasets.*

Algorithm	AVG ± STD	Classification rate(%)
Improved Grey Wolf Algorithm	**3.31E – 09 ± 0.000000**	**100**
Biogeography-based optimization [10]	3.65E – 07 ± **0.000000**	**100**
Particle Swarm Optimization [10]	0.084050 ± 0.035945	35
Algorithm Genetic [10]	0.000181 ± 0.000413	**100**

Table 6. Experimental results for balloon datasets.

Algorithm	AVG ± STD	Classification rate(%)
Improved Grey Wolf Algorithm	**9.1952E – 31 ± 4.8474E – 33**	**100**
Biogeography-based optimization [10]	8.09E – 27 ± **1.51E – 42**	**100**
Particle Swarm Optimization [10]	0.000585 ± 0.000749	**100**
Algorithm Genetic [10]	5.08E – 24 ± 1.06E – 23	**100**

Table 7. Experimental results for Iris datasets.

Algorithm	AVG ± STD	Classification rate(%)
IGWO	**0.015766** ± 0.0004059	**92**
BBO [10]	0.019150 ± **3.66E – 18**	90
Particle Swarm Optimization [10]	0.228680 ± 0.057235	37.33
Algorithm Genetic [10]	0.089912 ± 0.123638	89.33

Table 8. Experimental results for Cancer datasets.

Algorithm	AVG ± STD	Classification rate(%)
Improved Grey Wolf Algorithm	**0.001388** ± 3.1451e-05	**98**
Biogeography-based optimization [10]	0.002807 ± **0.000000**	95
Particle Swarm Optimization [10]	0.034881 ± 0.002472	11
Algorithm Genetic [10]	0.003026 ± 0.001500	**98**

5. Conclusion

This paper uses the methods ANN and DEA to cover the drawbacks of DEA especially in statistical noise. For that we started by identifying the objectives and the indicators affecting our system. Then, we applied the CCR model (oriented inputs) to evaluate the scores. As a result, we employed an improved grey wolf algorithm as new trainer for ANN to enhance the performance of the system. In addition, the algorithm was compared with the renowned benchmarks. Finally, the reported results proof that the improved grey wolf is able to outperform the Biogeography-based optimization, algorithm genetic and Particle Swarm Optimization on the collected port datasets and on almost of classification and approximation datasets.

References

[1] Douaioui K, Fri M, Mabrouki C and Semma E A 2018 Smart port: Dcsign and perspectives in *2018 4th International Conference on Logistics Operations Management (GOL)*, Le Havre pp. 1–6.

[2] Bentaleb F,Mabrouki C and Semma A 2015 Key Performance Indicators Evaluation and Performance Measurement in Dry Port-Seaport System: A Multi Criteria Approach *Journal of ETA Maritime Science* **3**(2) pp. 97–116.

[3] Bichou K and Gray R 2004 A logistics and supply chain management approach to port performance measurement *Maritime Policy & Management* **31**(1) pp. 47–67.

[4] Chul-hwan H 2018 *Assessing the impacts of port supply chain integration on port performance* | Elsevier Enhanced Reader.

[5] European Sea Ports *PPRISM Port PeRformance Indicators: Selection and Measurement Project Executive report (PPRISM WP4 D4.2)* European Sea Ports Organization, 2012.

[6] FRI M, Fedouaki F, DOUAIOUI K, Mabrouki C and SEMMA E A 2019 Supply Chain Performance Evaluation Models, State-of-the-Art and Future Directions *International Journal of Engineering and Advanced Technology* **9**(1) pp. 6336–6347.

[7] Fri M, Douaioui K, Lamii N, Mabrouki C and Semma E A 2020 A DEA-Based Hybrid Framework to Evaluate the Performance of Port Container in *Advanced Intelligent Systems for Sustainable Development (AI2SD'2019)* **1104** M. Ezziyyani, Ed. Cham: Springer International Publishing, pp. 1–12.

[8] Charnes A, Cooper W W and Rhodes E 1978 Measuring the efficiency of decision making units *European Journal of Operational Research* **2**(6) pp. 429–44

[9] Mirjalili S, Mirjalili S M and Lewis A 2014 Grey Wolf Optimizer," *Advances in Engineering Software* **69**, pp. 46–61.

[10] Mirjalili S, Mirjalili S M and Lewis A 2014 Let a biogeography-based optimizer train your Multi-Layer Perceptron *Information Sciences* **269** pp. 188–209.

Knitted Cotton Fabric Strain Sensor by In-situ Polymerization of Pyrrole

G B Tseghai[1,2,3,4], B Malengier[1], D A Mengistie[2], K A Fante[3] and L Van Langenhove[1]

[1] Department of Materials, Textiles and Chemical Engineering - Centre for Textile Science and Engineering, Faculty of Engineering and Architecture, Ghent University, Ghent,Belgium
[2] Ethiopian Institute of Textile and Fashion Technology, Bahir Dar University, Bahir Dar,Ethiopia
[3] Department of Electrical Engineering: Jimma Institute of Technology, Jimma University, Jimma, Ethiopia
[4] Author to whom any correspondence should be addressed

GranchBerhe.Tseghai@UGent.be

Abstract. The purpose of this research work was to develop a textile-based strain sensor. A conductive textile fabric was produced by the coating of knitted cotton fabric with in-situ polymerization of polypyrrole. The sensor consists of a conductive textile as an electrode, stainless steel yarn as interconnection, Arduino Nano as a controller, HC05 Bluetooth module, and a Lithium polymer battery as a power source. For the demonstration, the sensor was placed on the upper arm and bicep stretch was performed. It was observed that the contraction of the arm muscle causes a reduction in resistance of the electrode. Therefore, change in swelling was successfully detected from the increase and drop of resistance during contraction and relaxation of the muscle. This principle could be applied to determine the status of peripheral edema, where the increase in resistance in this work indicates edema is becoming severe.

1. Introduction

Nowadays, textile materials are relied upon to exhibit extra functionalities other than comfort and durability in accordance with the movement of electronic and computerized correspondence [1]. Stretchable hardware is constructed from yarns, fabric textures, or other articles of clothing and has received extensive interest. The construction of textile-based sensors i.e. the sensor is a device that acquires a physical quantity and converts it into a signal suitable for processing (mechanical, electrical, optical) to provide a usable output in response to a specific stimulus [2] is booming a buzzword in the field of smart textile materials. Textile based sensors are preferable for wearable applications due to their matching flexibility, light weight and possibility of washing. As a result, textile-based pressure sensors [3], [4], biopotential sensors [5], [6] and strain sensors [7], [8] have been constructed in different formats. Applications, where a polymer is applied to conventional textiles to make them conductive and use for an appropriate type of sensing function, are under investigation. Polypyrrole is one of the most explored conducting polymers because of its acceptable electrical conductivity and flexibility [9].

This work is aimed at developing a homemade strain sensor from a textile-based electrode produced by in-situ polymerization of lower pyrrole monomer concentration on a knitted cotton fabric under ammonium persulfate oxidizing agent. This type of sensors could have a typical use as stretching, swelling, relaxation and contraction sensor in biomedical and sports applications.

2. Material and Methods

A half bleached knitted cotton fabric having 140 GSM available at the Department of Materials, Textiles and Chemical Engineering, University of Ghent, was selected as a textile substrate. A reagent grade 98% pyrrole monomer obtained from SIGMA-ALDRICH, Belgium and a reagent grade, 98% Ammonium Persulfate oxidizing agent obtained from SIGMA-ALDRICH, Belgium were employed. The knitted cotton fabric was coated by in-situ polymerization of 0.3M of pyrrole monomer in the presence of 0.15M of ammonium persulfate. An aqueous liquor bath was prepared using a 2:1 molar ratio of pyrrole and ammonium persulfate. The fabric samples were batched to the prepared solution (figure 1-left) at 1:40 material to liquid ratio (MLR) and kept for two hours at room temperature and ambient atmospheric pressure. The sample was treated with ethanol in order to terminate polymerization and then with distilled water to remove unreacted components. Finally, the samples were dried in an oven drier at 100°C for 3 minutes to obtain the actual fabric (figure 1-right).

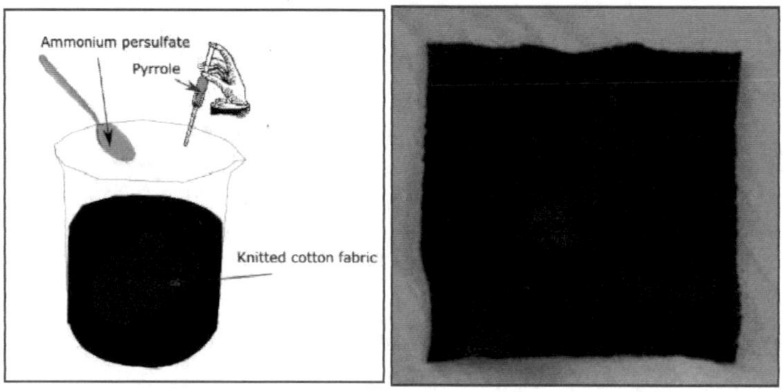

Figure 1. In-situ polymerization of pyrrole on knitted cotton fabric (left) and PPy coated knitted cotton fabric (right).

Thickness, bending length, tensile strength, elongation at break and SEM were tested to evaluate the mechanical property of the polypyrrole coated cotton fabric. The thickness was determined using ISO 5084:1996(E) (Determination of thickness of textile and textile products) using the Mitutoyo Digimatic Indicator. Bending length was analyzed following the BS 3356:1990 test method using bending length testing equipment. Using the appropriate bending length mean value, we calculated the flexural rigidity G, in mg cm, separately for the course and wale directions by the formula in equation 1.

$$G = 0.1MC^3 \quad (1)$$

Where C = Bending Length (cm), M= Mass/Area of Specimen (g/m^2)

The strength and elongation at break were tested using INSTRON universal strength tester. A tensile test according to ISO 13934-1 was used. The surface topology of the fabric before and after coating was studied using FEI Quanta 200 FFE-SEM. Images were taken with an accelerating voltage of 20 kV. The non-conductive samples were prepared prior to analysis by applying a gold coating using Balzers Union SKD 030 sputter coater.

The resistance of the fabric was measured by stretching of the samples using 3-7/8" MaxSteel Light Duty Drill Vise 83070 Stanley Hand Tool (figure 2-left) from 0 to 35% elongation. The surface resistance of all prototypes was measured by using a four-point probe measurement and a microcontroller (Arduino Nano). In the Arduino measurement, an "Arduino Nano" which was set up specifically for this particular purpose with the circuitry in figure 2-right was used. One end of polypyrrole treated fabric was connected to 5V of Arduino input and the other was connected in series to a 10 kΩ pulldown resistor, with the midpoint connected to the analog input (ADC) of the Arduino (figure 2-right). The ground of the Arduino was connected to another end of the resistor.

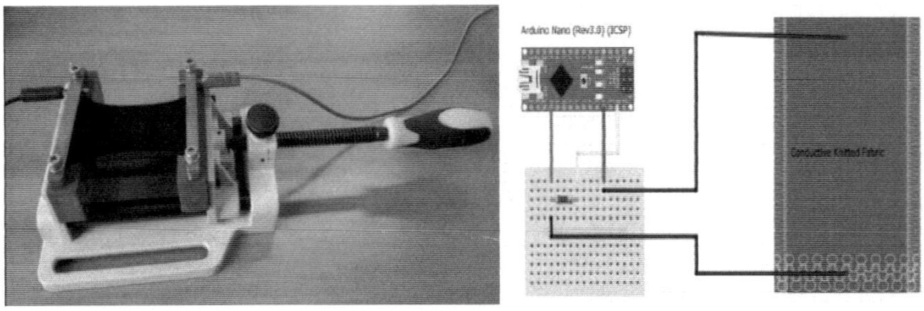

Figure 2. Electrical circuitry establishment and measurements: Duty Drill Vise Stanley Hand Tool (left), and Arduino Nano for dynamic test (right).

To test the performance of the sensor, a textile-based electrode was constructed by sewing a 5cm x 2cm conductive knitted fabric between two low extensibility strips of textile fabrics, see figure 3.

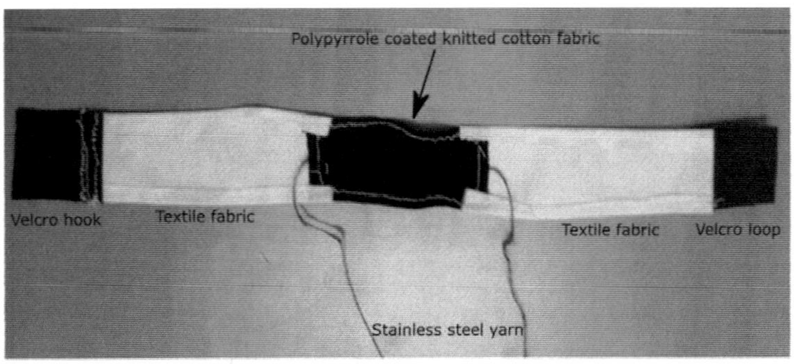

Figure 3. Actual textile-based strain electrode construction.

A stainless steel yarn was used as an interconnection between the textile-based electrode and the Arduino Nano controller. HC05 Bluetooth Module was used to transfer data from the controller to a Smartphone. A 6V battery was used to power the Arduino and the Bluetooth module.

By placing the strain sensor on the upper arm and subjecting the bicep muscle to physical relaxation and contraction, the effect on resistance was observed.

3. Results and Discussion

The thickness of the fabric has increased from 0.5 mm to 0.66 mm due to Polypyrrole coating. This increment in thickness imparted stiffness on the fabric. The fabric has gotten stiffer after treatment which is clear from the increase in bending length and weight of fabric. The bending length of the polypyrrole treated fabric has been increased from 1.225 mm to 1.45 mm in wale direction and 1.25 to 1.375 mm in course direction. The respective calculated flexural rigidity ($G = 0.1\ MC^3$) has also increased from 25.74 to 62.8 gm cm in wale direction and from 27.34 to 53.55 gm cm in course direction.

Both the tensile strength and strain at break of the fabric were reduced due to treatment from 107.3 to 100.7 N and 71.34 to 70%, see figure 4. This could be due to the effect of the coating process and drying after treatment that could cause very fabric degradation in addition to the presence of the polypyrrole.

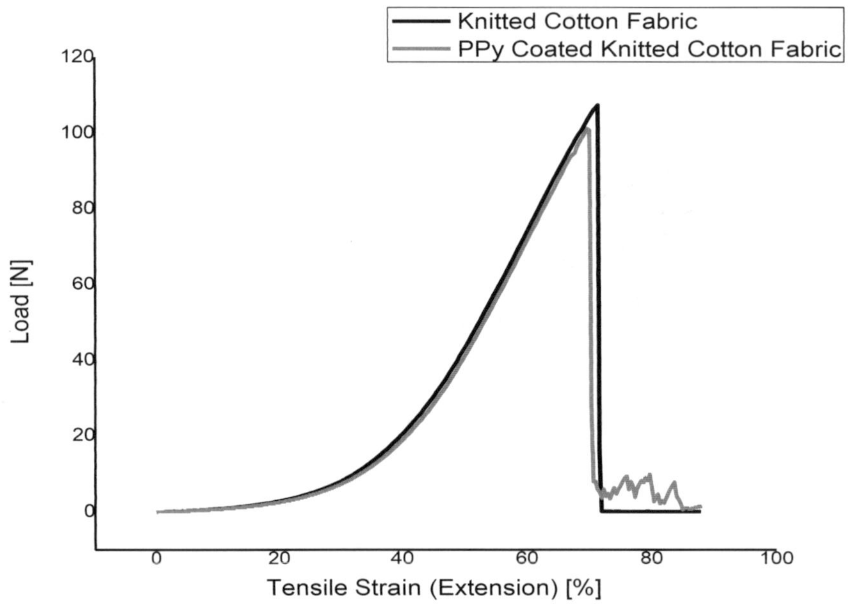

Figure 4. Example typical Load-Extension curve

The SEM results in figure 5 showed that the yarn loops interstices in the fabric were coved by the addition of the Polypyrrole. In addition, many protrudes loops were observed in the fabric but not after the polymers were applied. The presence of Polypyrrole made the smoothness better and led to coverage of the yarns.

Figure 5. SEM Results: Knitted cotton fabric (left), Polypyrrole coated knitted cotton fabric (right)

The four-point probe measurement showed the surface resistance of the polypyrrole coated knitted cotton fabric was 60Ω/sq and a linearly increasing function of the distance between the electrodes.

Of great importance to use the fabric as a sensor is to obtain a dependency of the resistance on elongation. This was the case, the surface resistance of the conductive fabric has increased to 350 Ω at 35% elongation. Next, a 5cm x 5cm stripe was used for demonstration, using the Arduino Nano to read out the resistance. The sensitivity of the coated fabric was found to be fairly stable, raising the resistance from 0.48 kΩ to 6.7 kΩ (+1,295.8 %) at 60% elongation. The typical dynamic response at 0.12 cm/s constant rate of stretching is shown in figure 6.

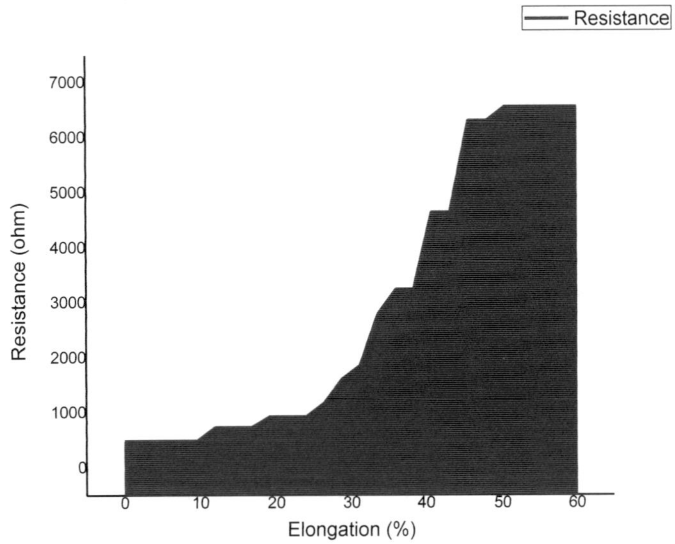

Figure 6. Dynamic testing of coated fabric resistance at constant rate of stretching.

It was noticed that the resistance of the textile-based strain sensor electrode (5cm x 2 cm) increases during muscle contraction (flexion) and vice versa during relation (extension), see figure 7. This is due to the increase in circumference of the upper arm during contraction causing stretching of the sensor placed on it.

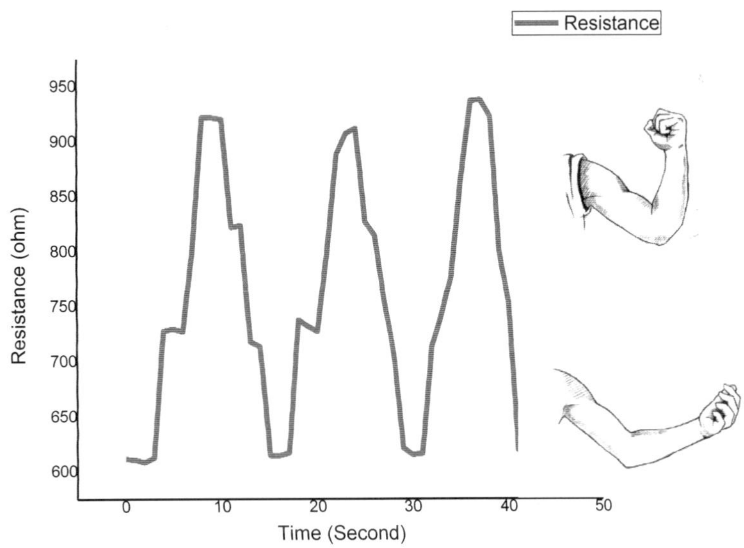

Figure 7. Performance testing of the strain sensor.

This principle could be applied to determine the status of peripheral edema, where the increase in resistance indicates edema is becoming severe. Peripheral edema is swelling of body parts due to excessive trapping of fluid which can happen in hands, arms, feet, ankles, and legs[10].

4. Conclusion

We have successfully developed a conductive knitted cotton fabric by in-situ polymerization of polypyrrole which was then used to produce a textile-based strain sensor. For demonstration, the strain sensor was placed on the upper arm and bicep flexion and extension were performed. It was observed that the swelling of the arm muscle due to bicep flexion causes an increase in resistance of the electrode. The swelling extent was noticed from the increase and drop of resistance during contraction and relaxation of the muscle. Therefore, the developed textile-based strain sensor can be used to determine the status of peripheral edema in biomedical applications and as a swelling/contraction and loading/unloading sensor in different areas.

Acknowledgment

The authors would like to express appreciation for the support of the NASCERE project. This project was funded with support from the Ethiopian Government. This publication reflects the views only of the authors, and the Ethiopian Government cannot be held responsible for any use which may be made of the information contained therein.

References

[1] Petrusic S and Koncar V 2016 Controlled release of active agents from microcapsules embedded in textile structures *Smart Textiles and their Applications* (pp. 89-114). Woodhead Publishing.

[2] Eke B O and Egbono F 2017 Designing Internet of Things System for Checking Cattle Rustling in Nigeria *International Journal of Computer Applications* **157**(7).

[3] Tseghai G B, Malengier B, Nigusse B, Bayrau A and Van Langenhove L 2018 Development and evaluation of resistive pressure sensors from electro-conductive textile fabric. In *Second International Forum on Textiles for Graduate Students* (pp. 651-7). Tianjin Polytechnic University.

[4] Lei K F, Lee K F and Lee M Y 2012 Development of a flexible PDMS capacitive pressure sensor for plantar pressure measurement *Microelectronic Engineering* **99** pp.1-5.

[5] Paul G M, Cao F, Torah R, Yang K, Beeby S and Tudor J 2013 A smart textile based facial EMG and EOG computer interface. *IEEE sensors journal* **14**(2) pp.393-400.

[6] Pani D, Dessì A, Saenz-Cogollo J F, Barabino G, Fraboni B and Bonfiglio A 2015 Fully textile, PEDOT: PSS based electrodes for wearable ECG monitoring systems *IEEE Transactions on Biomedical Engineering* **63**(3) pp.540-9.

[7] Zahid M, Papadopoulou E L, Athanassiou A and Bayer I S 2017 Strain-responsive mercerized conductive cotton fabrics based on PEDOT: PSS/graphene *Materials & Design* **135** pp.213-22.

[8] Seyedin S, Razal J M, Innis P C, Jeiranikhameneh A, Beirne S and Wallace G G 2015 Knitted strain sensor textiles of highly conductive all-polymeric fibers *ACS applied materials & interfaces* **7**(38) pp.21150-8.

[9] Weng W, Chen P, He S, Sun X and Peng H 2016 Smart electronic textiles. *Angewandte Chemie International Edition* **55**(21) pp.6140-6169.

[10] Cho S and Atwood J E 2002 Peripheral edema *The American journal of medicine* **113**(7) pp.580-6.

Improvement of enzymatic antifelting treatment of wool fiber and its effect on dyeing with juglone from a common walnut plant

Y Chemchame[1,3], H Benzbir[1], A Kachachi[1], M El Bouchti [2] and A Kharchafi[1]

[1]Department of Traditional Weaving, Academy of Traditional Arts, Foundation of Hassan II Mosque, Casablanca, Morocco.
[2]REMTEX Laboratory, High School of Industry Textile and Clothing (ESITH), PO Box 7731, Oulfa, Casablanca, Morocco.
[3] Author to whom any correspondence should be addressed

ychem2@gmail.com

Abstract. The objective of this study is to assess an enzymatic effect on the antifelting treatment of yarn and wool fabric. Enzymes, namely, the proteases cardosines A and B, are extracted from the flowers of the cardoon (*Cynara cardunculus L*) plant. Several conditions of enzymatic treatment are studied in order to optimize the antifelting process and improve shrink proofing. The optimal process is at 40 °C for 2 h in an alkaline medium at pH 10. The results show that the shrinkage of 16.8% for untreated yarn is reduced to 0%, while the shrinkage of 10% for the untreated fabric is reduced to 5%. In addition, the antifelting treatment increases the dye absorbance of a walnut stain (common walnut plant) on the yarn and wool fabric, although the fixation dye rate is decreased. Post oxidation is tested in order to improve the fixation dye rate and the washing fastness of juglone on the wool yarn and fabric.

1. Introduction

Antifelting treatment is significantly important for improving shrink proofing and preserving fabric armor. After washing, the whole orientation of the fibers on the fabric surface is conserved, thereby maintaining the original look.

Several studies have confirmed the efficiency of bioantifelting treatment and have shown that it can replace the oxidant agents such as sodium hypochlorite and peroxysulphuric, which can cause damage to the fibers and the environment. In the antifelting treatment of wool, cardosines were not used before this time or may be rarely. However, there were a lot of research in antifelting treatment with papain, savinase, cutinase. woolase, porcine trypsine and others. *Nurhan and Merih* studied enzymatic, hydrogen peroxide and sodium hypochloride, and the additive agents were based on polyurethane, polysiloxane, silicone, chitosan, a hydrophilic and a natural polymer oxidative, to improve the shrink-proofing and antifelting properties of wool. They performed the experiments with enzymatic treatments using commercial protease preparations, such as Perizym AFW, Alcalase 2.5 L, Savinase 16 L and Papain [1]. A biocomposite made from keratin polypeptides and waterborne polyurethane [2] was employed as a bioantifelting agent for wool fabric by *Zhuang D. et al.* [3]. Their results indicated that with an increasing content of keratin polypeptides from 0 to 6 wt.%, the area shrinking rate of the treated wool fabrics decreased from 4.55% to 0.47% [3].

Using others technics, a significant improvement in the wettability of chlorinated and plasma-treated wool fabrics has been observed, when compared with an untreated fabric sample [4]. The improved hydrophilicity of the wool fiber surface by nitrogen plasma could enhance and achieve a better polymer deposition on the surface for improving the antifelting properties of wool when compared with the chlorination process [4].

In this research, we seek to achieve a new zero-AOX [5] antifelting processing alternative to conventional processes, such as chlorine-Hercosett processing, and thus, use environmentally-friendly enzymes extracted from plants [6]. We choose the cardoon plant (*Cynara cardunculus L*) as a source of protease enzymes known as cardosines. They can break the peptidic bonds between the hydrophobic amino acids of the protein [7]. The maximum effect of these enzymes can be achieved in acid conditions. There are two principal types of cardosines: cardosine A and cardosine B. Like chymosin, cardosine A is considered to be responsible for the coagulant activity of proteins, whereas cardosine B has proteolytic activity (*Verissimo et al.* 1996).

The aim of this work is to study the enzymatic effect on the antifelting treatment of the wool. Hence, we tested the antifelting effect on yarn and wool fabric using cardosines extracted from the cardoon flower (*Cynara cardunculus L*). Various treatment conditions are studied in order to establish the optimal recipe. Moreover, the effect of antifelting treatment on the dyed yarn and wool fabric is studied to improve the absorbance dye of juglone extracted from a walnut stain (common walnut plant) [8–10].

2. Experimental

2.1. Materials

2.1.1. Wool fiber features. The wool fiber was from the Timahdit region in Morocco. Brown fleece was compacted and homogenized into a medium-weight fleece of 1.9 kg and the fiber fineness was 44–50 using the Bradford scale, with a fiber medium length of 9.6 cm.

2.1.2. Enzymes. The enzymes were extracted from the flowers of the cardoon plant (*Cynara Cardunculus L*). This plant grows in several regions of Morocco. The extraction method was based on the maceration in distilled water during 24 hours (h) under cover and room temperature.

2.1.3. Natural dye. Juglone dye was extracted from the walnut stain of the common walnut plant, which grows in the north of Morocco [11]. The extraction method was based on water solubilizing at high temperature of the dried and powdered walnut stain. The dye extraction can be achieved by macerating in water at low temperature, as described below.

2.1.4. Chemicals. Alkali reagents, sodium carbonate (Na_2CO_3) and the acidic reagent, acetic acid (CH_3COOH), were of analytical grade and were from Sigma Aldrich (Germany). H_2O_2 was from Solvapur (Casablanca, Morocco). A Marseille soap-type was prepared from vegetable oil and was purchased from a supermarket.

2.1.5. Spectrophotometry. An ultraviolet-visible spectrophotometer (Thermo, Helios Epsilon) was used at 325–1100 nm, with a spectral bandwidth of 1 nm.

2.1.6. Bath. A 250 mL flask was used. Heating was carried out with a thermostat hotplate (Scilogex MS-H280-Pro).

2.2. *Optimization of enzymatic treatment conditions*

2.2.1. Enzymatic extraction. A mass of 90 g of fresh flower cardoon was macerated in 450 mL of distilled water for 24 h under cover at room temperature. The blend was filtered using filter paper that was used in the antifelting treatment.

2.2.2. *Enzymatic treatment of wool yarn.* Nine yarns of 1 g were placed separately in nine baths containing 50 mL of enzymatic extract and acetic acid at pH 5. The first three baths were heated at 40 °C and kept at this temperature for 1, 2 and 3 h, respectively. The second and third sets of baths were heated at 50 and 60 °C, respectively, and kept at these temperatures for 1, 2 and 3 h, respectively, as for the first three baths. At the end of the treatment process, the yarns were rinsed in cold water.
2.2.3. *Enzymatic treatment of wool fabric.* A mass of 3 g of wool fabric was placed in a bath containing 150 mL of enzymatic extract and acetic acid at pH 5. The bath was heated at 60 °C and kept at this temperature for 2 h.
2.2.4. *Soaping test.* The treated yarns and fabric were soaped with 1 g/L Marseille soap at 60 °C for 15 min and a liquid ratio of 1/50. The same soaping process was repeated three times in order to test the enzymatic antifelting treatment of the yarns.

2.3. Dyeing process of yarn and fabric with walnut stain extract
2.3.1. Preparation of extract dyes. A walnut stain of the common walnut plant (3 g) was stirred in 50 mL of water at 40 °C for 2 h, with 10 g of sodium carbonate added to reach pH 10. The solution was filtered using filter paper and well preserved for the dyeing phase.
2.3.2. Dyeing. Two yarn samples of 0.5 g and two fabric samples of 1.5 g, with one of them antifelted (yarn and fabric) as described above, were placed in a bath containing 50 mL and 100mL of dye (extract prepared from 3 g of walnut stain in 50 mL of water), at 90 °C and pH 4 for 1 h at a ratio of 1/50 and 1/33, respectively.
2.3.3. *Soaping.* The dyed yarns were soaped with 0.5 g/L Marseille soap at 60 °C for 15 min and a liquid ratio of 1/50.
2.4. *Post oxidation of antifelted yarn and fabric*
The antifelted dyed yarn and fabric were placed in two baths containing 4 mL of peroxide hydrogen in 25 mL and in 37.5 mL of distilled water and acetic acid at pH 4 at 45 °C for 15 min.

2.5. Spectral analysis

2.5.1. Spectrophotometer calibration. Spectrophotometer calibration was achieved by using distilled water as a standard solution.
2.5.2. *Measurement of dye exhaustion and fixation rate.* One mL of solution was removed from each dye bath for measurement. Each sample of yarn and fabric was diluted to 20 and 30 mL, respectively, using distilled water. The absorbance measurements are shown in table 4. The absorbances were measured at 400 nm.

2.6. Washing fastness
The washing fastness was determined over 30 min at 40 °C, according to ISO 105-C6:A1S [12].

3. Results and discussion

3.1. Optimization of enzymatic treatment conditions
Table 1 shows the length variation of antifelted yarns after the soaping test. The initial length of the yarns is 127 cm.

Table 1. Measurement lengths (cm) of antifelted yarns at different temperatures and duration.

Duration (h)	40°C	50°C	60°C
1	122	121	121
2	120	124	127
3	127	127	127

The optimal antifelting conditions were established at 60 °C for 2 h of treatment and heating at 40 °C for 3 h. These conditions can ensure the perfect shrinking resistance after three soaping operations.

3.2. Enzymatic treatment of wool yarn

Table 2 shows the length variation of antifelted and non-antifelted yarns after the soaping test.

Table 2. Dimensional measurements (cm) of antifelted (AF) and non-antifelted (NAF) yarns.

	NAF yarn	AF yarn
Initial length (cm)	113	127
Final length (cm)	94	127
Shrinking rate (%)	16.8	0.0

The treatment with the optimal conditions described above ensured a high shrinking-resistance to the wool yarn, compared to the non-treated yarn.

3.3. Enzymatic treatment of wool fabric

The fabric is the wool canvas, which prevents the mobility of fibers. In order to assess better the antifelting effect on this fabric, the enzyme quantity was increased and the soaping test was intensified.

Table 3 shows the length variation of antifelted and non-antifelted fabric after the soaping test.

Table 3. Dimensional measurements (cm) of antifelted (AF) and non-antifelted (NAF) fabric.

	NAF fabric	AF fabric
Initial dimensions (cm)	10/10	10/10
Final dimensions (cm)	9.5/9.5	9/9
Shrinking proof (%)	5.0	0.0

As shown in table 3, the shrinking proof of antifelted fabric was lower than the non-treated fabric. This indicates that the intensified soaping test caused more important shrinking of the non-treated fabric.

3.4. Exhaustion and fixation dye rate in dye bath

The absorbances of the initial, final dye baths and residual Soaping bath are presented in table 4.

Table 4. Exhaustion and fixation rate of juglone in the dye bath.

	Initial dye bath (cm)	Final dye bath (cm)	Exhaustion rate (%) ((Absi - Absf)/Absi)*100	Residual Soaping bath	No fixed dye rate	Fixation rate ((Absi-Absf-Abss)/Absi)*100
Absorbance on AF yarn	1.795	1.349		0.093	5.1%	19.6%

Absorbance on NAF yarn	1.795	1.498	24.8	0.039	2.1%	14.3%
			16.5			
Absorbance on AF fabric	0.824	0.551	33.1	-	-	-
Absorbance on NAF fabric	0.824	0.564	31.5	-	-	-

The exhaustion dye rate on the antifelted yarn and fabric was higher than the non-antifelted yarn. This can be explained by the higher accessibility of the fiber to the dye, due to the partial destruction of the fiber scale. The effect of antifelting on the dye absorbance in fabric was not very important. As observed for the yarn, this can be due to the armor structure of the fabric, which is tight and dense.

The non-fixed dye rate was higher for the antifelted yarn than the non-antifelted yarn. This might be explained by some destruction of the active sites of the fiber.

3.5. Washing fastness after post oxidation

Measurements that were obtained for the antifelted dyed yarn and fabric are presented in table 5.

Table 5. Washing fastness of the dyed yarn and fabric, with and without post-oxidation (PO).

Samples	Washing fastness (assessing staining)						Washing fastness (assessing change in color)
	WO (wool)	PAC (poly-acrilic)	PES (polyester)	PA (poly-amide)	CO (coton)	CA (cellulose actetate)	
Dyed yarn with PO	4	4-5	4-5	4	4	4	4–5
Dyed yarn without PO	4	4-5	4-5	4	4	4	4–5
Dyed fabric with PO	3-4	3-4	4	4-5	4-5	4	4-5
Dyed fabric without PO	3-4	3	3-4	4-5	4-5	3-4	4-5

The values from these experiments confirm the solid fixation of dyes using the post oxidation. The washing fastness was improved on the dyed fabric, but still constant on the dyed yarn. The juglone colorant was converted to the non-soluble form inside the fiber after post oxidation. The good washing fastness on the dyed yarn was attributed principally to the high degree of hydrogens and Van der Waals bonds, established inside the fibers. Consequently, the post oxidation of dyed yarn did not influence the washing fastness.

4. Conclusions

The antifelting treatment with cardosine enzymes in the optimal conditions ensured a high shrinking-resistance to the wool yarn and fabric, compared to the non-treated yarn and fabric.

Also, the antifelting effect improved the dye absorbance on fabric, and more so on yarn. This improvement was due to the partial destruction of fiber scale, which increased the accessibility of the fiber to the dye juglone. The structure of the fabric is tight and dense; hence, the dye absorbance was lower than those observed on yarn.

In addition, no fixed dye rate on the antifelted yarn was higher than for the non-antifelted yarn. This can be explained by some destruction of the active sites of the fiber. The post oxidation was confirmed for improving the washing fastness of dyed wool fabric. The juglone colorant was converted to the non-soluble form inside the fiber after post oxidation.

References

[1] Onar N and Sariişik M 2004 Application of enzymes and chitosan biopolymer to the antifelting finishing process *J. Appl. Polym. Sci.* **93** 2903-08

[2] Cook H J R and Fleischfresser B E 1985 Shrink Resisting Wool with Synthappret BAP: The Effect of Drying Conditions *Tex. Res. J.* **55** 607

[3] Zhuang D, Chunyan H, Zhen X and Kelu Y 2017 Bio-antifelting agent based on waterborne polyurethane and keratin polypeptides extracted by protease from waste wool *Fibers Polym.* **18** 641

[4] Kan C W, Yuen C W M, Tsoi W Y I and Tang T P 2010 Plasma pretreatment for polymer deposition—improving antifelting properties of wool *IEEE Transa. Plasma Sci.* **38** 1505-11

[5] Wang Q, Wang P, Fan X, Cui L, Zhao X, and Gao X 2009 A comparative study on wool bio-antifelting based on different chemical pretreatments, *Fiber. Polym.* **10** 724

[6] Smith E and Shen J 2011 Surface modification of wool with protease extracted polypeptides. *J. Biotechnol.* **156** 134

[7] Leulmi I 2016 *Hydrolyse enzymatique des caséines bovines par la ficine et les cardosines en vue d'obtenir des peptides antimicrobiens* (Constantine : L'institut de la Nutrition, de l'Alimentation et des Technologies Agro-Alimentaires)

[8] Bôhmer H 2002 *Natural Dyes and Textiles (Koekboya)* ed Weppert (Schweinfurt) p 244

[9] Dument E 2010 *Teindre avec les plantes (les plantes tinctoriales et leur utilisation*) ed Ulmer p 9

[10] Marquet M 2011 *Guide des teintures naturelles* ed Belin pp 41-42

[11] Bellakhdar J 1997 *La Pharmacopée marocaine traditionnelle : Médecine arabe ancienne et savoirs populaires (*Paris) ed Ibis Press

[12] ISO 105-C06 2010 Textiles *Tests for colour fastness Part C06: Colour fastness to domestic and commercial laundering* (Basel: ISO)

IOP Conf. Series: Materials Science and Engineering **827** (2020) 012043 doi:10.1088/1757-899X/827/1/012043

Empirical assessment of critical success factor of lean and six sigma in the Moroccan automotive industry

Raja Elboq[1,4], Mustapha Hlyal[2], Jamila El Alami[3]

[1] LASTIMI Laboratory, Mohammedia School of Engineers (EMI) and High School of Technology, Mohammed V University in Rabat, Morocco
[2] CELOG, School of Textile and Clothing (ESITH), Casablanca 7731, Morocco
[3] LASTIMI Laboratory, High School of Technology (EST), Mohammed V University in Rabat, Morocco
[4] Author to whom any correspondence should be addressed

rajaelboq@research.emi.ac.ma

Abstract. Lean manufacturing (LM) and Six sigma (SS) are two methods of continuous improvement that became essential in several industrial sectors. These approaches interest the researchers and also the business managers. Thus, LM and SS success that can be involved simultaneously and systematically is based on a set of Critical Success Factors (CSF). First, we extracted the CSFs discussed in the literature. Then, these CSFs have been projected to those used in the Moroccan automotive industry. For this purpose, we adopted a qualitative research methodology using structured interviews with 12 experts from the Moroccan automobile industry. Through this study, LM and SS implementing characteristics were revealed. Thus, we performed, based on CSF importance and maturity, a bi-dimensional scan that describe Lean manufacturing (LM) and Six sigma (SS) implementation within Moroccan automotive industry.

1. Introduction:

Companies are constantly investigating the improvement of their competitiveness. LM and SS are among the most effective ways to optimize management systems and business processes.

Actually, hundreds of papers produced trace their origins, techniques and highlight the benefits of their implementation. While little researches has attacked the pre-implementation phase through the necessity of taking into account the CSF, and still few have conducted their importance and rarely tested the integration level required to their success. All these elements have to be assimilated before integrating each of them [1].

The main objective of this study is to explore the status of LM and SS's critical success factors (CSF) in terms of importance and integrating levels within automotive industry in Moroccan context.

Thus, through this empirical study we will define:

- The difference of importance degree between each CSF for the same approach and for the same CSF relatively to the other approach: To list the top factor for success and help company to switch easily from one approach to the other by simply reinforcing certain CSF
- The maturity level of CSF required to success LM and SS implementing: To highlight from any level the CSF integrated is becoming functional.
- A Comparative matrix of between the two approaches according to the factors

The remainder of this paper is organized as follows: First, we present an overview of the literature to extract the CSF. Next, we describe the data, sample, and measures. Results are analyzed in the after next section.

2. Literature review:

Critical Success Factors can be seen as one of the important methods that aims to achieve effective quality management [2], organizational target and goal [3] [4].

The CSF extraction is based on a literature review of papers and researches that had been collaborated in this topic of continuous improvement methods especially LM and SS.

Indeed, the exploration of the literature consists in examining articles resulting from a wide search and a careful sorting from different sources [5]. This allows discovering the conceptual field of research and contributing thereafter to its theoretical development [6] [7].

A review of the literature revealed that there are a set of attributes that are prerequisites for the successful implementation of both methods: management commitment, culture change in organizations, good communication in the hierarchy, new approaches to production and customer service and a high level of training and education of employees [8].

These axes are more frequently reported in the literature as success factors for Six sigma, and are considered critical for effective and efficient implementation of six sigma [8].

Through the literature also, it has been revealed that there are a number of critical success factors for the implementation of LM and SS: Management commitment and support, Education and Training, Communication, Employee Involvement, Culture Change, Skills and Expertise, Linking the SS Method to Customers, Linking the Six Sigma Methodology to Business Strategy, Linking the Six Sigma Method to Vendors, Linking the SS Method to Human Resources, Reward System, Competency in project management. In 2017, a slight modification was made to these axes by adding the "understanding of tools and techniques" applied to clerks [9].

Key success factors identified include: communication, culture change, management commitment [10]. The common axes between the two studies are: “Commitment and support of management”, “Training”, “Communication”, “Employee involvement”, “Culture Change”, “Understand the tools and techniques” and “Skills & expertise". They are judged more critical and will therefore serve to refine our comparison.

Many companies often fail to integrate the above thrusts in their organizational systems and do not identify them well before their continuous improvement initiatives, so they never achieve the desired results [11].

This could be explained by the fact that LM does not include the organizational infrastructure. In contrast, SS understands the organizational infrastructure, and thus facilitates communication between employees, since one of the benefits of a SS program is the creation of a common language [10]. Henderson and Evans [12] also mentioned that the organizational infrastructure of SS is as follows: "Have all members of the organization speaking the same language".

Lean tools are geared towards cultural change. However, some organizations believe that it is possible to implement LM only in a specific culture, such as Japanese [13], while SS is based only on statistical analysis. It has been shown that culture is less important for the implementation of SS than for the implementation of LM.

The literature review also found that education and training are of high importance or weight for both Six Sigma and LM methodology. The results of this study showed that education and training were more important for LM than for SS.

This is explained by the fact that an operator can understand LM easily. Also because LM consists of individual methods that provide quick improvements [14].

Hence, the remainder of this paper will examine the level of achievement of each CSF described below, in the Moroccan companies that succeeded or try to implement LM and SS.

3. Material and Methods

This study undertook a combination of two research approaches. It comprises literature review, and interviews with 12 practitioners of Lean and Six sigma.

Through the literature review conducted at the initial session, we had identified specific CSFs of Lean and Six sigma.

The data was collected through interviews covering automotive manufacturer industry based in Morocco. We have selected companies that are considered leaders in implementing LM and SS.

The interviewers are top or middle management who have a good overview of LM and SS methodologies. To analyze the collected data, we had used the content analyses approach. Thus we had extracted significant elements. The objective of the data collection was to explore the maturity and importance of the same specific CSF in the implementation of lean manufacturing and six sigma within these companies.

The survey questionnaire was split into the following categories with several or more questions in each category to determine:

- Integration characteristic of LM and SS: chronology, maturity, due dates
- Importance assigned to each CSF for LM and SS implementing
- Integration level accomplished for each CSF In comparison with LM and SS maturity

4. Results and discussion

By exploring the extracted data, we got a preliminary picture of the company's involvement in the two methodologies. It was revealed that 58% of the companies had implemented Lean Manufacturing methodology only, (0%) none company had implemented six sigma only, while 42% had implemented the combination of the two methodology (figure1). It means that 100% of the companies are implementing LM, of which 42% had implemented LM and Six sigma.

This was explained by the interviews by adopting Toyota strategy in their companies and then the necessity to implement Lean manufacturing to avoid wastes and reduce Lead time. More so, if they tend to take their process to a greater level and meet their customer requirement, then they add Six sigma methodology to which perform their continuous improvement system.

This was consistent with the finding in literature which claim that implementing LM and SS together yield better results [7].

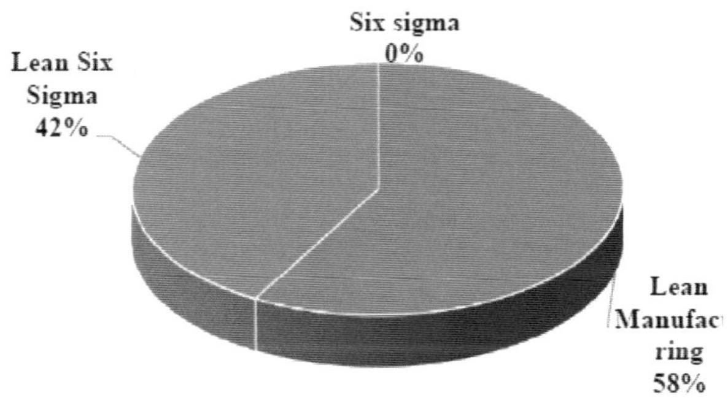

Figure 1. Repartition of implementing Lean, six sigma.

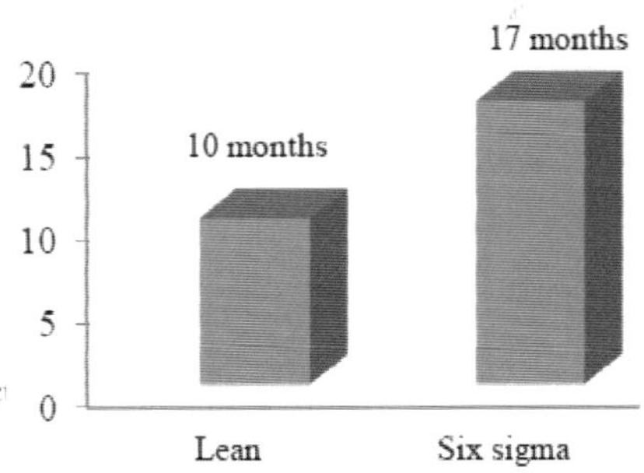

Figure 2. Duration implementing before achieving first improvements.

The maturity level achieved of implementing LM and SS within these companies is respectively 33% and 25%. It took 10 months for LM and 17 months for Six sigma to mark these levels (Figures2 & 3). It explains that Six sigma is requiring more time to achieve expected results, while Lean allows quick improvements. According to [1], this deduction is confirmed, claiming that implementing lean guarantee a rapid improvements in processes with simple actions. Then, they can focus on deeper issues that require statistical analysis through Six sigma [1].

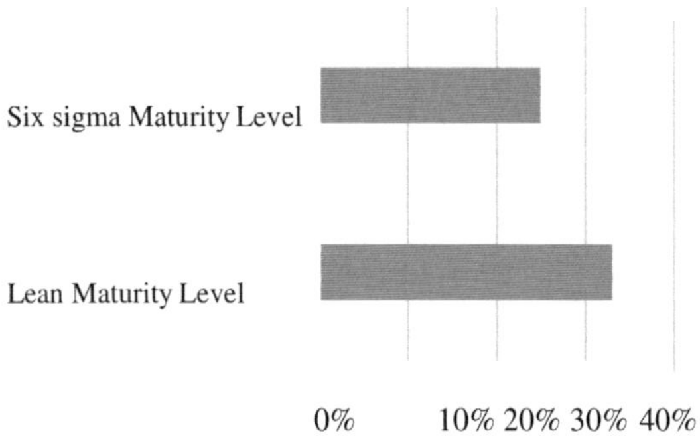

Figure3. Maturity level achieved within studied companies.

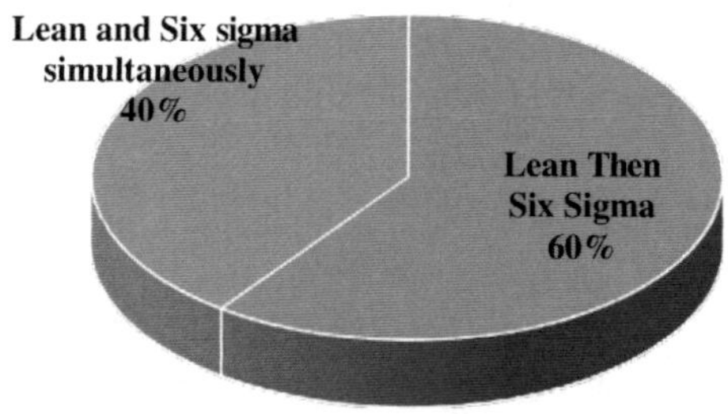

Figure4. Chronology of implementing Lean & Six sigma.

Furthermore, the main goal of this paper is to draw by CSF importance and maturity a bi-dimensional sheet-root to companies need to success their process of continuous improvement. As mentioned above, the CSF include: "Commitment and support of management", "Training", "Communication", "Employee involvement", "Culture change", "Understand the tools and techniques", "Skills and expertise".

Unsurprisingly, for the entire companies investigated, the specific Critical Success Factors had been converted and processed directly or indirectly while implementing within their organizations.

By collecting the importance assigned to each CSF and analyzing the differences between the two approaches. For LM, the results state that the top factors which are slightly rated are "Understanding tools and Techniques" and "Training". While the most important considered are "Management commitment", "Employee Involvement" and "Communication". In contrast, for Six sigma, "Understanding tools and techniques" and "Training" exceed other factors. The lowest ranked for successful implementing is "Culture change" (Figure5).

Comparatively, "Management commitment" and "Training" are considered as the most important factors for both LM and SS. Whereas, "Skills and Expertise" is the lowest ranked common factor.

Figure5. Comparing CSF importance between Lean and Six sigma.

In conclusion, the successful implementation of the two approaches rely on "Management commitment" reinforcement as well as the "Training" of the improvement team, while balancing between "Understanding tools and techniques" and "Communication".

Equally significant, "Training" can facilitate success when it is provided to both the LM team and the SS staff involved with the implementation. This training can reduce staff hesitation generated from misunderstanding and hence increase their involvement.

Furthermore, if the leaders don't understand and believe in LM and SS philosophies enough to take the time and remove resistance through employee's qualification and involvement and strong communication, is less likely to be embraced and the miraculous gains through the two approaches are less likely to be reached.

Conversely, the study depicts a high integration of the majority CSF within these companies development system (65%) (figure6) while the LM and Six maturity levels, as mentioned before, still modest 25% for SS and 33% for LM. (Figure2).

By linking the integration level and importance assigned to each CSF we can conclude that: Despite the high importance of "Employee involvement" and "Culture change" as success factor, but it appear they was not properly handled into the improvement program. Also, the "Understanding tools" was considered among of the most important factors, but it wasn't enough take in account. Thus, it explain the modest LM and SS maturity levels within these companies.

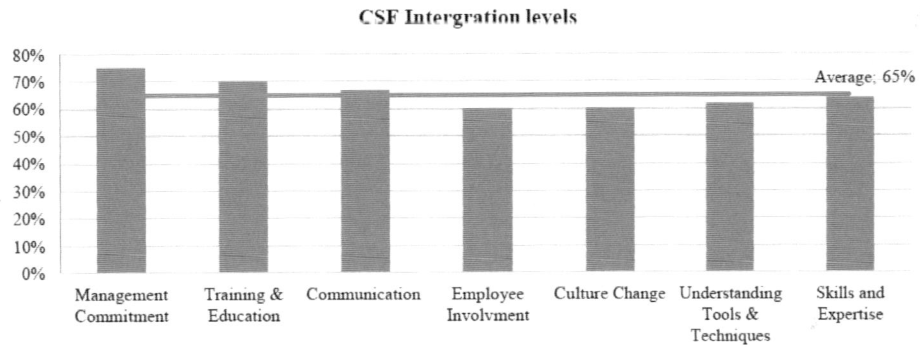

Figure6. Integration levels accomplished for each CSF.

5. Conclusion & Perspectives:

In a dynamic environment, LM and SS are used as a strategic methodologies and has been implemented as a means of achieving a competitiveness by Moroccan automotive industry.

The most implemented approach is LM, while SS is implemented only when it will be combined to LM. This combination is almost starting with LM.

In order to foster these approaches, it is important for companies to develop an implementation roadmap based on CSF. This will enable the organization to achieve quickly and effectively the gains expected. "Management commitment", "Communication", "Training" and "Understanding tools" are among the most important factors as assigned by the interviews. Thus, the provision of the necessary resources to properly could enable LM and SS to succeed.

Thus, integrating properly these factors could enable LM and SS to succeed. Otherwise, will delay intended changes and results, and consequently, their low level of integration in the system will stifles the LM and SS maturity.

The above measures are simple and direct tactics that can be undertaken to improve the results of LM and SS further by encouraging a CSF based strategy

Finally, we have investigated how LM and SS are integrated and what could make LM and SS successful in a Moroccan automotive industry. For further researches, it is recommended to study additional industries to draw up an exhaustive framework in order to ensure success of LM and SS deployment and also to measure their maturity in different industries.

References

[1] Antony J 2011, "Six Sigma vs Lean: Some perspectives from leading academics and practitioners," *Int. J. Product. Perform. Manag.* **60**(2), pp. 185–190.

[2] Saraph J V, Benson P G, and Schroeder R G 1989 An instrument for measuring the critical factors of quality management *Decis. Sci.*, **20**(4), pp. 810–829.

[3] Hardaker M and Ward B.K 1987 How to make a team work *Harv. Bus. Rev.*

[4] Hayes B J 2000 Assessing for lean six sigma implementation and success, *Six Sigma Advant.*

[5] Garza-Reyes J A 2015 Lean and green–a systematic review of the state of the art literature *J. Clean. Prod.* **102**, pp. 18–29.

[6] Seuring S and Müller M 2008 From a literature review to a conceptual framework for sustainable supply chain management *J. Clean. Prod.* **16**(15) pp. 1699–1710.

[7] Alhuraish I, Robledo C and Kobi A 2017 A comparative exploration of lean manufacturing and six sigma in terms of their critical success factors *J. Clean. Prod.* **164**, pp. 325–37.

[8] Qu L, Ma M and Zhang G 2011 Waste analysis of lean service," in *Management and Service Science (MASS) International Conference on* pp. 1–4.

[9] Meredith J 1993, Theory building through conceptual methods," *Int. J. Oper. Prod. Manag.* **13**, (5), pp. 3–11.

[10] Cheng C Y and. Chang P Y 2012 Implementation of the Lean Six Sigma framework in non-profit organisations: A case study*Total Qual. Manag. Bus. Excell.* **23**(3–4), pp. 431–447.

[11] Mulrow C D 1994 Systematic reviews: rationale for systematic reviews *Bmj*, **309**(6954), pp.597–9

[12] . Henderson K M and Evans J R 2000 Successful implementation of six sigma: benchmarking General Electric Company *Benchmarking Int. J.* **7**(4) pp. 260–282.

[13] Tranfield D Denyer D and Smart P 2003 Towards a methodology for developing evidence-informed management knowledge by means of systematic review *Br. J. Manag.* **14**(3) pp. 207–222.

[14] Bożek M and Hamrol A 2012 Analysis of efficiency of Lean Manufacturing and Six Sigma in a production enterprise *Manag. Prod. Eng. Rev.* **3**(4) pp. 14–25.

IOP Conf. Series: Materials Science and Engineering **827** (2020) 012044 doi:10.1088/1757-899X/827/1/012044

Physical Internet: Literature review and research opportunities

K Jharni [1,2,3,4], M Hlyal[2] and J El Alami[1,3]

[1]Mohammedia School of Engineers / Mohammed V University, Rabat, Morocco
[2]CELOG, Superior School of Textile and Clothing - (ESITH), Casablanca, Morocco
[3]LASTIMI, High School of Technology, Sale, Morocco
[4]Author to whom any correspondance should be adressed

kjharni@gmail.com

Abstract. With an increasingly rich and personalized product offering, the digitalization of the commercial relationship and the opening of world borders, commercial transactions have never been this important. And so are the requirements in terms of time and availability that logistics and supply chains should fulfill. These latters still support the growth of the economic exchanges but struggle to find a disruptive model capable of meeting more requirements with sustainable economic, environmental and social approaches. Several initiatives are under consideration, including the Physical Internet, which offers an abstract, generic and global framework for managing the logistics chain of physical objects. In this article, we review the literature of the Physical Internet following the qualitative content analysis method. We found that although the number of contributions is limited, the contribution grows at a rate of 25% every year on average. We analyzed the contributions after categorizing them to find that the governance (management and community) of the Physical Internet from a global perspective is not covered enough by the current literature which can be an opportunity for future research.
Keywords: Physical Internet, Literature review, Qualitative content analysis

1. Introduction

The technological evolution and particularly moving everything to its digital representation is transforming the way people act and behave. The commerce is not an exception where the physical contact between the buyer and the seller or its products is no longer necessary. People can move from one product to another and even change stores with a few clicks instead of minutes or hours. They can do it without leaving their place instead of walking hundreds of meters. In such a new context, there seem to be no physical nor temporal constraints. This adds more constraints to current supply chains mainly because customers may believe that they can physically get the goods they are buying with just a few clicks. In fact, they don't know whether the goods are stored in the neighborhood or are located on another continent but still expect them to be delivered as quickly as possible. This challenge is what today's and future logistics and supply chains must handle.

The Council of Supply Chain Management Professionals defines the logistics [1] as "The process of planning, implementing, and controlling procedures for the efficient and effective transportation and storage of goods including services, and related information from the point of origin to the point of consumption for the purpose of conforming to customer requirements". If we can replace the goods by the data in this definition we will end up with data logistics which is continuously evolving with the Digital Internet.

In fact, in the digital world, user experience is continuously enhanced. Users can share big amounts of data easily (images, videos, etc.). Data exchanges are increasingly growing at a very fast rate but paradoxically the costs are dropping. The new infrastructures support even more volumes and further enhance service quality. So, we could ask ourselves if the data logistics is that powerful, why not to be inspired by its underlying principles to disrupt goods logistics? How similar are the digital exchanges compared to physical exchanges? And if these two areas are similar, what approach to reproduce the success of the digital exchanges onto the physical ones?

2. The Physical Internet

The introductory questions suggest a concept that physical objects can be moved with as much simplicity, security, and efficiency as digital data. Such a concept appealed to the authors of a logistics study published in a special edition of The Economist [2] in June 2006 under the title "Physical Internet". The authors suggest the word implicitly by describing the similarity between a goods warehouses and Internet routers, physical packets, and information packets, and logistic networks compared to computer networks. Thus the word "Physical Internet" (abbreviated PI or π [3]) was used for the first time.

This new term interested researchers [4] who have found in the physical Internet not only a metaphor but a new angle of view that could consolidate the various singular initiatives to improve the performance of supply chains. Thus was born the concept of the physical Internet as it will be presented in this article with the publication of the 1st "Manifesto for a Physical Internet" *by B. Montreuil* in 2009 [5]. Then, *B. Montreuil* gives a first version of the vision of the Physical Internet [6] by describing its characteristics as a response to the inefficiency and the unsustainability of current logistics. Flowing the analogy with the Digital Internet, the Physical Internet was then defined as "an open global logistics system founded on physical, digital and operational interconnectivity through encapsulation, interfaces, and protocols. It is a perpetually evolving system driven by technological, infrastructural and business innovation." [7].

This work catalyzed and federated researches aimed at optimizing supply chains around the Physical Internet. By addressing the three issues (economic, environmental and social), the paradigm was adopted by the European Union in 2013, which is ALICE (Alliance for Logistics Innovation through Collaboration) with the ambition to fully implement the Physical Internet in Europe by 2050 [8]. In the same way, several academic institutions have launched research branches to meet the needs of their government and/or industry [5].

Although the Physical Internet concept comes from the Digital Internet, they both remain related. The recent digital and data management concepts can be very useful to the Physical Internet and even be concrete enablers for its real implementation. Some technologies are particularly very tied to the Physical Internet such as IoT [9], big data [10], AI [11] and blockchain [12].

3. Review methodology

The main goal of this work is to give an up to date overview of the research publications about the Physical Internet and to "summarize the state of the art " [13] in that subject. Following the guidelines from *Seuring and Gold* [14] which are based on *Mayring* [15] works, we used the qualitative content analysis method with its four-steps illustrated in Figure 1.

Figure 1. Four steps of the qualitative content analysis.

3.1. Data Collection
The Physical Internet is a recent topic. In order to give the best picture possible about the Physical Internet, we used different databases: ScienceDirect, IEEE, and ResearchGate. To select the most relevant publications up to 2018, we looked for "Physical Internet" as an exact expression in titles, keywords, and abstracts (Table 1). The publications were then post-processed to eliminate non-relevant work and especially where "Physical Internet" means the physical infrastructure of the Internet.

Table 1. Search Expressions and post processing.

Database	Search Expression	Post Processing	Publications removed
ScienceDirect	"Physical Internet"		5
IEEE	("Publication Title":"Physical Internet") OR ("Author Keywords":"Physical Internet") OR ("Publication Title":"Physical Internet") OR ("Document Title":"Physical Internet")	Title, keywords, abstract, screening	3
ResearchGate	"Physical Internet"		4

3.2. Data Analysis
Using the above criteria, we found 96 publications of different types: conference papers, journal articles, books, magazines, etc. From the overall publications, 62 were published in the last 3 years (2016-2018) which represents 65% of the publications with a progress average of 25% every year as illustrated in Figure 2. ResearchGate has the most important share with 64% of the publications followed by ScienceDirect with 21% and IEEE with 15% as illustrated in Figure 3. Most contributions are published as articles (54%) then as conference papers (35%) as illustrated in Table 2.

As to authors, 141 contributors were found. The top 10 authors are sited 87 times which is 31% of the overall citations. Eric Ballot is the most cited author (21 times) followed by Shenle Pan (16 times) and Benoit Montreuil (15 times).

Through the research process, we noticed that the publications on the Physical Internet are very limited. We found only four literature reviews. In the first [4], the authors focus on the theoretical foundation of the Physical Internet. The second [5], is a special issue to identify critical challenges and to stimulate research contribution. The authors of the third review [16] focus on the adoption of the Physical Internet. The last one [17] is a systematic literature review of the scientific literature around the Physical Internet.

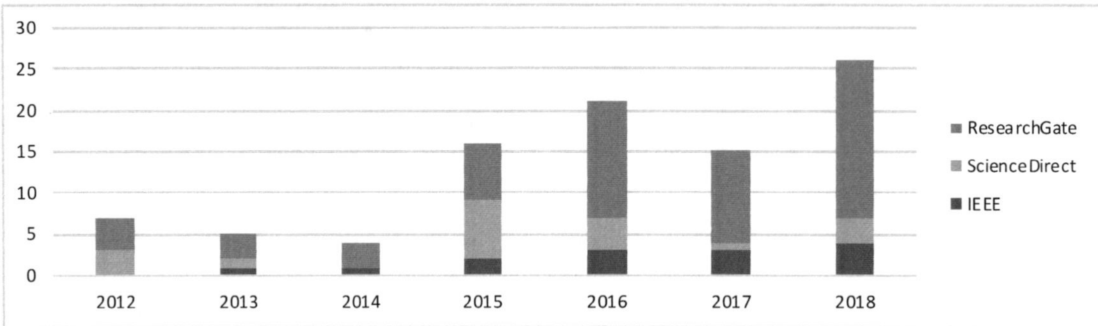

Figure 2. Physical Internet publications over the years.

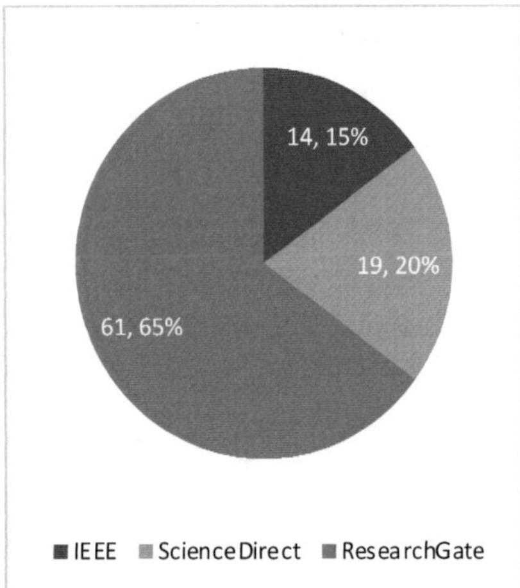

Figure 3. Physical Internet publication sources.

Table 2. Physical Internet publication types.

Publication Type	Number of Publications	Share
Article	51	54%
Conference Paper	33	35%
Book Chapter	3	3%
Poster	2	2%
Preprint	1	1%
Thesis	1	1%
Book	1	1%
Data	1	1%
Magazines	1	1%
Total	94	100%

With respect to the qualitative content analysis, the categorization process and the findings are explained in the next section.

4. Results and discussion

The purpose of this section is to highlight the least covered areas of the Physical Internet in order to give researchers the opportunity to contribute with original works. Starting from the digital Internet as a basic metaphor, we have identified some categories to qualitatively analyze publications and frame their evaluation.

4.1. Data Categorization

The literature about the Internet reveals four different aspects that make its history [18]: the technical/technological evolution, the operations and management, the community and the commercialization. The Internet disrupted the way people and systems communicate. To reach this goal or at least approach it on the physical world, we can start with what made the success of the Internet. In this journey, we want to check whether the Physical Internet is following the same pattern or a different one. We used the four aspects behind the success of the Internet as our main categories. Then we analyzed the contributions and classified them to the category that best matches their content (Table 3).

Table 3. Physical Internet publication categories.

Category	Number of publications
Technical/technological evolution	92
Commercialization	2
Total	94

4.2. Data Evaluation

The final analysis shows that almost all contributions are related to the Physical Internet itself. They give answers to how PI should be implemented with models, simulations and uses cases.

Only two contributions cover commercialization and business model. Unfortunately, no contribution specially addresses how to manage the Physical Internet from a worldwide perspective and how to accelerate and engage further contributions, how to manage and build a community that will support and guarantee the Physical Internet evolution.

The success of the Digital Internet is due among others to the community of its users and the way it was governed trough time [18]. To complete the analogy and if we target a wide adoption of the Physical Internet, we think that the identification of the governance and the potential business models of this disruptive concept are required. These topics are not covered enough by the current literature which opens opportunities for contribution. Future research may particularly answer the following questions:

- The relationship between the Digital and the Physical Internet is only a metaphor or a deep analogy? We think that the relationship between the Digital and the Physical Internet is not clearly defined. From our perspective, they are analog and parallel [19]. Thus, we can infer most of the knowledge from the Digital to the Physical Internet with few adaptations.
- How to develop a community of scientists, professionals, industrials and users that will contribute to the implementation, management, and improvement of the Physical Internet? Recent collaborative models such as crowdsourcing and open sourcing show that the bigger the community and the more heterogeneous it is, the quicker is the progress and the adoption. We believe that the adoption of the Physical Internet and its development can be further enabled and accelerated by demystifying and popularizing its underlying concepts to wider community.
- What can be the disruptive business models for the Physical Internet? Is it possible to transpose SaaS, PaaS and IaaS concepts to the supply chain? Companies have different needs and maturity levels. Big ones want more control with less investment and small ones struggle to manage their logistic service. We think that the Physical Internet can serve different kinds of needs thanks to the anonymization and the optimization in every layer. We can, for example, offer infrastructure services like logistics centers rental or platform services like cross-docking operations.

5. Conclusion

In this article, we reviewed the literature around the Physical Internet following the qualitative content analysis method. We categorized the results using the historical aspects (technical / technological, management, community and commercialization) behind the success of the Digital Internet to find similarities and suggest new research contributions.

The findings show that the Physical Internet is still a local initiative and that a worldwide management of its operations is not covered by the literature. We also found that the current contributions do not work on building a large community trough popularizing of the fundamental concepts of the Physical Internet. Thus, we suggest some research perspectives by first inferring the knowledge from the Digital Internet and its history. Then, work on a worldwide vision of the Physical

Internet through a governance system and a wider community. And last, looking for disruptive and affordable business models inspired by the digital ones and enabled by new advancements in automation, autonomous navigation and IoT.

References

[1] Kate V 2013 *CSCMP Supply Chain Management Definitions and Glossary*. [Online]. Available:https://cscmp.org/CSCMP/Academia/SCM_Definitions_and_Glossary_of_Terms/CSCMP/Educate/SCM_Definitions_and_Glossary_of_Terms.aspx. [Accessed: 08-Sep-2009].

[2] The Economist 2006 *The physical internet: A survey of logistics* The Economist.

[3] Montreuil B, Meller R D, and Ballot E 2010 Towards a Physical Internet: the impact on logistics facilities and material handling systems design and innovation Prog. *Mater. Handl. Res.*

[4] Treiblmaier H, Mirkovski K and Lowry P B 2016 Conceptualizing the Physical Internet: Literature Review, Implications and Directions for Future Research . *In 11th CSCMP Annual European Research Seminar, Vienna, Austria, May*

[5] Pan S, Ballot E,Huang G Q and Montreuil B 2017 Physical Internet and interconnected logistics services: research and applications I*nternational Journal of Production Research*

[6] Montreuil B 2011 Toward a Physical Internet: meeting the global logistics sustainability grand challenge *Logist. Res.* **3**(2–3), pp. 71–87.

[7] Montreuil B, Meller R D, and Ballot E 2012 Physical Internet Foundations *IFAC Proc*. **45**(6) pp. 26–30.

[8] Ambra T, Caris A and Macharis C 2019 Towards freight transport system unification: reviewing and combining the advancements in the physical internet and synchromodal transport research *Int. J. Prod. Res*. **57**(6), pp.1606-23

[9] Tran-Dang H and Kim D S 2018 An Information Framework for Internet of Things Services in Physical Internet *IEEE Access*, **6**, pp.43967-77.

[10] Zhong R Y, Xu C, Chen C and Huang G Q 2017 Big Data Analytics for Physical Internet-based intelligent manufacturing shop floors *Int. J. Prod. Res*. **55**(9), pp.2610-21

[11] Klumpp M 2017Artificial Divide: The New Challenge of Human-Artificial Performance in Logistic in *Innovative Produkte und Dienstleistungen in der Mobilität* (pp. 583-593). Springer Gabler, Wiesbaden.

[12] Treiblmaier H 2019 Combining Blockchain Technology and the Physical Internet to Achieve Triple Bottom Line Sustainability: A Comprehensive Research Agenda for Modern Logistics and Supply Chain Management *Logistics*, **3**(1), p.10.

[13] Rowley J and Slack F 2004 Conducting a literature review *Management Research News*.

[14] Seuring S and Gold S 2012 Conducting content-analysis based literature reviews in supply chain management *Supply Chain Management: An International Journal.*

[15] Mayring P 2000 Qualitative Content Analysis *FQS - Forum Qual. Soc. Res*. **1**(2)

[16] Sternberg H and Norrman A 2017 The Physical Internet – review, analysis and future research agenda *International Journal of Physical Distribution and Logistics Management.*

[17] Domański R, Adamczak M and Cyplik P 2015 Physical Internet (PI) : a Systematic Literature Review *Logforum* **14**.

[18] Leiner B M et al., 2009 A Brief History of the Internet *ACM SIGCOMM Comput. Commun. Rev.*

[19] Bartha P 2010 *By Parallel Reasoning: The Construction and Evaluation of Analogical Arguments* (Oxford Scholarship Online: May 2010)

IOP Conf. Series: Materials Science and Engineering **827** (2020) 012045 doi:10.1088/1757-899X/827/1/012045

New textile-based hybrid photovoltaic/thermal (PV/T) system

B Pause[1,2]
[1] Textile Testing & Innovation, LLC, Longmont, Colorado, USA
[2]Author to whom any correspondence should be addressed

Pause@textile-testing.com

Abstract. This paper introduces a new textile-based hybrid PV/T system utilizing a spacer fabric treated with phase change material (PCM). The PCM absorbs excess heat from the PV modules, which keeps the modules in their optimal temperature range so that their power output is enhanced. The thermal energy stored in the PCM is then used for hot water generation. Cold water, which is pumped through thin capillaries integrated into the spacer fabric, absorbs the heat stored in the PCM and is heated up. The warm water can be used immediately or stored in an intermediate storage tank. Comparing the proposed textile-based hybrid PV/T system to conventional hybrid PV/T systems, the proposed system is less expensive and provides a higher overall efficiency. The textile-based collector is lightweight, thin and flexible.

1. Introduction

Combining photovoltaic and solar thermal components into a single system can quadruple the energy output of a conventional photovoltaic array. Furthermore, such a system addresses the problem of overheated PV modules. Current PV/T collectors are typically bulky and heavy PV panels in combination with solar thermal collectors. It needs to be considered that the hybrid PV/T collectors are currently more expensive than a combined use of a conventional PV panel and a thermal collector based on the same performance [1], [2].

In order to overcome some of the obstacles related to the use of the commercially available hybrid PV/T systems, a textile-based hybrid PV/T system has been developed. In this system, phase change material (PCM) is used for intermediate heat storage.

2. Phase change material

A phase change material (PCM) possesses the ability to change its physical state in a certain temperature range. When the melting temperature is obtained in a heating process, the phase change from the solid state to the liquid state occurs. During this melting process the phase change material absorbs and stores a large amount of so-called "latent heat". The temperature of the phase change material remains nearly constant throughout this process. In a reverse cooling procedure, the stored latent heat is released into the environment and a phase change from the liquid state to the solid state takes place. This return to the previous state also happens without a change in the material's temperature. The fact that a large amount of latent heat is absorbed or released without any temperature change makes a PCM highly desirable as a means of heat storage [3].

PCMs are characterized by the temperature ranges in which the phase change takes place and their latent heat storage capacities. Common PCMs are paraffin waxes, salt hydrates, bio-based materials, and eutectics.

3. Containment of the PCM

Before a PCM can be applied to a textile substrate, it needs to be durably contained in a carrier material in order to prevent dissolution while in its liquid state. For the application in the textile-based hybrid PV/T system, the selected PCM is directly integrated into a silicone rubber compound and durably contained therein by a cross-linking procedure. The cross-linking prevents leakage of the PCM out of the silicone rubber compound while liquid [4].

4. Textile-based hybrid PV/T system

4.1 Collector

The textile-based hybrid PV/T collector consists of a spacer fabric in which capillaries are integrated. The back side of the spacer fabric is covered with a protective coating in order to withstand the elements. The front side of the spacer fabric faces thin film PV cells and is coated with the silicone rubber compound containing the PCM as a thermal storage medium. During the coating procedure, the silicone rubber compound penetrates into the spacer fabric and encloses the capillaries completely. The thin film PV cells are then laminated onto the surface of the silicone rubber coating layer equipped with the PCM. The textile-based hybrid PV/T collector is shown in figure 1.

Figure 1. Textile-based hybrid PV/T collector.

4.2. Photovoltaic/thermal system

The entire system used for the hot water generation is shown in figure 2. The system comprises the textile-based hybrid PV/T collector, a cold water supply system consisting of a pump and a valve as well as a temperature sensor, and a water tank for intermediate storage of the generated hot water.

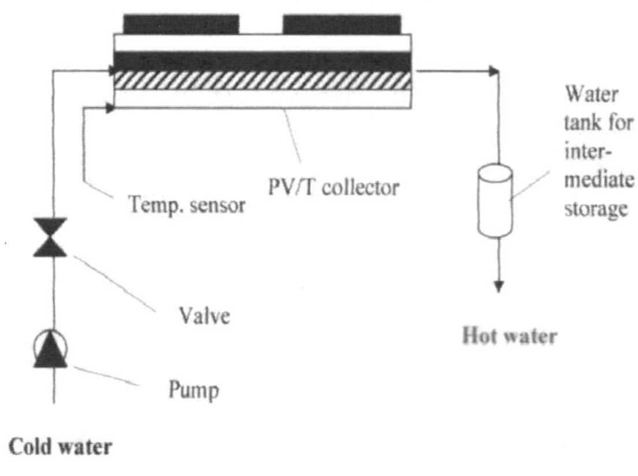

Figure 2. Photovoltaic/thermal system.

4.3 Operation of the system

The thin film PV cells of the textile-based hybrid PV/T collector convert the solar radiation into electricity. As soon as the temperature of the PV cell reaches a given trigger temperature, the excessive heat is absorbed by the PCM integrated in the silicone rubber coating compound. In this way, the PV cell is cooled and can operate at its optimal temperature, staying at its highest possible efficiency and maximum electrical output. Tests have shown that the PCM cooling used in the textile-based hybrid PV/T system is more efficient than the forced liquid or air cooling used in commercialized PV/T systems, because the construction of the textile–based hybrid PV/T collector provides direct contact between the PV cell and the heat absorber.

The temperature measurements carried out by the temperature sensor at the collector location deliver an indication at which point the PCM is ready for a recharge, i.e. when the temperature exceeds a certain value the sensor provides a signal to the valve which normally shut off the cold water supply. The valve opens and cold water with a temperature of about 20°C is pumped through the capillaries which are integrated into the spacer fabric.

Because of a high temperature gradient between the PCM's temperature and the water temperature at the beginning of the recharge cycle, there is a fast transfer of the latent heat stored in the PCM to the water, and so the water is quickly heated. The water which is flowing out of the textile composite has a temperature of about 40 °C.

The water flow through the capillaries can be automated through a thermostat or operated manually. The warm water can be used instantly and/or can be stored in an intermediate hot water tank. This eliminates the need for a heat exchanger and its components, as well as for a large storage tank which, in turn, reduces costs and saves space.

5. PCM selection

In order to select the PCM which operates in a suitable temperature range, field tests were carried out. In these field tests the temperature development on top of the PV cells were measured over the course of the day. The test results indicate that the temperature rises substantially in the afternoon hours when the PV cell faces in the southwest direction with a maximum temperature of about 60 °C.

It needs to be taken into consideration that the optimal operation temperature of a PV module is about 25 °C. It has been calculated that for every 1 °C above 25 °C, the electrical output drops by about 0.5% [5], [6].

Taking these operation requirements into account, an eutectic PCM has been developed and selected which absorbs latent heat in a temperature range between 25 °C and 40 °C and possesses a high latent heat storage capacity of about 270 J/g. Figure 3 shows a DSC thermograph of the developed PCM.

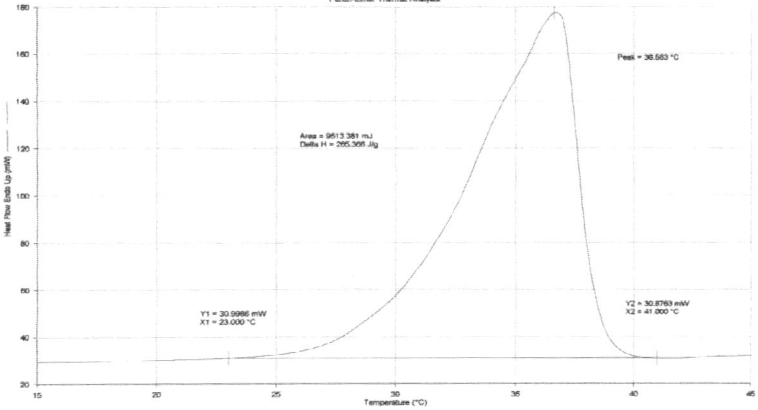

Figure 3. DSC thermograph of the developed PCM.

A concentration of 50 wt. % of the selected PCM in the silicone rubber compound leads to a latent heat storage capacity of about 540 kJ per one m^2.

6. Test results

6.1 Cooling of the PV cells

The main purpose of the PCM is the cooling of the PV cell in order to ensure that they remain in the optimal temperature range during the window of operation. Field tests were carried out where the temperature on top of the PV cells was monitored utilizing collectors with and without PCM treatment which were placed side by side. The test results are summarized in figure 4.

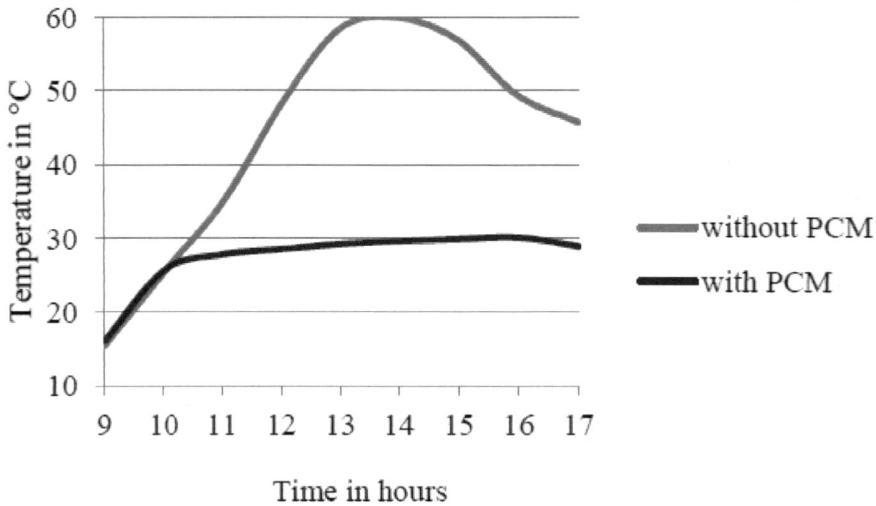

Figure 4. Temperature development at the PV cell with and without the PCM treatment.

The results summarized in figure 4 indicate that the latent heat absorption of the PCM starting at a trigger temperature of 25°C and the recharge of the PCM in consecutive time intervals reduces the PV cell's temperature by up to 30 degrees.

6.2 Power generation by the PV cells

The cooling feature provided by the PCM leads to a substantial improvement in the power output of the PV cell as it is shown in figure 5. The power output from both arrangements was measured simultaneously with the temperature measurements.

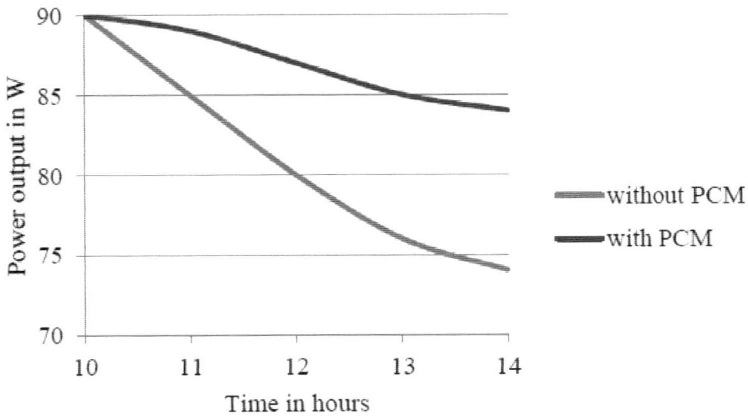

Figure 5. Power output of the PV cells with and without the PCM-cooling.

For instance, the cooling of the particular PV cell provided by the latent heat absorption of the PCM reduced the power output by only 4 %. In contrast, without the PCM's cooling feature the power output of the same PV module dropped by 18 % due to the overheating of the PV cell.

6.3. Hot water generation
The latent heat storage capacity of 540 kJ/m^2 allows for the heating of about 6 liter of water from 20°C to 40 °C referred to a collector size of one square meter in a single recharge. Multiple recharges are usually carried out over the course of the day.

7. Applications
The textile-based hybrid PV/T collector is lightweight, thin, flexible and weatherproof. The collector fits any roof shape and can be used not only on roofs of residential and commercial buildings, but also for swimming pool and greenhouse covers. Another application can be emergency shelters set up in disaster areas. The system would provide shelter as well as electricity and hot water at the same time and, therefore, can address three major problems in one arrangement.

8. Comparison with conventional hybrid PV/T systems
The textile-based hybrid PV/T collector is roughly 75-80 % lighter than a comparable conventional PV/T system. Furthermore, the price of the textile-based PV/T collectors are about 50% of the price of the commercialized PV/T collectors based on the same performance. Moreover, maintenance requirements of commercialized PV/T systems are high and the requirement of continuous pumping of fluid or air though the system leads to high operational costs. The textile-based hybrid PV/T system maintenance requirements and operational costs are lower and its overall efficiency is higher, compared to common PV systems, meaning that the break-even period for a PV/T installation will be much shorter.

9. Conclusions
The newly-developed textile-based hybrid PV/T system provides a solution for enhancing the efficiency of PV modules substantially. The heat which is absorbed and stored by PCM when cooling the PV modules is then used for hot water generation. The textile-based hybrid PV/T collector is much lighter when conventional collectors and fits various roof shapes which make applications other than roofs of buildings possible. In comparison to conventional hybrid PV/T systems, the proposed solution provides substantial cost and space savings.

IOP Conf. Series: Materials Science and Engineering **827** (2020) 012045 doi:10.1088/1757-899X/827/1/012045

References

[1] Davis, T.A. and Cappelle, M.A., University of Texas System, 2016. *Photovoltaic-Thermal (PV-T) system for desalination*. U.S. Patent 9,278,315.

[2] Othman M Y, Yatim B, Sophian K, Bakar, M. N.A., 2008 Advances in Hybrid Photovoltaic Thermal (PVT) Solar Air Heaters, *World Renewable Energy Congress* in Glasgow, UK.

[3] Pause B 2017 Smart textiles created by the application of phase change material (PCM) *International Conference on Intelligent Textiles & Mass Customization ITMC2017* in Ghent, Belgium.

[4] Pause B 2019 Smart functional textiles created by the application of macro-encapsulated phase change material *AUTEX conference* in Ghent, Belgium.

[5] Tangsiriratana E, Skolpap W, Patterson R, Sripapha K 2019 Thermal properties and behaviour of microencapsulated sugarcane wax phase change material *Helion* **5** p.e02184

[6] McCabe J 2004 *Optimization of photovoltaic/thermal collectors Solar conference* in Portland, USA.

An enhanced power harvesting from woven textile using piezoelectric materials

N Chakhchaoui[1,2,6], H Jaouani[2], R Farhan[3], A Eddiai[3], M Meddad[4], O Cherkaoui [2], L Van Langenhove[5]

[1] BGIM Laboratory, Higher Normal School (ENS), Hassan II University, Casablanca, Morocco

[2] REMTEX Laboratory, Higher School of Textile and Clothing Industries (ESITH), Casablanca Morocco

[3] Laboratory of Physics of Condensed Matter (LPMC), Faculty of Sciences Ben M'Sik Hassan II University, Casablanca, Morocco

[4] LAS Laboratory of Setif, Mohamed el Bachir el Ibrahimi BBA University, Algeria

[5] Ghent University, Department of Textiles, Technologiepark 907, 9052 Zwijnaarde, Belgium

[6] Author to whom any correspondence should be addressed

nabil.chakhchaoui@esith.com

Abstract. The field of power harvesting has experienced significant growth over the past few years due to the ever-increasing desire to produce portable and wireless electronics with extended lifespans. The present work aims to introduce an approach to harvesting electrical energy from a mechanically excited piezoelectric element and investigates a power analytical model generated by a smart structure of type polyvinylidene fluoride(PVDF) that can be stuck onto fabrics and flexible substrates. Moreover, we report the effects of various substrates and investigates the sticking of these substrates on the characterization of the piezoelectric material.
Keywords: Power harvesting, Wireless electronics, Piezoelectric energy, Technical textile, PVDF

1. Introduction

Over the past decade, the number of wireless sensors and portable electronics carried by individuals had a rapid increase [1, 2]. Each of these devices is typically powered using a traditional electrochemical battery. However, the use of batteries involves many drawbacks, such as limited energy storage capability, potential damage to environment and a limited life time, which have limited further development and applications of wearable electronics. To overcome this issue, many efforts have been focused on wearable energy harvesters, allowing the conversion of ambient energy surrounding the system to a usable electrical energy [3, 4] such as the mechanical energy dissipated from human motion [5, 6], including electromagnetic [7], electrostatic [8], thermoelectric [9], and piezoelectric harvesters [10]. Compared with electromagnetic, electrostatic and thermoelectric methods, the piezoelectric approach to power harvesting provides several advantages including: higher energy density, the ability to be fabricated in custom shapes and higher flexibility of being integrated into a system.

The focus here is on piezoelectric materials, which accumulate electrical charge in response to applied mechanical stress. Representative piezoelectric materials can be categorized into piezoceramics and piezopolymers. However, PZT (lead zirconate-titanate) is a commonly used piezoelectric ceramic. On the other hand, the piezopolymer commonly used is polyvinylidene fluoride (PVDF). Compared with rigid, brittle, and heavy PZT [11,12], PVDF is particularly interesting due to its low cost, considerable flexibility, durability, good stability and easy integration into elements such as clothes and shoes [13]. Recently, a great amount of researches have been conducted to develop simple and efficient energy harvesting devices from vibration by using piezoelectric materials based on smart textiles which harbor potential in various applications such as energy harvesting, sensing and actuation [14]. Smart Textiles are defined as textile products, such as woven, knitted or non-woven structures made from fibers, filaments or yarns, which are able to sense and respond to changes in their environment [15]. Furthermore, recent developments in the field have led to the design of a number of mechanisms that can be used for harvesting electrical energy, from a variety of textiles substrates including textile plus an interface layer such as Kapton and Alumina, and so on [16]. In addition, *Almusallam. A, et al* [17] presented an experimental study on clamping effect on the piezoelectric responses of screen-printed low temperature PZT/polymer films on flexible substrates. Later, *Almusallam. A, et al* [18] made a series of studies about the use of a flexible piezoelectric nano-composite films for kinetic energy harvesting from textiles. In this context, the study has shown a piezoelectric patch that could recharge a small device within few hours when excited by a kneepad located on the knee due to walking [19].

This paper has investigated the development of novel energy harvesting systems, which are a clean and durable solution, based on the thin film PVDF sticking on flexible substrates such as fabrics. The smart structure shows greater piezoelectric and dielectric properties compared to our previous work [19]. These properties are explored and discussed in this paper.

2. General theory of kinetic energy harvesting

2.1 Standard approach to energy harvesting

The simplest method to recover energy is to directly connect the electrical circuit to the piezoelectric elements. This device is shown in figure 1(a) where the resistance *R* represents the input impedance of the supplied electric circuit. In this case, the load voltage is alternative. The waveforms associated with this technique are shown in figure 1(b), and in this case, a permanent sinusoidal stress has been performed.

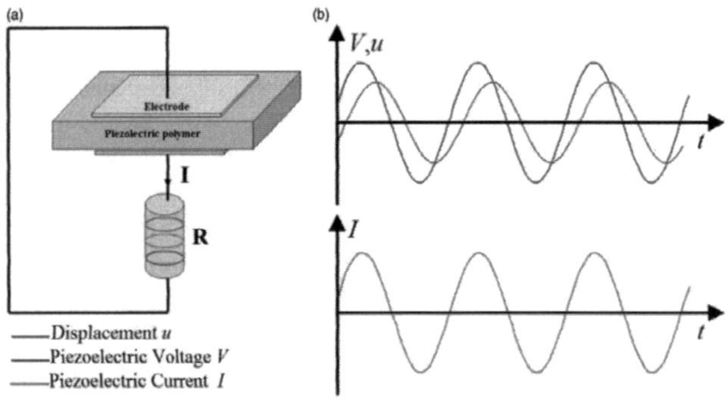

Figure 1. Standard approach Alternating Current (AC) to energy harvesting: (a) standard network and (b) standard waveforms.

2.2 Analytical and modelling of power harvesting system

As one of the important vibration-based energy harvesting methods, piezoelectric conversion has received much attention because of the simple structure of piezoelectric converter and the ease of application characteristic of piezoelectric materials. There are two types of piezoelectric effects that can be used for technological applications: the direct piezoelectric effect that describes the ability of a given material to convert mechanical strain into electricity. On the other hand, the converse effect, which is the ability to convert an applied electrical solicitation into mechanical energy.

Two equations are used to depict the piezoelectric nature of a material and they have been considered from *Jaffe and Cook* [5]. Equations (1) and (2) are defined as the piezoelectric constitutive equations:

$$S_\alpha = s_{\alpha\beta}.T_\beta + d_{i\alpha}.E_i \quad (1)$$
$$D_i = d_{i\alpha}.T_\alpha + \varepsilon_{ij}.E_j \quad (2)$$

$$\alpha = \beta = 1,2,\ldots,6 \qquad i,j = 1,2$$

The first equation defines the mechanical response of the material, while the second equation defines the electrical response. Where S is the strain, T is the stress, E is the electric field, D is the electric displacement, d is the piezoelectric electromechanical coupling coefficient, s is the compliance, which relates stress and strain at constant electric field, ε is the dielectric permittivity, which indicates the charge stored in the capacitive element of the piezoelectric material at constant stress, and the subscripts represent the direction of each property.

$$S_1 = s_{11}.T_1 + s_{12}.T_2 + s_{13}.T_3 + d_{31}.E_3 \tag{3}$$

$$S_3 = s_{13}.(T_1 + T_2) + s_{33}.T_3 + d_{33}.E_3 \tag{4}$$

$$D_3 = d_{31}.(T_1 + T_2) + d_{33}.T_3 + \varepsilon_{33}.E_3 \tag{5}$$

When placed on an isotropic substrate (the fabric plus interface layer), the lateral strain is the same in each direction $\mathrm{S}_1 = \mathrm{S}_2$ and the stress $\mathrm{T}_1 = \mathrm{T}_2 = T$, equation (1) can be re-written as

$$S_1 = (s_{11} + s_{12}).T_2 + s_{13}.T_3 + d_{31}.E_3 \tag{6}$$

Thus, the value of S1 and S2 must be determined as a function of T_3 using Hooke's law

$$S_1 = S_3 = (-\frac{v_{sub}}{\gamma_{sub}}).T_3 \tag{7}$$

Also, the mechanical compliance matrix parameters were calculated using the following equations

$$s_{11} = \frac{1}{\gamma_{piez}}; \quad s_{12} = -\frac{v_{piez}}{\gamma_{piez}}; \quad s_{13} = -\frac{v_{piez}}{\gamma_{piez}} \tag{8}$$

Relating the strain and the ratio of the generated charge to the applied force and combining with equations (4), (5) and (6) gives

$$\left(-\frac{v_{sub}}{\gamma_{sub}}\right).T_3 = \frac{1-v_{piez}}{\gamma_{piez}}.T - \frac{v_{piez}}{\gamma_{piez}}.T_3 + d_{31}.E_3$$

It is possible to express the T stress as a function T_3 and E_3 , given by

$$T = \frac{\gamma_{piez}}{1-v_{piez}}.\frac{v_{piez}.\gamma_{sub} - v_{sub}.\gamma_{piez}}{\gamma_{piez}.\gamma_{sub}}.T_3 - \frac{\gamma_{piez}}{1-v_{piez}}d_{31}.E_3 \tag{9}$$

Hence the expression of the electric displacement D; Substituting equation 7 into 3 and rearranging it results in

$$D_3 = d_{33}.\left(\frac{-2.\gamma_{piez}}{1-v_{piez}}.\frac{v_{piez}.\gamma_{sub} - v_{sub}.\gamma_{piez}}{\gamma_{piez}.\gamma_{sub}} + 1\right).T_3 + (\varepsilon_{33} - \frac{2.\gamma_{piez}.v_{piez}^2.d_{33}^2}{1-v_{piez}}).E_3 \tag{10}$$

The current supplied in the polymer in the case of transverse mechanical stress is expressed in the form

$$I = A.\frac{\partial D}{\partial t} = A.\{\, d_{33}.\left(\frac{-2.\gamma_{piez}}{1-v_{piez}}.\frac{v_{piez}.\gamma_{sub} - v_{sub}.\gamma_{piez}}{\gamma_{piez}.\gamma_{sub}} + 1\right).\frac{\partial T_3}{\partial t} + (\varepsilon_{33} - \frac{2.\gamma_{piez}.v_{piez}^2.d_{33}^2}{1-v_{piez}}).\frac{\partial E_3}{\partial t}\,\} \tag{11}$$

A is the active surface of the polymer.

$\frac{\partial T_3}{\partial t}$ and $\frac{\partial E_3}{\partial t}$ are the time derivatives of the electrical field and strain, respectively.

IOP Conf. Series: Materials Science and Engineering **827** (2020) 012046 doi:10.1088/1757-899X/827/1/012046

$$\alpha = d_{33} \cdot \left(\frac{-2.\gamma_{piez}}{1 - \upsilon_{piez}} . \frac{\upsilon_{piez}.\gamma_{sub} - \upsilon_{sub}.\gamma_{piez}}{\gamma_{piez} \cdot \gamma_{sub}} + 1 \right)$$

$$\beta = \varepsilon_{33} - \frac{2 \cdot \gamma_{piez}.\upsilon_{piez}^2.d_{33}^2}{1 - \upsilon_{piez}}$$

The electric field E is equal to the alternating electric field E_R across the resistor: $E = E_R$. The voltage output of the system across the load resistance is defined by the term figure 1

$$E_R = -\frac{R.I}{t_p}$$

Where I is the current which crosses the load R, e is the thickness of the polymer film.
So:

$$E = E_R = -\frac{R.I}{t_p} \quad (12)$$

The general expression the current I is

$$I = A.\alpha.\frac{\partial T_3}{\partial t} - \frac{A.\beta.R}{t_p}.\frac{\partial I}{\partial t} \quad (13)$$

In the frequency domain, the current is expressed in the form

$$I = j.w.A.(\alpha.T_3 - \frac{\beta.R}{t_p}.I) \quad (14)$$

$$I = \frac{j.w.A.\alpha.T_3}{1 + j.w.A.\frac{\beta.R}{t_p}} \quad (15)$$

In the case of a resistive load connected in series with the piezoelectric polymer working in piezoelectric mode, the objective is to take the expression of the power dissipated in the resistive load, on the basis of the equation that links the power, current, and load $P = R \cdot I^2$.

The power dissipated in the resistor R is expressed by the equation

$$P = \frac{R.(w.A.\alpha.T_3)^2}{1 + \left(w.A.\frac{\beta.R}{t_p}\right)^2} \quad (16)$$

3. Simulation results and discussion

3.1 Flexible textile substrate

A smart structure of type PVDF was stuck on three different substrates, Kermel and thick woven fabric (Polyester 65%–Cotton 35%) and Cotton 100%. The Kapton substrate was used commonly as reference substrate material for investigating the dielectric and piezoelectric properties of the composite. However, this material is used as a substrate for flexible electronic devices. The PVDF polymer layer was sandwiched between two silver electrodes that were used to enable film poling and perform study of the different parameters. The investigated textiles substrates are polyester-cotton, cotton, and polyamide-imide (Kermel) were plain weave, 65% polyester with a warp and weft.

3.2 Design mode study

This section focuses on the effect of substrates parameters textiles in order to increase the power recovered by the PVDF polymer film. The optimization of these substrates is implemented in order to ensure good performance of this electroactive polymer for energy harvesting.

The theoretical results are presented with an analysis of the 33-mode for the piezoelectric generator to verify the validity of the model developed for application in textile.

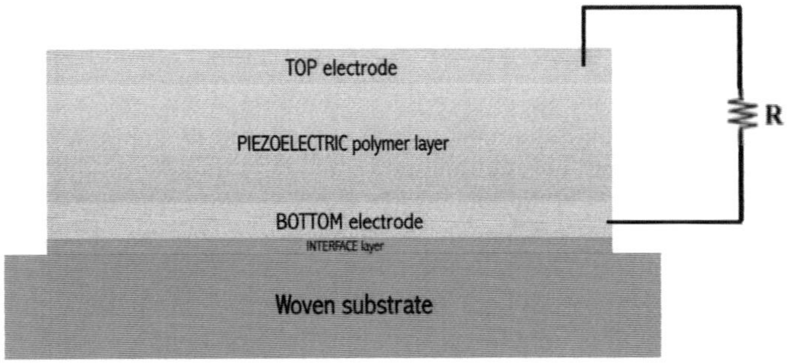

Figure 2. The smart structure stuck on woven fabric.

3.3 Effet of substrats on energy harvesting under compressive force

Figure 3 shows the theoretical results obtained with piezoelectric polymer PVDF film stucked on substrats (cotton; polyster/cotton; kermel) having a thickness of 500 µm, the frequency 6 Hz, and the applied force 80 N. The theoretical results show that the harvested power increases while increasing the resistance load, up to an optimal value in which the power decreases.

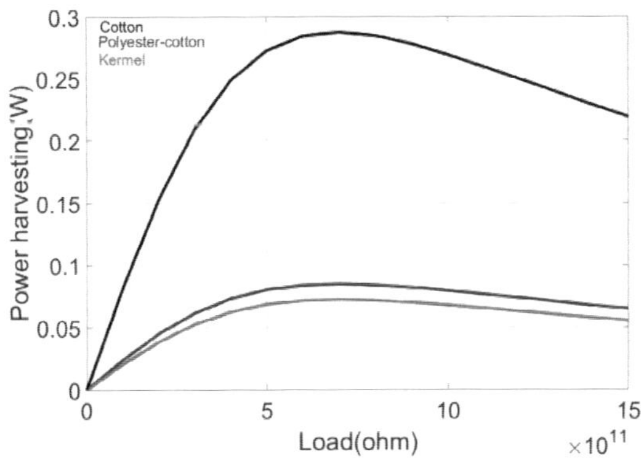

Figure 3. The power harvested as a function of resistive load for cotton, polyester-cotton and Kermel substrates when applying force F=80N at 6Hz.

4. Conclusion

Textile substrate bears one of the most promising form factors for future electronic applications due to its wide degree of freedom in shapes with flexible and stretchable characteristics. This work investigates the energy harvesting performance of PVDF polymer material stuck on woven-fabric substrates. Clearly the energy harvesting performance is fundamentally linked to the piezoelectric properties of the stuck film. The theoretical investigations confirmed an increase on the power harvesting values by 258mW, 96mW and 95mW for the cotton, Polyester-cotton and kermel substrates, respectively.

References

[1] Patel S, Park H, Bonato P, Chan L, Rodgers M A 2012 review of wearable sensors and systems with application in rehabilitation *Journal of Neuroeng. Rehabil* **9** 1–17

[2] Meddad M, Eddiai A, Cherif A, Hajjaji A and Boughaleb Y 2014 Model of piezoelectric self-powered supply for wearable devices *Superlattices and Microstructures* **71** 105-116

[3] Park J, Lee S, and Kwak B M 2012 Design optimization of piezoelectric energy harvester subject to tip excitation *Journal of Mechanical Science and Technology* **26** 137-143

[4] Priya S 2007 Advances in energy harvesting using low profile piezoelectric transducers *Journal of electroceramics* **19** 167-184

[5] Starner T 1996 Human-powered wearable computing *IBM systems Journal* **35** 618-629

[6] Silva H R, Afonso J A, Morim P C, Oliveira P M, Correia J H and Rocha L A 2007 Wireless

hydrotherapy smart-suit network for posture monitoring in 2007 IEEE International Symposium on Industrial Electronics (pp. 2713-2717) IEEE

[7] Saha C R, O'donnell T, Wang N and McCloskey P 2008 Electromagnetic generator for harvesting energy from human motion *Sensors and Actuators A: Physical* **147** 248-253

[8] Kornbluh R D, Pelrine R, Pei Q, Heydt R, Stanford S, Oh S and Eckerle J 2002, July Electroelastomers: applications of dielectric elastomer transducers for actuation, generation, and smart structures Smart *Structures and Materials* **4698** 254-270

[9] Ramadass Y K and Chandrakasan A P 2010 A battery-less thermoelectric energy harvesting interface circuit with 35 mV startup voltage *IEEE Journal of Solid-State Circuits* **46** 333-41

[10] Pozzi M and Zhu M 2012 Characterization of a rotary piezoelectric energy harvester based on plucking excitation for knee-joint wearable applications *Smart Materials and Structures* **21** 055004

[11] Shenck N S and Paradiso J A 2001 Energy scavenging with shoe-mounted piezoelectrics *IEEE micro* 30-42

[12] Almusallam A, Yang K, Zhu D, Torah R N, Komolafe A, Tudor and Beeby S P 2015 Clamping effect on the piezoelectric responses of screen-printed low temperature PZT/Polymer films on flexible substrates *Smart Materials and Structures* **24** 115030

[13] Kymissis, J, Kendall C, Paradiso J and Gershenfeld N 1998, October Parasitic power harvesting in shoes In *Digest of Papers Second International Symposium on Wearable Computers (Cat. No. 98EX215)* 132-139 IEEE

[14] Tognetti A, Lorussi F, Bartalesi R, Quaglini S, Tesconi M, Zupone G, de Rossi D 2005 Wearable kinesthetic system for capturing and classifying upper limb gesture in post-stroke rehabilitation *J. NeuroEng. Rehabil.* **2** 1–16

[15] Ma R, Lee J, Choi D, Moon H and Baik S 2014 Knitted fabrics made from highly conductive stretchable fibers Nano letters **14** 1944-1951

[16] R N Torah S P 2004 Experimental investigation into the effect of substrate clamping on the piezoelectric behaviour of thick-film PZT elements *physics d applied physics* **37** 1074–1078

[17] Almusallam A, Yang K, Zhu D, Torah R N, Komolafe A, Tudor J and Beeby S P 2015 Clamping effect on the piezoelectric responses of screen-printed low temperature PZT/Polymer films on flexible substrates Smart Materials and Structures **24** 115030

[18] Almusallam A, Luo Z, Komolafe A, Yang K, Robinson A, Torah R and Beeby S 2017 Flexible piezoelectric nano-composite films for kinetic energy harvesting from textiles *Nano Energy* **33** 146-156

[19] Chakhchaoui N, Ennamiri H, Hajjaji A, Eddiai A, Meddad M and Boughaleb Y 2017 Theoretical modeling of piezoelectric energy harvesting in the system using technical textile as a support *Polymers for Advanced Technologies* **28** 1170-1178

IOP Conf. Series: Materials Science and Engineering **827** (2020) 012047 doi:10.1088/1757-899X/827/1/012047

Application of make/use platform for sustainable fashion design

A Scerbaka[1,2]

[1]Faculty of Materials Science and Applied Chemistry Design Technology Department, Riga Technical university, Riga, Latvia
[2] Author to whom any correspondence should be addressed

anda.scerbaka@gmail.com

Abstract. The article explores the challenges of the clothing industry by focusing more on textile and the clothing industry interactions to reduce industry fabric waste and promote sustainability. The main research focus is to gather and analyze various possibilities for fashion designers to achieve zero waste designs in nowadays context. The introduction gives insights on the problems of the fashion industry and formulates the purpose and tasks of the research work, identifies and unpacks problems and also justifying the relevance of this study. Attention is drawn to the analysis of pre-consumer waste generation in the clothing production processes and the ways to drive the industry towards sustainability. The rich source of zero waste design ideas can be found in the ethnographic costumes' solutions. Integrating historical experience with the methods and approaches defined by the 21st century designers can significantly reduce or even avoid fabric pre-consumer waste. The paper will reflect the collected principles of achieving zero waste fashion by using Make/Use platform, an open source system for zero waste user, modifiable fashion practice. The methods of Make /Use platform are described, and its practical application is illustrated through the zero-waste coat design. Premeditate changes are based on the fashion designer's thinking encouraging minimizing the waste in the early stages of the costume idea development, providing a sensible use of the fabric with its silhouette and design specificity.

1. Introduction

Although most consumers are aware of the amount of waste that they generate themselves, relatively few are aware of the waste generated in the course of producing the goods that they consume. Fashion is one of the most resource-intensive industries in the world, which is complex and takes a leading role in world economics. *Michael F. Ashby* describes the last two centuries as being an era of material abundance, pointing out our ability to locate, mind, refine and use materials increased and, with it, material availability and affordability. Claiming the dominant business model, in material terms, was one of take - make – use – dispose, focusing more on economic growth resulting in increasing consumption [1]. This system has created sustainability issues including fabric waste at the pre and post-consumer stage, it is in this part of the process where industry has the opportunity to take more responsibility.

- The fashion and textile industry is one of the largest environmental pollutant on earth [2];
- Garment waste is a growing problem in the clothing industry: in pre- and post-consumer stage;
- In traditional clothing cut-sew production system, at least 15% of the cut off fabric is thrown away; [3] it is a waste of energy from invested resources, which could be used more rationally to not generate environmental pollution;
- One of the most preferred steps in the waste management hierarchy when it comes to sustainable

fashion is prevention [4];

- There is a growing interest in zero waste fashion to obtain sustainability in the textile and clothing industry [5];
- Industry is looking for solutions to discover applicable zero waste possibilities and for designers to understand and develop how to incorporate such techniques and methods intoproduction.

There are different strategies and direction to reduce waste in production stage. One of them - zero waste pattern making, involving designer, patternmaker and their integration or close collaboration. As an example, the rich source of zero waste design ideas can be found in the ethnographic costumes' solutions. Integrating historical experience with the methods and approaches defined by the 21st century designers can significantly reduce or even avoid fabric pre-consumer waste [6].

This study is a result of collected examples of ways to eliminate pre-consumer waste and it will question if in practice an engagement with Make/Use platform methods can lead the industry towards sustainability.

As well as Make/Use platform which is focused on user modifiable zero waste fashion to consumer to be aware of waste, will be tested in its application and relationship with user. Identifying this kind of methods allows designers to transform the way how to look at pattern making not only to minimize waste but also to develop new ideas to a sensible use of the fabric with its silhouette and design specificity.

2. Materials and methods – Make/Use platform application

Make/Use is freely available online platform developed by designer and researcher *Holly McQuillana* and her team. [7] Patterns are made considering the width and length of fabric, as well as fixed areas of it. *Holly McQuillan* emphasizes that the purpose of her designed patterns is not to reduce the consumption of fabric, but to eliminate the concept of waste [7]. The Make/Use system is simplified and accessible to different level skilled users, offering more sophisticated changes what can be made for experts.

The system features seven ready-made garments patterns, templates, fabric prints. There are available cut-out blocks, which gives modification options for users, and other tools and tips to help the user to change design (Figure 1) [7]. It states on the platform that the tools and concepts have been embraced by a diverse range of users, from retail clothing consumers and makers, to large international brands, including sustainability pioneers Patagonia and fast-fashion giant H&M [7] [8].

In Figure 2 shows seven garment collections and their patterns of Make/Use platform: crop T-shirt, long T-shirt, long coat, tube dress, wrap dress, wrap skirt, spiral trousers.

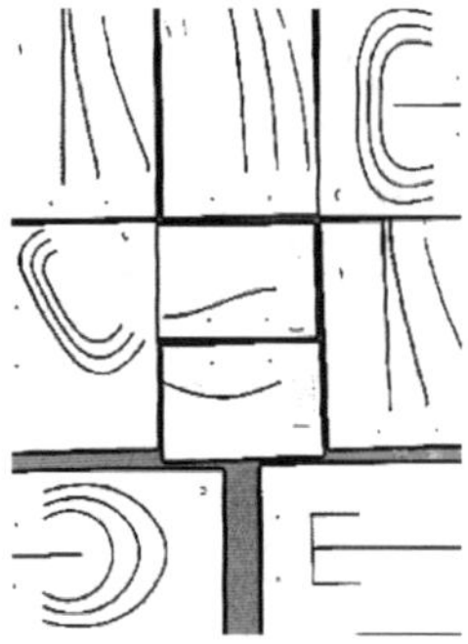

Figure 1. Make/Use platform templates [8].

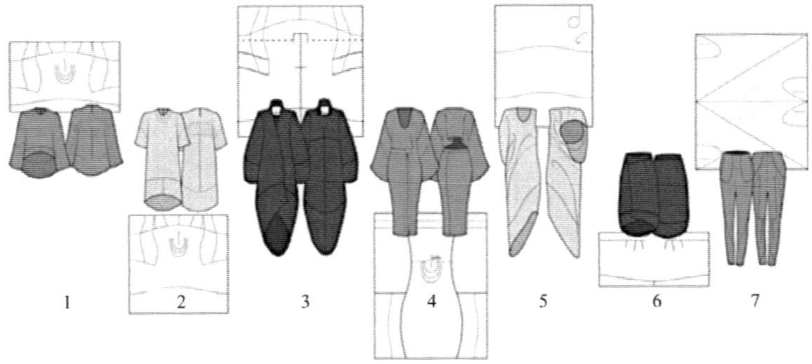

Figure 2. Make/Use platform seven garment collection: 1 – crop T-shirt; 2 – long T-shirt; 3 long coat; 4 - tube dress; 5 – wrap dress; 6 – wrap skirt; 7 – spiral trousers [7].

The key concept of the system patterns is to create volume by joining two edges of a piece of fabric to make a tube shape where is space for the body to pass through – as a wrap or two conjoined tubes. One of the main principles of Make/Use is to let the user understand how the modifications can be done, allowing the designer to find new solutions. There are offered four types of modifications:

- manipulation of the tubular volumes' structures;
- rotation of parts affecting the silhouette of the garment; rotation is achieved by cutting the tubular constructions with a curved line and rotating fabric 180 degrees. It can improve the fit and freedom of movement;
- changing and swapping parts to places that make asymmetric shapes in clothing;
- resetting fabric parts – removing them off and resetting back to its original place [7].

3. Results and discussion

The Make/Use open-source was tested by making coat prototype to understand how user experiences is relative. Based on the practical application and analysis of results, challenges were perceived and introduced to design, new insights, knowledge were gained and ideas for further development were then formulated.

The platform application in this research was tested with in online accessible pattern for a free silhouette long coat with the shawl lapel (Figure 3). The coat consists of two tubular constructions; horizontal tube is for the sleeves and shoulders, vertical for the torso; the coat opening is at the front that transforms into a collar neckline, in the back forming a back detail - a rectangular opening. The pattern can be adjusted to different lengths and widths depending of the used fabric.

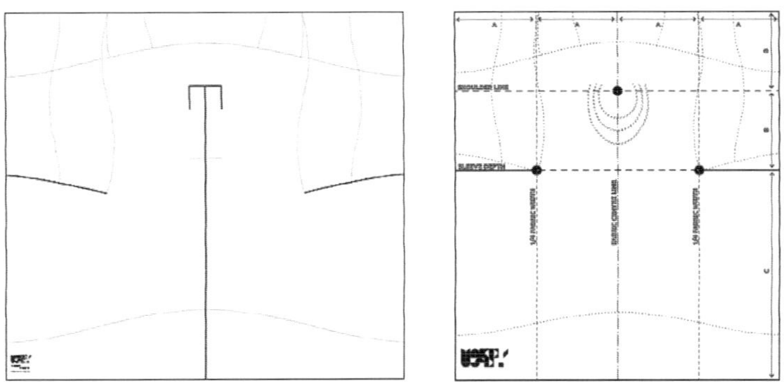

Figure 3. Make/Use platform long coat grid and pattern [7].

The coat was made taking in account all considerations and recommendations found on the platform as well as fabric specifics and design ideas. The final look of garment was influenced by a lot of factors such as designer's choice of details for joining edges, the drape of the cloth. Used material: 100% wool fabric, which was steamed, ironed, cut off not even edges of fabric bought in shop.

Pattern was printed according to instructions on website; transferred on to the fabric. Before cutting the fabric, considering the possibility of inaccuracy the pattern was checked with the base grid available in the system. The length of the coat was increased taking in account fabric length.

The fabric was cut in the indicated lines: in the central part vertically, the collar mark, the horizontal line which separates sleeves from torso part. The joining edges of the collar's back was sewn together, to form a shawl lapel and reveal a rectangular opening in the back of the coat and is seen as design detail. The cut edges were joined in the back vertical seam, horizontal back seam which forms sleeves. It was decided to reduce the size of the sleeves by folding cuff, thereby creating bigger volume on elbow part of sleeve.

Cut edges were joined in creative way, examples of stiches are shown in Figure 4. Needle felting, hand stitching as the insertion stitch, machine stitch has been given as example in platform.

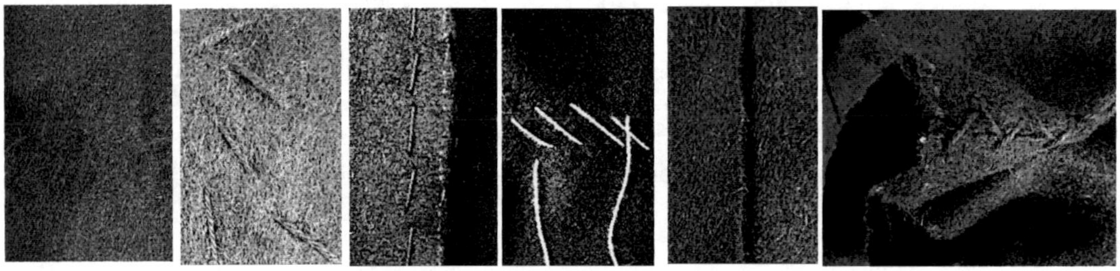

Figure 4. Examples of cut edges joining's. From the left: needle felting technique; needle felting technique with large stroke stitches; the selvedge stitched with straight hand stitches; diagonal hand stitches over needled felt; straight stitch; the insertion stitch.

In traditional fashion system exist 3 main evaluation criteria: appearance, fit and cost. [9] Usually these elements are viewed together as they depend on each other. To evaluate zero – waste design *Timo Rissanen* and *Holly McQuillana* have developed five primary criteria that should be taken into account when designing zero-waste patterns: appearance, fit, costs that can compete with traditional fashion cost, additionally should be valuated fabric waste and sustainability criteria, as well as the criteria for manufacturability. The weight of these criteria depends on the imposed privileged preferences.

It was detected that for non-experienced user achieving well-constructed garment could be problematic without guide of expert who understands how to get garment flat to form, even thou all instructions are given. Problems was detected on printing the given pattern in home environment, but solution was found by given possibility to recheck the pattern with the base grid available in the system. The available grid was helpful to understand how to adjust the pattern for specific fabric length and width for using all the fabric as was tested in prototype creation.

As one of main platform missions is to provide zero waist patterns, this is achievable if the fabric edges are accurate straight. During the creation of the prototype the only left-waste were pieces of the fabric what end up from the edges of the fabric, when not cut straight when the fabric was brought directly for the wholesaler. What is shown is the waste that was created due to the way fabric is sold to the consumer (Figure 5(2)).

Different practical solutions were made to join cut edges, the aim was to understand and valuate how many possibilities in this particular case can exist. The conclusions were that the design has the potential for future development and variability by adding design value such as functional details. As well as fabric manipulating possibilities were detected to the design of the original prototype; as in upgraded prototype was customized sleeves to match desirable silhouette. Also, the edges of the savage are integrated as a design component, not hidden or cut out.

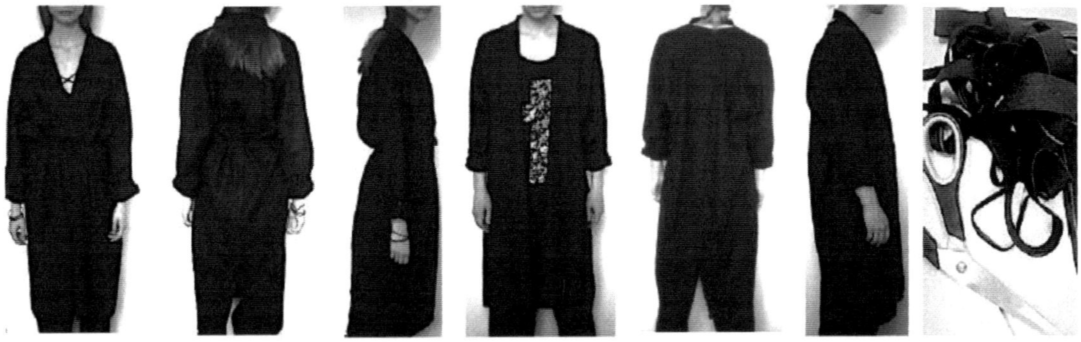

1 2

Figure 5. 1 Long coat from Make/Use platform on female (front, back and side view) and male figures (front, back and side view); 2 – The waste from the edges of the fabric.

Many criteria were considered to evaluate zero waste coat prototype result that includes appearance, design, fitting, sustainability and future possibility for production, zero waste approach. This can be seen in the table 1. All criteria were found approached and future possibilities have been observed.

After analysing the coat, observations show that the coat's appearance and fit are good, appearance can be considered as unisex, equally well suited to both genders (Figure 5(1)), its loose fit allows the garment to cover large range of sizes and it's easily adaptable to different body shapes.

Table 1. Characteristics of the coat.

Product description	Black 100% wool one-size unisex loose-form wrap coat with a shawl lapel, different design solutions can be found in the seams; sleeve size decreases in the wrist making fabric fold detail; opening is located on the front of the coat; a rectangular opening and split in the seam is in the back of the coat. Disadvantages: no functional details (pockets, clasps, snaps etc.); integration possibilities exist.
Fit	Balanced silhouette; loose fitting; freedom of movement is not restricted; comfortable to wear; the silhouette does not distort the wearer's figure.
Sustainability	No fabric waste in creating coat; design is sustainable because the silhouette is classic, addressing to a wide range of wearers. Disadvantages: the durability of seams should be increased to make sure that the joints are mechanically resistant; missing functional parts that can be integrated in further study.
Production possibilities	It is possible to produce a model or modified model in industrial production; possible problems may be due to the need for straight edges of the layout, to have the exact alignment of the joining edges. From the economic side, the production of such coats could be economically advantageous. Cutting requires less time for separating edges.
Conclusions	Pattern and design possibilities has potential of the future application for zero waste design development and production.

4. Conclusions

1. Zero waste design is a new way of thinking about the possibilities of the fashion design industry, it offers new opportunities to develop design ideas.
2. Make/Use Platform allows to discover, approach and successfully presents patterns for

interested users. It's a good source to get familiar to zero waste clothing pattern making.

3. The practical part of research - the coat prototype - uncovered a lot of positive qualities and reached objectives, for example zero waste result.
4. Understanding Make/Use platform application can help understand how to incorporate such techniques and methods into production that drives industry closer to the aim of the prevention of waste. Further studies should be made.

Acknowledgements
The authors gratefully acknowledge the funding by Riga Technical university grant, Latvia, PVS ID 3926.

References

[1] Ashby M F 2015, *Materials and Sustainable Development*, Butterworth Heinemann p.312
[2] Climate Works Foundation Quantis 2018 *Measuring Fashion: Insights from the Environmental Impact of the Global Apparel and Footwear Industries study* p.7
[3] Hayes S, McLoughlin J and Fairclough D Cooklin's 2012 *Garment Technology for Fashion Designers* 2nd edition, UK: A John Wiley and Sons, Ltd. p,210
[4] Gertsakis J and Lewis H 2003 Sustainability and the Waste Management Hierarchy: a discussion paper on the waste management hierarchy and its relationship to sustainability. *RMIT University, Melbourne*, pp.1-15.
[5] TED, The Textile Environment Design *The Ten*, [online] [01.03.2019]. Available at: http://www.tedresearch.net/teds-ten/.
[6] Timo Rissanen H M 2015 *Zero waste fashion design,* London: Bloomsbury Publishing Plc p. 224.
[7] *Make/Use research project* led by Holly McQuillan, [online] [03.03.2019]. Available at: makeuse.nz.
[8] McQuillan H, Archer-Martin J, Menzies G, Bailey J, Kane K and Fox Derwin E 2018 Make/Use: a system for open source, user-modifiable, zero waste fashion practice. *Fashion Practice,* **10**(1), pp.7-33.

IOP Conf. Series: Materials Science and Engineering **827** (2020) 012048 doi:10.1088/1757-899X/827/1/012048

Investigating the effect of air permeability and moisture on 2.5GHz textile microstrip patch antenna performance

A Wahab Memon[1,2,4], P Van Torre[3], B Malengier[1] and L Van Langenhove[1]
[1]Department of Materials, Textiles and Chemical Engineering - Centre for Textile Science and Engineering, Faculty of Engineering and Architecture, Ghent University, Ghent, Belgium
[2]Department of Textile, Mehran University of Engineering & Technology, Jamshoro, Pakistan
[3]Department of Information Technology, Ghent University, Ghent, Belgium
[4] Author to whom any correspondence should be addressed

AbdulWahab.Memon@UGent.be

Abstract. When discussing textile antenna and its comfortable use, the effect of air permeability and atmospheric moisture are eminent factors which have direct influence on textile antenna transmission characteristics and comfort of the wearer when used practically. The rate of airflow under certain air pressure passing through any surface is termed as air permeability of that surface and amount of moisture regained by a material when exposed to varying environmental humidity ranges compared to dry weight of that material is its moisture regain. Fleece and spacer fabrics were selected each having different air permeability and moisture regain values but both fabrics constitute polyethylene terephthalate - PET. Microstrip patch antennas that resonate at 2.45GHz were constructed utilizing both textile fabrics. Spacer fabric has high air permeability and low moisture regain compared to fleece fabric. Spacer fabric is a knitted 3-dimensional structured fabric whereas fleece is nonwoven in structure. FlecTron® is utilized as the conductive part of the microstrip patch antenna. Antenna made off spacer fabric showed better transmission characteristics and are least altered due to change in humidity compared to antenna made off fleece fabric. In this research, we conclude that the selected textile spacer fabric, with its high air permeability and low moisture regain, is the most favourable textile material in the construction of textile antennas.

1. Introduction

Wearable textile systems create an improved quality of life by enhancing functionality of clothing. This new generation of garments has the ability to monitor bio-signals from the wearer and from the environment around the wearer of these garments. By wireless means, information is transferred to a base station through wearable antennas integrated into the outer garment of the wearer [1]. Microstrip planar antennas are low profile, comfortable to wear, light in weight, simple in design and robust to harsh environmental conditions. These antennas utilize electro textiles as conductive and dielectric textile material as insulating part of the antenna [2] as illustrated in Figure 1.

Antenna performance is affected when exposed to high humid conditions especially by altering relative humidity. By selecting proper textile materials, this effect can be minimized. Textile materials with the lowest moisture regain and highest air permeability least alter, when employed to construct antennas, electromagnetic properties and hence better performance can be obtained from antennas constructed from those materials. Textile material contains air cavities and their relative permittivity is situated between 1 and 2, whereas water has a much higher relative permittivity of 78 at 2.45 GHz frequency and 25°C [3]. Textile material used for wearing purpose, having high air permeability is advantageous, therefore substrates like foam are not advantageous. Also, low moisture regain value and low relative permittivity can be effective in the antenna construction and for its performance.

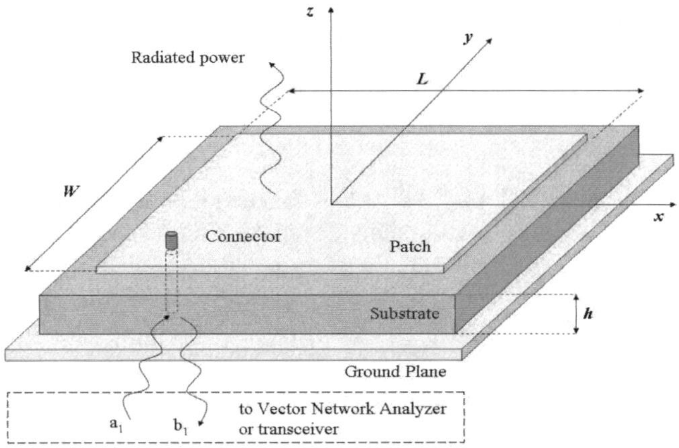

Figure 1. Geometry of Microstrip patch Antenna.

2. Materials and Method

In order to design an efficient microstrip patch antenna, selection of suitable textile materials was done. Textile materials having lowest possible permittivity, close to permittivity of air that is 1 together with high air permeability which allows any moisture if present in that material to evaporate more quickly and with low moisture regain are most suitable material properties to get high efficiency from antennas [4]. Spacer fabric which is a knitted 3-dimensional structure is chosen against fleece fabric which is nonwoven in structure as was used before in antenna construction [5]. Both are quite thick with sufficient difference in their air permeability values, their moisture regain values should be close to each other and they are suitable for the construction of microstrip patch antennas. Meanwhile, difference in air permeability allowed to check the effect on transmission characteristics for both antennas when relative humidity remains equal during transmission characteristics measurements.

FlecTron® is used as conductive electro textile to construct microstrip patch antennas. It is a copper plated nylon ripstop woven fabric with a surface resistivity of less than 0.1 Ω/m^2 and air permeability of 97mm/sec. Both the knitted spacer fabric and fleece fabric constitute of the same PET material. The thickness of the textile materials was measured using *Digimatic Indicator* based on the ISO 5084 standard whereas the air permeability is measured under ISO standard 9237.

The extent to which a material is sensitive to moisture is defined as its moisture regain. Both textile fabrics were put in an oven for 5 hours under 105°C and weighed. Afterwards, both fabrics were put in conditioned room for 24 hours at 20°C and 65% relative humidity for conditioning, later these fabrics were weighed again. This procedure is followed under ISO standard 6471 (1-4). The moisture regain is then calculated by,

$$Moisure\ Regain\% = \frac{Conditoned\ weight - Dry\ weight}{Dry\ weight} \times 100 \qquad (1)$$

Figure 2. Anechoic chamber room.

Two materials (fleece and spacer fabrics) chosen for microstrip patch antenna construction are moisture dependent, but one is less dependent than the other as moisture regain is different for both materials.

Reflection co-efficient (S_{11}) is measured in an anechoic chamber shown in figure 2 by employing a *Vertical Network Analyzer* (VNA). Relative permittivity and loss-tangent of textile materials are determined by comparing simulated and measured resonance frequency of microstrip antennas. The constructed spacer fabric-based antenna is illustrated in figure 3. Antennas are simulated using *CST Studio Suite* 2017 and constructed manually utilizing both fleece and spacer textile fabrics.

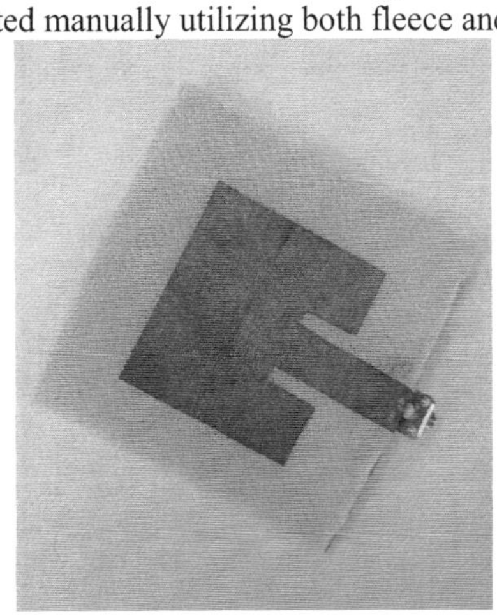

Figure 3. Spacer fabric-based microstrip patch antenna.

Both the antennas were put to conditioning again for 24 hours at 65% relative humidity and 25°C temperature and later measured their transmission characteristics.

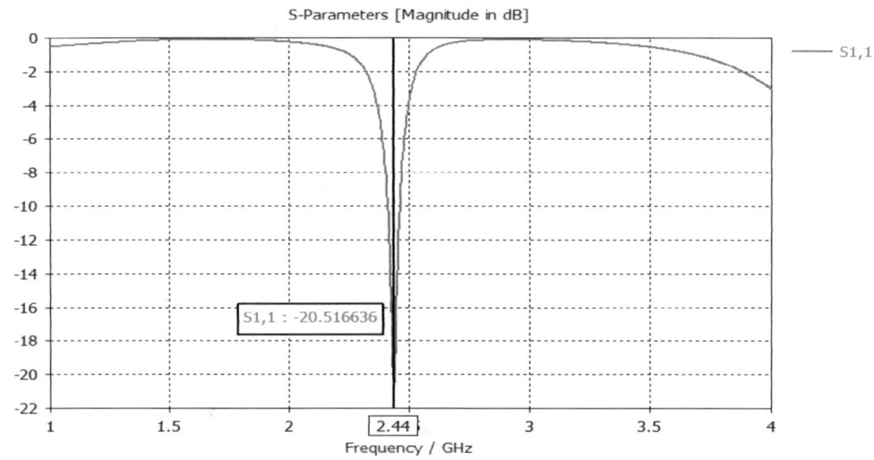

Figure 4. Measured return loss for spacer fabric-based antenna.

3. Results and discussion

The dimensions of simulated spacer fabric-based antenna are given in table 1. The simulated dimensions for spacer fabric-based antenna proved accurate when antenna was constructed using spacer fabric as constructed antenna also resonates at 2.45GHz.

Table 1. Dimensions of simulated spacer fabric-based antenna.

Length of patch (mm)	Width of patch (mm)	ength of ground plane (mm)	Vidth of ground plane (mm)	ength of ed line (mm)	Width of eed line (mm)	Gap between eed line and patch (mm)
53	60	100	100	15	13.36	5

The transmission characteristics of spacer fabric-based antenna which is designed to resonate at 2.45GHz are illustrated in figure 4. SMA connector is employed to construct microstrip patch antennas. Antennas resonate at 2.45GHz when provided with 50Ω of impedance. The properties of FlecTron® conductive fabric, spacer fabric and fleece fabric are mentioned in table 2.

Table 2. Properties of fleece and spacer fabrics.

Materials	Permittivity (εr)	Moisture regain (%)	Thickness (mm)
Fleece	1.25	0.65	2.56
Spacer	1.05	0.52	3

Air permeability was measured as an average or mean value of five different test results taken from five different positions of one fabric sample. Five different positions within the sample were chosen to see the difference in air permeability values within one sample. This procedure was followed for both fleece and spacer fabrics. An average or mean value with standard deviation for air permeability of the spacer and fleece fabrics are mentioned in table 3.

Table 3. Mean and standard deviation of air permeability test.

Material	Mean	Standard deviation
Spacer fabric air permeability (mm/sec)	2618.4	30.41
Fleece fabric air permeability (mm/sec)	610	14.61

Antennas constructed from both different textile materials resonate at 2.45 GHz ISM band. Main focus is kept on the sharp resonance peak rather than covering the entire communication frequency band because the peak position is related to the relative permittivity of textile material which changes when the relative humidity in the atmosphere changes. Influence of relative humidity at its particular value on antenna performance is examined by measuring return loss characteristics of both antennas.

It was observed that resonance peak for conditioned (65% relative humidity and 25°C) spacer fabric-based antenna is least altered when measuring its transmission characteristics compared to that of conditioned fleece fabric-based antenna at the same values of relative humidity and temperature. The results for the influence of increasing relative humidity ranging from 10% to 90% on transmission characteristics for both antennas are forthcoming. High air permeability of spacer fabric-based antenna helps to let the moisture in it evaporate more quickly, hence resonance peak is least altered at particular value of relative humidity and temperature. Whereas, due to the low air permeability of fleece fabric-based antenna, moisture remains for long in fleece fabric, due to this when transmission characteristics were measured, resonance peak is more altered.

4. Conclusion

Because of the fact that high air permeability of spacer fabric-based antenna takes the least time to let the moisture from the antenna substrate evaporate more quickly, this will help achieve less shift of resonance frequency of that antenna. Whereas, fleece fabric-based antennas cannot allow moisture to dissipate in the atmosphere as quick as spacer fabric based antenna does. Hence resulting in more shift of its resonance frequency. Another important factor is the relative permittivity of both antenna substrates. Spacer fabric has relatively lower permittivity than fleece fabric, almost as close to the permittivity of air. Water has relatively higher permittivity that is 78 at 2.45GHz and tangent loss is 0.15 at 25°C. So, the influence of humidity on transmission characteristics of antenna made from spacer fabric is less because spacer fabric has low moisture regain and high air permeability. Whereas, the influence of humidity on transmission characteristics of antenna made from fleece material is slightly more because of slight high moisture regain and low air permeability.

Acknowledgment

The authors acknowledge the scholarship financed by Higher Education Commission – HEC of Pakistan for the first author of this paper. Being the first author, I would like to thank my promotor and co-promotors for their support in all aspects.

References

[1] C Hertleer, A Tronquo, H Rogier, L Vallozzi and L Van Langenhove 2007.Aperture-coupled patch antenna for integration into wearable textile system *IEEE Antenn Wirel Pr* **6** pp.392-395

[2] L Vallozzi, C Hertleer and H Rogier 2016 Latest developments in the field of textile antennas *Smart textiles and their applications* **178**, ed V Koncar UK: Elsevier. pp 599-626

[3] C Hertleer, A Van Laere, H Rogier and L Van Langenhove 2009 Influence of relative humidity on textile antenna performance *Text. Res. J.* **80**(2) pp. 177-183

[4] A Tronquo, H Rogier, C Hertleer and L Van Langenhove 2006 A robust planar textile antenna for wireless body lans operating in the 2.45 GHz ISM band *Inst. Elect. Electro. Lett.* **42**(3) pp.142- 143

[5] A Tronquo, H Rogier, C Hertleer and L Van Langenhove 2006 Applying textile materials for the design of antennas for wireless body area networks *IEEE Antenn Wirel* France.

Needleless electrospinning of PAN/SBA-15 for the preparation of nanofibers membranes

M H Rafe[1,2,3,5], C Delaite[1], B Lebeau[2], M Bonne [2], E Mohsenzadeh[4] and D Adolphe[3]

[1]Université de Haute Alsace (UHA) Laboratoire de Photochimie et d'Ingenierie Macromoléculaires (LPIM), F-68100 Mulhouse France
[2]Université de Haute Alsace (UHA), CNRS, Institut de Science des Matériaux de Mulhouse (IS2M) UMR 7361, F-68100 Mulhouse, France
[3]Université de Haute Alsace (UHA), Laboratoire de Physique et Mécanique Textiles (LPMT), F-68100 Mulhouse, France
[4]UCL, Hautes Etudes d'Ingénieur (HEI), GEMTEX, 13 rue de Toul. F-59000 Lille, France
[5] Author to whom any correspondence should be addressed

Muhammad-hassan.rafe@uha.fr

Abstract. Polyacrylonitrile is one of the few waterproof polymers that can be spun from relatively safe solvents, facilitating the use of PAN nanofibers membranes in medical, filtration and biological applications. In this study PAN nanofiber membranes containing SBA-15 nanoparticles were prepared by the process of electrospinning in needless electrospinning machine and DMSO was used as solvent. A small amount of dispersing agent BYK W-9010 was added for a better deagglomerating of SBA-15 nanoparticles in the PAN/DMSO solution. The influence of rod like SBA-15 type ordered mesoporous silica particles as filler on the morphology and textural parameters of electrospun PAN nanofibers was studied. The effect of concentration, applied voltage, distance and feed rate on nanofibers diameter was investigated. The results obtained demonstrated that by increasing the concentration, applied voltage and feed rate, the number of nanofibers was increased and thicker nanofibers were obtained. Viscosity increased by increasing the concentration of polymer and it increased further by imparting small amount of silica nanoparticles. The SEM images of nanofibers were taken and their morphology explains that presence of silica nanoparticles in the nanofibers increases the specific surface area and makes the surface rough.

1. Introduction

Nanofibers can be produced by different methods like jet blowing, melt blowing, co-extrusion, interfacial polymerization, electrospinning and many others. Among these, electrospinning is a versatile and well organized technique that is used to produce very fine homogeneous micro fibers or nanofibers. There are many advantages of the electrospinning process including being a relatively quick, simple, economical and efficient way to produce nanofibers. The applications of electrospun products are mainly in electronics, automotive, cosmetics, filtration and healthcare sectors [1].

In the process of electrospinning, usually polymer is dissolved in a solvent to make a solution and the selection of solvent is very important as it should be not toxic, corrosive or dangerous to be used [2]. In this work polyacrylonitrile (PAN) is the selected polymer with ordered mesoporous silica SBA-15 particles as inorganic fillers dissolved and suspended, respectively, in dimethyl sulphoxide (DMSO) to make a suspension to be electrospun. PAN is one of the few water resistant polymers which is semi crystalline and well resistant to degradation by sunlight, and in addition fibers of PAN are thermal insulators [3].

The addition of SBA-15 particles in the nanofibers provides increase in specific surface area of nanofibers [1]. Mesoporous silica materials have a great interest that continues to grow in emerging applications such as drug delivery system, catalysis, sensors, filtration etc., that benefit from the properties of these materials [4]. There are many specific cases where ordered mesoporous technology is urgently needed. These include ultrapure water purification and reuse in semiconductor industry, treatment of pharmaceutical waste water discharge and recovery of toxic homogeneous catalyst used in fine chemical production [5]. Nanocomposites filled with nanofibers show enhanced mechanical properties in contrast to unfilled composites [6].

Dispersion of silica nanoparticles by adding some surfactant implies the modification of surface energy of the silica particles that results in disrupting the Van der Waals forces that attract the silica particles to each other and thus hindering the SBA-15 particles to reaggregates [7].

This study gives an overview of the effect of the different PAN concentrations in DMSO, the effect of incorporating SBA-15 particles and the effect of dispersing agent in PAN/SBA-15/DMSO solutions, variation of electrospinning parameters on the process and resulting nanofibers.

2. Material and Methods

2.1. Materials
Polyacrylonitrile with Mw = 150,000 g/mol and DMSO (99.9% pure) were purchased from Sigma-Aldrich. SBA-15 has been synthesized according to S. Almuhamed et al [6] and Y. Belmoujahid et al [7] procedure. Dispersing agent BYK-W 9010 was obtained from BYK Company (Germany).

2.2. Preparation of solution
PAN solutions of 7, 8.5 and 10wt% were prepared in DMSO and 2wt% SBA-15 was added in all three solutions. A dispersing agent (1 wt% with respect to SBA-15) was also added in the SBA-15 containing suspension. Each SBA-15 containing suspension was stirred for 30 minutes by IKA® T18 digital Ultra Turrax to make the suspension homogeneous and obtain well deagglomeration of (SBA-15) silica particles.

2.3. Electrospinning
For electrospinning, the needleless nanospinning machine "Nanospider Lab" (Elmarco, Czech Republic) was used. The spinning parameters such as voltage, electrode-substrate distance, nozzle diameter, carriage speed etc. were varied to find optimum conditions. The optimized parameters are given in table 1.

Table 1. Electrospinning parameters.

PAN Concentration in DMSO (wt %)	Voltage (KV)	Distance (mm)	Feed rate (mL/h)
7%	20	200	0.5
8.5%	30	200	0.6
10%	40	190	0.6

2.4. Electrospinning conditions
The prepared PAN and PAN/filler solutions were electrospun directly by means of nanofibers machine "nanospider" by (Elmarco, Czech Republic). Temperature of the spinning chamber was 22-24 °C and

relative humidity was 38% to 40%. Carriage speed was 190 mm/sec and each sample was electrospun for 20 minutes. The electrospinning process is carried out between two high voltage wires, i-e the bottom rotating wire (electrode) in which polymer solution carriage is moved and it produces nanofibers, whereas the top wire (electrode) that is connected to nanofibers collector sheet.

2.5. Characterization
The viscosity of PAN/DMSO, PAN/SBA-15/DMSO and PAN/SBA-15/BYK-W 9010/DMSO solutions were measured using Anton Paar Modular Compact Rheometer 502 on "plan plan mode". SBA-15 particles sizes were measured by ZETASIZER nano series in DMSO. The surface morphology of nanofibers was examined using scanning electron microscopy JEOL JSM-IT100. Images obtained by SEM were analyzed by Image J 1.45S software to obtain the average fibers diameter. Three images were taken by sample to get the mean values of diameter of fibers. On each image, 25 measurements of fibers diameters were taken and average value of all measurement were determined.

3. Results and Discussion

3.1. Effect of PAN concentration on PAN/DMSO solution viscosity
The viscosities of all the solutions were measured at constant temperature of 20 °C and before each measurement solutions were mixed by Ultra Turrax at 1×10^4 r.p.m for 20 minutes to make solutions homogeneous. The viscosities are mentioned in table 2.

Table 2. Viscosity measurement of different PAN/ DMSO solutions.

Sr. No.	PAN concentration in DMSO (wt%)	Viscosity mPa/sec
1	7	$9.62\times10^2 \pm 45$
2	7+2%SBA-15	$1.5\times10^3 \pm 145$
3	7+2%SBA-15+1% BYK	$1.65\times10^3 \pm 212$
4	8.5	$2.55\times10^3 \pm 294$
5	8.5%+2%SBA-15	$3.25\times10^3 \pm 217$
6	8.5+2%SBA-15+1% BYK	$3.47\times10^3 \pm 188$
7	10	$3.93\times10^3 \pm 161$
8	10+2%SBA-15	$5.87\times10^3 \pm 408$
9	10+2%SBA-15+1% BYK	$6.3\times10^3 \pm 175$

The results show that an increase in concentration of PAN in DMSO will increase the viscosity. The increase in viscosity is ascribed to the increase of intermolecular interactions. In the same way, dispersion, stronger polarization and hydrogen bonding between the solvent and the polymer affects the solubility and the viscosity, where continuously increase in viscosity is observed by increasing the concentration from 7%wt to 8.5%wt and then 10%wt. Similarly by imparting the 2%wt of silica nanoparticles increases the viscosity of solution to much extent and it is observed that same behavior of increasing the viscosity in small impact is seen by adding BYK W-9010 (1%wt of total weight of silica particles) (figure 1).

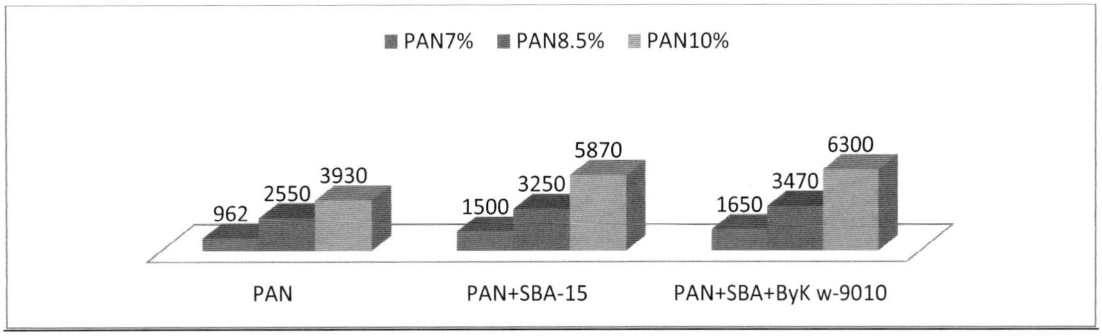

Figure 1. Impact of SBA-15 & BYK W-9010 on PAN solution viscosity.

The viscosity of solution has great impact on the electrospinnability of the solution and on the morphology of the resulted nanofibers. If the solution has less viscosity only beads and beads on nanofibers will be obtained. This is due to the number of chains entanglements of polymers that are lower than the critical point. This lack in viscosity gives the place to the solvent surface tension to be dominant and thus to break up the jet into individual droplets and electrospying or beads on nanofibers takes place [7]. Further increase in the solution concentration yields higher number of chains entanglement which is high enough to resist the jet breakage and also to withstand further elongation the solvent evaporation. Thus bead free nanofibers were collected.

3.2. Particles size measurement

The particles size of SBA-15 in DMSO was measured by light scattering, with and without BYK-W 9010, and results are given in table 3.

Table 3. Particles size measurement.

Sample Name	Poly dispersity index (PDI)	Mean Diameter (nm)
2wt%SBA-15+ BYK,W-9010 (1wt% vs SBA-15 wt) in DMSO	0.945	110
2wt%SBA-15/DMSO	0.641	250

The dispersing agent causes deagglomeration of silica particles in the solution by coating the surface of the inorganic filler. Dispersing or deagglomeration of silica nanoparticles by introducing the dispersing agent was a physical treatment which was pretreated by mechanical shear mixing. Mechanical shear mixing had the role of creating high local shear stress, which produced the tinny gap in the aggregates and the dispersing agent diffused into those gaps and started to disperse the nanoparticles by steric or electrostatic repulsions and as a result the previously formed gaps propagate, leading to ultimately separating the silica nanoparticles from their aggregates.

3.3. Nanofibers surface morphology

The nanofibers morphologies are shown in figure 1. For all three different solution concentrations, voltage was increased gradually 20, 30 and 40kV and feed rate of 0.5mL/h for low concentration (7wt% PAN) was used whereas for higher PAN concentrations (8.5, 10wt% PAN) it was increased to 0.6mL/h to make the feed rate convenient with electric field and concentration, so that droplets suspended at the needle tip must be drawn away with equilibrium mass balance and as a result nanofibers with uniform diameter were obtained. The distance was fixed at 200mm for 7, 8.5wt% and 190mm for 10wt% of PAN in DMSO concentration. SEM images of different PAN concentrations with different diameters are shown in figure 2.

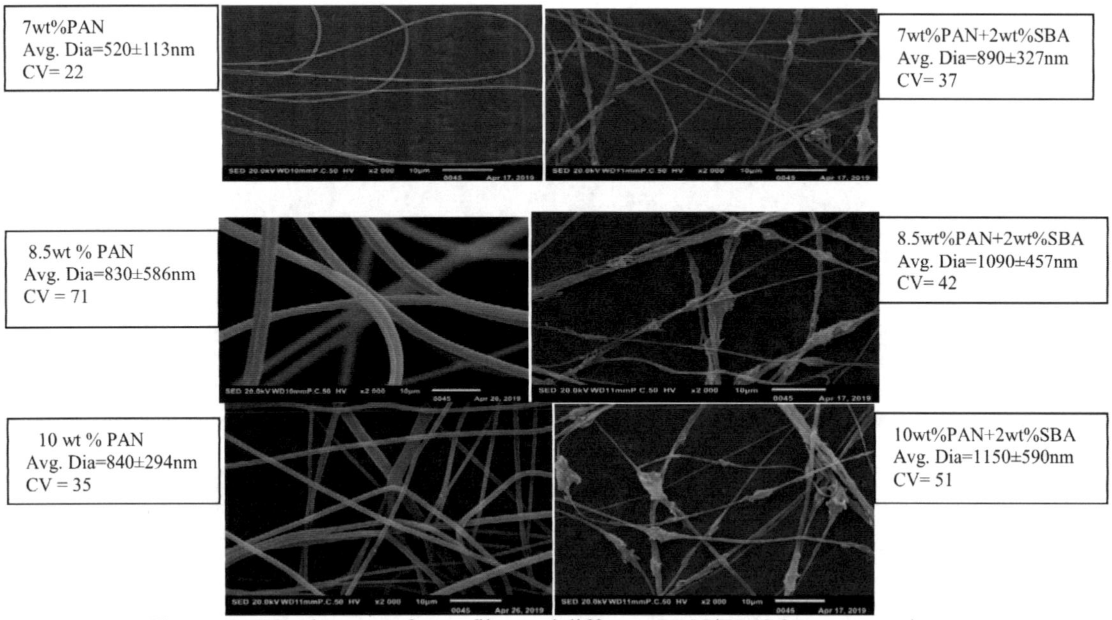

Figure 2. SEM images of nanofibers of different PAN/DMSO concentrations.

The SEM images show that at low PAN concentration, there are very few nanofibers and with small average diameter of 520nm. By increasing the polymer concentration, voltage and feed rate, the amount of nanofibers increased and their diameter also increased to 830 nm and 840 nm for 8.5 and 10wt% of PAN in DMSO, respectively. Similarly, by introducing silica nanoparticles in PAN/DMSO solution the viscosity behavior is much increased because the introduction of silica nanoparticles increased the resistance in flow of solution. And increase in viscosity resulted in thick nanofibers of 890nm, 1090nm and 1150nm for 7%wt, 8.5%wt and 10%wt of PAN concentration respectively, so high viscosity resulted in nanofibers of larger diameter. The impact of PAN concentration on nanofibers diameter is shown in figure 3.

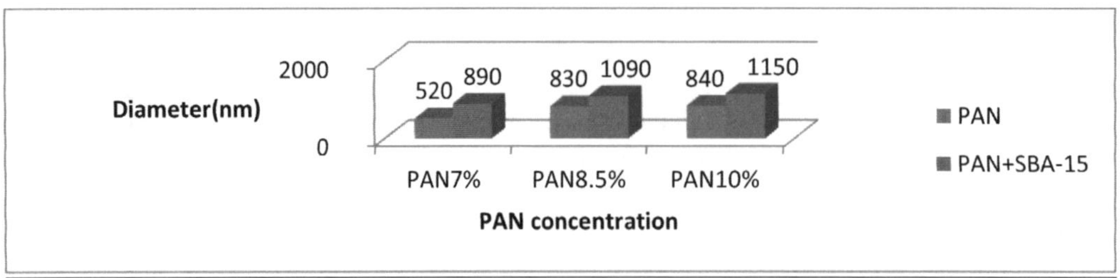

Figure 3. Impact of PAN concentration on nanofibers diameter.

In the case of applied voltage, as it is increased for 7%wt to 8.5%wt and then for 10%wt concentration from 20kV to 30Kv and 40kV, the quantity of nanofibers is increased and their diameter is also increased. The reason behind it is that low voltage electrospinning does not occur because the magnitude of generated electric field has not enough power to overcome the surface tension of the solution, whereas at higher voltages, the electric field overcomes the surface tension of the solution and droplets takes the shape of typical Taylor cone and it will accelerate the jet, reduce the flight time. Shorter flight time will offer less time for the fiber to be stretched and the defect free cylindrical thick nanofibers are obtained [7]. The decrease in distance between electrodes to 190mm in case of 10wt% PAN concentration resulted in an increased in the electrical field and a reduction in flight time, which caused thick nanofibers. By adding the silica particles in the nanofibers, a few agglomerates are observed on the surface of nanofibers and their diameter also increased. Silica particles make the surface of nanofibers rough and increase the specific surface area.

4. Conclusion

For needleless electrospinning, finding the best parameters is very important, as voltage, PAN concentration and feed rate give the significant effect on electrospinning results. High voltage accelerated the jet and reduced the flight time, which allowed less time for fibers stretching and elongation, resulting in thicker nanofibers. High concentration of polymer solution resulted in higher number of polymer chains entanglement which led to the formation of larger diameter nanofibers. Whereas higher feed rate, due to large nozzle, coated more polymer solution on the electrical wire, which resulted in more fibers and increased fibers' diameter. The addition of silica particles increased the viscosity of PAN/DMSO solution. The dispersing agent BYK W-9010 was successfully used to promote the deagglomeration of silica particles in the PAN/DMSO solution. BYK W-9010 has also small impact on solution viscosity. The introduction of silica particles make the surface of nanofibers rough and increase the nanofibers diameter.

References:

[1] Faccini M, Borja G, Boerrigter M, Morillo Martín D, Martìnez Crespiera S, Vázquez-Campos S, Aubouy L and Amantia D, 2015. Electrospun carbon nanofiber membranes for filtration of nanoparticles from water. *Journal of Nanomaterials* **2015**.

[2] Sabantina L, Mirasol J R, Cordero T, Finsterbusch K and Ehrmann A 2018 *Investigation of needleless electrospun PAN nanofiber mats*. In AIP Conference Proceedings (Vol. 1952, No. 1, p. 020085). AIP Publishing LLC.

[3] Properties of polyacrylonitrile, *Institute Garanado de Technologia da polyacrilonitrila*. 2016. http://www.igtpan.com/Ingles/propriedade-poli.asp

[4] Almuhamed S, Khenoussi N, Bonne M, Schacher L, Lebeau B, Adolphe D and Brendlé J 2014. Electrospinning of PAN nanofibers incorporating SBA-15-type ordered mesoporous silica particles *European polymer journal* **54** pp.71-78.

[5] Ka Y H, Gordon M and King L Y 2003 *Langmuir* Selective Adsorbents from Ordered Mesoporous Silica **19** pp.3019-24.

[6] Wei L and Sabit A, 2014 Properties of Electrospun Polyacrylonitrile Membranes and Chemically-activated Carbon Nanofibers *Tex. Res. J.* **80**(2) pp.124-34

[7] Almuhamed S 2015 Study and development of nonwovens made of electrospun composite nanofibers (Doctoral dissertation, Mulhouse).

A project to improve awareness for cyber security in smart textile applications

Manuela Bräuning[1,3], Tobias Scheible[2], Martin Rieger[2], Holger Morgenstern[2], Désirée Göbel[1], Nele Schäfer[1]

[1]Faculty of Engineering, Textile and Clothing Technology and Management, Albstadt-Sigmaringen University, Germany
[2] Faculty of Computer Sciences, IT Security and Digital Forensics, Albstadt-Sigmaringen University, Germany
[3] Author to whom any correspondence should be addressed

braeuning@hs-albsig.de

Abstract. The textile and clothing industry is facing structural changes and will enter into the interdisciplinary market of Smart Textiles. This market has a high market potential, but there is also plenty new challenges. That is why our project systematically investigates the integration and IT security of electronic communication systems in textile products. The main objectives are the scientific investigation of the area of IT security of smart textile products, the transfer of existing security concepts as well as the development of secur innovative prototypes. Based on the knowledge gained, guidelines for further action are deduced and courses for the textile and clothing industry will be created, ensuring an intensive multidisciplinary and cross- industry know-how transfer. This challenge is to be solved by an interdisciplinary team of scientists and companies from the fields of textile and clothing as well as information technology.

1. Introduction

The textile industry is one of the oldest industries in the world. In addition to the automation of the loom, the development of synthetic fibers was one of the greatest revolutions in the industry. Synthetic fibers have opened up a multitude of new possibilities in addition to classic clothing. The field of technical textiles has been on the advance in the last decades and has opened up a multitude of industries with product innovations. From underwear to interior fittings and lightweight construction which makes e-mobility possible, to the rotors of wind power edges which support the energy revolution, textiles make an important contribution in our lives and are often hidden despite everything. Electronic components are integrated into textiles and trigger the next revolution in textile and clothing technology. These textile products with extended functionalities are summarized under the term Smart Textiles. Although the development of Smart Textiles has been going on for decades, their use in everyday life is still in its infancy and there are numerous definitions. In the following, Smart Textiles are defined as textile products that interact with their environment and can thus actively support the user. The terminology used in this paper is divided into intelligent textiles, so-called I-textiles, and textiles with integrated electronic components, so-called e-textiles. For this research work we concentrate exclusively on the field of e-textiles and divide them into adaptable, integrated and modified integration levels. Adapted means that the electronic components are located in a closed housing and the textile is the carrier. On the next level, the connections of an electronic circuit with textiles are realized, for example with conductive yarn. At the highest level, modified, electronic components such as batteries and temperatures are realized directly as textiles.

1.1. Market evolution

From automated communication with a washing machine and intelligent anti-theft protection to vital-sign monitoring – technology has the ability to make textile products more intelligent and this way "smarter". Notably, textiles and clothing are one of the oldest industries and are essential to our everyday life. Whilst our ancestors used garments to protect themselves against environmental impacts, thereby facilitating the highly developed human civilization; textiles and clothing also paved the way for globalization in the shape of the Silk Road, connecting different societies across different continents. In addition, the field of technical textiles has been on the rise in recent decades and is opening up a wide variety of industries with product innovations; here Germany is regarded as an innovation engine and world market leader [1][2]. The market for Smart Textiles products is a young and dynamic area. The Centre for European Economic Research (ZEW) expects the global market volume to have increased thirtyfold by 2030. According to the study, the greatest growth is expected in the medical, industrial, military and fashion sectors [3]. This trend can also be deduced from some megatrends such as connectivity, new work, health and individualization [4].

1.2. Cybersecurity

The report "The situation of IT security in Germany 2018" by the German Federal Office for Information Security (BSI) states that the threat situation for IT attacks has intensified and the focus of attacks is clearly spreading. The Smart Home and IoT areas are now also among the targets for attacks (see Figure 1). Smart Textiles are assigned to the Internet of Things category and are therefore also a potential target for attackers. However, IT security in this area has not yet been systematically researched [5].

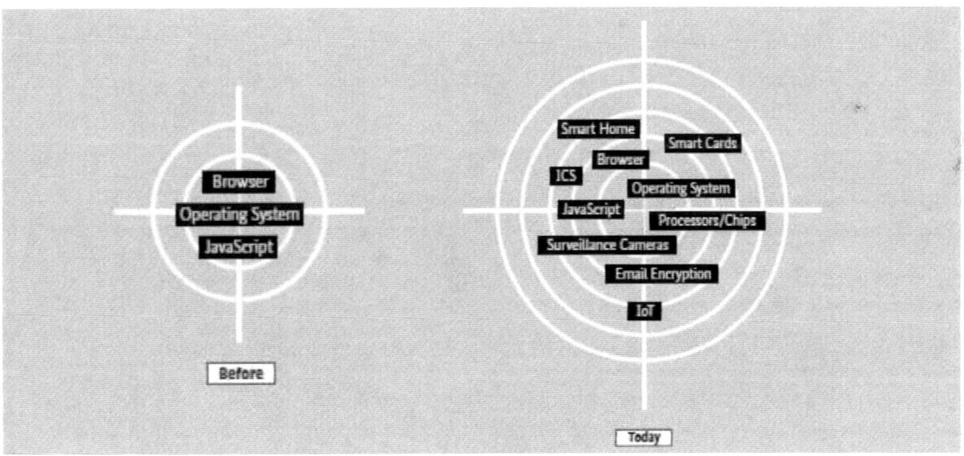

Figure 1. Focus of attacks broadening [5].

2. The project SEKT

Therefore, the interdisciplinary research project SEKT of the Faculties of Engineering and Computer Science at the Albstadt-Sigmaringen University takes up this approach and intensively combines the fields of textile and clothing technology and IT security. The SEKT project systematically investigates the integration and IT security of electronic communication systems in smart textile products and develops practical security concepts and smart demonstrators. This challenge is to be met by an interdisciplinary team of scientists and companies from the fields of textile and clothing technology and IT security between December 2018 and November 2021. In addition to three professors a team of three employees and twelve project partners – ranging from Start-Ups to associations to large cooperations –

will be involved. The core objective is the scientific investigation of the area of IT security of smart textile products and the transfer of existing security concepts as well as the development of secure, innovative prototypes. Recommendations for action are derived from the knowledge gained and courses for the textile and clothing industry are created, thus ensuring intensive interdisciplinary and cross-sector know-how transfer.

The project is divided into the following five phases:

1. Investigation of current system solutions via survey and analysis of the latest product developments
2. Examination of specific systems with which the integration of the electronic components into the textile products is evaluated as well as the safety concepts used are investigated.
3. Development and adaptation of practice-relevant IT security concepts to smart textiles products
4. Development, design and manufacture of three selected smart demonstrators
5. Knowledge and technology transfer by means of guidelines and a teaching module in the blended learning process [6].

2.1. Use of communication technologies in Smart Textiles products

The main focus in the analysis of the systems was on the wireless communication technologies used. When a transmission of data takes place, this is relevant for IT security. Since an attacker does not have to have direct access to the product, this is also the most likely attack scenario. The following Figure 2 shows the distribution of different technologies in different categories.

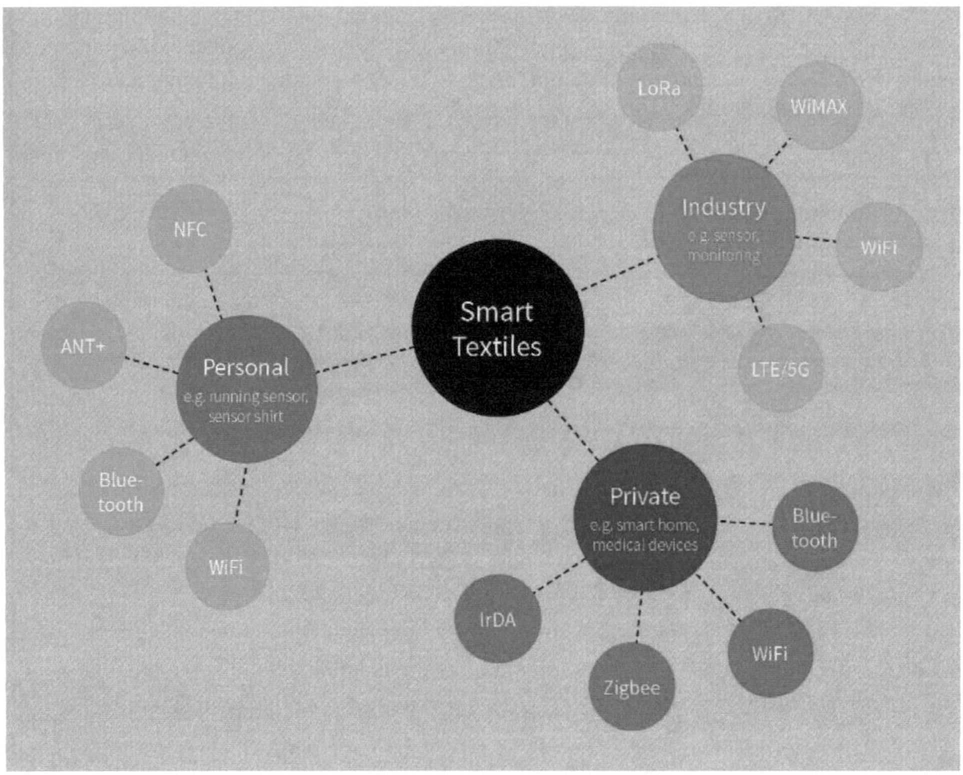

Figure 2. The four most common wireless communication technologies in the three main areas.

The next steps will include a deeper analysis of these communication technologies. Performing this, we will use well established methods as e.g. penetration testing in general and communication capturing followed by deciphering and replay attacks in particular. We will develop functional prototypes fulfilling as well a reasonable functionality as a good IT security.

This will result in a recommendation as to which protocols should be used for safeguarding the individual technologies. An important part of this will be how the compliance with the requirements of textile companies that have purchased these components can be systematically checked.

3. Conclusions

The textile industry is and will remain an innovative industry dedicated to progress in all sectors; an enormous development is predicted in the near future, especially in the Smart Textiles sector. As an interdisciplinary project, SEKT combines these two innovative areas of textiles and clothing and information technology and thus creates the basis for future innovative product development. It has been shown that wireless transmission methods are the most important core points for IT security. Effective protocols exist for all technologies to secure communication. However, these must also be activated or used. Therefore, checking the mechanisms used is one of the most important factors.

Acknowledgement

The Federal Ministry of Education and Research (BMBF) funds interdisciplinary research and development projects in the disciplines of engineering, natural sciences and economics, in which universities of applied sciences cooperate with partners from industry and science. The current research project SEKT (funding code FKZ:13FH180PX6) is the first research team to focus on evaluating the cyber security of electronic communication systems in smart textile products. The project started in December 2018 and will last until November 2021. The project team works with well-known brands with backgrounds in textiles from different areas such as fashion, sports, military or medicine and cooperates with various partners from the IT sector. In addition, the group cooperates with various associations and the universities of Erlangen-Nuremberg and Hamburg – a consortium of 12 partners is involved. The SEKT project is funded by the FHprofUnt2016 initiative with approx. 500,000 Euros.

References

[1] Textil, T.V.Q., 2004. Gesamtverband der deutschen Textil-und Modeindustrie eV Eschborn (in Vollmacht für seine Mitgliedsverbände). *IG Metall Vorstand: Tarifvertrag zur Förderung von Aus-, Fortund Weiterbildung der Beschäftigten der Textilindustrie und der Bekleidungsindustrie nach Ziffer*, **4**.

[2] Gesamtverband textil+mode, *Textilindustrie für ein starkes Europa*, Gesamtverband textil+mode, 02 04 2019. [Online]. Available: https://www.textil mode.de/presse/detailseite/news/textilindustrie-fuer-ein-starkes-europa. [Accessed 14 04 2019].

[3] Ohnemus D J and Rasel D F, 2018 *Fashion Tech–Smart Textiles: Kurzexpertise im Auftrag des BMWi,* ZEW - Zentrum für Europäische Wirtschaftsforschung GmbH, Mannheim.

[4] Zukunftsinstitut GmbH, *Megatrends*, 2018. [Online]. Available: https://www.zukunftsinstitut.de/dossier/megatrends/. [Accessed 14 04 2019].

[5] Federal Office for Information Security (BSI), *The State of IT Security in Germany 2018*, Federal Office for Information Security (BSI), Bonn.

[6] Bräuning M,2019, *Project Sekt: Cyber security of electronic communication systems in smart textile products*, Albstadt-Sigmaringen University, project team SEKT. [Online]. Available: https://projekt-sekt.de/en/. [Accessed 07 02 2020].

IOP Conf. Series: Materials Science and Engineering **827** (2020) 012051 doi:10.1088/1757-899X/827/1/012051

Development of 3D warp interlock fabrics based on moroccan natural fibers

Z Samouh[1,2,3,4], O Cherkaoui[2], D Soulat[3], A R Labanieh[3], F Boussu[3] and R El moznine[1]

[1] Laboratory LPMC, Faculty of Science El Jadida, Chouaib Doukkali University, EL Jadida, Morocco [2]

[2] Laboratory REMTEX, ESITH (Higher School of Textile and Clothing Industries), Casablanca, Morocco

[3] Laboratory GEMTEX, ENSAIT, Univ. Lille North of France, F-59100 Roubaix, France

[4] Author to whom any correspondence should be addressed

zineb.samouh@gmail.com

Abstract. The aim of this research is the valorization of Moroccan natural resources through the manufacturing of innovative 3D interlocks textiles. In order to investigate the potential of the sisal yarns used into 3D warp interlock fabrics, the 3D fabrics were studied at multi-scales (micro-scale, meso-scale and macro-scale). Before starting the manufacture of the 3D fabrics, different physical properties of the sisal fibers were analyzed according to standards. At the meso-scale, the mechanical and thermal properties of the sisal yarns constituting the 3D warp interlock fabric were studied according to standards. After these yarns and fibers characterizations, different types of architecture of the 3D warp interlock fabric have been designed and produced. Some difficulties have been revealed during the weaving process, which highlights the difficulty to handle the sisal yarns in order to obtain 3D fabrics to be used as a future fibrous reinforcement for composite material.

1. Introduction

Environmental issues, such as the greenhouse effect, the CO_2 emission and climate change have led researchers to develop alternative biomaterials to replace conventional ones. Scientifics have found that natural fibers are one of the alternatives to synthetic fibers, offering several advantages including renewability, low density and biodegradability. Natural fibers, such as flax, bamboo, can be prepared into different forms (non-woven, unidirectional, bidirectional...etc.) for different applications mostly in composite materials [1, 2].

Among the fibrous structures applied in composite materials, 2D fabrics are characterized by their good mechanical properties in plane of the fabric. However, stacking of bidirectional woven structures without any through-the-thickness linking points provides low delamination resistant fibrous reinforcement for composite material. For this reason, the introduction of three-dimensional woven structures as reinforcement of composite materials seems to be interesting to improve the mechanical properties of the preforms applied to aerospace and automotive [3].

Several scientific works are carried out to develop and study the 3D warp interlock woven fabrics based on synthetic fibers used in the fields of aeronautics, space and/or defense as protective structures.

IOP Conf. Series: Materials Science and Engineering **827** (2020) 012051 doi:10.1088/1757-899X/827/1/012051

The complex behavior of 3D warp interlock woven structures is due to the multi-scale appearance of these textiles. In the literature, several stages of the elaboration of the fibrous elements are necessary to manufacture the 3D warp interlock textile. In other words, the elaboration of the 3D fabric is based on the development of the fibrous element (from the fiber to the yarn, then from the yarn to the elementary pattern and finally from the elementary pattern to the 3D fabric) [4, 5].

In general, fibers used for the manufacture of 3D woven structures are synthetic fibers such as glass fibers, carbon fibers...etc. [3]. However, the exploitation of the natural fibers to develop the 3D fabric is limited, e.g. flax [6]. The current work is a Franco-Moroccan bilateral cooperation program. This collaboration aims to broaden the field of application of Moroccan natural fibers and more precisely in the development of multi-layered fiber reinforcements.

Among the natural Moroccan fibers, pine cone fibers [7] and doum fibers [8] are the ones used to develop biocomposites. Sisal fiber, which we used, is originally from the northern regions of Morocco [9] and is currently applied mainly in traditional medicine [10]. Sisal plant is one of the most widely used natural fiber sources and is very easy to cultivate. In our study, it has been extracted from the leaves of one agave specie plant (*Agave sisalana*).

This work was carried out to develop the 3D warp interlock textiles based on Moroccan sisal fiber. The 3D fabrics were studied in multi-scale approach (micro-scale, meso-scale and macro-scale) to better understand their mechanical behaviors. Additionally, the study unveiled the difficulties to produce the 3D warp interlock fabrics at each step of the production.

2. Materials and Method

The preforms elaborated in this work are described under the micro-scopic, meso-scopic and macro-scopic scales in order to understand the behavior of these structures.

2.1 Characterization of 3D warp interlock textiles at microscale

The description of the 3D interlock structure in the microscopic scale corresponds to the smallest entity of the final preform, which is the fiber. In our work, the fiber used is a Moroccan sisal fiber. The tensile properties of the sisal fiber were analyzed according to ISO 507 standard, using universal load machine Zwick Z005.

2.2 Characterization of 3D warp interlock fabrics at mesoscale

At this scale, the properties of the yarn are identified. The yarn placement and their interlacement pattern inside the structure are studied. The sisal yarn used in this work were chosen from several types of twisted sisal yarns marketed by the Moroccan company *Sonajute*. The chosen type was 300/1. The properties of sisal yarns were analyzed according to the standard norms. The geometrical representation of the 3D fabric in the mesoscale is based on the definition of the 3D warp interlock fabric and the types of yarns constituting the preform. *Boussu et al.* [11], offers a general definition of the different components of the 3D warp interlock fabric (number of warp yarns, number of weft yarns, number of layers, woven pattern ...) to better describe the position of the different yarns located inside the woven structure. The studied 3D warp interlock fabrics of this work belong to the classification of 3D warp interlock fabrics. Based on the general definition given by *Boussu et al.* [11], the 3D interlock fabric is defined in Equation 1 [11]:

$$\text{3D warp interlock X1–X2 \{N\} Y1k–Y2k Binding Wbk\{Bki\}–Surface Ws\{Ci\}–Stuffer\{Si\}} \quad (1)$$

Where:

- X1 represents the type of angle of binding warp yarn, O (orthogonal) or A (angle)
- X2 corresponds to the type of depth of the binding warp yarn, L (layer to layer) or T (through the thickness)
- {N} corresponds to the repetitive sequence of the different number of weft layers for each column of the 3D warp interlock fabric elementary pattern.
- Y1k is equal to the path of the binding warp yarn of group k
- Y2k is equal to the depth of the binding warp yarn of group k

- Binding term corresponds to binding warp yarns
- Wbk is related to the type of weave diagram on fabric surface of binding warp yarns of group k
- Bk i contains the numbering of binding warp yarns of group k with inter-ply i position
- Surface term corresponds to surface weave warp yarns and disappears if surface weave
- warp yarns are not included
- Ws is related to the type of weave diagram of surface weave warp yarns
- Ci contains the numbering of surface weave warp yarns with inter-ply i position
- Stuffer term corresponds to stuffer warp yarns and disappears if stuffer warp yarns are not included
- Si represents the numbering of stuffer warp yarns with inter-ply i position.

In order to describe the architecture and geometrical properties of the preform at the mesoscale the definition of the 3D warp interlock fabric and the REV are introduced.

2.2.1 The Representative Elementary Element (REV) of 3D warp interlock fabrics. The Representative Elementary Volume represents a unit cell of the preform. The unit cell is the elementary repeated unit pf the structure. It describes the geometry of the structure, the interlacement pattern between the constituting yarns and number of layers. Four structures were designed and woven on the loom.

2.3 3D warp interlock textiles at macroscale

Based on the design of the 3D warp interlocks structures, textile structures were produced using the sisal yarn. The weaving process followed in this current study is illustrated in Figure 1.

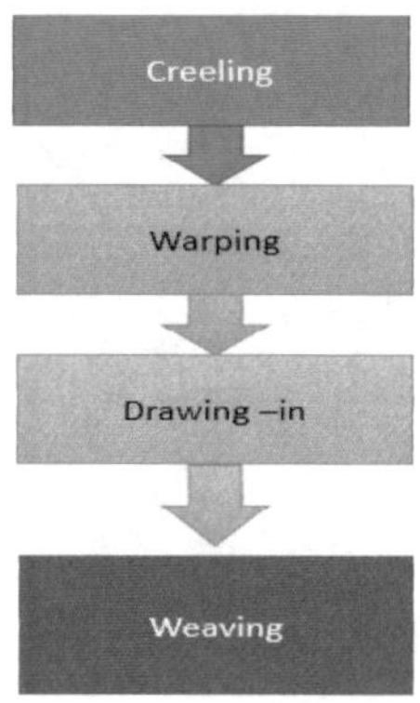

Figure 1. Weaving Process.

On the loom, the warp yarns are extended but the weaving of the 3D warp interlock fabrics is based on an area located between a set of warp yarns arranged parallel to one to another on a beam with a constant tension.

3. Results and Discussions

3.1 3D warp interlock fabrics at micro-scale

Characterization of the preform at the microscale involves the identification of the sisal fiber properties which is the smallest entity of the preform. The mechanical properties of the sisal fiber are summarized in Table 1. The average diameter of sisal fiber was 239,00 ± 80,2 µm. The average tensile strength of sisal fiber was 340,02 ± 70,4 Mpa. The average Young's modulus of sisal fiber was 12,5 ± 7,8 Gpa. The average elongation at break of sisal fiber is 3,2 ± 0,8%. The mechanical properties of sisal fiber in this current study are in agreement with the works studying sisal fiber in the literature [12,13].

Table 1. Mechanical Properties of Moroccan Sisal Fiber.

Mechanical Properties of Moroccan Sisal Fiber	
Average diameter [µm]	239.0 ± 80.18
Elongation at break [%]	3.2 ± 0.8
Tensile strength interval [MPa]	186.9.7-409.4

IOP Conf. Series: Materials Science and Engineering **827** (2020) 012051 doi:10.1088/1757-899X/827/1/012051

Tensile strength [MPa]	340.02 ± 70.4
Young's modulus [GPa]	12.5 ± 7.8

3.2 3D warp interlock textiles at mesoscale

At this scale, the sisal yarns were characterized according to standard norms. The Representative Elementary Element (REV), which represents the repeated unit of the textile architecture characterizing the yarns interlacement pattern of the 3D fabric, was described.

3.2.1 Characterization of Moroccan Sisal Yarn. The linear density of the twisted sisal yarn was 3300 Tex. Table 2 illustrates the physical and mechanical properties of the sisal yarns. Thermogravimetric analysis of sisal yarns (Table 3) shows that the thermal degradation of sisal yarns is related to the degradation of the chemical components of sisal fiber (lignin, cellulose, etc.).

Table 2. Physical and Mechanical properties of Moroccan sisal yarns.

Sisal Yarn	Diameter (mm)	Linear Density (Tex)	Tenacity (cN/Tex)	Twist level (tr/m)
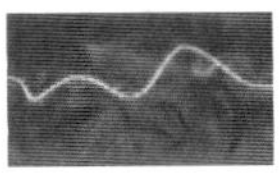	1,75	3300	13,1	80

Table 3. Thermal characterizing temperature of sisal yarns.

Sisal Yarns	Thermal Decomposition Temperature of Hemicellulose of Sisal Yarns (° C)	Thermal Decomposition Temperature of Cellulose of Sisal Yarns (° C)	Thermal Decomposition Temperature of Liginin of Sisal Yarns (° C)
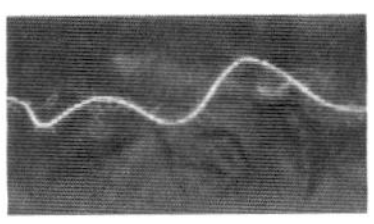	230	350	385

3.2.2 The Representative Elementary Volume (REV) of 3D warp interlock textiles. Four 3D fabric architectures with the same number of 4 layers have been designed to be produced on the same weaving loom. The definition of the four designed fabrics according to the Equation 1 [11] is as follows in Table 4:

Table 4. The designed unit cell of the four 3D warp interlock fabrics.

3D Representation	**Loom Pattern**
3D Warp Interlock O-L 2-3 4 Binding {Twill 3 weft effect left shift }{1 5 9 - 2 6 10 - # - # - #} – Stuffer {# – 3 8 – 4 11 – 7 12 - #}	
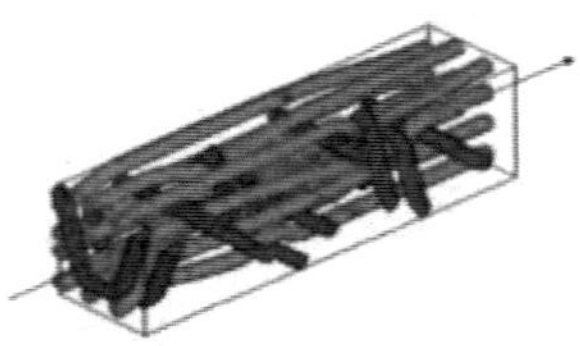	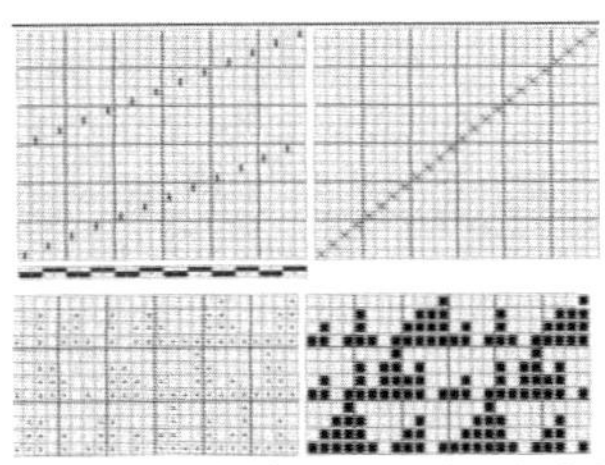
3D Warp Interlock O-L 1-3 4 Binding {Twill 3 warp effect left shift }{1 5 9 - 2 6 10- # - # - #} – Stuffer {# – 3 8 – 4 11 – 7 12 - #}	

3D Warp Interlock O-T 2-4 4 Binding {Basket 4-2 }{1 2 5 6 9 10 -# - # - # - #} – Stuffer {# – 3 8 – 4 11 – 7 12 - #}

3D Warp Interlock A-L 5-3 4 Binding {Twill 6 weft effect left shift} {1 3 5 7 9 11 – 2 4 6 8 10 12 - # - # - # - # - #}

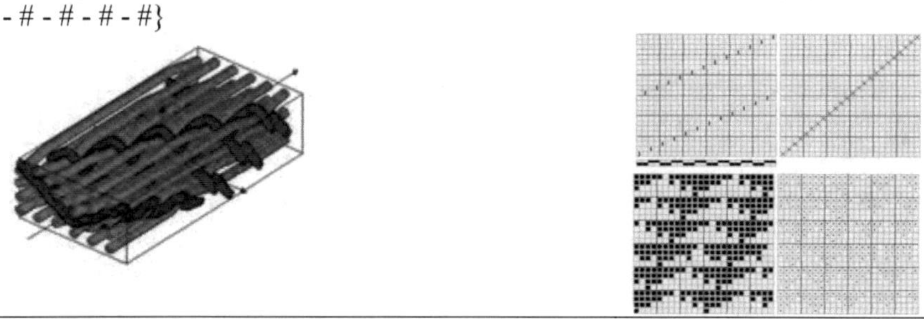

3.2.3 3D warp interlock preform at macroscale. After the production of the four fabrics, the main descriptive elements have been defined in the following table (Table 5) according to standard norms.

Table 5. Characterization of 3D warp interlock textiles.

3D Warp Interlock Textiles manufactured							
3D Warp Interlock O-L 2-3 4 Binding {Twill 3 weft effect left shift }{1 5 9 - 2 6 10 - # - # - #} – Stuffer {# – 3 8 – 4 11 – 7 12 - #}		**3D Warp Interlock O-L 1-3 4 Binding {Twill 3 warp effect left shift }{1 5 9 - 2 6 10- # - # - #} – Stuffer {# – 3 8 – 4 11 – 7 12 - #}**		**3D Warp Interlock O-T 2-4 4 Binding {Basket 4-2 }{1 2 5 6 9 10 -# - # - # - #} – Stuffer {# – 3 – 4 811 – 7 12 - #}**		**3D Warp Interlock A-L 5-3 4 Binding {Twill 6 weft effect left shift} {1 3 5 7 9 11 – 2 4 6 8 10 12 - # - # - # - # - #}**	
Warp Yarn count(yarn/cm)	3	Warp Yarn Count(yarn/cm)	3	Warp Yarn Count(yarn/cm)	2	Warp Yarn Count(yarn/cm)	3
Weft Yarn Count(yarn/cm)	3	Weft Yarn Count(yarn/cm)	2	Weft Yarn Count(yarn/cm)	3	Weft Yarn Count(yarn/cm)	3
Width (cm)	48	Width(cm)	48	Width(cm)	48	Width(cm)	48
Length(cm)	150	Length(cm)	150	Length(cm)	150	Length(cm)	150

The warp yarns are subjected to stresses at the different phases of the weaving and its tension is regulated manually. The increase of the tension of the warp yarn will affect the embossing of the yarns. The linear density and the structure of the sisal yarn allow to have significant friction between the yarns at the time of opening the shed. After the insertion of weft yarn by shuttle, the packing of the weft yarn modifies the weft and warp reduction of the fabric. The structure of the sisal yarns makes the packing of the middle weft yarns supported. Thus, it is concluded that the process of weaving sisal yarns will influence the properties of the yarns and the properties of the fabrics.

4. Conclusion

This work aims to valorize the Moroccan natural resources for the development of 3D warp interlock fabrics. The Moroccan fiber exploited is the sisal fiber. The manufacturing of four architectures of the 3D interlock woven fabric is evaluated. The properties of the 3D woven preform are characterized at three scales: micro (fiber), meso (yarn) and macro (fabric). The sisal fiber shows interesting properties. However, weaving the used rigid yarn of this type of fiber represents a challenge and requires adjustment to the weaving loom and/or special treatment to the used yarn.

Acknowledgements

This work was funded by the Franco-Moroccan Hubert Curien partnership (PHC) with the reference Toubkal/19/83 – Campus France: 41511ZF.

References

[1] Shekar H S and Ramachandra M 2018 Green composites: a review *Materials Today: Proc*, **5**(1), pp.2518-2526.

[2] Saheb D N and Jog J P 1999 Natural fiber polymer composites: a review. *Adv Polymo Tech* ,**18**(4) pp351-63.

[3] Mouritz A P,Bannister M K, Falzon P J and Leong K H 1999, Review of applications for advanced three-dimensional fibre textile composites .*Compos Part A-Appl S*, **30**(12) pp.1445-61.

[4] Corbin A C, Boussu F, Ferreira M and D Soulat 2018 Influence of 3D warp interlock fabrics parameters made with flax rovings on their final mechanical behaviour *Journal of Industrial Textiles* p.1528083718808790

[5] Dufour C, Pineau P, Wang P, Soulat D and Boussu F 2015 Three-dimensional textiles in the automotive industry ,*Advances in 3D Textiles*, (Sawston:Elsevier) pp. 265-91.

[6] Lansiaux H, Soulat D, Boussu F and Labanieh A R 2018 Manufacture and characterization of 3D warp interlock fabric made of flax roving ,*IOP Conference Series: Materials Science and Engineering*,**406**(1) .

[7] Arrakhiz FZ, El Achaby M, Benmoussa K, Bouhfid R, Essassi EM and Qais A 2012 Evaluation of mechanical and thermal properties of pine cone fibers reinforced compatibilized polypropylene *Mater Des* ,**40** ,pp 528-35.

[8] Arrakhiz FZ, El Achaby M, Malha M et al 2013 Mechanical and thermal properties of natural fibers reinforced polymer composites: Doum/low density polyethylene, *Mater Des*,**43,** pp 200-05.

[9] El-Hilaly J, Hmammouchi M and Lyoussi B 2003 Ethnobotanical studies and economic evaluation of medicinal plants in Taounate province (northern Morocco) *J Ethnopharmacol* **86**(2-3) pp 149-58.

[10] Fennane M and Rejdali M 2016, Aromatic and medicinal plants of Morocco: richness, diversity and *threats Bulletin de l'Institut Scientifique, Rabat, Section Sciences de la Vie* **38**.

[11] Boussu F, Cristian I and Nauman S 2015 General definition of 3D warp interlock fabric architecture *Composites Part B: Engineering*, **81** pp 171-88.

[12] Silva FDA, Chawla N and Filho RDDT 2008 Tensile behavior of high performance natural (sisal) fibers *Compos Sci Technol* **68**(15–16) pp3438-43.

[13] Belaadi A, Bezazi A, Bourchak M and Scarpa F 2013 Tensile static and fatigue behaviour of sisal fibres *Mater Des* **46** pp 76-83.

Review of literature on existing models about the impact of continuous training on business performance

Asmaâ MOUSSAID[1,2,3], Mohamed TKIOUAT[2] and Mustapha HLYAL[1]

[1] Laboratory CELOG, Higher School of Textile and Clothing Industries (ESITH), Casablanca, Morocco
[2] Mohammedia School of Engineering (EMI), Laboratory LERMA, Rabat, Morocco
[3] Author to whom any correspondence should be addressed

moussaid@esith.ac.ma

Abstract. The development of a company requires the development of its human resources. Research has shown that overall performance is measured not only by the economic dimension but also by the social dimension and actions on development processes, in particular, the continuous training process which is the constructor of adequate skills for improvement. The below analysis will be divided into two axes:

The first axis aims to define the overall performance through a review of literature. It demonstrates a relationship between the individual's performance and continuous training as a process of individual development, and cites some performance measurement tools.

The second axis examines the literature following specific inclusion criteria, its impact on overall performance, and highlights existing assessment models.

Keywords: Continuous training, Performance, Training plan, Models, Evaluation, Knowledge and interviews.

1. Introduction

Nowadays, companies are facing several challenges such as intensive competition, the market turbulence as well as technological progress, which requires constant monitoring.

The company can only reduce these threats if it has qualified human resources and can achieve overall performance if they enhance individual performance. Despite technological and industrial progress, all over the world, the human factor is the driving force behind any company's growth. Without developing the human resources, we can never achieve performance.

Many studies show that continuous training is the main tool for human development, and it guarantees growth of the company's income [1].

What effects does the training have on the individual, and subsequently on the company's performance? What are the models in the literature that study the impact of continuous training? These are the two axes that will be addressed in the rest of this work.

2. Material and Methods

This article is a literature review. We have extended the time range in order to cover the evolution of the addressed problem. The articles selected for analysis are from 1997 to 2019.

In the first axis, the research defines the impact of individual performance on the company's overall performance, the factor that affects the individual's performance, and the key indicator that measures it.

In the second axis, the study investigates the articles or theses that measure the impact of continuous training on the company's overall performance, and proposes models to assess this impact.

The context of the studies selected varies from one reference to another, from Moroccan, French and Anglo-Saxon contexts.

3. Results and discussion

3.1. Axis 1: Overall and individual performance

According to *Bourguignon 1997* [2], performance revolves around three essential points: action, its result and its success.

Action performance within a company is based on processes acting on the individual, the material and the organization.

The process that directly influences the individual's performance is continuous training, which is considered a key in skills production.

Furthermore, performance measurement according to *Bouquin 2004*, means measuring three components: economy, efficiency and efficacy [3].

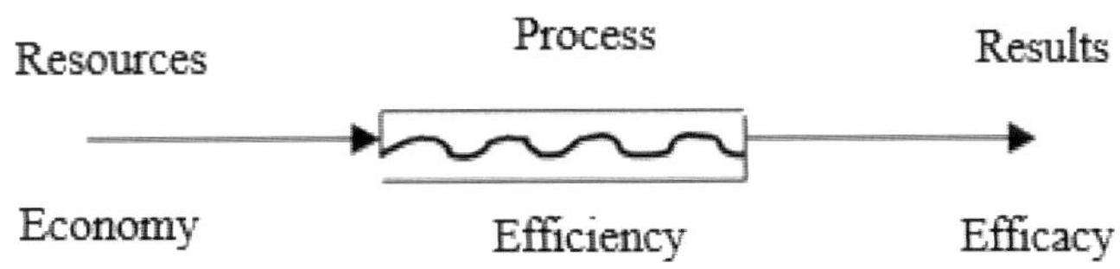

Figure 1: Performance (*Bouquin, 2004*)

The efficiency component is strongly correlated with the continuous training process. We conclude that a properly implemented and controlled continuous training process generates high efficiency.

For *Bouquin (2004)*, what makes it possible to evaluate individual performance is measuring productivity before and after a training.

According to *Movahedkhah (2005)* [4], the success of an organization is associated with the performance of people, which can be assessed by motivation, skills development and performance.

He proposed a model for the concept of overall performance of the business, taking into consideration the most important indicators, namely the financial indicator, Customer indicator, Internal Processes and organizational training.

The key indicator for training impact is measuring the productivity. This confirms that measuring the performance of the individual that influences overall performance is similar to measuring productivity.

The definition of performance according to *Maurell 2014*[5],which is based on *Reynaud's (2003)* definition of performance, is an aggregation of three dimensions; economic, social and environmental.

This performance is measured using several tools, namely the Balanced Score Card (BSC) which extends to both social and environmental dimensions, or the Triple Bottom Line.

Maurell 2014[5] proposes a structural equation model, based on data from a sample of 121 firms, from various sectors.

Among the results of his research, social performance has twice as much impact on overall performance compared to the economic and financial component.

According to *ISSOR 2017*[6], performance is a complex concept that integrates several dimensions (strategic, competitive, organizational, financial, business and human performance), and it is necessary to measure all its dimensions and rely on its determining factors to improve it.

Human performance is impacted by the management practices of human resources (HR).

Continuous training, which is aimed at developing initiative, human resources creativity, and the ability to monitor technical progress, is the most critical practice for the development of employees within a company.

Hence we conclude that in order to ensure overall performance, it is necessary to improve the performance of the people working within the company, and that a continuous training process guarantees the production of competences in a way that meets the constantly evolving requirements.

3.2. Axis 2: Impact of training on performance

A study by *Carriou and Jeger* published in 1997[7], deals with the return on investment of continuous training, through an analysis of productivity data from the tax returns of a sample of 10,000 companies, with a size of more than 50 employees.

The impact of continuous training was tested using three production function models, and each model is based on a combination of assumptions (table 1):

Table1. Results of continuous training impact

Hypothesis 1: the delayed impact effect on productivity / Hypothesis 2: the accumulation effect Hypothesis 3: the obsolescence effect / Hypothesis 4: the saturation phenomenon

Model 1 of the Cobb-Douglas type	Model 2	Model 3 of type TRANSLOG
H1	H1 H2 H3	H1 H4
A basic model confirms a positive impact of the continuous training rate of year n-1 on the productivity of year n, and measures the elasticity of the company's added value compared to the training rate of 2%,	Shows an obsolescence of continuous training of about 50% per year, and confirms the elasticity of 2%.	Also attests to the positive impact of continuous training on productivity.

In addition, *Arcimoles* (2012) addresses the effect of evaluation on the effectiveness of training, through a study of the impact of training on employee productivity [8].

The study used a quantitative approach of evaluation practices done after the training (on SPF120: sample of companies in the region of Paris, France), then measured the impact of these evaluation practices on the effectiveness of the training, by observing two studied populations, to one of which the evaluation was announced and the other wasn't.

The study showed positive consequences of the implementation of a learning assessment process on employees' attitudes and on the effectiveness of training, and it increases performance orientation.

Aubert (2009) [9] studied the apparent performance of continuous training in companies, and analysed the impact on productivity through the processing of productivity data from tax returns of a sample of 1,605 companies with a size of more than 1,000 employees. The effect of continuous training was tested using the Cobb-Douglas production function model, with constant yield, and with only two production factors: capital and efficient labour. The study estimates confirm the existence of significant productivity gains associated with continuous training provided by employers. Investment in continuous training would also be beneficial to companies.

Another study published in 2010 by *Dyane* [10], analyses the systems and practices of continuous training in France and Morocco for the development of the necessary skills among companies, and discusses public responsibility in this process.

It was based on a quantitative and qualitative survey, for a sample of 40 respondents out of 110 companies in various sectors, the data collected were then statistically processed.

The perceived effects of training vary mainly between production yield and increased revenue.

This proves that continuous training has a positive impact on the development of the company's human resources.

Additionally, *Zaim (2017)* confirms the effect of continuous training on the improvement of employees' skills, through a questionnaire, with a sample of companies (46 respondents out of 93 companies in Morocco in the Rabat-Sale region) [11].

The final conclusion of this article is that the majority of companies are aware of the importance of continuous training: a factor of stability and security by guaranteeing a staff always qualified and ready for change.

Lazzarini (2019), for his part, addressed the issue of university studies for future engineers, and proposed the integration of a module on sustainable human development within companies [12].

He proceeded through focus groups with engineering students and semi-interviews with coordinators. The integration of this module was appreciated by the students, who are the bearers of change within companies, and the professors have brought added value to the training curriculum which will help to spread this notion of sustainable human development, which increases performance within companies.

According to *Muehler and Beckmann* [13], when comparing the evolution of wages of employees who had undergone firm-specific training and those who had general training, we notice that general training is linked to larger wage increases than firm-specific training.

They use for demonstrating a Difference in differences (DID) , which is a statistical technique used in quantitative research, and they conducted the training as well as observed data on the span of 3 years. Which leads us to conclude that general training impacts productivity more than firm-specific training.

Nevertheless, and on the same orientation of observing wage effects and productivity by training, *Gallen, Peneder, and Robinson* insist that investing in continuous training is highly recommended and important for the increase and growth of the overall performance of the European Union labour Force's sample of study [14].

Gallen, Peneder, and Robinson have also found in their statistics of EU Labour Force Survey that general education and continuous training are both positively correlated.

Through this set of studies we develop a conceptual model of performance, its components, its impact factors, and its measurement models (Figure 2).

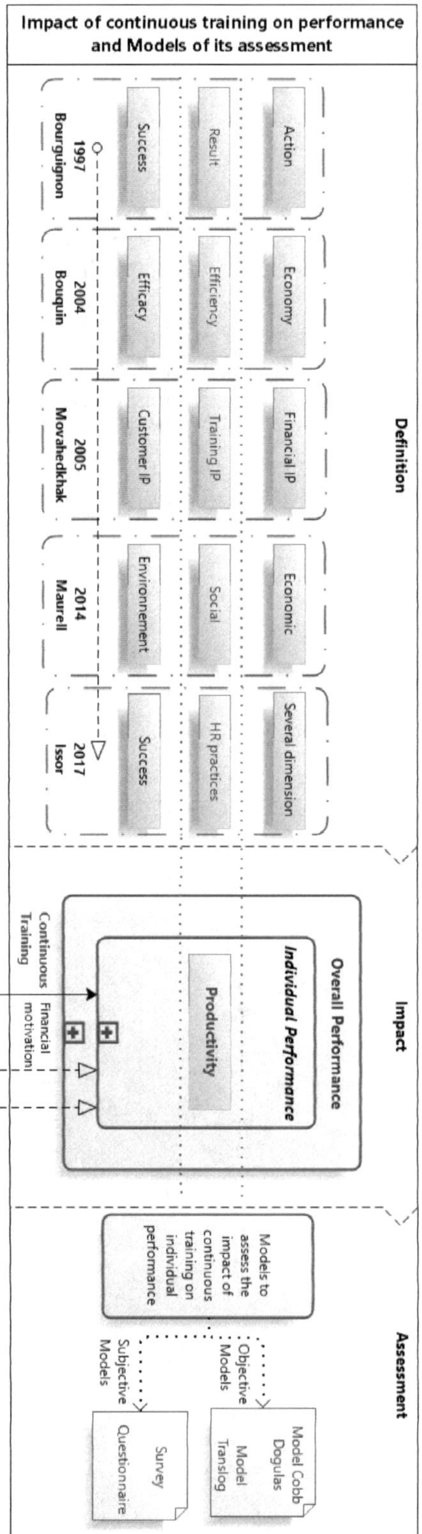

Figure 2*: Impact of continuous training on performance and Models of its assessment*

Among the main findings, the components of the company's overall performance include the results of actions directly related to the individual.

The definition of performance over the years, since 1997, has been based on components. The common measure is individual productivity, which provides information on the overall state of the company's performance.

According to the above-mentioned studies, the process of continuous training has a direct influence on the individual's productivity. This is proven through models that measure the impact of continuous training on individual performance.

These models are divided into two categories: objectives through an analysis of individual productivity data for a sample of companies, and subjective through questionnaires and surveys (Figure 2).

4. Conclusion

This review of literature shows the positive effect of continuous training on the development of individuals and, consequently, on the development of the overall business performance. This confirms that continuous training is an essential tool for the constant improvement of companies.

Questions that might be areas of research later on are: What are the practices of continuous training in Moroccan companies today? Does the state as a stakeholder support this process of upgrading the skills of employees within a company? If so, what are the mechanisms that exist? And to what extent do these mechanisms affect the real needs of Moroccan companies?

References

[1] Sal A. and Raja M., 2016. The impact of training and development on employees performance and productivity. *International Journal of Management Sciences and Business Research*, **5**(7).

[2] Bourguignon A., 1997. Sous les pavés la plage... ou les multiples fonctions du vocabulaire comptable: l'example de la performance. *Comptabilité contrôle audit*, **3**(1).

[3] Douhou A and Berland N 2007, "Mesure de la performance globale des entreprises," *28e Congrès l'Association Francoph. Comptab* in Poitiers, France.

[4] M. Movahedkhah, P. Barillot, and D. Thiel, 2005, Modèle conceptuel causal de la performance de systèmes industriels, *6ème Congrès Européen de Science des Systèmes in* Paris, France.

[5] Maurel C and Tensaout M, 2014. Proposition d'un modèle de représentation et de mesure de la performance globale. *Comptabilité-Contrôle-Audit*, **20**(3), pp.73-99..

[6] Issor Z 2017 La performance de l'entreprise: un concept complexe aux multiples dimensions. *Projectics/Proyectica/Projectique*, (2), pp.93-103..

[7] Carriou Y and Jeger F, 1997. La formation continue dans les entreprises et son retour sur investissement.

[8] Monnot A, 2013. L'évaluation de la formation professionnelle: Analyse des pratiques, de leurs déterminants et des conséquences d'une évaluation des apprentissages sur l'efficacité de la formation *(Doctoral dissertation, Atelier national de reproduction des thèses)*.

[9] Aubert P, 2009, Le rendement apparent de la formation continue dans les entreprises : effets sur la productivité et les salaires.

[10] Dyane S 2010. Formation continue et développement des compétences: etude des systèmes et des pratiques de formation en France et au Maroc *(Doctoral dissertation, Artois)*.

[11] Zaim M, Zaim F, Sbiti M and Benazza H 2016 L'intérêt de la Formation Continue dans les Entreprises Marocaines. **6**(8), p.103.

[12] Pérez-Foguet A, and Lazzarini B 2019. *Journal of cleaner production*, **218**, pp.772-781.

[13] Muehler G, Beckmann M and Schauenberg B 2007. The returns to continuous training in Germany: new evidence from propensity score matching estimators. Review of Managerial Science, **1**(3), pp.209-235.

[14] O'Mahony M A R Y 2012. Human capital formation and continuous training: Evidence for EU countries. *Review of Income and Wealth*, **58**(3), pp.531-549.

The esterification and dyeing properties of raw wool fibers

Fatine Akoh[1,2,3], El Bouchti Mehdi[2], Omar Cherkaoui[2] and Mohamed Tahiri[1]

[1] Geoscience laboratory, Faculty of Sciences Ain Chock, Hassan II University. Km 8 Route d'El Jadida, B.P 5366 Maarif 20100. Casablanca, Morocco

[2] Remtex Laboratory, Higher school of textile and clothing industries. km 8, Oulfa, BP 7731, Casablanca, Morocco

[3] Author to whom any correspondence should be addressed

akoh.fatine@gmail.com

Abstract. The reaction of raw wool fibers (RWFs) with methanol, ethanol and n-propanol catalyzed by HCl has been found to modify carboxyl side chains. The effect of esterification of wool on its behavior toward acid dye has been studied. Esterified wool fibers (EWFs) showed an increase in acid dye affinity, and dyed esterified wool shows an increase in dye fastness to washing. The process of esterification with alcohols/hydrogen chloride caused some modification of carboxyl and other groups. The surface modification of wool fibers was determined by FTIR, modification rate of carboxyl groups and the acid dye distribution of RWFs and EWFs with various alcohols was evaluated by Datacolor.
Keywords: Carboxyl groups, acid dye, methanol, ethanol, n-propanol

1. Introduction

Blackburn & Phillips (1944) informed that a combination of methanol and acetic anhydride esterified the carboxyl groups, in part at least, but that some of the terminal amino groups in lysine and the peptide imino linkage were simultaneously acetylated. *Fraenkel-Conrat & Olcott* (1945) attested that anhydrous alcohols containing a catalytic quantity of mineral acid reacted precisely with the carboxyl groups in proteins without modifying any other group[1, 2]. Methanol containing 0,1M hydrochloric acid was used by *Blackburn & Lindley* (1948) to introduce 0,8% of methyl groups into wool; this is equivalent to masking 65 % of the carboxyl groups [3, 4].

The present investigation shows an increase in dye affinity with esterified wool fibers, it appears that the basis of this increase, in this case, was the conversion of anionic carboxylate groups into ionizable ester groups and it was also confirmed by the increased rate of dye uptake [1, 4], especially from a neutral dye bath. Furthermore, a hypothesis was made by *Alexander* that by blocking the carboxylate groups in wool, acid dyes should be more resistant to washing[1, 5]. *Kilpatrick and Maclaren (1969)* found that amide groups, in addition to carboxyl groups, are modified during treatment of wool with boiling methanol containing hydrogen chloride, which cast doubts over the specificity of the esterification procedure by *Alexander* [4, 6, 7]. Esterification with boiling methanol has been used by numerous workers, who measured the changes in fiber properties (such as, water sorption stability and dyeing behavior) which resulted on esterification and then rationalized them in terms of changes in carboxyl content[1, 3]. It is therefore important to know the nature and extent of any side reactions accompanying the modification of carboxyl groups by esterification. We have studied the esterification of wool with three alcohols and this work describes the side reactions resulting from the modification of carboxyl groups.

2. Materials and Methods

2.1 Esterification and dyeing

The raw wool fibers (RWFs) were cleaned by a washing process in a Soxhlet apparatus with distilled water for 6 cycles. RWFs were esterified with methanol (65°C), ethanol (75°C) and n-propanol (97°C), containing hydrogen chloride (introduced as concentrated HCI); the liquor/wool ratio was 50/1. Reactions were carried out in a water bath with a reflux heating from 3h to 48h and the EWFs were then rinsed with acetone.

The RWF and EWFs dyeing was performed by a commercial acid dye (blue BEMACID) in the water bath; the samples were introduced in pots, under heating and manual stirring to a temperature from 45°C to 98°C in acid medium, then the samples were washed and dried.

2.2 Modification rate of carboxyl groups

50mg of each sample (RWFs and EWFs with various alcohols) was added to 50 ml of NaOH (0.05M). The mixte was stirred for 1h, then filtered and titrated with HCl (0.05M). (HCl and NaOH are purshased from Sigma Aldrich). The consumed NaOH by samples should be corresponding to the extent of fibers carboxylic groups. Thus, the modified amount of carboxylic groups is calculated as below (1) [8] :

$$\frac{C_b(M)\ V_b(ml) - C_a(M)\ V_a(ml)\ (mmols)}{m\ (g)} \qquad (1)$$

Where C_b, V_b, C_a and V_a are the concentration and volume of the initial base and the acid solution (titration), and m represents the amount of samples used in the titration. To calculate the modification rate of the carboxylic groups, we proceed as follows (2):

$$\frac{mmols\ (RWFs) - mmols\ (EWFs)}{mmols\ (RWFs)} \times 100 \qquad (2)$$

2.3 FTIR analysis

Fourier transform infrared spectra of the investigated samples were realized using Nicolet iS10FTIR-ATR spectrophotometer, connected with the Golden Gate single attenuated total reflection (ATR) accessory.

2.4 Color Measurement

For the study and characterization of the dyed wool samples with blue BEMACIDE, measurements are made using DATACOLOR spectrocolorimeter and they are systematically conducted along with visual observations, the experimental results are then listed in form of tables.

3. Results and discussions

3.1 Modification rate of Carboxyl groups

Although esterification with methanol containing 0.1 M hydrogen chloride took place rapidly at 3h and 2h with carboxyl groups modification of 38% and 73% respectively. Meanwhile, the esterification with higher alcohols, especially n-propanol, was slower (Table 1); it took 48h to esterify only 33% of carboxylic groups. Therefore, the modification rate of carboxyl groups significantly decreased as the size of the alcohol increased. As a result, methanol is preferable over ethanol or n-propanol when it comes to esterification with a maximum modification of carboxyl groups [1, 5].

Table 1. The extent of modification of carboxyl group in RWFs during esterification with various alcohols

Alcohol	Reaction time (h)	Carboxyl groups %
Methanol	6	38
	24	73
Ethanol	3	23
	6	42
	24	47
n-propanol	6	17
	24	31
	48	33

3.2 FTIR analysis

In the following spectrum (Figure 1), the additional bands are attributed to the newly formed esters groups indicating the successful surface modification of RWFs with various alcohols especially the methanol. Therefore, the modification rates of carboxyl groups significantly decreased as the size of the alcohol increased [1, 5]. Thus, methanol is preferred over ethanol and propanol for esterification of carboxyl groups.

An analogy is noted between the IR spectrum of RWFs and methanol EWFs, with the appearance of a peak at 1730,88 cm^{-1}, this peak corresponds to the carbonyl group[2, 9]. This observation allows us to say that the treatment of wool with an alcohol, such as methanol, leads to the penetration of the latter and creating a new function "-CO"[2, 9]. It should be noted that the carbonyl peak also appears using other alcohols, but it is not as significant as the methanol's.

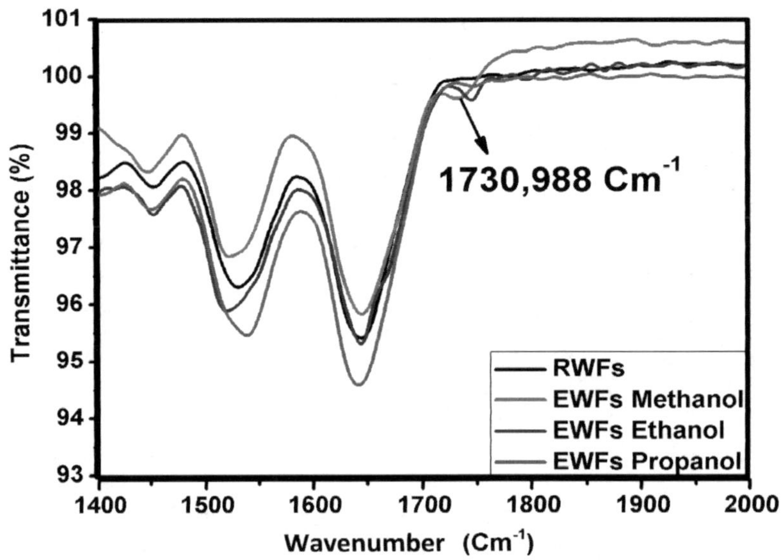

Figure 1. Infrared-Raw and Esterified wool fibers with various alcohols.

3.3 Color differences DATA

The spectropolarimeter allowed us to determine the parameters L * a * b * of the colored samples[10, 11]. Knowing the properties, L*, a* and b* corresponding to the two samples [11], it is possible to evaluate the difference between the color 1 and the color 2 as below (3):

$$\Delta E = [(\Delta L^*)^2 + (\Delta a^*)^2 + (\Delta b^*)^2]^{1\backslash 2} \quad (3)$$

if the color (1) is considered as reference, (4), (5) and (6) will be:

$$\Delta L^* = L^*2 - L^*1 \quad (4)$$
$$\Delta a^* = a^*2 - a^*1 \quad (5)$$
$$\Delta b^* = b^*2 - b^*1 \quad (6)$$

The signification of color differences is shown in (Table 2)[11]. In our case, the sample reference is the raw wool fibers (Table 3).

Table 2. Color differences interpretation.

Color Differences	Positive	Negative
ΔL*	The sample is clearer	The sample is darker
Δa*	The sample is redder or less green	The sample is greener and less red
Δb*	The sample is yellower or less	The sample is bluer and less yellow

The color differences (Table 4, 5 and 6) in the samples shows that the wool treated with the various alcohols has a greater affinity for the acid dye than the RWFs. Hence, EWFs are 6, 4, 2 times darker than RWFs for methanol, ethanol and propanol respectively (Figure 2). These results indicate that the ester functions are not largely reactive compared to the carboxyl groups of wool.

Table 3. Color Differences obtained for RWFs and EWFs.

Parameters	RWFs	Methanol EWFs	Ethanol EWFs	Propanol EWFs
L^*	30.31	26.19	27.39	29.98
a^*	-4.50	-1.30	-1.19	-2.83
b*	-30.99	-33.50	-34.20	-32.78

Table 4. Comparison with methanol EWFs.

Differences	Value	Comment
ΔL*	-4.12	Methanol EWFs is darker than RWFs
Δa*	3.2	Methanol EWFs is redder
Δb*	-2.51	Methanol EWFs is bluer
ΔE	5.82	Methanol EWFs is almost 6 times darker than RWFs

Table 5. Comparison with ethanol EWFs.

Differences	Value	Comment
ΔL*	-2.92	Ethanol EWFs is darker than RWFs
Δa*	3.31	Ethanol EWFs is redder
Δb*	-0.21	Ethanol EWFs is bluer
ΔE	4.41	Ethanol EWFs is 4 times darker than RWFs

Table 6. Comparison with n-propanol EWFs.

Differences	Value	Comment
ΔL*	-0.33	n-propanol EWFs is darker than RWFs
Δa*	1.67	n-propanol EWFs is redder
Δb*	-1.79	n-propanol EWFs is bluer
ΔE	2.47	n-propanol EWFs is 2 times darker than RWFs

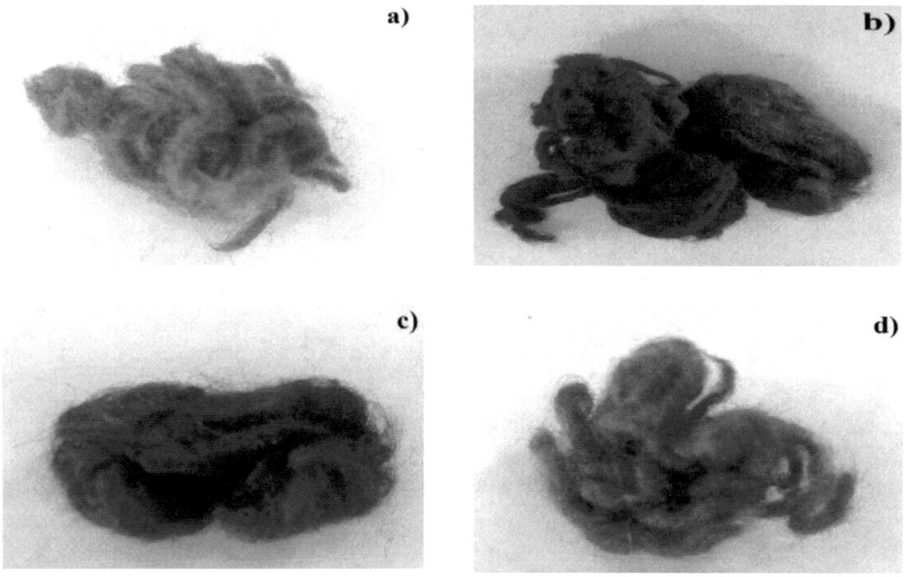

Figure 2. a) RWFs, b) Methanol EWFS, c) Ethanol EWF and d) n-propanol EWFs colored with acid dye.

4. Conclusion

The effect of the wool esterification and its behavior towards the acid dye was studied. EWFs with various alcohols have an increased affinity for the dye especially Methanol EWFs. It seems that the basis of this affinity in the wool was the conversion of anionic carboxylate groups into nonionizable ester groups. In contrast to the acidic groups, these ester groups are no longer in competition with the coloring anions for the basic cationic groups of the wool and, therefore, the increase of the dye affinity was successfully confirmed by the modification rate of carboxyl groups.

Acknowledgments

The authors would like to thank the entire REMTEX laboratory team at Higher School of Textile and Clothing Industries and Professor Said Gmouh for their scientific and technical contribution.

References

[1] Holt L A, Leach S J and Milligan B 1969. *Textile Research Journal* **39**(3) pp.290-293..

[2] Pasquini D, Belgacem M.N and Gandini A, 2006. Aprigio da Sliva Curvelo A. *J. Colloid Interface Sci*, **295**, pp.79-83..

[3] Kilpatrick D J and Maclaren J A, 1969. The Dyeing Properties of Esterified Wools. *Textile Research Journal* **39**(3) pp.279-283..

[4] Robert S 1943 Dyeing cellulose esters. *U.S. Patent* 2,328,682.

[5] Alexander P, Carter D, Earland C and Ford O E 1951 Esterification of the carboxyl groups in wool. *Biochemical Journal* **48**(5) p.629..

[6] Ammayappan L 2013 Eco-friendly surface modifications of wool fiber for its improved functionality: An overview *Asian journal of textile* **3**(1) pp.15-28..

[7] Kilpatrick D J and Maclaren J A 1969 The Dyeing Properties of Esterified Wools. *Textile Research Journal* **39**(3) pp.279-283..

[8] Moaseri E, Baniadam M, Maghrebi M and Karimi M 2013 A simple recoverable titration method for quantitative characterization of amine-functionalized carbon nanotubes *Chemical Physics Letters* **555** pp.164-7.

[9] Gomez N, Julia M R, Lewis D M and Erra P 1995. The use of FTIR to investigate modifications to wool treated with sodium sulphite and cationic protein hydrolysate. *Journal of the Society*

of Dyers and Colourists **111**(9) pp.281-284.
[10] Equipment: 2012 Datacolor launches handheld spectrophotometer for colour measurement, *Additives for Polymers* **2012**(4) p.10.
[11] Pathare P B, Opara U L and Al-Said F A J 2013 Colour measurement and analysis in fresh and processed foods: a review. *Food and bioprocess technology* **6**(1) pp.36-60.

IOP Conf. Series: Materials Science and Engineering **827** (2020) 012054 doi:10.1088/1757-899X/827/1/012054

Sustainable Supply Chain Management: review of triggers, challenges and conceptual framework.

D Saidi[1,3], J El Alami[1] and M Hlyal[2]

[1] LASTIMI Laboratory, Mohammed V University in Rabat, Rabat, Morocco, Graduate School of Technology of SALE,
[2] CELOG, ESITH, Morocco, Casablanca
[3] Author to whom any correspondence should be addressed

dounia.saidi@um5.ac.ma; dounia.saidi@gmail.com

Abstract. An urgent need for organizations arises during the last nine years to incorporate social and environmental concerns into their Supply Chain. The adoption of new sustainable practices is no more a choice neither an option, it becomes one the company worries. This review aims to take a fresh look into the consideration of Sustainability in Supply Chain Management patterns for the last nine years (from 2010 to 2019) with a critical point of view. A growing awareness toward years of sustainability has contributed to the development of the new Sustainable Supply Chain (SSC) practices. The literature review provides an overview about the area of Sustainable Supply-chain Management (SSCM), it also outlines practices and patterns that need to be developed, with a focus on the current issues and emphasizing on used methodologies. A time line of relevant papers is also provided as a reference by using the rich and targeted body of researches. Moreover, a conceptual framework is proposed and the role of stakeholders is clearly defined. Finally, a summary of results and interpretations is provided, and the future research perspectives and issues are highlighted.
Keywords: Green supply chain management; literature review; social supply chain Introduction

1. Introduction

Over the last 9 years, Sustainable Supply Chain Management (SSCM) has been a major focus of business concerns. It transforms the whole value chain and consumer's behavior. Sustainability drivers should not be considered solely emanating from organizations, it comes from each part of the global supply chain. Many companies have noticed significant changes in the economic environment. The ecosystem has become more complicated, with an intensified competition, new consumption patterns emerged and demand prediction has become more difficult and consumer's power has raised.

Agility is the main feature of unstable and turbulent markets. The control of the supply chain is vital and necessary for the survival of the latter, in this context, implementing sustainability in companies is not easy.

One of the success keys of an overall literature review, is to formulate precise and clear questions. The purpose is to guide the reader throughout the review and inform him of what the present review might answer and what results are to be expected from it.

Q1: What is the state of art of Sustainable Supply Chains? What are their outcomes of the last ten years? How has this research area progressed?

Q2: What are triggers and challenges of a Sustainable Supply Chain Management?

Q3. What are perspectives and gaps that can orientate future research studies?

2. Material and methods

Before starting any study, it is necessary to trace the path to follow and define further steps to take. ''A literature review is a systematic, explicit, and reproducible design for identifying, evaluating, and interpreting the existing body of recorded documents'' [1].

The purpose of the present literature review is to bring an overview about the research subject area, summarize current reviews, identify issues and patterns, and finally to define the conceptual framework of the field.

2.1. Search scoop identification

Defining clear boundaries and delimitating the search scoop is particularly essential for a literature review [2]. In this step, papers to be selected and analyzed are delimited and criteria of our analysis is defined. Paper identification process concerns mainly literature reviews of the last 10 years ranged from 2009 to 2019. We opted for SCOPUS as a database source for our study.

A total of 171 reviews resulted from the first filter of keywords, period and paper category review using keyword strings as sustainable, supply chain, management.

To narrow the field search, and to have more reliable results the engineering subject area was targeted. Thus, the number of papers was reduced to 59.

2.2. Research result analysis

2.2.1. Contributions by year. The descriptive analysis is an important section of this study. It aims to present the insights of 59 reviews in Sustainable Supply Chain objectively and accurately selected from SCOPUS database. It provides a background for the category identification and evaluation of papers. The distribution of papers within a 9-year horizon is carried out as shown in the figure 1 below.

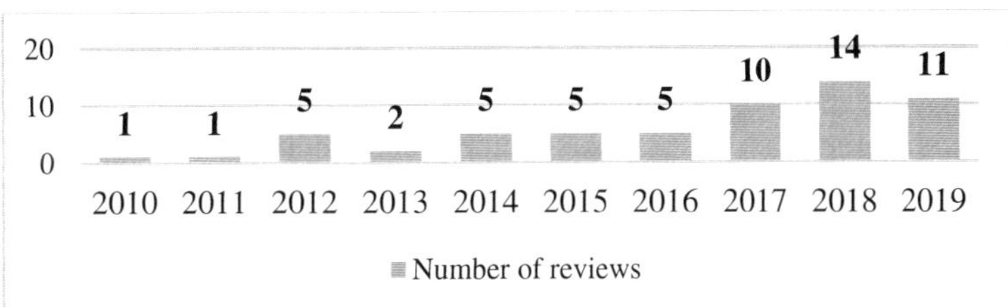

Figure 1. Evolution of SSCM review papers for the period ranged from 2010 to 2019.

The figure 1 clearly represents the evolution of SSCM state of art papers during the last ten years. Number of publications has raised and has reached 11 until April 2019 compared to 2010, which implies that the literature papers on SSCM are still developing and growing. Recently in 2018, 14 reviews were published, it provides us with a global idea about the total number of publications that might be reached on December 2019.

3. Results and discussion

3.1. SSCM Conceptual Framework

With regard to supply chain sustainability considerations, to draw a conceptual framework is a milestone of this state of art. The inclusive methodology was detailed previously [3]. The first step towards building the theory before analyzing results, is to conceptualize the field [3], [4]. The consolidation of multiple research papers into a unified scoop leads to reviews such as [2], [5], [6]. The Figure 2 summarizes the conceptual framework of a sustainable company, triggers and challenges in an agile ecosystem [2], [5]–[13].

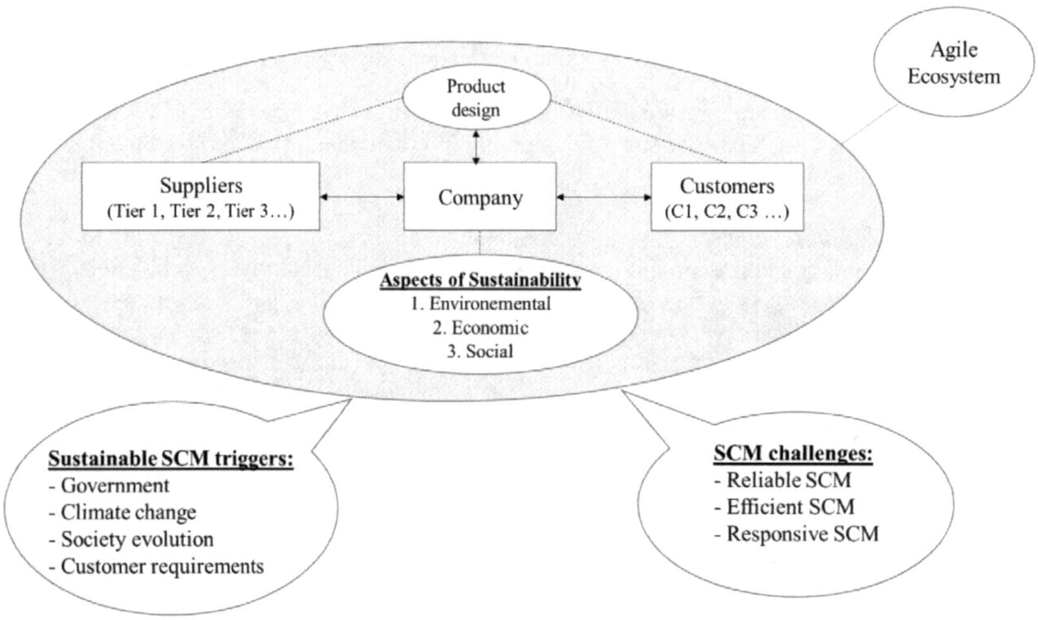

Figure 2. The conceptual framework of a sustainable company, triggers and challenges in an agile ecosystem.

Global supply chains are characterized by the complexity of interaction between stakeholders [14]. We find several physical flows (material), information and financial flows between the suppliers and the company and between the latter and its customers [15], [16].

This review of the literature will go through the question to see current research trends and authors who have already treated the topic focusing on the triggers of the SSCM and the challenges to overcome.

3.2. Sustainability aspects

3.2.1. First aspect: environmentally friendly. This is the most popular aspect. Sustainable Development is often wrongly limited to the environmental dimension. When we look to natural resources they are decreasing considerably otherwise they are unlimited and essential to our survival [17]. To ensure this mission several issues must be considered; to limit waste and foster the use of renewable energy, to protect biodiversity and maintain the variety of animal and plants, to fight climate change reduce CO2 emissions. It is true that in industrialized countries, they produce too much, consequently they pollute too much and generates too much waste. Hence waste management and recovery is an important focus [18].

3.2.2. Second aspect: socially inclusive. The social aspect is attracting more and more interest from researchers and practitioners in the business world. But what is it really about? It's about the ability to ensure the well-being of citizens. To allow everyone, regardless of their standard of living, to meet the basic needs of life in this case: food, housing, health, education, heritage culture and so on [19], fight against exclusion and discrimination especially of disabled people, elderly, minority,to limit disparities of gender equality, salary leveling, accessibility for all and foster solidarity [18] and more importantly to contribute to the well-being of everyone by promoting social cohesion.

3.2.3. Third aspect: economically efficient and responsible. The economic aspect is the one that interests the most practitioners and people in the business world. For many years, being efficient and effective was directly related to making profit and increasing the turnover of a company, or on a larger scale contributing to the economic development of a country.

Concretely, it is a question of reconciling the viability of a project, a company or an organization with ethical principles, such as the protection of the environment and the preservation of social cohesion by developing innovative and ethical business practices and fairly distribute benefits and wealth. Other alternatives are emerging to mitigate the effects of capitalism such as the circular economy or the collaborative consumption [18].

3.3. SSCM triggers

The 21st century has been marked by the effects of globalization, which has spread to new forms. New practices and behaviors arise. The opening of barriers between countries has led to economic and cultural change. In such context, the evolution of society is notable, an awareness of the need for reactivity and comfort emerged. It is now the customer who has the power to influence the market.

The demand prediction is very complex, new customer requirements are set up such as the customization and originality of services and products [20], [21]. Therefore, it is necessary to arm oneself with an agile work organization and review supply chain configurations [22] as pointed out by *E. Koberg & A. Longoni*. As a result, the supply chain becomes an essential component that influences customers' decisions. They are looking for multiple possibilities and more comfort during their shopping trip. Agility is not possible without investing in new technologies [20], [21]! The advent of e-commerce powered by NITCs brings productivity gains and new opportunities for customers. As a result, traditional distribution schemes are disrupted [23]. Moreover, consumer's awareness of the importance of environment respect and corporate social responsibility is increasing.

Since pre-industrialization, the concentration of greenhouse gases in the atmosphere has increased due to human activities. The UN is sounding the alarm and the emergency to reduce CO_2 emissions. The public authorities concerned must be in line with this perspective. Hence the importance of thinking globally and acting locally, everyone at their own level. Governments, local authorities and universities are encouraged to join this approach and contribute each within their own scope of action.

3.4. SSCM Challenges

It is important to know the triggers for the adoption of a sustainable supply chain. But it is even more important to know the challenges facing a SSCM.

3.4.1. Efficient SSCM. The first challenge of SSCM is to be efficient. It inevitably involves the Quality, Cost and Delivery (QCD) and resource optimization. We can say that a SSC is efficient if it does not consider customer's Quality by meeting its requirements, Cost control, because it is critical to any company or have high rate of Delivery Schedule Adherence. It also helps as an excellent starting point for being lean [24], and limit or eliminate wastes for all links of supply chain. However, lean management sometimes does not foster the sustainable part.

3.4.2. Responsive SSCM. In an agile ecosystem, it is difficult to predict demand, and to plan ahead for supplies, production and deliveries. Consumer's behavior has changed, requiring customization and speed of service. Some organizations integrate the customer into their product design and development process. This involves a variety of products to be designed, produced and delivered. An alignment of the supply chain configuration is required. For the production part, mass-customization is a partial solution to overcome this heterogeneity [25]. The rest is the delivery part, with the problems of the last mile, in order to optimize the costs of deliveries and to be able to have a reactive and sustainable supply chain.

3.4.3. Reliable SSCM. A supply chain reliability is the ability to deliver perfect orders in accordance with customer expectations. It is considered as the balance of the supply chain links where information flows must be in perfect synchronization in real time to allow readjustments and facilitate decision-making. In an environment where uncertainties are managed, it is even more difficult to be sustainable and reliable at the same time. The NICTs, the Internet of Things and big data [16], are there to overcome its constraints and help to keep the chain stable and synchronized between these different links anticipate potential upcoming issues.

3.5. Future research perspectives

In this paper, we provide an overview about SSCM research trends, and contribute to the development of upcoming literature reviews. More often, research of this field are more focused on the green or environmental side. The social dimension is not treated or rarely; only few papers focused on the social side of sustainability applied on supply chains. Interests are more oriented on the green aspect of it and a lot of research papers are produced in this sense.

Through this literature review, we were able to identify related research areas with the SSCM. The circular economy based on recovery, recycling and repair instead of producing is a research field that needs to be developed in line with sustainability of supply chains. Another interesting niches that was raised in this study is the collaborative supply chain, mutualization and Cross-channel and omnichannel logistics which can be extraordinary opportunities for companies to optimize their distribution configurations and be part of a sustainable development approach. Another research niche that jumped out at us was cobotics, where humans and robots are led to collaborate in a supply chain. Is this an opportunity to better concretize the social part of sustainable development in supply chains? Or is it a threat given that we are marginalizing the human being in the value chain?

Many other questions remain to be asked. But to sum up, supply chains still need to integrate the three aspects of sustainability, to be economically efficient, socially oriented, and environmentally friendly at the same time. This research field still needs to be developed taking into consideration an integrated vision of sustainable development.

References

[1] FINK, A 2005 *Conducting Research Literature Reviews: From the Internet to Paper*, 2nd edition. Thousand Oaks, London, New Delhi: SAGE Publications.

[2] Seuring S and Müller M 2008 From a literature review to a conceptual framework for sustainable supply chain management *J. Clean. Prod.* **16**(15) pp. 1699- 1710.

[3] Meredith J 1993 Theory Building through Conceptual Methods *Int. J. Oper. Prod. Manag.* **13**(5) pp. 3- 11.

[4] Weick K E 1995 What Theory is Not, Theorizing Is *Adm. Sci. Q.* **40**(3) pp. 385-90,

[5] Martins C L and Pato M V2019 Supply chain sustainability: A tertiary literature review *J. Clean. Prod.*

[6] Brindley C and Oxborrow L 2014 Aligning the sustainable supply chain to green marketing needs: A case study *Ind. Mark. Manag.* **43**(1) pp. 45- 55

[7] Srivastava S K 2007 Green supply-chain management: A state-of-the-art literature review *Int. J. Manag. Rev.* **9**(1) pp. 53- 80

[8] Ahi P and Searcy C 2013 A comparative literature analysis of definitions for green and sustainable supply chain management *J. Clean. Prod.* **52** pp. 329- 41.

[9] Tseng M L, Islam M S, Karia N, Fauzi F A and Afrin S 2019 A literature review on green supply chain management: Trends and future challenges *Resour. Conserv. Recycl.* **141** p. 145-62

[10] Brandenburg M, Govindan K, Sarkis J and Seuring S 2014 Quantitative models for sustainable supply chain management: Developments and directions *Eur. J. Oper. Res.* **233**(2), pp. 299-312.

[11] Jadhav A, Orr S and Malik M 2018 The role of supply chain orientation in achieving supply chain sustainability *Int. J. Prod. Econ.* ***217*** pp.112-25.

[12] Dallasega P and Sarkis J 2018 Understanding greening supply chains: Proximity analysis can help *Resources, Conservation and Recycling* **139** pp.76-77.

[13] Bastas A and Liyanage K 2018 Sustainable supply chain quality management: A systematic review », *J. Clean. Prod.* **181** pp. 726- 744.

[14] Mayring P 2015 Qualitative Content Analysis: Theoretical Background and Procedures », in *Approaches to Qualitative Research in Mathematics Education*, Éd. Dordrecht: Springer Netherlands pp. 365- 380.

[15] Agarwal A, Shankar R 2007 and Tiwari M K Modeling agility of supply chain », *Ind. Mark. Manag.* **36**(4) pp. 443- 457.

[16] Klötzer C and Pflaum A 2017 Toward the Development of a Maturity Model for Digitalization within the Manufacturing Industry's Supply Chain *Hawaii Int. Conf. Syst. Sci. 2017 At: Hilton Waikoloa Village, HI, USA* **50**.

[17] Redclift M 1987 *Sustainable development: Exploring contradictions*. Routledge.

[18] Giddings B, Hopwood B and O'Brien G 2002 Environment, economy and society: fitting them together into sustainable development *Sustain. Dev.* **10**(4) pp. 187- 196.

[19] Eltantawy R A, Fox G L and Giunipero L 2009 Supply management ethical responsibility: reputation and performance impacts *Supply Chain Manag. Int. J.*

[20] Christopher M 2000 The Agile Supply Chain: Competing in Volatile Markets *Ind. Mark. Manag.* **29**(1) pp. 37- 44.

[21] Christopher M and Towill D 2001 An integrated model for the design of agile supply chains *Int. J. Phys. Distrib. Logist. Manag.* **31**(4) pp. 235- 246.

[22] Koberg E and Longoni A 2019 A systematic review of sustainable supply chain management in global supply chains *J. Clean. Prod.* **207** pp. 1084- 1098.

[23] Croom S 2001 Restructuring supply chains through information channel innovation *Int. J. Oper. Prod. Manag.* **21**(4) pp.504-15

[24] Garza-Reyes J A 2015 Lean and green – a systematic review of the state of the art literature *J. Clean. Prod.* **102** pp. 18- 29.

[25] Dubey R, Gunasekaran A and Childe S J 2015 The design of a responsive sustainable supply chain network under uncertainty *Int. J. Adv. Manuf. Technol.* **80**(1), pp. 427- 445.

[26] Wu K J, Liao C J, Tseng M L, Lim M K, Hu J and Tan K 2017 Toward sustainability: using big data to explore the decisive attributes of supply chain risks and uncertainties *J. Clean. Prod.* **142** p. 663- 676.

IOP Conf. Series: Materials Science and Engineering **827** (2020) 012055 doi:10.1088/1757-899X/827/1/012055

Wastewater treatment using flocculation method and water re-use for dyeing of polyester fibers

P Forte Tavčer [1, 2,3] and K Prelog[2]

[1]Department of Textiles, Graphic Arts and Design, Faculty of Natural Sciences and Technology University of Ljubljana, Snežniška 5, 1000 Ljubljana, Slovenia.
[2] Donar d.o.o., Kamniška 41, 1000 Ljubljana, Slovenia
[3] Author to whom any correspondence should be addressed

petra.forte@ntf.uni-lj.si

Abstract. Textile industry uses and pollutes large quantities of water, especially for chemical treatments of textile materials, like bleaching, desizing, dyeing, printing and finishing. After each process, the water is usually drained into the sewage system and fresh water is used for new process of textile treatment. Re-using of wastewater for further processing would be of great importance in term of saving fresh water and diminishing the environmental burden. The goal of our research was to find out whether treatment of dye-house wastewater with flocculation could be efficient enough for water re-use in further dyeing of polyester fabric.
Wastewater collected in Slovenian dye-house was treated with a cationic flocculant. Treated water was analysed for chemical oxygen demand (COD), total organic carbon (TOC), spectral absorption coefficient (SAC) and dry matter and further used for laboratory dyeing of polyester fabric. The fabric was dyed comparatively using technological and treated water under the same conditions in light, medium and dark shade. The colour of dyed samples was evaluated on spectrophotometrically. Wet fastness and colour fastness to perspiration (acid and alkaline) of differently dyed samples were investigated. Change of colour was acceptable for all dyed samples. Fastness to washing and perspiration were good, very similar when comparing samples dyed in technological and those dyed in cleaned water. The colorimetric and fastness results showed a high efficiency of flocculation for dye-house wastewater treatment and reuse of treated water in production.

1. Introduction

Textile industry uses and pollutes large quantities of water, especially for chemical treatments of textile materials, like bleaching, desizing, dyeing, printing and finishing. A large problem of wastewater represents the dyes, which even in small quantities are already very visible and disturbing for the environment. During dyeing most of the dyes are exhausted from the dye-bath and bound to the fibers, but a part of unbounded dye always remains in the dye-bath and consequently finishes in the effluent. After each process the water is usually drained into sewage systems and fresh water is used for new process of textile treatment. Re-using of wastewater for further processing would be of great importance in term of saving fresh water and diminishing of environmental burden. The composition of dye-house wastewater is very complex and varies daily because of different dyes and additives which are used in the dyeing processes [1]. Most dyes have a complex aromatic structure, consistent to light, bio decomposition, oxygen and other environmental influences. That is why it is difficult to remove them

IOP Conf. Series: Materials Science and Engineering **827** (2020) 012055 doi:10.1088/1757-899X/827/1/012055

from water with classical treatment methods [2]. It is also impossible to find a process that would be suitable for all wastewaters [3].

Flocculation is a simple and economical process to remove the dyes from wastewater [4,5]. The goal of our research was to find out whether treatment of dye-house wastewater with flocculation could be efficient enough for water re-use in further dyeing of polyester fabric.

Wastewater collected in Slovenian dye-house was treated with cationic flocculants. Treated water was analysed for chemical oxygen demand (COD), total organic carbon (TOC), spectral absorption coefficient (SAC) and dry matter and further used for laboratory dyeing of polyester fabric. The fabric was dyed comparatively using technological and treated water under the same conditions in light, medium and dark shade. The colour of dyed samples was evaluated on spectrophotometer Datacolor Spectraflash SF-600X. Wet fastness and colour fastness to perspiration (acid and alkaline) of differently dyed samples were investigated.

2. Experimental

2.1. Flocculation
The representative wastewater from the dye-house was treated using different flocculants; Flocculant CHT-CV (Bezema, Switzerland), Flocculant CHT-VEL (Bezema, Switzerland), Rucofloc (Rudolf Group, Germany) and Sevofloc (Textilcolor AG, Switzerland). From 0,2 to 2,0 g/l flocculant at pH 6, 7, 8 and 9 was added to the water samples pH was adjusted with Na_2CO_3 or CH_3COOH. The laboratory flocculator Velp Scientifica FP4 was used. The solutions were mixed for 2 min. at 100 o/min following 20 min at 20 o/min. After 120 min of sedimentation the best results were found at 0,3 g/l of Flocculant CHT-CV at pH 8, at 0,5 g/l of Flocculant CHT-VEL at pH 7, at 0,7 g7l of Rucofloc at pH 8 and at 0,7 g/l of Sevofloc at pH 8.

2.2. Dyeing
The treated water was used for dyeing of 100 % polyester yarn (PES) with the fineness of 25 tex in light (L), medium (M) and dark (D) shade. Before dyeing the yarn was washed at 70 °C for 30 min with 1 g/l Foryl OV (Pulcra). After dyeing the yarn was reductive cleaned with 1% Rucorit RAL (*Rudolf Group GmbH)* and 2 g/l NaOH 50% for 20 min at 85 °C. The same material was also dyed with regular technological water at the same conditions. The dyeing was done in a laboratory dyeing machine Ahiba Nuance ECO at 130 °C for 60 min at bath to liquor ratio of 1:10

Dyes and additives: Foron dyes (*Archroma, Switzerland),* Alwiron PWE (*Textilcolor, Germany);* (dispersive agent), Tanacid SAB (*Tanatex, Germany);* (buffer).

Table 1: Recipes for dyeing of polyester, a) Quantity of dyes, b) Quantity of auxiliaries

a)

Dye	Concentration (%)		
	L[a)]	M[a)]	D[a)]
Foron brilliantgelb SWF	0	0	0,1500
Foron goldgelb SWF	0,0040	0,2480	0,5000
Foron rot SWF	0,0001	0,0003	0
Foron blau SWF	0,0002	0,0030	0
Foron marineblau SWF	0	0	0,5000

b)

Chemicals and additives	Concentration (gl^{-1})		
	L[a)]	M[a)]	D[a)]
CH_3COOH 80% (pH 4,5)	1,10	1,10	1,10
Alwiron PWE	1,50	1,50	1,50
Tanacid SAB	0,75	0,75	0,75

[a] Color tone

2.3. Analysis
Chemical oxygen demand (COD) - SIST EN ISO 6060:1996
Total organic carbon (TOC) - SIST EN ISO 8245:2000

IOP Conf. Series: Materials Science and Engineering **827** (2020) 012055 doi:10.1088/1757-899X/827/1/012055

Spectral absorption coefficient (SAC) - SIST EN ISO 7887:1996
Colour measurement – spectrophotometer Spectraflash SF-600X (Datacolor).
Colour fastness to washing - SIST EN ISO 105-C06:1999 .
Colour fastness to perspiration - SIST EN ISO 105-E04:1999

3. Results and discussion

Table 2 represents the environmental parameters of representative sample of wastewater and water cleaned with different flocculants.

Table 2: Environmental parameters of representative sample of wastewater and water cleaned with different flocculants.

Flocculant	Dry matter (%)	COD ($mgO_2\,l^{-1}$)	TOC (ppm)	IC (ppm)	SAC (m^{-1}) 436 nm	525 nm	620 nm
Untreated	0,23±0,01	680±10	615±10	92±1	420,0	219,0	190,0
CHT- CV	0,25±0,01	580±10	597±10	113±1	1,6	0,6	0,5
CHT- VEL	0,26±0,01	650±10	770±10	77±1	2,4	1,2	1,0
RUCOFLOC	0,25±0,01	630±10	671±10	110±1	10,6	5,3	2,8
SEVOFLOC	0,28±0,01	640±10	784±10	112±1	3,7	2,1	1,7

It can be seen that all flocculants thoroughly decreased the SAC values, while other parameters; dry matter, COD, TOC and IC, remained high or even increased. COD value of representative sample is 680 mg O_2/l. The best results of treatment were obtained for Flocculant CHT-CV, which decreased it for 15% to a value of 580 mg O_2/l. Even less has changed the TOC value, from 615 ppm for untreated water to 597 ppm for water treated with Flocculant CHT-CV. The content of dry matter even increased during treatment from 0,23% to 0,25%. On the other hand, very good results were obtained for SAC values, where the reduction in light absorption was almost 100% and the colour of the water was completely disappeared. This means that the flocculation removed almost all dyes from the wastewater although other components remained there in a considerable amount.

Polyester yarn was dyed in a water treated with Flocculant CHT-CV. Table 3 represents the CIE L*, a*, b* values for samples dyed with technological and treated water and colour differences between them.

Table 3: CIE L*, a*, b* values of polyester dyed in light (L), medium (M) and dark (D) tone and colour difference (ΔE_{ab}*) between samples dyes with technological and treated water.

Water	Colour tone	L*	a*	b*	ΔE_{ab}*
Technological	L	82,85	2,59	14,00	/
	M	55,48	9,37	21,87	/
	D	24,77	6,53	3,48	/
Treated	L	83,36	3,33	14,05	0,90
	M	55,30	9,70	22,13	0,45
	D	25,01	6,57	3,47	0,24

The highest colour difference of ΔE_{ab}* = 0,9 was obtained for light shade. It is low enough to be accepted for further production. ΔE_{ab}* for medium shade is 0,45 and for the dark shade 0,24. Both values are very low and show that the material can be dyed with treated wastewater.

Table 4 represents the colourfastness's values to wash and acid and alkaline perspiration of polyester samples dyed with technological and treated water.

Table 4: Colourfastness to wash and acid and alkaline perspiration of polyester.

Water	Colour tone	Evaluation		
		Acid perspiration	Alkaline perspiration	Washing
Technological	L	5	5	5
	M	5	4-5	5
	D	4-5	4	5
Treated	L	4-5	4-5	5
	M	5	4-5	5
	D	4	4	5

The fastness to washing are the same on polyester yarn dyed with treated or with regular water, excellent in both cases. Very good are also the fastness properties to acid and alkaline perspiration at both dyeing conditions.

4. Conclusions

In the first part of the investigation, it was found that the pollution of the wastewater from a dye-house oscillates during the day and during the week. The average pollution is high, the TOC values are above 900 ppm and the coloration is high. To obtain the average polluted wastewater, the samples of the water were taken from the collection pool once a day for one week and consequently all seven samples were mixed together. In this way the representative wastewater was obtained.

The wastewater from the textile dye house was treated with 0,3 g/l CHT Flocculant-CV at pH 8. The SAC decreased for 99%, COD for 15%, whereas the amount of dry matter increased during the flocculation process.

The treated water was used for dyeing of polyester yarn in light, medium and dark shade. For comparison the polyester yarn was dyed in the same shades by using fresh process water. The dyeing results were appropriate for all three shades when using the treated wastewater and when using fresh process water. Change of colour was acceptable for all dyed samples. The lowest colour difference was obtained at dark shade, the highest at light shade. Fastness to washing and perspiration were good, very similar when comparing samples dyed in technological and those dyed in cleaned water. The colorimetric and fastness results showed high efficiency of flocculation for dye-house wastewater treatment and a possibility of reusing of treated water in further production.

Flocculation has proven to be a process, which successfully removes mainly dyes from the wastewater, while a part of other components remained there together with a part of flocculant. Despite high COD and TOC values the treated water is appropriate for preparing new dyebaths, especially for dyeing of synthetic polyester fibers in dark tones. Polyester is namely very hydrophobic, without functional groups, absorbs only few chemicals and is hard to dye. This is a reason why the flocculants and other chemicals that remained in treated water do not enter into the fibers during dyeing but remain in the dye-bath while small dispersive dyes enter into the fibers at appropriate conditions and successfully dye them.

Acknowledgement
This work was supported by the Slovenian Research Agency (P2-0213).

References

[1] Joshi M and Purwar R 2004 Developments in new processes for colour removal from effluent. *Rev. Prog. Color.* **34** 58–71
[2] Skelly K.J 2003 Water Recycling: Textile Wet Processing. *Society of dyers and colourists*.
[3] Menezes E 2003 Developments in waste water treatment methods. *Energy and waste water*, 58–65.
[4] Holkar C et all. 2016 A critical review on textile wastewater treatments: Possible approaches. *J. Env. Man.* **182**, 351–66.
[5] Forte Tavčer P 2003 Treatment of wastewater from pigment printing. *Tekstilec*, **46** 120–126.

Microencapsulation for improved mosquitoes' repellent efficacy of cotton fabrics

A M Grancaric[1,5], K Laird [2,3], L Botteri[1], J Shen[3] and K Laatikainen[4]

[1]Department of Textile Chemistry and Ecology, Faculty of Textile Technology, University of Zagreb, Prilaz baruna Filipovića 28a, HR-10000 Zagreb, Croatia
[2]Infectious Disease Research Group, School of Pharmacy, De Montfort University, Leicester, UK
[3]Textile Engineering and Materials Research Group, School of Design, De Montfort University, The Gateway, Leicester, UK
[4]Lappeenranta-Lahti University of Technology LUT, Lappeenranta, Finland
[5]Author to whom any correspondence should be addressed

amgranca@ttf.hr

Abstract. In recent years, mosquitoes that can transfer viruses causing vector-borne diseases, such as dengue fever, chikungunya, Zika and West Nile virus have dramatically increased and reached Europe. In 2018, a higher number of 1503 human cases were reported in the EU/EEA and EU neighbouring countries. The current research was involved in the development of mosquito repellent cotton fabrics with natural essential oils and further improvement of mosquitoes repellent efficacy by microencapsulation of repellents on cotton Fabrics. The repellents efficacy for Anopheles spp. is calculated by using the results of WHO modified test method CTD/WHO PES/IC/96.1. Mosquito-repellency of the treated cotton fabrics against Aedes aegypti mosquito species were tested by using Y-tube Olfactometer. SEM images of treated cotton fabrics were also represented in this paper.

1. Introduction

A recent outbreak of West Nile virus in Central Europe (2018) demonstrates the need for increased research on repellents against invasive mosquitoes which carry the West Nile virus. In 2018 a higher number of 1503 human cases were reported in the EU/EEA and EU neighbouring countries [1].

Many infectious diseases are vector-borne including malaria, Zika virus, dengue and chikungunya virus, West Nile virus and plague, all of which are spread via the bites of infected mosquitoes, and are prevalent in subtropical regions to which South European countries belong. Recent studies indicate that Dengue viruses infect more than 50 million people per year and cause more than 20,000 deaths annually [2]. During the last 15 years Aedes albopictus (Asian tiger mosquito) was introduced into Europe and has spread to several countries, which increases the possibility of its transmission [3,4]. There is limited published data on the prevention of these mosquito borne diseases. Therefore, the development of repellents new targets and related technology becomes paramount importance. The current research was involved in the development of mosquito repellent cotton fabrics with natural essential oils and further improvement of mosquitoes repellent efficacy by microencapsulation of repellents on cotton fabrics.

2. Material and Methods

Scoured and bleached samples of woven cotton fabric were used in the current study. The first set of cotton fabrics was treated with immortelle oil, vibroactivated zeolites (VZ) and water glass, by using traditional Pad-Dry laboratory device [5]. In the continuous impregnation process, the cotton fabric wet pick-up of the treatment liquor was 100%, followed by drying at 120 °C for 2 min. The concentrations of repellents used in the treatment and their corresponding sample labelling code are listed in Table 1. The repellents efficacy for Anopheles spp. is calculated by using the results of WHO modified test method CTD/WHO PES/IC/96.1, which essentially is a chamber method, where mosquitoes are released onto treated fabric. In this method, the mosquito will be repelled to the bait through the gap provided. their effective mosquito repulsion i.e. their ability to migrate away is observed and calculated as Percentage Repellency (resp. efficacy). The observations were recordered over 10 and 30 minutes after mosquitoes were released [6].

Table 1. The treatment of **c**otton fabric samples with different repellents.

Sample code	Treatment with repellents
UN	Untreated
VZ_20	20 g/l vibroactivated zeolite
I_5	5 g/l immortelle essential oil
I_5_WG_10	5 g/l immortelle essential oil + 10 g/l water glass

The second set of cotton fabrics, was treated with immortelle oil microcapsules (ME), by using Pad-Dry impregnation system, the same system used for the first set samples. Microcapsules were produced in laboratory scale by coacervation phase separation technique, whereby immortelle essential oil (EO) (30%) was emulsified with 1% w/v chitosan and 0.025% w/v sodium alginate. Mosquito-repellency of the treated cotton fabrics against Aedes aegypti mosquito species were tested by using Y-tube Olfactometer [6]. The physical structure and elemental composition of cotton fibers were analyzed by Hitachi SU3500 scanning electron microscopy equipped with energy dispersive spectrometer detector (SEM-EDS). The imaging was made with 3 kV accelerating voltage. The samples were coated with a 10 nm thick layer of gold before analysis.

3. Results and discussion

Cotton samples used in the current research were conventionally treated with new natural mosquito repellents: Immortelle essential oil and micro- to nano- vibroactivated zeolite. Furthermore, waterglass were also used to incorporate with immortelle essential oil (see Table 1 and Figure 1).

Figure 2 shows SEM images of the surfaces of the cotton fibers after treatment with different mosquito repellents. It can be seen that untreated cotton shows fibrilar surface structure. Cotton fibers surface looks smoother and uniform when impregnated with immortelle oil, Vibroactivated zeolite or water glass. vibroactivated zeolite (VZ_20) and water glass (I-WG_10) particles are clearly seen on the fiber surface. The zeolites particles are uniform in the range from 5μm to 10 μm. It is found that water glass (sodium silicates) particles are smaller and located only in the spaces between fibers.

IOP Conf. Series: Materials Science and Engineering **827** (2020) 012056 doi:10.1088/1757-899X/827/1/012056

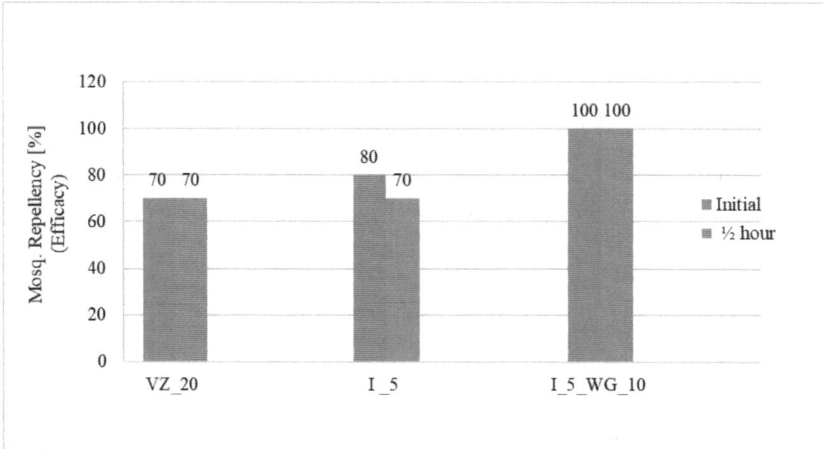

Figure 1. Repellents efficacy of conventionally treated cotton fabrics against Anopheles spp mosquitoes.

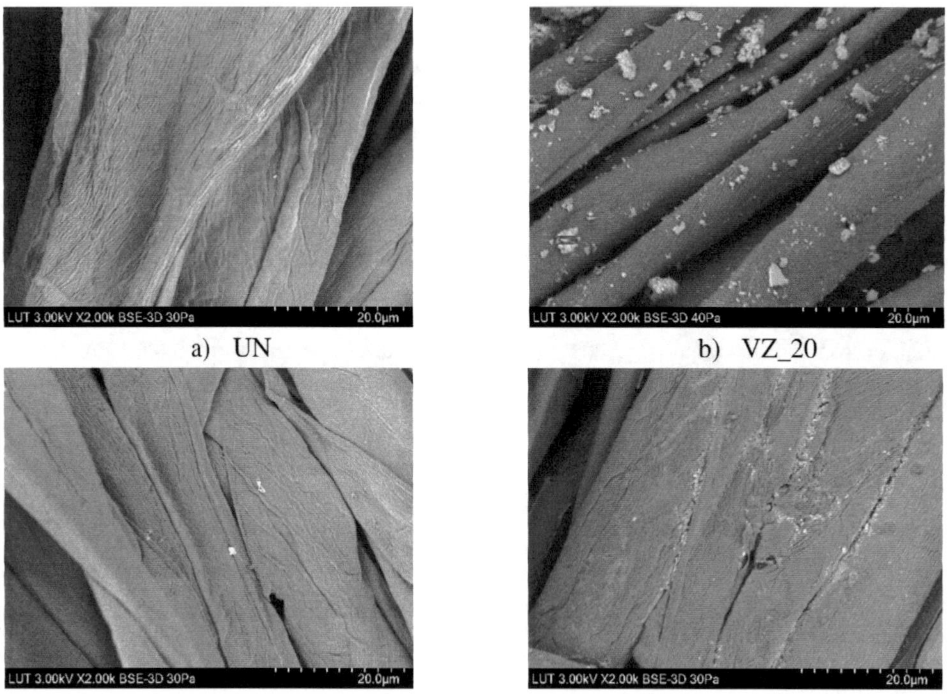

a) UN b) VZ_20

c) I_5 d) I_5_WG_10

Figure 2. SEM images of cotton fabrics: a) untreated, b) treated with vibroactivated zeolite, c) immortelle oil, and d) immortelle oil and water glass - signals detected by the backscattered electrons (BSE) (LUT).

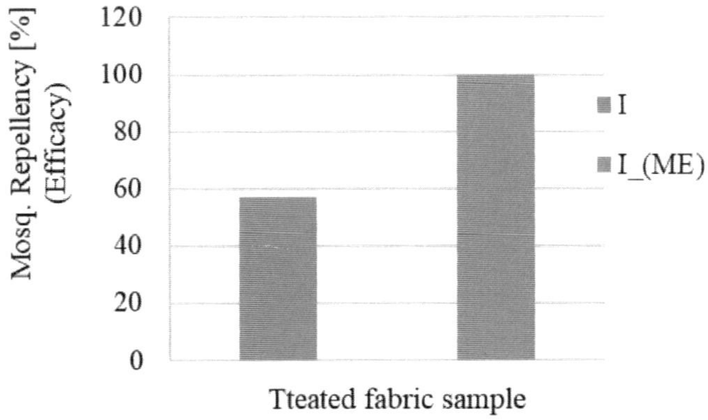

Figure 3. Repellency for conventional finishing with immortelle oil (I) and microencapsulated immortelle oil (I_ME) on cotton samples against Aedes aegypti by using Y-tube olfactometer; ME-microcapsules.

Efficacy results for microencapsulated immortelle oil are high giving a 100% efficacy against Aedes aegypti (Figure 3), compared with the immortelle oil alone and against the same mosquito, only a 57% efficacy rate against Aedes aegypti was achieved but 70% to 80% against Anopheles spp.

The repellent with the greatest efficacy was cotton treated with immortelle oil and water glass and will be considered for future research work on this area. Further studies are being undertaken to optimize the concentration of immortelle oil to be used and to improve the efficacy of microencapsulation of immortelle oil for the controlled release of bioactive oils to achieve the long lasting of mosquito-repellency. This is a pilot study demonstrating initial findings on potential repellents against mosquitos further work on synergistic effects of different repellents, fabric construction, repellent concentration, wash fastness properties and technology application still need to be assessed.

4. Conclusion

The natural bio repellent used in this paper, immortelle oil, showed good to excellent repellent efficacy, when applied using conventional technology or novel microecapsulation procedure. For Anopheles, the highest efficacy is obtained with immortelle oil and water glass but not so high for vibroactivated zeolite. Microencapsulated immortelle oil against Aedes aegypti showed the highest efficacy, compared with immortelle oil alone when applied using conventional technology for the same mosquito.

Acknowledgment

This paper is a part of the COST Action project CA16227 (2017-2021) entitled: Investigation and Mathematical Analysis of Avant-garde Disease Control via Mosquito Nano-Tech-Repellents (IMAAC). The authors would like to thank the European Commission for funding this COST-Action project.

References

[1] ECDC Report: *West Nile virus infections by affected areas, in the EU/EEA Member States and EU neighbouring countries, 2018 transmission season,* https://www.ecdc.europa.eu/en/publications-data/west-nile-virus-infections-affected-areas-eueea-member-states-and-eu-neighbouring
[2] Devaux CA 2012 Emerging and re-emerging viruses: A global challenge illustrated by Chikungunya virus outbreaks, *World J Virol* **1**(1), pp 11-22
[3] Scholte E J and Schaffner F 2007, Waiting for the tiger: establishment and spread of the Aedes

albopictus Mosquito in Europe, Takken W Knols BGJ eds. *Emerging pests and vector-borne disease in Europe*. Wageningen Academic Publishers, pp 241-60.

[4] Enserink M 2008 Entomology. A mosquito goes global, *Science*, **320**(5878), pp 864-866

[5] Grancaric A M, Botteri L and Thakar V 2018 Textile Treatments with a new Mosquito Repellents Based on The Natural Vibroactivated Zeolites, *Ninth Workshop Dynamical Systems Applied to Biology and Natural Sciences DSABNS*, Turin

[6] Anuar A A and Yusof N 2016 Methods of imparting mosquitoes repellents agents and the assessing mosquito repellency on textiles, *Fashion and Textiles* **3**(1)

Comfort-related properties of the seersucker woven fabrics

Malgorzata Matusiak[1,2]

[1]Institute of Architecture of Textiles, Faculty of Material Technologies and Textile Design, Lodz University of Technology, 116 Zeromskiego str., 90-924 Lodz, Poland
[2] Author to whom any correspondence should be addressed

malgorzata.matusiak@p.lodz.pl

Abstract. Clothing creates a barrier between the human body and surroundings. Dependably on the predestination, it protects the human organism against outer factors dangerous for human health and safety. From the point of view of human health and well-being, the comfort-related properties of clothing are very important. The aim of presented work is to investigate the seersucker woven fabrics in the range of their comfort-related properties and to analyse an influence of the structure of fabrics on their properties. 9 variants of the seersucker woven fabrics were the objects of the investigation. The fabrics were manufactured from cotton yarns. In the experimental fabrics, 3 kinds of repeat of the seersucker pattern and 3 kinds of weft yarns were applied. Measurements of the comfort-related properties of the investigated fabrics have been performed using the following testing devices: Alambeta, Permetest and Moisture Management Tester. Results showed that both structural factors of fabrics: repeat of the seersucker effect and a kind of weft yarn influence the thermal insulation properties of the fabrics being investigated. Especially, the repeat of the seersucker effect influences the thermal conductivity, thermal absorptivity and thermal resistance in the statistically significant way at the probability level $p = 0.05$.

1. Introduction

Protection of human being is primal and one of the most important functions of clothing. Clothing creates a barrier between the human body and surroundings. Dependably on the predestination, it protects the human organism against outer factors dangerous for human health and safety. From the point of view of human health and well-being, the comfort related properties of clothing are very important. There are many factors influencing an ability of clothing to ensure the thermo-physiological comfort of clothing usage. An appropriate selection of textile materials from the point of view of their comfort-related properties plays a significant role in ensuring the thermal comfort. Thermal resistance, moisture transport and air permeability are considered as the most important comfort-related properties of textile materials [1].

Some researchers [2, 3] consider that the seersucker fabrics have good comfort properties. It results from their puckered structure. A typical seersucker structure is characterized by occurring the puckered strips in warp direction (figure 1). The seersucker fabrics are considered as the three-dimensional (3D) fabrics. Such structure is usually received on loom by an application of two warps of different tension [3].

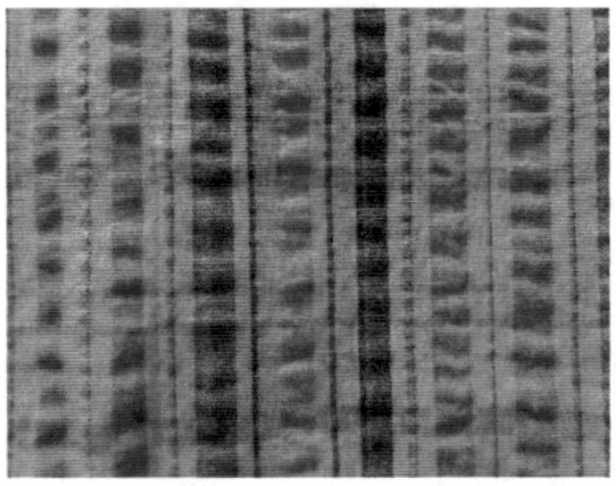

Figure 1. Example of seersucker woven Fabric.

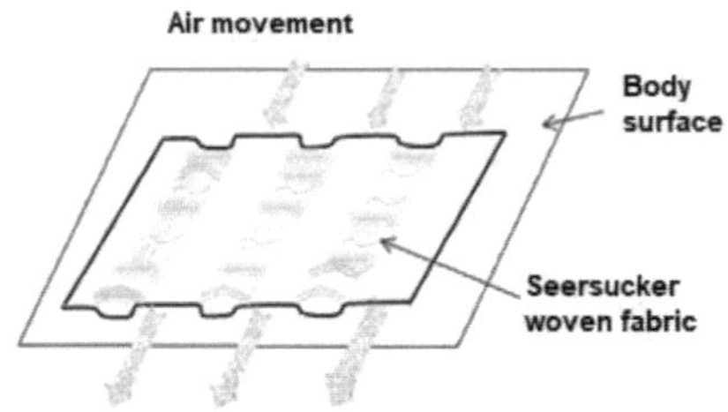

Figure 2. Schematic diagram of air circulation between the seersucker woven fabric and human skin.

The puckered effect generates air spaces between body and fabric. It ensures cooling of the user's body in hot conditions, because the puckered area of the seersucker fabric holds the fabric away from the skin (figure 2). It facilitates the circulation of air and water-vapor.

Matusiak and Fracczak [4] showed up that the seersucker woven fabrics are characterized by low values of the thermal absorptivity in a wet state. According to *Hes et al.* [5, 6], it provides an excellent physiological comfort conditions while using the clothing made of such kind of fabrics.

The aim of the presented work is to investigate the seersucker woven fabrics in the range of their comfort-related properties and to analyse an influence of the structure of fabrics on their properties.

2. Materials and methods

Totally, 9 variants of the seersucker woven fabrics were the objects of the investigation. The fabrics were manufactured from cotton yarns. In the experimental fabrics, 3 kinds of repeat of the seersucker pattern and 3 kinds of the weft yarns were applied. Both warps: basic and puckering were made from the same yarn: 20 tex x 2 cotton. As a weft the following yarns were used: 20 tex x 2, 25 tex x 2 and 30 tex x 2.

The following variants of pattern of the seersucker effect were applied (figure 3):

- variant MM1 – width of puckered and flat strips appropriately: 4 mm and 8 mm,
- variant MM2 – width of puckered and flat strips appropriately: 9 mm and 17 mm,
- variant MM3 – width of puckered and flat strips appropriately: 10 mm and 37 mm.

The fabrics were finished by a tensionless method. The finishing process included washing, rinsing and drying. Because the fabrics were manufactured from 2 ply yarns, the warps did not have to be sized. Due to this fact, the desizing process was unnecessary. The fabrics being investigated were measured in the range of their basic structural parameters. The results are presented in table 1.

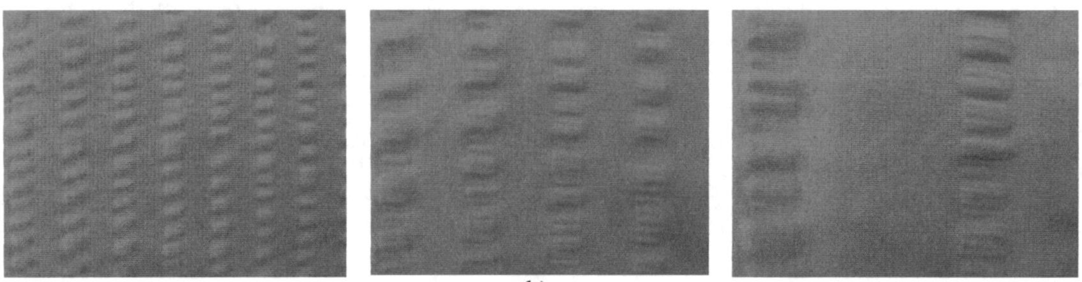

a) b) c)

Figure 3. Pictures of the seersucker woven fabrics being investigated: a) variant MM1, b) variant MM2, c) variant MM3.

Table 1. The basic structural parameters of the seersucker woven fabrics being investigated.

Variant of repeat of the seersucker effect	Weft yarn	Mass per square meter g/m^2	Warp density threads/dm	Weft density threads/dm
	20 tex x 2	221.88	137.6	120.2
MM1	25 tex x 2	253.03	134.6	116.0
	30 tex x 2	262.62	137.8	108.0
	20 tex x 2	216.96	136.8	118.0
MM2	25 tex x 2	236.76	138.0	109,6
	30 tex x 2	254.75	134.0	105.6
	20 tex x 2	205.43	127.8	116.8
MM2	25 tex x 2	227.06	129.8	111.6
	30 tex x 2	243.50	131.0	104.6

Measurements of the comfort-related properties of the investigated fabrics have been performed using the following testing devices: Alambeta, Permetest and Moisture Management Tester M290 (MMT). The Alambeta is a computer-controlled instrument for measuring the basic static and dynamic thermal characteristics of textiles [4, 5, 7, 8]. This method belongs to the so called "plate methods". The principle of measurement by means of the Alambeta relies on the convection of heat emitted by the hot upper plate in one direction through the measured sample to the cold bottom plate. It is two plate measuring device developed by *Hes* (Sensora, Czech Republic). The Alambeta measures: thermal conductivity, thermal diffusivity, absorptivity, thermal resistance and fabric thickness. For each fabric variant 10 repetitions were performed. Measurement was done in the standard climatic conditions.

The Permetest was also developed by *Hes* (Sensora, Czech Republik). This device can be considered as a portable "skin model". By means of the Permetest, the following comfort-related properties are determined: thermal resistance, water-vapor resistance and relative water-vapor permeability [1]. By means of the Permetest, 3 repetitions of measurement were performed for each fabric variant being investigated.

The Moisture Management Tester (SDL Atlas) is an instrument designed to measure the dynamic liquid transport properties of textiles such as knitted and woven fabrics in three aspects [9]:

- absorption rate – moisture absorbing time for inner and outer surfaces of the fabric,
- one-way transport capability – one-way transfer of liquid moisture from the inner surface to outer surface of fabric,
- spreading/drying rate – speed of liquid moisture spreading on the inner and outer surfaces of fabric.

The device is controlled by a PC and MMT290 SOFTWARE. Measurement is performed for samples cut into 80 mm x 80 mm squares. For each fabric 5 repetitions of measurement are performed. During the test, a pre-defined amount of test solution (synthetic sweat) is introduced onto the upper side (skin side) of the fabric, and then the test solution is transferred onto the material in three directions [10]:

- spreading outward on the upper surface of the fabric,
- transferring through the fabric from the upper surface to the bottom surface,
- spreading outward on the bottom surface of the fabric.

In order to asses a statistical significance of the influence of structural parameters of the seersucker woven fabrics on their comfort-related properties, a statistical analysis of the results was performed using the multifactor ANOVA available in STATISTICA software. The repeat of the seersucker effect and a kind of weft yarn were taken as the independent variables (main factors), whereas the comfort related properties were the dependent variables. Statistical significance was assessed at the significance level 0.05.

3. Results and discussion

Results of measurement using the Almbeta, Permetest and Moisture Management Tester showed that both structural factors of fabrics; repeat of the seersucker effect and a kind of weft yarn, influence the thermal insulation properties of the fabrics being investigated.

3.1. Results from the Alambeta

Statistical analysis using the ANOVA shows that the repeat of the seersucker effect and a kind of weft yarn influence the thermal conductivity, thermal absorptivity and thermal resistance of the fabrics in statistically significant way at the 0.05 significance level. The highest thermal conductivity was stated for the fabrics representing the MM3 variant of the seersucker effect repeat (figure 4). However, it is statistically observed a significant interaction between two main factors: seersucker effect repeat and a kind of weft yarn. Influence of the repeat of the seersucker effect on the thermal conductivity is modified by a kind of weft yarn (figure 5). In the case of fabrics containing the 25 tex x 2 and 30 tex x 2 weft yarns the lowest thermal conductivity was stated for the fabric representing the MM2 pattern variant.

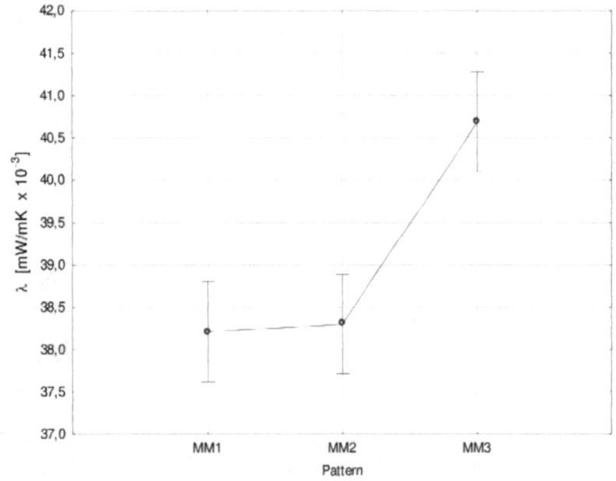

Figure 4. Thermal conductivity of the seersucker woven fabrics of different repeat of the seersucker effect.

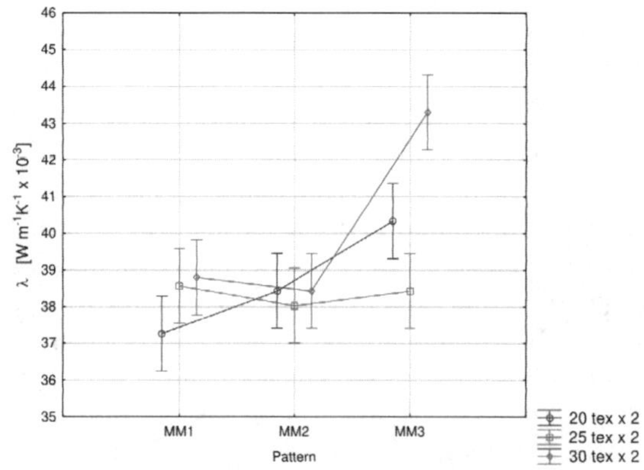

Figure 5. Thermal conductivity of the seersucker woven fabrics in function of repeat of the seersucker effect and a kind of weft yarn.

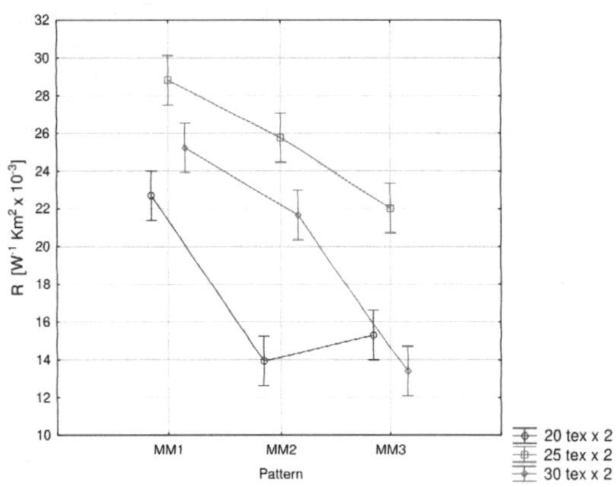

Figure 6. Thermal resistance of the seersucker woven fabrics in function of repeat of the seersucker effect and a kind of weft yarn.

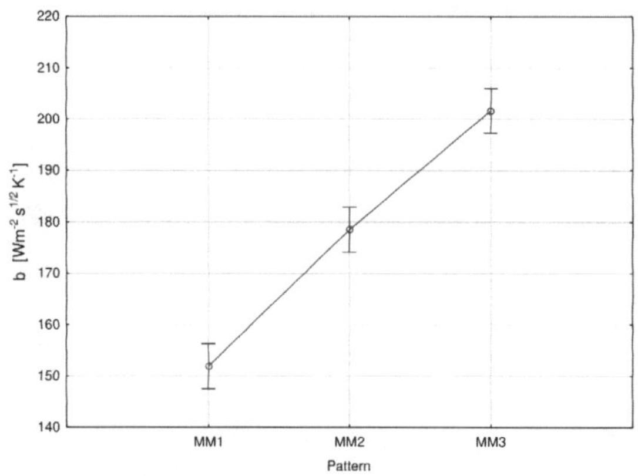

Figure 7. Thermal absorptivity of the seersucker woven fabrics of different repeat of the seersucker effect.

In the case of the thermal resistance and thermal absorptivity it was also stated that the influence of the seersucker effect pattern and a kind of weft yarn as well as the interaction between both independent variables (main factors) are statistically significant at the significance level 0.05. However, thermal resistance of the seersucker fabrics is decreasing from the MM1 to the MM3 variant of pattern (figure 6), whereas for the thermal absorptivity the tendency is opposite (figure 7). The thermal absorptivity is the surface property. It expresses the warm/cool feeling while first contact of fabric with human skin. In the case of the MM3 variant the share of the puckered strips in total area of fabrics is the lowest in comparison to the MM1 and MM2 fabric variants. It means that in the case of the MM3 fabric variant the number of direct contact points between the fabric and human skin is the lowest. It influences the heat absorption by the fabric.

3.2. Results from the Permetest
The results from the Permetest confirmed the relationships between the thermal resistance and structure parameters of the investigated fabrics. The thermal resistance decreases from the MM1 variant to the MM3 variant (figure 8). However, it should be mentioned here that in the measuring channel of the Permetsst, it is an air movement of the velocity 1.0 m/s. The puckered strips can disturb the air movement. The question was how should the samples of the seersucker woven fabrics be placed on the measuring channel of the device: puckered strips parallel or perpendicular to the direction of air movement? Performed investigations showed that the thermal resistance measured by means of Permetest at parallel to the air movement placement of puckered stripes is lower than that measured at perpendicular placement (figure 9) and the difference between them is statistically significant. It should be taken into consideration while assessing such kind of patterned fabrics by means of the Permetest.

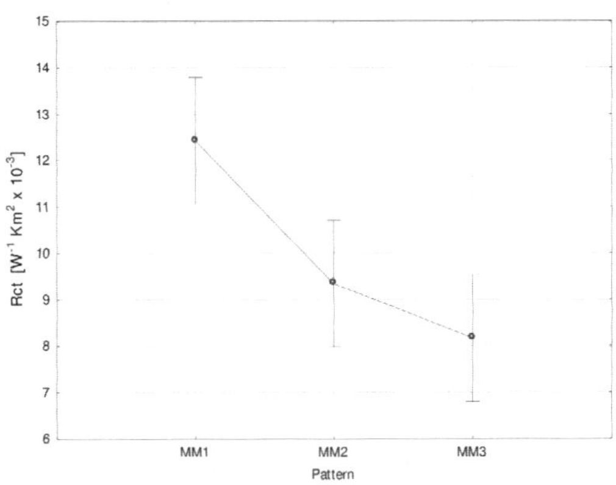

Figure 8. Thermal resistance according to the Permetest measured at parallel to air movement placement of puckered strips.

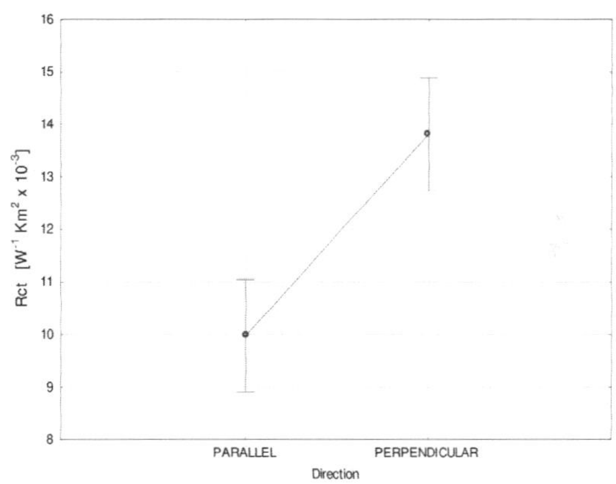

Figure 9. Thermal resistance according to the Permetest measured at different direction of sample placement.

3.3 Results from the Moisture Management Tester
The Moisture Management Tester provides 10 different parameters characterizing the fabrics from the point of view of their ability to transport of the liquid moisture. It is impossible to present all of them in this paper. Due to this fact, the detailed results from the MMT will be the presented in separate publication. Generally, the results showed that the seersucker fabrics being investigated should be classified as fast absorbing and slow drying fabrics. The average values of the Overall Moisture Management Capacity (OMMC) was not higher than 0.4. It is a medium level because the range of this parameters is from 0 to 1. The higher value of the OMMC parameter is the better ability of fabric to transport the liquid moisture. It was also observed that in the majority of cases the pattern of the seersucker effect influences the values of parameters assessed by the MMT.

Obtained results allowed drawing two important conclusions from the point of view of the assessment of such kind of patterned fabrics as the seersucker woven fabrics in the range of their moisture transport capability. First of all, it was stated that the puckered strips disturb the spreading of liquid moisture in the warp direction of the fabric. Secondly, while measuring the seersucker woven fabrics by means of the MMT the placement of sample is very important. During the test a pre-defined amount of test solution (synthetic sweat) is introduced onto the upper side of the fabric in one, central point, and then the test solution is transferred onto the material in three directions. Depending on whether the droplets fall on a flat or puckered area of the sample, a moisture spreads in different ways. In the presented investigations, the placement of samples was done in such a way that the synthetic sweat droplets were falling on the puckered surface. It is due to the fact that while using the clothing made of the seersucker woven fabrics, the puckered area of the fabrics adheres to the user's skin and absorbs sweat from the skin.

4. Conclusions

Performed investigations confirmed that the structure of the seersucker woven fabrics significantly influences the majority of comfort-related properties of the fabrics. The repeat of the seersucker effect influences the thermal conductivity, thermal resistance, thermal absorptivity as well as the moisture transport ability. In some cases, there is also statistically significant interaction between the main structural factors: repeat of the seersucker effect and a kind of weft yarn. In the case of measurement by means of the Permetest, the placement of the sample in measurement channel is also important for the results of thermal resistance measurement. The values of thermal resistance determined when the direction of the puckered strips is parallel to the air movement are lower than that determined when the direction of the puckered strips is perpendicular to the air movement.

Acknowledgments

This work is financed by National Science Centre, Poland in the frame of the project titled „Geometrical mechanical and biophysical parameterization of the three-dimensional woven structures", project No. 2016/23/B/ST8/02041.

References

[1] Hes L 2006 Alternative methods of determination of water vapour resistance of fabrics by means of "Skin Model", *3rd European Conference on Protective Clothing (ECPC) and NOKOBETEF* ***8***, CIOP-PIB and ESPC, Gdynia

[2] Ashraf W, Nawab Y, Maqsood M, Khan H, Awais H, Ahmad S, Ashraf, M. and Ahmad S 2015 Development of seersucker knitted fabric for better comfort properties and aesthetic appearance, *Fibers and Polymers*, **16**(3), pp 699-701.

[3] Gandhi K 2012 *Woven Textiles Principles, Technologies and Applications*, Woodhead Publishing, New Delhi, pp 142-158.

[4] Matusiak M and Fracczak L 2017 Comfort-related properties of seersucker fabrics in dry and wet state, *International Journal of Clothing Science and Technology* **29**(3), pp 366-379

[5] Hes L, Boguslawska-Baczek M and Geraldes J 2014 Thermal comfort of bedsheets under real conditions of use, *Journal of Natural Fibers*, **11**(4), pp 312-321

[6] Bajzik V, Hes L and Dolezal I 2016, Changes in thermal comfort properties of sportswear and underwear due to their wetting, *Indian Journ. of Fibre & Textile Res.*, **41**(2) pp 161-166.

[7] Özçelik G, Çay A and Kirtay E (2007), A study of the thermal properties of textured knitted fabrics, *Fibres & Textiles in Eastern Europe*, **15(**1(60)), pp 55-58

[8] Matusiak M and Sikorski K 2011 Influence of the structure of woven fabrics on their thermal insulation properties, *Fibres and Textiles in Eastern Europe,* **19**(5(88)), p 46-53.

[9] Matusiak M 2019 Moisture Management Properties of Seersucker Woven Fabrics of Different Structure, *Fibres & Textiles in Eastern Europe* **27**(3(135)) p 43-50

[10] SDL Atlas *MMT M 290 manual*

IOP Conf. Series: Materials Science and Engineering **827** (2020) 012058 doi:10.1088/1757-899X/827/1/012058

Exponential success through integrated supply chain optimization, ecomotional intelligence and reputation-based leadership: Zara model

N Berbiche[1,2,4], M Hlyal[1,2] and J El Alami[1,3]

[1]Laboratory LASTIMI, Higher School of Technology, Salé/Mohammed V University in Rabat, Rabat, 10080, Morocco

[2] Center of Logistics Excellence CELOG, ESITH, Casablanca, 20190, Morocco

[3] National Center for Science and Technology Research, Rabat, 10102, Morocco

[4] Author to whom any correspondence should be addressed

nawfalber@gmail.com

'You need to have five fingers touching the factory and five touching the customer.' Amancio ORTEGA, Inditex Chairman

Abstract. Fast fashion figures among the most lucrative using industries of the modern economies. With $25 billion annual sales in 2018, Zara that pioneered fast-fashion art has achieved an outstanding rapid surge, conquering 96 markets worldwide. The proposed paper presents the results of a critical thinking on how Zara applied innovation and continuous improvement in its operations and strong business model to maintain its unique competitive advantage over time. Our study shows that the secrets behind its exponential success are mainly a sophisticated sustainable supply chain, integrating efficiently the layers of design, sourcing, manufacturing, and distribution to stores for customers. It has to do with the best trade-off between low storage and shorter time response to customers in a contracted volatile market. Concisely, this research on specialized studies done over recent years seeks to help professional leaders grasp the best managerial and relational practices, along with developed operations' research tools based on demand forecasting, inventory-distribution and clearance pricing optimization models. It aims also at enriching their knowledge with some non-quantitative methods related to clients' preferences, '*glocal strategy*' for adaptation, and social responsibility. As such, in view of the frequent conjunctional use of *economic intelligence* and *emotional intelligence* by Zara, we coined the neologism '*ecomotional intelligence*', as a guide to belief and action.

1. Introduction

Fast fashion figures among the most lucrative industries of the modern economies. However, textile companies confronted with new constraints of rapid customer satisfaction, challenging environment and complexity of sustainment systems are bearing fierce competitiveness to preserve their benefits and respective market shares, if not to survive. In this context, Zara has achieved outstanding rapid surge, conquering 96 markets [1]. Inditex group where Zara is its largest subsidiary by more than 60%, propelled Ignacio to the 2nd fortune in Europe and 6th in the world [2], contributing by more than 2% of total social taxes to Spain economy [1]. By the eve of the 21st century, Zara's average annual growth has extraordinarily accelerated to 34.8 %, leading since then fast-fashion killers, as illustrated in the figure 1[3]. So, what are the secrets behind Zara's exponential success, in such a very short period?

IOP Conf. Series: Materials Science and Engineering **827** (2020) 012058 doi:10.1088/1757-899X/827/1/012058

In fact, the keys to Zara's huge success are mainly a sophisticated sustainable supply chain, integrating efficiently the layers of design, sourcing, manufacturing, distribution to stores for customers [4], within the strong business model on figure 2. It has to do with the best trade-off between low storage and shorter time response to customers in a contracted volatile market [5]. To learn from the prosper Zara's experience, the proposed paper presents the results of a critical thinking on the factors to its exponential success through analyzing and synthetizing a large number of related scientific and specialized studies done over recent years. It seeks to help professional leaders and decision-makers grasp the best managerial and relational practices, along with operations' research analytical tools applied for the continuous improvement of its efficient resilient supply chain. It aims also at enriching their knowledge with some implemented non-quantitative methods related to clients' preferences and social responsibility. As such, due to the frequent conjunctional use of various forms of economic and emotional intelligence by Zara and other killers in low manner, we come up with a new paradigm, for which the neologism of 'ecomotional intelligence' has been coined. In short, a guide to belief and action for a lasting competitive advantage [6].

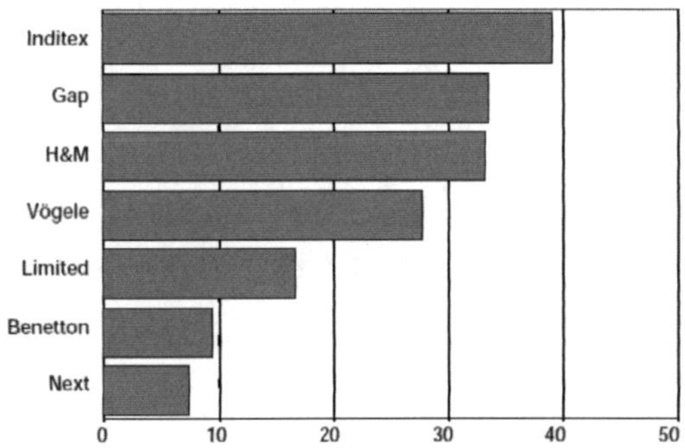

Figure 1. Annual Growth 1994-2000 [3].

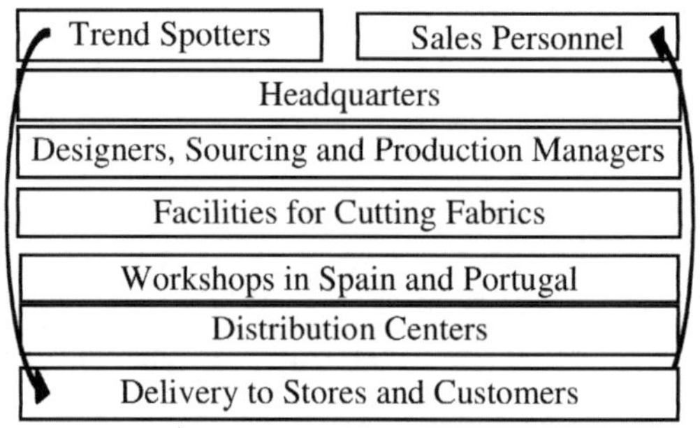

Figure 2. SCM Organization of Zara [4].

After presenting the best practices of Zara's integrated supply chain, this research paper will examine the climax role of applied demand forecasting and optimization models as a catalyst to durable competitiveness, to follow on with ecomotional intelligence and reputation-based leadership, expressive of corporate marketing and social responsibility.

2. Best practices of integrated supply chain management

For its continuous improvement, Zara focused on developing smart efficient supply chain management through the vertical integration of layers and key functions, just-in-time strategy, diversification and externalization.

2.1. Vertical integration and just in time strategies

Zara's success highlights the decisive importance of high degree vertical integration and just-in-time approach to maintain tight control over the different stages of its supply chain for low storage levels and flexibility to speed response to customers and current fashion trends. We distinguish functional, geographical and informational integrations [7], contributing to the synchronization of all the layers with real-life visibility over on the shelf and online stores stocks [8], [9]. Zara's ownership of its supply-chain steps displayed in figure 3, allows for more rapid product turnover and full capability to produce and deliver in less than three weeks worldwide, empowered by eliminating limitations and not having risks. Controlled by a centralized command server, Zara also counts on its well-elaborated ICT system using essentially the RFID for identification and tracking, in addition to DBMS and ERP that enable both inter-functional and organization's control, and improve planning at different levels [1]. Thanks to just-in-time strategy installed by Toyota along with the elaborated inventory model detailed in figure 4, the company schedules materials, goods and labor to arrive, then replenishes them when needed, avoiding idle-time loss and increasing benefits. In typical seasons, it produces 11,000 different items.

2.2. Diversification, externalization and low price sourcing

Zara is a brand that sells myriad types of fashion: basic, fast and seasonal. Besides textile, it extends its activities to housing equipment and other accessories. Concentrating on core functions, it externalizes high cost and workers' consuming activities like merchandises' transport, loading from warehouses and unloading to stores to less expensive specialized logistics providers. However, for basic, easy to forecast products, Zara uses cheaper suppliers with longer lead times.

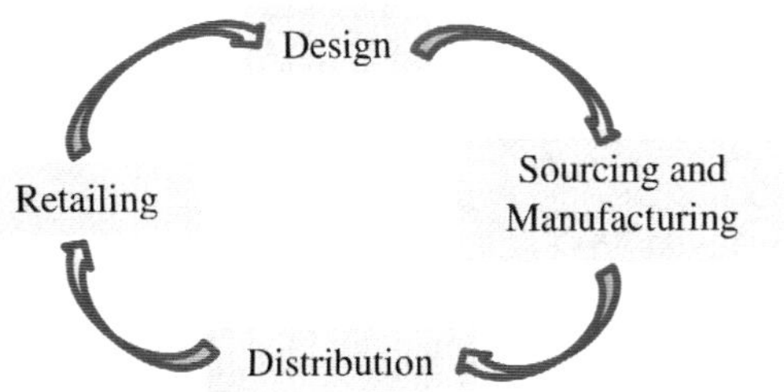

Figure 3. Zara Business Model [5].

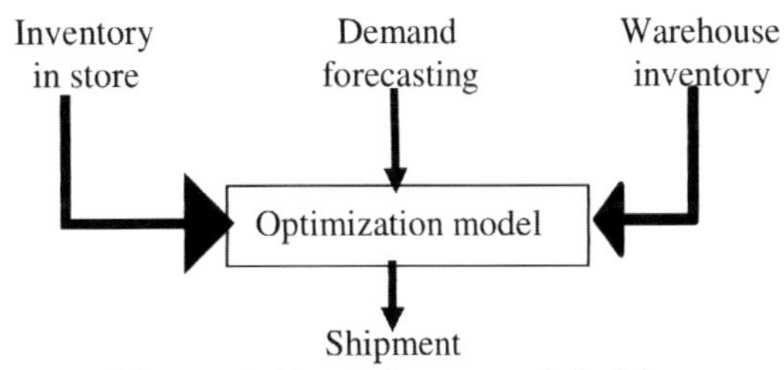

Figure 4. Zara's Inventory Model.

As such, while backward integration reduces cost and improves quality, forward integration reduces the bullwhip effect avoiding costly overproduction and asking scientists for uncertainty mitigation methods where demand forecasting probes significant impact [5].

3. Supply chain optimization and uncertainty mitigation

The huge rise of Zara relies predominantly on the use of sophisticated operations research tools and algorithms, in partnership with eminent scientists from cutting edge MIT's operations research department and London Business School a few years ago. Hence, the company developed many optimization models alongside the global supply chain, but the most famous and impactful ones in terms of competitive advantage extension, are those dedicated to optimize the allocation of inventory across the two warehouses and the store network to increase sales, summed up hereafter in table 1.

It consists of inventory distribution and clearance pricing optimization on one hand, and uncertainty mitigation through demand forecasting on the other. Developed models propose for leaders how to determine key drivers and decision variables like shipment and storage quantities, markdown quantities, retail size-assortment or pricing amount of garments, subject to the general prevailing constraint of customers' satisfaction. Pricing models help commercials find out if a manager is making mistakes [4]. Fast-paced clearance sales and pricing decisions apply to more than 11,000 different fashion designs introduced each year by Zara, considering the 70 countries' different market conditions.

The proposed project consists in the development of a pricing mathematical model based on sales forecasting that estimates consumer's reactions to price discounts and a linear optimization model that paves the way for profit-maximizing optimal price assignments.

Table 1. Zara's Optimization and Uncertainty Mitigation Models

Zara Goal / Developed Model	Method	Outcomes / Benefits
- **Demand Forecasting and Inventory Distribution Optimization.** Developed distribution management algorithms to increase sales. - Determination of each inventory shipment it sends from its two central warehouses to its 1,500 stores worldwide [10].	- Retail size-assortment view of a store's inventory, linking stock levels/ demand to select store replenishment quantities [11]. - Regression demand forecasting for each product at every store location and auto-correction of store manager predictions.	- Increased sales by 3–4% additional revenues for 2007 and 2008 [12].
- Clearance Pricing Optimization	- Pricing mathematical model based	- Increase of 6% to

Models and reduced markdowns for 110000 garments churned out annually in 96 countries, (less than 15%).	on a sales forecasting that estimates consumers' reactions to price discounts and a linear optimization model leading to profit-maximizing optimal price assignments.	Zara's clearance sales profits in 2009 [13].
- Initial Shipment Demand Forecasting and Distribution Optimization. Best trade-off between large initial shipments to stores/stock at warehouse [14].	- Data-driven model of forecast updating and a dynamic optimization formulation for allocating limited stock by location over time, reducing lost sales once initial ones observed.	- Increase in season sales by 2% and reduction in unsold units by 4% in 2012 [15].

4. Leading by Social Responsibility and Ecomotional Intelligence

4.1. Zara's social responsibility

Inditex-owned Zara confirms its innovative green engagement. Just recently, it announced a courageous transitional 2020-2025 plan regarding eco-friendly fabrics and recycling, that follows the Strategic Plan for a Stable and Sustainable Supply Chain 2014-2018 [16]. Besides organic and recycled, when asked what "Sustainable" means, a manager from Zara explained, '... prioritizing more energy- and water-friendly processes in the growing and manufacturing of these textiles.' [17]. This ambitious plan also includes partnership with MIT scientists to research how to recover fibers from old garments using clean energy. Eco-efficient lighting, internal transport using bikes and electric vehicles, recycled and recovered materials and cardboards, besides eco-stores confirm the innovative green willingness and engagement of Zara, constantly enforced through auditing the conformity to the Code of Conduct.

In 2014, Zara achieved 95% of its sustainability supply chain plan thanks to innovation coupled with improvement, traceability and compliance for individual and common lasting wellbeing, product quality and safety status. The company also promotes its reputation and image by highlighting its citizenship through the application of good governance practices with an INDITEX investment of 48 million $ in education, social well-being and humanitarian programs, along with good workers conditions. As such, the 700 designers, alongside the 170,000-plus people working at the Company, produce quality fashion using their creative passion and put it at customers' fingertips. It explains how Zara achieved the record performance of 26.15 billion euros in net sales in 2018, and why it continues to share that growth with its employees.

4.2. Ecomotional intelligence and 'glocal strategy'

Zara's correct prediction of buying behaviors and trends' analysis helps in taking quick decisions and inventory control. Instead of setting the trends, Zara follows them as it combines fashion reconnaissance, spotting trends everywhere from the street to movies to couture fashion shows, and information from its customers to keep its merchandise fresh. In addition, it follows 'glocal strategy,' to think globally but act locally dealing with cultural awareness to adapt worldwide-distributed clothes to local customers' traditions. It also influences clients through their open pleasant architecture.

Among Zara's key success factors, our analysis of its marketing strategy shows that its professionals apply both 'economic intelligence' and 'emotional intelligence' that we can merge in the neologism we have just coined 'ecomotional intelligence.' It helps guess and communicate the right customer's needs, along with new fashion trends of competitors and attractiveness. The idea to propose this new concept began to germinate when we have observed that many companies tend increasingly to combine almost necessarily the two techniques in their marketing strategy and business models, to maximize customers' satisfaction and minimize lost sales, seeking to last their competitive advantage. By the way, designers pay close watch on trends and buying behaviors through:

- Markets research on competitors' stores and fashion shows through economic intelligence ;
- Investigations on places with customers' presence as university campuses and discos ;

- Other feedbacks coming from internal stores through sales report and qualitative comments of customers immediately transmitted to designers.

4.3. Customer's loyalty versus unconventional booming low-budget advertising
The Company publishes two catalogs twice a year, along with of course attractive and familiar online stores. Zara managers reserve small budget for advertisement but it is effective through its aggressive presence near competitors in big places and famous streets of grand cities. The short shelf life for trendy clothes creates psychological attraction on customers [16]. They usually visit Zara stores more than 12 times a year fearing of they would not find them anymore on the shelves, compared with its competitors by 5 to 7 times a year. Hence, the policy of customer loyalty of client portfolio is in full evolution as a heavy and satisfied customer you keep is ten times less expensive than acquiring a new one. Furthermore, collaborators in Zara stores show interest to customers' wants and preferences, and transmit them immediately to warehouses.

5. Conclusion and future research
Fast With its accelerated growth attaining more than $25 billion annual sales in 2018 [16], we have seen through the critical analysis how Zara that pioneered and developed fast-fashion art, insisted to seek innovation and continuous improvement in its operations and business model in order to maintain the uniqueness of its huge competitive advantage over time. Most of analysts sum up the secrets behind its success within the efforts and investments led in the human, structural and relational capitals, as illustrated in table 2.

Table 2. Secrets behind Zara's Huge Success.

Human Capital	Structural Capital	Relational Capital
Development of analytical tools in partnership of cutting-edge universities scientists resulting in innovative demand forecasting and optimization models.	Vertical integration and just in time strategy throughout global supply chain, besides externalization and diversification.	Leading by ecomotional intelligence, customer loyalty and corporate social policy.

We find out that Zara's production system follows a pull system to control alignment of demand to customers' expectations. Besides elaborated optimization models, vertically integrated supply chain is the keystone of its success as it contributes to total control of the production process and needed flexibility to be able to react quickly to the market trends, so that it is extremely difficult to benchmark them by competitors, called fast-fashion killers.

Finally, many scientists and fashion decision-makers remain interested in how Zara will extend its competitive advantage in a market increasingly immersed in digital revolution, and willingness for best trade-off between sustainability and design.

References
[1] Industria de Diseno Textil INDITEX S.A 2019 Industry of textile design 2017-2018 *Inditex Annual Report* (Coruna: INDITEX S.A) from https://www.inditex.com/investors/investor-relations/annual-reports
[2] Costa L 2015 Inditex abre su primer outlet en Coruna e integra a 9 empleados con discapacidad *La Opinión de la Coruña* (Septienbre 2015) 11 from https://www.laopinioncoruna.es/economia/ 2015/09/11/inditex-abre-primer-outlet-coruna/993716.html
[3] Witter M S D 2000 *21st Century* strategy *Speciality Clothing Report*, London and New York
[4] Stevens G C 1989 Integrating the supply chain *Int. J. Physical Distribution and Materials Management* **19**(8) pp 3-8
[5] Ferdows K, Lewis M and Machuca J 2004 Rapid fire fulfillment *Harvard Business Review* **82**(11)
[6] Bean J C 2011 *Engaging Ideas: The Professor's guide to integrating writing, critical thinking and active learning in the classroom*, San Francisco: Jossey-Bass, pp 10-55
[7] Stadtler H and Kilger C 2010 *Supply Chain Management and Advanced Planning: Concepts,*

Models, Software, and Case Studies 4th ed, Berlin: Springer Publishing Company, pp 37-68

[8] Bhatnagar R, Chandra P and Goyal S K 1993 Models for multi-plant coordination *European Journal of Operational Research* **67**(2) pp 141-160

[9] Chen F, Federgruen A and Zheng Y S 2001 Coordination mechanisms for distribution system with one supplier and multiple retailers, *Management Sci.* **47**(5), pp 693-708

[10] Gallego G and Özer O 2001 Integrating replenishment decisions with advance demand , *Management Sci* **47**(10), pp 1344–60

[11] Fisher M and Raman A 1996 Reducing the cost of demand uncertainty through accurate response to early sales, *Operations Research* **44**(1) pp 87-99

[12] Caro F, Gallien J, Diaz M, et al. 2010 Zara uses operations research to reengineer its global distribution process (Recipient of the Franz Edelman Award) *Interfaces* **40**(1) pp 71-84

[13] Carboni Borrasé R 2009 *Clearance pricing optimization at Zara* MBA and MS thesis Massachusetts Institute of Technology (Cambridge: MIT)

[14] Garro A 2011 *New product demand forecasting and distribution optimization: A case study at Zara,* MBA and MS thesis Massachusetts Institute of Technology (Cambridge: MIT) pp 14-72

[15] Gallien J, Mersereau A J, Garro A, Mora A D and Vidal, M. N 2015 Initial shipment decisions for new products at Zara *Operations Research* **63**(2) pp 269-286

[16] Industria de Diseno Textil INDITEX S.A 2019 Strategic Plan for a Stable and Sustainable Supply Chain 2014-2018 *Inditex Annual Report* (Coruna: INDITEX S.A) in https://www.inditex.com/documents/10279/619384/Inditex+Annual+Report+2018.pdf/25145dd4-74db-2355-03f3-a3b86bc980a7

[17] Segran E 2019 Zara built a $20B empire on fast fashion. Now it needs to slow down, *Fast Company* **7** (July 2019) 90379824 ed Mehta S (New York: Fast Company, Inc)

IOP Conf. Series: Materials Science and Engineering **827** (2020) 012059 doi:10.1088/1757-899X/827/1/012059

Dynamic colour in textiles: combination of thermo, photo and hydrochromic pigments

I Cabral[1,2], **A P Souto**[1]

[1] 2C2T – Centro de Ciência e Tecnologia Têxtil, University of Minho, School of Engineering, Campus Azurém 4800-058 Guimarães, Portugal
[2] Author to whom any correspondence should be addressed

diascabral@gmail.com

Abstract. Colour Change Materials are a class of smart materials that present distinct characteristics relative to conventional pigments and dyes. When exposed to a given stimulus, these materials exhibit reversible chromatic changes, allowing the introduction of dynamic and interactive qualities to textiles. Currently, the combination of various chromic materials presents an under-developed area. The objective of this research is to study the integration and combination processes of different types of dynamic colourants in textiles, namely thermo, photo and hydrochromic pigments. The experimental work conducted demonstrates the effect of screen printing variables in textile chromatic qualities, being the variables set: screen mesh count, magnetic field of the screen printing table, printed layers (number and process) and pigments' combination method – overprinting and combination in the screen printing paste. The results attained also highlight design possibilities to further research colour change effects and outline a methodology to explore and design screen printed textiles with multi-sensitive qualities.

1. Introduction

The emergence of Colour Change Materials (CCMs) has enabled the introduction of dynamic chromatic qualities to textiles by sensing and reacting reversibly in response to an external stimulus. This chromic behaviour is based on the variation of the substances microstructure or electronic state, affecting their optical characteristics; absorptance, reflectance, scattering or transmittance [1-2].

CCMs are denominated according to the stimulus that triggers the changes, also presenting different properties and behaviour [3].

Thermochromic (TC) leuco dyes change colour upon heat. They fade away above their activation temperature and return to the predefined colour below it. TC pigments can be combined with conventional pigments, according to the defined binder type. In this case, the effect of thermal variation reaches variations between colours [4, 5].

Photochromic (PC) pigments react to Ultraviolet (UV) radiation, being colourless in the stimulus absence and acquiring their predefined colour when exposed to it. Transition between colours can be also attained through combination of PC and conventional pigments [6].

The presence of moisture or water changes the optical characteristics of hydrochromic (HC) pigments. These materials are usually white and opaque in dry conditions, changing to transparent when exposed to water [3-7].

Currently, limited research has focused on systematic methods to develop colour changing textiles and the combination of various chromic materials presents an under-developed area. This research explores colour as a dynamic variable of textile design through three materials' types: TC, PC and HC pigments. The main objectives are to study the integration and combination processes of these colourants in textiles and to develop a methodology to create chromic textiles with multi-sensitive qualities.

2. Materials and methods

The chromic pigments handled in this research were supplied by SFXC: water based TC dispersion black, blue, magenta and yellow, with 31°C activation temperature; water based PC ready formulated ink blue, magenta and yellow; water based HC ready formulated ink white. Conventional pigments were ATUSMIC Magnaprint yellow HG, pink H5B, blue HG and black H3B. Screen printing pastes were formulated with Gilaba vinyl acrylic binder.

Samples were screen printed on a Zimmer Mini MDF R541 table on a cotton substrate. The selection of the table's magnetic field level defines the pressure that the metallic rod-squeegee applies during paste application on the substrate. The levels tested were 1; 3 and 6 from a range of 1 to 6 available, low to higher pressure. The diameter of the rod-squeegees used was 12 and 6 mm and the mesh count of the screens were 46; 89 and 107 TPI.

After being screen printed, each sample completed a process of drying and thermo setting in a Werner Mathis AG laboratory oven at 160°C during 2 minutes for TC; 130°C during 3 minutes for PC and HC pigments.

Based on experimental practices in textile engineering and design, the experimental work consisted on three phases focused on a) individual use of each chromic material; b) combination of each stimulus-sensitive colourant with conventional pigments; c) combination of chromic materials.

The first phase studied the effect of printing parameters in the chromatic qualities of TC, PC and HC textiles. The variables set were screen mesh count, magnetic field level and printed layers (number and process). Considering that the HC colour was commercially available in white, HC samples were developed with the substrate previously screen printed with 3% conventional pigment black.

The second phase studied the combination of each CCM with conventional pigments, applied to textile substrate by different processes: overprinting of conventional and chromic pigments in different pastes and screen printing with the pigments combined in the same paste.

The third phase explored the possibility to combine different typologies of chromic materials, mixed in the screen printing paste or screen printed with overlapped layers. The combinations studied were TC & PC; TC & HC; PC & HC and TC, PC & HC.

Samples analysis encompassed qualitative and quantitative methods through laboratory studies, namely colour measurement using a spectrophotometer Datacolor International SF600 Plus – CT with Datacolor TOOLS software, direct observation, photographic and video record.

3. Results and discussion

3.1. Individual use of each chromic material

The analysis of printing parameters' influence on chromatic qualities of each pigment type was conducted with a set of sample frameworks. Screen printed samples were developed with the 12 mm rod and encompassed a relationship between screen mesh counts – 46, 89 and 107 TPI – and magnetic field levels (Mf) – 1; 3 and 6. A framework of samples screen printed with each pigment type is presented in figure 1 with samples in colourized state: TC blue below 31°C, PC blue under UV radiation and HC in dried state.

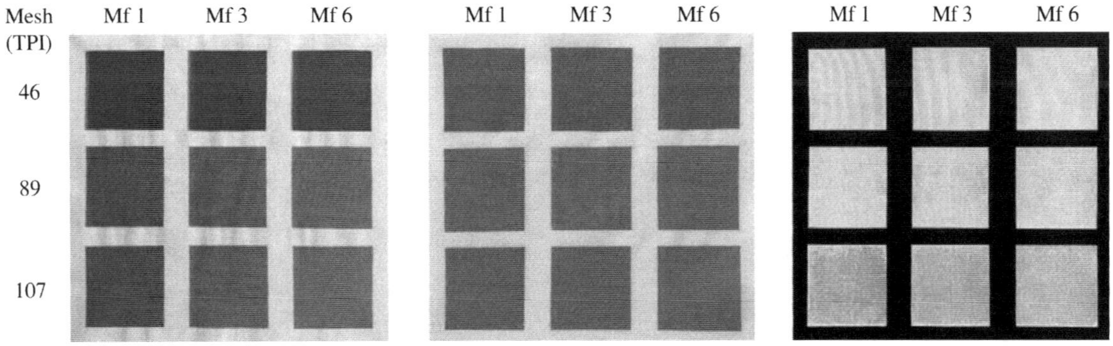

Figure 1. Samples framework: screen mesh count vs. magnetic field level. Left: TC blue samples below 31ºC. Centre: PC blue samples under UV radiation. Right: HC samples in dried state.

For the three chromic pigments handled, screen mesh count variable showed a greater effect on textile printed results than the magnetic field level. Screens with lower mesh count present larger open areas, allowing more paste to flow onto the substrate and with TC and PC pigments, samples developed with 46 TPI screen are darker than other samples, but they present uneven printing. This effect suggests an excessive paste layer thickness printed. In samples developed with higher mesh counts, the printing was more uniform, although a subtle moiré pattern was perceived. An additional experiment was conducted with a lower diameter rod – 6 mm – that reduced this effect.

HC samples screen printed with the 46 TPI screen present a thicker layer in relation to samples developed with higher mesh counts, which was obviously perceived through the white layer coverage level above the black background and through touch. Mean values of samples thickness varied from 0.45, 0.35 and 0.33 mm for 46, 89 and 107 TPI screen, respectively. Besides colour qualities, the printing can change textile characteristics, such as thickness, weight, stiffness and handle, becoming an important design consideration as these variables also build up the textile expression(s).

The change of magnetic field levels produced slight variations between the samples' qualities. TC and PC textiles screen printed with the lowest pressure (Mf 1) have a subtle darker nuance than samples with higher levels, whereas for HC a more uniform printing was attained with Mf 6.

To study the variable printed layers, number and process, a framework for each chromic material and colours was developed with samples screen printed with one layer (1L), one layer applied on top of a previously printed and dried layer (1+1L) and two consecutive rod-squeegee passages of printed layers (2L). Figure 2 presents an example of the results, for each pigment type.

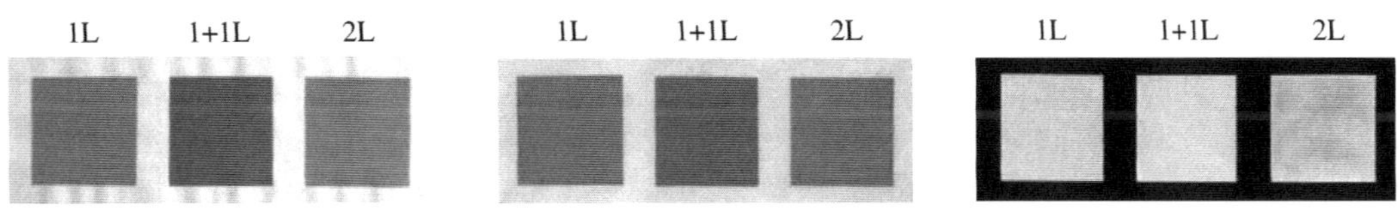

Figure 2. Samples screen printed with different layer number and process. Left: TC blue samples below 31ºC. Centre: PC blue samples under UV radiation. Right: HC samples in dried state.

Number and process of the printed layers showed a similar effect along the chromic materials tested. In relation to one layer application (1L), the overprinting of a second layer (1+1L) increases colour saturation and thickness, also contributing to the evenness of the printed surface. Samples with two layers consecutively applied (2L) attained a blurred effect with areas of different colourization.

When chromic pigments are in colourless state, printing parameters were not expected to influence textile colours. Nevertheless, an incomplete decolourization was observed in samples with all pigments used and thus, the subtle colours or opacity can also present slight differences, particularly comparing samples with the smaller screen mesh count and overprinted layers, as well as samples of different colours or large concentration ranges. Colour difference (dE*) measured of HC samples wet in relation to the black background were 26,20 for 1L and 33,55 for 1+1L (samples developed with 89 TPI screen and Mf6). At a temperature above 31°C, the sample screen printed with 2% TC blue presented 10,48 dE* in relation to 10% TC blue (samples developed with 107 TPI screen and Mf3). The residual colouration or opacity of chromic pigments can affect how colours or colour change ratios with pigments combination will behave, being an important variable to assess.

3.2. Chromic materials and conventional pigments

Combinations of chromic materials with conventional pigments were studied through two processes: overprinting and pigments mixed in the paste. Samples developed by overprinting present the static pigment at the background and different colours at various concentrations were tested. A relationship between these parameters was applied for paste elaboration with mixed pigments. For example, if a sample was developed with 1% conventional pigment magenta at the background and overprinted with 100% HC, the corresponding samples' paste has 1% conventional pigment magenta with 99% HC.

Figure 3 presents an example of each chromic set, developed by different processes.

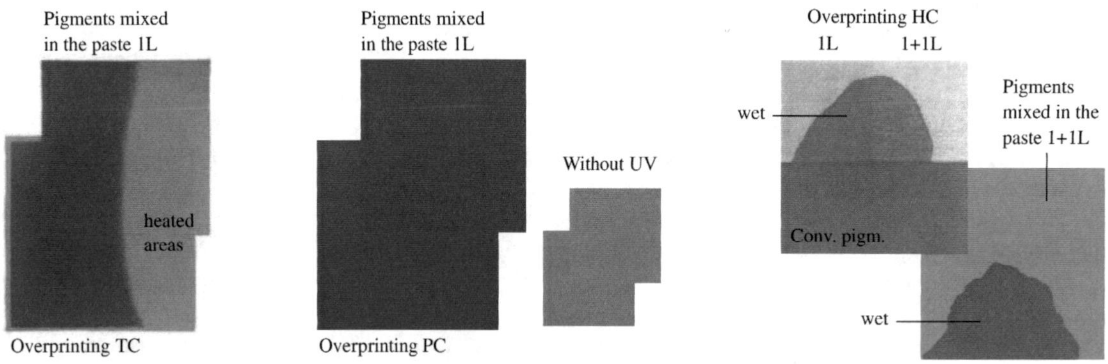

Figure 3. Combination of chromic and conventional pigments through different processes.

Considering the relationship between dynamic and static colourants, colour change behaviour can encompass a transition between colours and tones. Besides colours and concentration of the pigments combined, application processes can also influence textile chromatic behaviour. TC and PC examples presented in figure 3 attained a slightly darker colour in samples where pigments were combined in the paste than by overprinting. Although, by decreasing the conventional pigment concentration, the colour of the chromic pigment was more apparent in samples developed by overprinting. HC pigment is white and also displays a lighter surface through overprinting process.

Furthermore, when the conventional pigment selected presents a dark colour or is applied in high concentration, TC and PC colour change effect is subtle or not perceived. The same occurs for HC pigment when combined with lighter colour or low concentration of the static colourant.

Definition of screen printing paste colours, concentrations and application processes involves a set of relationships. The results attained also highlight the importance of a systematic practice-based approach to explore these materials' potential for designing colour change effects.

3.3. Combination of chromic materials

The results discussed in the previous sections set a basis to study the combination of different chromic materials types, which initially focused on two pigment types at a time, followed by an experiment with the three chromic materials.

When combining different CCMs, colour change effect of each chromic pigment type is not just influenced by the stimulus that it reacts to, but also by the stimuli of other pigments as they also influence its colour. In this sense, the relationship between several stimuli is an important parameter in textile colour change behaviour. For example, if a TC blue with 31°C activation temperature is combined with a PC yellow, below 31°C the textile is green when exposed to UV radiation (also depending on concentrations applied) and blue without UV; above 31°C the textile colour is yellow with UV radiation and colourless without.

In addition, when mixing chromic materials in the paste, printing parameters such as screen mesh count and magnetic field, are defined for an individual paste with different pigment types mixed. In overprinting processes, each pigment type can be screen printed with selected printing variables for each layer, using for example a different screen mesh count for each chromic paste.

The combination of three pigment types was researched with pigments mixed in the paste and overprinting process. The pigments selected for the study were 10% TC yellow, PC blue and HC. Sample 1 was screen printed with pigments combined in the paste. Through overprinting, sequence order of printed layers encompasses six options. For the following analysis, just four samples were considered (samples 2 to 5), excluding the two samples with the HC background. Figure 4 presents an image of each sample at a selection of different stimuli conditions.

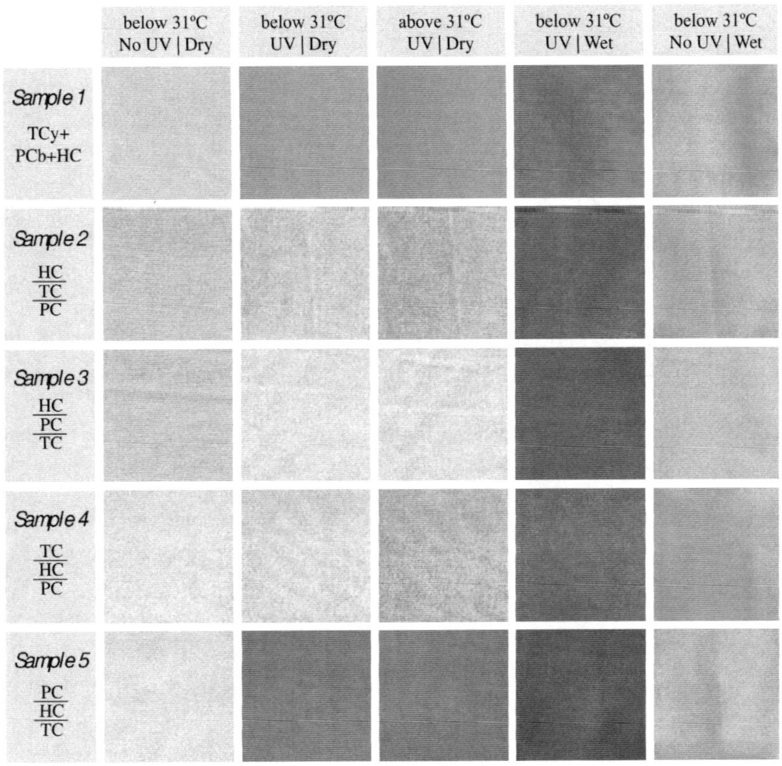

Figure 4. Framework of chromic samples that combine the same pigments by different processes. Each line presents the sample description and an image at a specific stimuli condition, as identified at the framework top.

Considering the intrinsic dynamic behaviour of the pigments researched, screen printing decisions play a crucial role in colour change effects. Samples presented in figure 4 applied the same pigment types vs. colours and similar concentrations between mixture in the paste and overprinting, although their chromatic behaviour at each stimuli condition tested have varied significantly, as can be analysed through samples' colours at each framework column.

When HC pigment is printed at the top layer (Samples 2 & 3), its opacity in colourized state does not allow an evident perception of changes between hues of the pigments below it. The effect observed relies on a variation between colour nuances. By mixing the HC in the paste, colour changes can be observed in HC dry state, varying between yellow, green and blue.

Layer order sequence of HC pigment also affects how the textile dries and becomes wet, particularly when PC layer is above the HC, when it was perceived that sample 5 did not become wet as fast as the other samples.

TC yellow concentration applied displays a light colour. In this sense, when PC is at decolourized state, the change between TC and HC states does not create an evident change between yellow and white hues. The pigments' incomplete decolourization discussed also contribute for this effect.

PC pigment selected has a dark colour and the transition between UV absence and exposure displayed a clear influence on textile chromatics. During experimentation, PC sensibility to heat was observed. When heating up the textile, PC colour saturation has decreased but returned to colourized while TC pigment was still colourless. This behaviour will be further researched.

In addition, chromic materials studied present gradual colour transitions at different rates in colourizing and decolourizing processes, according to each stimulus-sensitive type. According to the work developed, chromic textiles may exhibit different chromatic and dynamic qualities in response to the external stimuli combination and printing parameters.

4. Conclusion

This research studied processes to combine and integrate chromic pigments in textiles, exploring colour as a dynamic variable of textile design.

The results demonstrate the interdependency of textile colours and colour change behaviour in relation to pigment types, colours, concentrations, fabrication processes and stimuli parameters.

When working with chromic materials, decisions on pigments' combination and printing processes through paste mixture or overprinting are critical, as they significantly affect textile expressions and performance. Printing processes present designers with enhanced opportunities to explore colour change effects, namely to work with textile patterns, where each pigment type printed can add to more than a colour change effect, interacting with other pigments in the composition.

The systematic approach conducted to study the colour change effects through independent variables creates an understanding of TC, PC and HC behaviour and proposes a design methodology to explore and create interactive textile surfaces capable of displaying a wide range of chromatic changes, under diverse external stimuli.

The samples framework developed demonstrates colour change effects and can be used to assist design decisions when creating chromic textiles for expressive and functional purposes.

Acknowledgements

This work is financed by Project UID/CTM/00264/2019 of 2C2T – Centro de Ciência e Tecnologia Têxtil, funded by National Founds through FCT/MCTES.

References

[1] Talvenmaa P 2006, Introduction to chromic materials *Intelligent Textiles and Clothing* H Mattila, Cambridge

[2] Bamfield P and Hutchings M G 2018 *Chromic phenomena: technological applications of colour chemistry,* Cambridge: Royal Society of Chemistry

[3] Ferrara M and Bengisu M 2014 Materials that change color: smart materials, intelligent design, Springer, Cham

[4] Christie R M 2013 Chromic materials for technical applications *Advances in the dyeing and finishing of technical textiles* ed M L Gulrajani, Oxford: Woodhead Publishing

[5] Cabral I and Souto A P 2014 Thermochromic textile structures: a dynamic ambient light design *International Journal of Designed Objects* **6**(2), pp 55-64

[6] Viková M 2019 Type of chromic materials *Chromic Materials: Fundamentals, Measurements and Applications* ed M Viková, Waretown: Apple Academic Press

[7] Merati A 2018 Application of Stimuli-Sensitive Materials in Smart Textiles *Advanced Textile Engineering Materials* eds Shahid-ul-Islam and B S Butola, Beverly: Scrivener Publishing

Research regarding electromagnetic shielding achieved by the fabrics support

L Surdu[1,2], R M Aileni[1], R I Radulescu[1], L Chiriac[1]

[1] National Research & Development Institute for Textiles and Leather, Romania
[2] Author to whom any correspondence should be addressed

lilioara.surdu@yahoo.fr

Abstract. Electromagnetic shielding achieved by flexible materials has gained an increased significance in today's radiation polluted environment. One of the modern approaches for obtaining electromagnetic textile shields with good surface electrical conductivity is the deposition of metallic thin films onto the fabrics. The use of the magnetron sputtering technique ensures some advantages, e.g. improved reflections of the fabric once metallic coatings are deposited on both sides of the fabric, as well as lightweight and excellent flexibility due to nanoscale thickness of layers. In the present work, cotton and polyester fabrics with various structural parameters, such as yarns count, fabric density, and specific mass were used as samples to be coated. PVD magnetron sputtering technique was utilized for the deposition of metallic layers with various thicknesses, either on one side or on both sides of the fabrics, in order to ensure good conductivity of the fabric. Besides, fabric samples with inserted conductive yarns of silver and stainless steel were considered as substrates, as well. The electromagnetic shielding tests using TEM cell evidenced good shielding effectiveness of 10-34 dB within 0.1-1000 MHz.

1. Introduction

Human health and wellbeing are affected by non-ionizing radiation of electromagnetic waves [1, 2], such as the radiation of mobile GSM communication or internet, WiFi networks [3]. Electromagnetic waves cause the heating of cellular tissues with adverse long-term effects [5]. Moreover, the functioning of electronic devices may interfere with electromagnetic waves from various sources (such as power transmission lines, telecommunication, broadcasting), causing ElectroMagnetic Interference (EMI) [4].

This problem requires adequate solutions. The shielding of electromagnetic waves belongs to ElectroMagnetic Compatibility (EMC) discipline. EMC's main aim is to ensure the proper functioning of electronic devices and integration with other electronic components [6]. Thus, in order to avoid interference with other electronic devices functioning at various frequencies, a rigorous set of measures has to be followed within design of electronic devices. Shielding materials are based on good electrical conductivity, being conventionally made out of metallic plates and out of fabrics with metallic insertion or coatings [4]. When compared to metallic plates, electrically conductive fabrics present some advantages, such as flexibility, lightweight, mechanical resistance, and shape-ability, well described in the scientific literature [6]. Magnetron plasma is a novel technology for coating fabrics with metallic layers on nanometre scale [7, 8], while inserting conductive yarns into a fabric structure is already a conventional technology. The main aim of this paper is a comparative study on manufacturing methods for electromagnetic shields out of fabrics, based on coating using magnetron sputtering (PVD in argon) and inserted conductive yarns. The paper focuses mainly on both manufacturing methods and the evaluation result based on the shielding effectiveness of the fabrics.

2. Experimental part and Methods

In our experimental part, we used three samples made from 100% polyester with mass 210g/m^2 that were obtained by coating with successive metallic layers using a magnetron sputtering equipment. For coating, titanium (Ti) was used for the first sample (p1), nickel (Ni) for the second one (p2), and the third sample (p3) chromium (Cr) was used. Besides, we obtained two woven structures based on stainless steel yarns (p4 Bekinox) and polyamide yarns coated with silver (p5 Ag) from STATEX that were coated with a thin copper layer by PVD magnetron sputtering.

For samples coated by magnetron sputtering, physico-mechanical tests in INCDTP laboratories were performed and is presented in Table 1.

Table 1. Values for physico-mechanical tests - samples P1-P3.

No. art.	Tests		Results obtained			UM	Reference documents
			P1	P2	P3		
1	**Mass per unit area**		213	236	210	g/m^2	SR EN 12127:2003
2	**Density (yarns number per length unit)**	U	320	378	382	no. yarns/ 10 cm	SR EN 1049-2:2000 – Metoda A,B
		B	160	220	220		
3	**The maximum force of breaking through the procedure with the sample type tape**	U	888	2140	2328	N	SR EN ISO 13934-1/2013
		B	380	551	1274		
	Elongation at maximum force of breaking through the procedure with the sample type tape	U	7,76	38,6	38,3	%	SR EN ISO 13934-1/2013
		B	11,94	11,61	36,5		
4	**Tearing force**	U	18,43	12,24	89,1	N	SR EN ISO 13937-3:2002
		B	13,55	12,04	58,2		
5	**Thickness**		0,531	,437	0,342	mm	SR EN ISO 5084:2001
6	**Water vapor permeability**		31,0	30,6	30,9	%	STAS 9005:1979
7	**Air permeability**	268,2	98,15	76,86		l/m^2/sec	SR EN ISO 9237:1999

For samples obtained by weaving technology with conductive yarns (stainless steel, respective Ag) followed by a thin layer copper deposition by PVD magnetron sputtering, tests presented in Table 2 were performed in our laboratories.

Table 2. Physico-mechanical properties for samples P4 and P5.

Tests		P4	P5
Electrical resistance (Ohm / m)	**Measured**	2200	220

Relative magnetic permeability	**Calculated**	40	1
Length density (dtex)	**Specification**	400 dtex	296 dtex
	Measured	200,5x2 dtex	148x2 dtex
Yarn structure		Spun yarn, 80% cotton, and 20% inox	PA6.6 coated with Ag
Apparent diameter (µm)		273 / 54,6	228
Total yarn twist (twist/m)		593,6	-
TorsionTors/m)		827,6	-
Breaking tenacity (N/tex)		0,1365	0,4564
Breaking force (N)		5,46	13,69
Relative elongation (%)		5,8	27,36

To test the electromagnetic effectiveness, according to standards: ASTM ES7-83, ASTM D4935-89, and IEEE Std 299-2006, all samples (P1-P5) were cut in the form of a sword with outer diameter 100 mm and inner diameter 30 mm. After this stage, all samples were covered on the edges with a conductive layer based on Ag and were mounted inside the measuring cell used. For this experiment an Osciloscop Tektronix MDO 3102, a coaxial cell TEM 2000, a power amplifier SMX5 and signal generator E8257D were used.

Electromagnetic effectiveness shielding (SE [dB]) is the difference between the signal measured without sample and the level of the signal measured with sample mounted inside the measuring cell used. The results of the SE [dB] tests, according to frequency, are presented in table 3.

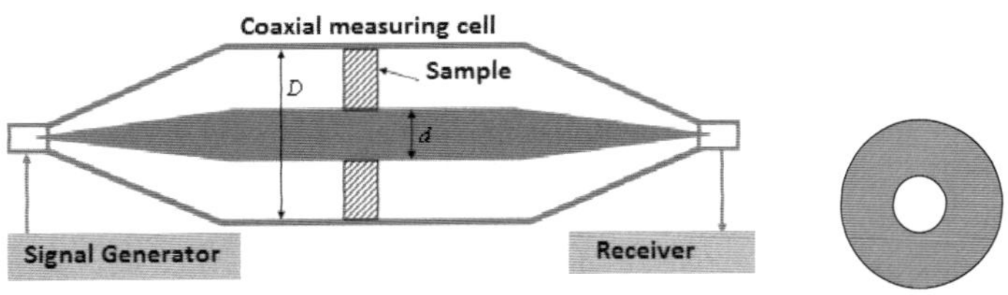

Figure 1. Schematic representation of the sample and coaxial measuring cell TEM 2000.

Table 3. Electromagnetic effectiveness shielding values.

Frequency (f [MHz])	P1	P2	P3	P4	P5
0.1	0	0.3	0	9.9	33.9
0.2	0.3	0.4	0.1	9.4	33.7
0.3	0.2	0.4	0.1	9.9	33.6
0.5	0.2	0.4	0.1	10	34.5
1	0.2	0.3	0.3	10.3	34.3
5	0.3	0	0	10	33.1
10	0.2	0	0	7.7	29.1
50	0.1	0	0	18.2	34
100	0	0	0.1	19.7	28.9
200	0.2	0	0	21.4	22.6
300	0.1	0.2	0.2	18.8	14.9
400	0.7	1.1	1.1	21.7	13.4
500	1	0.9	1.3	23.7	14.6
600	0.7	0.5	0.8	20.1	11.1
700	1.4	1.6	1.6	19.6	7
800	1.6	1.8	2	25.7	9.9
900	0.9	1	1.2	25.8	8.1
1000	1.8	2.3	1.9	24	3.6

3. Results and discussion

By using the experimental data of the analyses of the electromagnetic shielding effectiveness, a bivariate analysis between the frequency and electromagnetic shielding effectiveness was conducted (figures 2-6). Also, the goodness of fit of the statistical bivariate model that describes how well it fits a set of observations was analyzed by the R^2 (R-squared), SSE (sum of squared errors) that is the standard deviation of residuals, Adjusted R-squared, and RMSE (root mean square error) in Table 4.

$$\mathbf{SSE} = \sum_{i-1}^{n} w_i (y_i - \hat{y}_i)^2 \quad (1)$$

Table 4. Goodness of fit.

Sample	R^2	SSE	Adjusted – R^2	RMSE
P1	0.9912	0.04856	0.9516	0.1252
P2	0.9984	0.01305	0.9912	0.06489
P3	0.9961	0.03436	0.9787	0.1053
P4	0.9988	0.3149	0.9942	0.2968
P5	1	0.009361	1	0.05117

In Table 5 presents the conductivity of the metals used for coating the woven structure or yarns.

Table 5. Electroconductivity vs. resistivity of the metals used for coating the samples P1-P5.

Sample	Conductivity (σ [S/m])	Resistivity ([Ω·m])
P1 (Ti)	2.38×10^{6}	4.20×10^{-7}
P2 (Ni)	1.43×10^{7}	6.99×10^{-8}
P3 (Cr)	5.10×10^{6}	1.96×10^{-7}
P4 (Bekinox)	1.00×10^{7}	1.0×10^{-7}
P5 (Ag)	6.30×10^{7}	1.59×10^{-8}

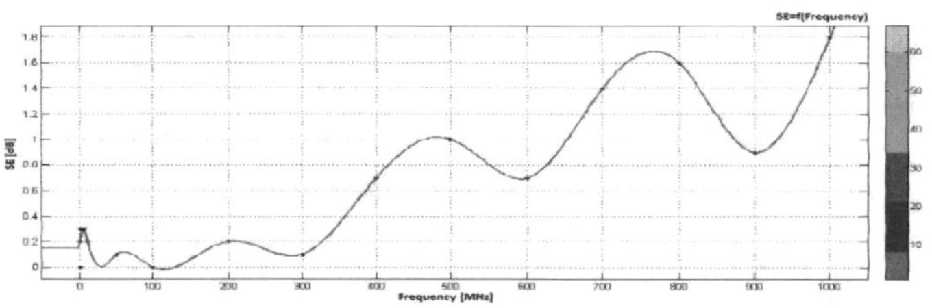

Figure 2. Bivariate analysis SE=f(Frequency) for sample P1.

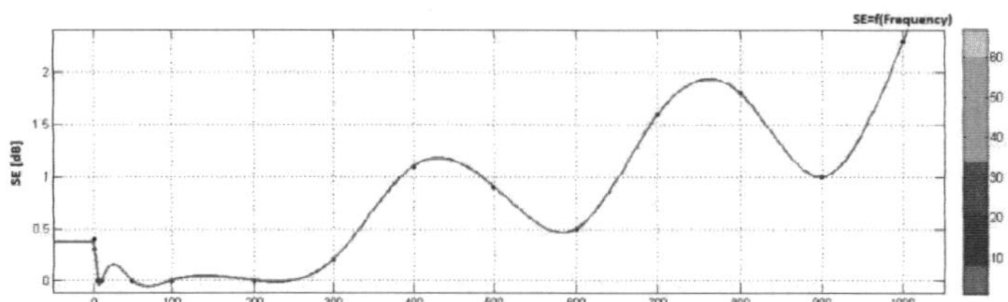

Figure 3. Bivariate analysis SE=f(Frequency) for sample P2.

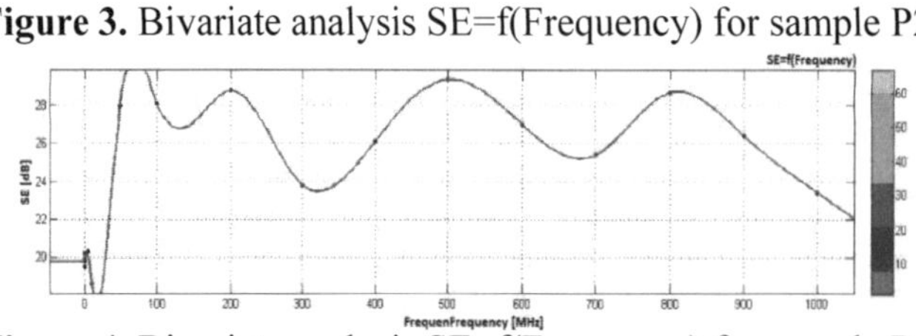

Figure 4. Bivariate analysis SE=f(Frequency) for sample P3.

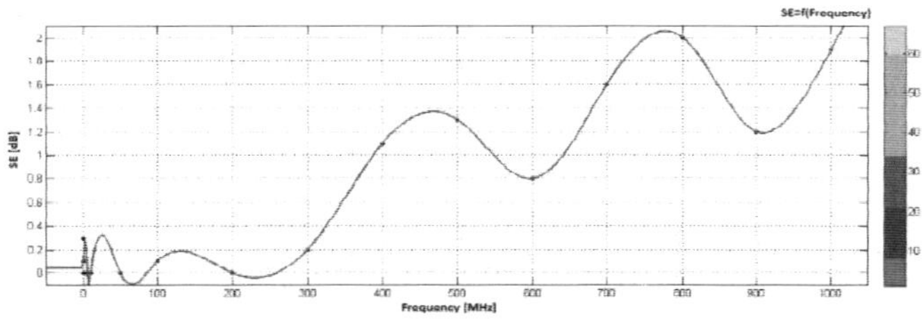

Figure 5. Bivariate analysis SE=f(Frequency) for sample P4.

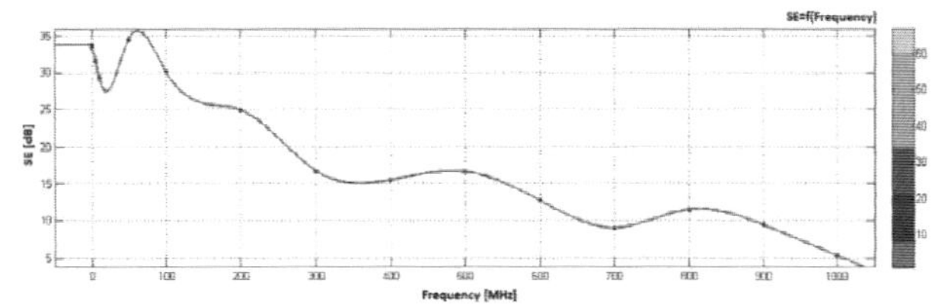

Figure 6. Bivariate analysis SE=f(Frequency) for sample P5.

By analysing the covariation of the electromagnetic shielding effectiveness (SE) and frequency(f) we obtained:

$$\text{cov}\,(f,SE_{P1}) = 1.0e{+}0.5* \begin{vmatrix} \mathbf{1.2534} & \mathbf{0.0018} \\ \mathbf{0.0018} & \mathbf{0.0000} \end{vmatrix} \tag{2}$$

$$\text{cov}\,(f,SE_{P2}) = 1.0e{+}0.5* \begin{vmatrix} \mathbf{1.2534} & \mathbf{0.0021} \\ \mathbf{0.0021} & \mathbf{0.0000} \end{vmatrix} \tag{3}$$

$$\text{cov}\,(f,SE_{P3}) = 1.0e{+}0.5* \begin{vmatrix} \mathbf{1.2534} & \mathbf{0.0024} \\ \mathbf{0.0024} & \mathbf{0.0000} \end{vmatrix} \tag{4}$$

$$\text{cov}\,(f,SE_{P4}) = 1.0e{+}0.5* \begin{vmatrix} \mathbf{1.2534} & \mathbf{0.0191} \\ \mathbf{0.0191} & \mathbf{0.0000} \end{vmatrix} \tag{5}$$

$$\text{cov}\,(f,SE_{P5}) = 1.0e{+}0.5* \begin{vmatrix} \mathbf{1.2534} & \mathbf{-0.0392} \\ \mathbf{-0.0392} & \mathbf{0.0000} \end{vmatrix} \tag{6}$$

By analysing the correlation between frequency and the electromagnetic shielding effectiveness (corr(f,SEp_x) (7)), we obtained the following correlation coeficients corr(f, SEp_1) =0.9013, corr(f, Sep_2)=0.8542, corr(f, Sep_3) =0.9217, corr(f, SEp_4)=0.8368; corr(f, SEp_5)=-0.9552

$$\text{corr}(f,SEp_x) = \boldsymbol{corr\left(f, SE_{p_x}\right)} = \frac{\boldsymbol{cov(f,SE_{P_x})}}{\sqrt{\boldsymbol{var(f)}}\sqrt{\boldsymbol{var(SE_{P_x})}}} \tag{7}$$

The analysis of the correlations and covariation shows that in the case of the sample based on Ag (Silver) and Cr (Chromium) thin layer deposited by PVD magnetron sputtering (p5), the correlation between frequency and shielding effectiveness is negative. This aspect means that the values of the frequency are in inverse proportionality report, and at a small frequency (approx 0.1 MHz), the values of the shielding effectiveness are high, whereas at high frequencies (approx 1GHz) values of the shielding have a decreasing trend.

Besides, there is a positive correlation between the frequency and shielding effectiveness of the samples coated with titanium (p1), chromium (p3), nickel (p2), the sample coated with Cr and based on Bekinox yarns inserted (p4). This means that the values of the shielding effectiveness are in direct proportionality report with the frequencies.

4. Conclusion

Both manufacturing technologies of achieving electric conductivity on fabrics were applied within this research study: insertion of conductive yarns (stainless steel/silver) and coating with metallic nanoscale films by physical vapor deposition (PVD) in argon. The shielding effectiveness tests were conducted according to ASTM ES-07 via TEM cell with values depending on the structure of conductive fabrics.

The results of the samples (100% cotton and 100% PES fabrics) coated by PVD magnetron sputtering in argon had satisfying results only in the case of PES with values of 6-8 dB within 0.1-1000 MHz. The fabrics coated with Cu by PVD magnetron sputtering in argon and based on woven structures with inserted conductive yarns of stainless steel (p4) and silver (p5) had values of 10-20 dB, respectively 30-34 dB in the specified frequency range. For both these fabrics (P4 and P5) by copper thin layer deposition by PVD magnetron sputtering in argon, the shielding effectiveness has been improved with 6-10 dB for fabrics with stainless steel yarns (p4) and only with 2-3 dB for fabrics with Ag yarns (p5). However, the sample p5 had the best overall shielding effectiveness, with 34.3-34.5 dB in the frequency ranges of 0.5-1 MHz.

Each of the two technologies renders surface conductivity and electromagnetic shielding effectiveness to fabrics, and by achieving the application of both technologies on a single fabric, the values of shielding effectiveness are improved.

Acknowledgments

The research presented in this paper was prepared in the INCDTP laboratories. The scientific paper is funded by The scientific paper is funded by ERA-NET Manunet project TexEMFiRe under the contract 28/2018 and Program 1 - Development of the national RD system, Subprogram 1.2 - Institutional Performance - RDI excellence funding projects, Contract no. 6PFE/2018.

References

[1] Schwab A, Kuerner W 2013 *Electromagnetic compatibility, Chapter 6–Electromagnetic shielding* ed AGIR
[2] Jan Z and Maciej J, 2011,EMI shielding using composite materials with plasma layers, *Electromagnetic waves,* p 425
[3] Shahin S, Banerjee S, Singh S P and Chaturvedi C M 2015,*Toxicological Sciences* **148**(2) pp 380-99
[4] Mashevich M, Folkman D, Kesar A, Barbul A, Korenstein R, Jerby E and Avivi L 2003, *Bioelectromagnetics: Journal of the BEMS, SPRBM,* **24**(2) pp 82-90
[5] Paul C R 2006 *Introduction to electromagnetic compatibility* **184** ed John Wiley & Sons
[6] Miclăuş S, Bechet P, Paljanos A, Aron A M, Mihai G, Pătru I and Baltag O 2016 Shielding Effectiveness Of Some Conductive Textiles And Their Capability To Reduce The Mobile Phones Radiation *Int. conf. Knowledge-based organization* Romania, **22**, (3) pp 524-530
[7] Brzeziński S, Rybicki T, Malinowska G, Karbownik I, Rybicki E and Szugajew L 2009, *Fibres & Textiles in Eastern Europe,* **1**(72) pp 60-65
[8] Ziaja J, Koprowska J and Janukiewicz J 2008 Using plasma metallisation for manufacture of textile screens against electromagnetic fields ,*Fibres & Textiles in Eastern Europe,* **16**(5) pp 64-66

Thermal resistance of double jersey fabric knitted by different yarn raw material

D Kopitar[1,2], Z Pavlovic[1], Z Skenderi[1] and Z Vrljicak[1]

[1] University of Zagreb, Faculty of Textile Technology, Department of textile design and management, Prilaz baruna Filipovica 28a, 10000 Zagreb, Croatia
[2] Author to whom any correspondence should be addressed

dragana.kopitar@ttf.hr

Abstract. The research was carried out on five double jersey knitted fabrics, knitted from cotton, Viscose®, Tencel®, Modal® and polyester ring spun yarns finesses of 20 tex. A circular double bed knitting machine gauge of E17 with the same production parameters was used for knitting all the fabrics. A change in all knitted fabric structure parameters was reflected through fabric mass per unit area, which ranged from 142 g m^{-2} to 165 g m^{-2}. The minimum and maximum knitted fabric mass per unit area difference is up to 16 %. The lowest thermal resistance have polyester, while the largest have cotton knitted fabric. Thermal resistance of cotton fabric is higher for 36.7 % related to polyester knitted fabric. Considering basic knitted fabric parameters, knitting with same constructional parameters but different yarn raw materials doesn't provide knitted fabric with same mass per unit area or thickness, i.e. thickness factor. It can be concluded that beside basic knitted fabric parameters, yarn raw material influence on thermal resistance.

1. Introduction

More and more yarns are being made in the world that will replace classic cotton yarns produced by the ring spinning system. For a number of reasons the increasing number of population does not cause a proportional increase in the production of cotton yarns. Therefore, yarns by different raw materials are manufactured which will replace or complement cotton yarns in different areas of application [1]. Investigations on textile properties carried out by physiologists showed that raw material composition does not significantly affect comfort parameters since humans can not feel the difference between different raw material compositions or type of fabric [2]. Contrary to physiologists, researches done by textile experts indicate that there are relevant differences in heat and water vapour resistance of fabrics made by different raw materials [3]. In research where knitted fabrics produced from cotton and angora fibres in different ratios were analysed, it was indicated that 25 % of angora fibre caused a significant difference in relative water vapour permeability values [4]. In another research, untreated fabric made of spun polyester yarn has higher water vapour resistance than the one made of cotton, where after fabric treatment, the decrease in water vapour resistance for polyester fabric is significant, but it is not for cotton fabric [5]. Recent research showed that different types of fibres and yarns are being used to improve the heat and moisture management, and therefore the comfort of the fabric wearable next to the skin [6]. In most of the studies of thermo-physiological properties, synthetic fibres are compared to natural fibres. Fabrics made by different yarn structure or different fabric parameters where researchers did not manage to produce yarns or fabrics that was identical, differ only in one single factor [7-11]. The present research aim is to analyse knitted fabric parameters and influence on thermal resistance of double jersey knitted fabrics. The fabrics were made on the same knitting machine with same production parameters by yarn finesses of 20 tex. Yarns were produced in one factory differing only in raw material, namely cotton, Viscose®, Tencel®, Modal® and polyester.

1.1. Knitted fabrics

A circular double bed knitting machine gauge of E17 was used for knitting the fabrics (Figure 1 and Figure 2). It is used to knit plain or plated double knit jersey fabrics, and due to its construction (Table 1), it is recommended to knit single cotton yarns from 12 to 36 tex or ply yarns from 10 tex x 2 to 17 tex x 2.

This type of machine has 8 knitting systems so it was necessary to prepare 8 individual yarn spools for each yarn group (cotton, Viscose®, Tencel®, Modal® and polyester). The tensile force of the yarn fed to the knitting system was regulated with Coni positive feeders.

Table 1. Construction features of the double bed circular knitting machine.

Machine gauge, E	Cylinder diameter, mm (e")	Number of knitting systems, S	Number of needles, N_i	Cylinder working speed, rpm
17	200 (8")	8	432 x 2	60

Twenty meters of each sample were knitted. The fabric take-down was performed by two pairs of rollers located 70 cm away from the knitting zone. The fabric was not wound onto a fabric roll but it was plaited down on the tray below the take down rollers. The research work was carried out on five double jersey knitted fabrics, knitted from cotton, Viscose®, Tencel®, Modal® and polyester ring spun yarns finesses of 20 tex.

Figure 1. Machine that was used for knitting the samples.

Figure 2. Close up of the double-bed cylinder during the knitting process.

1.2. Knitted fabric structure parameters

Basic structure parameters such as horizontal (D_h) and vertical fabric density (D_v), fabric thickness (t) and mass per unit area (m) are tested. Based on the fabric thickness and mass per unit area, porosity and fabric thickness cover factor were calculated.

From the width of the knitted fabric and number of needles in the cylinder, horizontal fabric density (D_h, number of loops in course direction) can be determined using following equation:

$$D_h = \frac{N_i}{S_p} \tag{1}$$

Where: Dh is the horizontal fabric density in cm-1, Ni is the number of needles, Sp is the width of the fabric in cm.

Vertical fabric density (Dv, number of loops in wale direction) was measured on the sample, counting the loops in one course at a certain length. Thickness (t) was measured using the precision Thickness Gauging testing device model 2000-U, from HESS MBV GmbH, according to the standard DIN EN ISO 5084, where the standard recommend to apply pressure of 1 kPa by a stamp of 50.5 mm diameter.

The most important technological and economical parameter of the knitted fabric structure is the fabric mass per unit area (m, g m-2). It can be determined by weighing the mass of the fabric (mu) and calculating its surface area (Pu) with this equation:

$$m = \frac{m_u}{P_u} \tag{2}$$

Where: m is mass per unit area in g m^{-2}, m_u is the weight of the knitted sample in g, P_u is the surface area in m^2.

Thickness cover factor (T_{cf}) is calculated with the ratio of mass per unit area (m) and fabric thickness(t):

$$T_{cf} = \frac{m}{1000 \cdot t} \tag{3}$$

Where. T_{cf} is thickness cover factor in g m^{-3}, m is mass per unit area in g m^{-2}, and t is fabric thickness in mm.

The porosity of the knitted fabric (P) is analysed as the ratio of the substance mass that makes the fabrics and the yarn volume of the knitwear. For the porosity of knitted fabric, the main data is the average density of fibres (γ) that builds the knitted structure. Porosity was calculated according to the following equation [12]:

$$P = \frac{\gamma - T_{cf}}{\gamma} \cdot 100 \tag{4}$$

Where: P is porosity, %, γ is the density of the fibre in g cm^{-3} where density of the cotton fibre is 1.52 gcm^{-3}, viscose fibre is 1.48 g cm^{-3}, tencel fibre is 1.50 g cm^{-3}, modal fibre is 1.53 gcm^{-3} and polyester fibre is 1.39 g cm^{-3} [13,14], T_{cf} is the thickness cover factor, g m^{-3}.

1.3. Thermal resistance

Thermal resistance of the knitted fabric were determinate by measuring thermal resistance under steady-state conditions using the Sweating Guarded Hot Plate according to standard ISO 11092:2014. The Sweating Guarded Hot Plate simulates the heat and moisture transfer from the body surface through textile fabric to the environment under specified environmental conditions. Standard environment for testing thermal resistance is air temperature of 20 ± 0.1 °C and relative humidity of 65 ± 3%. The thermal resistance of knitted fabric gives quantitative evaluation of fabric as thermal barrier to the wearer. The thermal resistance of knitted fabrics are required to assess the heat exchange of the human body with the environment, and are related to human perceptions of comfort [15].

2. Results and discussion

The results of the knitted fabric parameters, i.e. mass per unit area, knitted fabric thickness, horizontal and vertical knitted fabric density, calculated knitted fabric porosity and thickness cover factor are presented in Table 2. The results of thermal resistance are shown in Table 3. All tested yarns were ring spun yarns with nominal yarn count of 20 tex from 1.3 dtex fibres with a length of 38 mm. Ring-spun yarns were produced using the carding manufacturing process, comprising fibre preparation phases (opening, blending and carding), spinning preparation (drawing, pre-spinning and ring spinning), winding and cleaning. A Zinser 351 ring spinning machine connected to an Autoconer X5 winding machine was used for the ring spinning process [16,17]. The knitted fabrics were made on one machine and under the same conditions, i.e. without machine operation control. A change in all parameters of knitted fabric structure was reflected through fabric mass per unit area, which ranged from 142 g m^{-2} (knitted fabric made of polyester ring spun yarn) to 165 g m^{-2} (knitted fabric made of Viscose® ring spun yarn) (Table 2.). The minimum and maximum knitted fabric mass per unit area difference is up to 16 %.

Table 2. Basic knitted fabrics properties [18].

Sample	m, g m^{-2}	t, mm	D_v, cm^{-1}	D_h, cm^{-1}	P, %	T_{cf}, g m^{-3}
Cotton	157	0.64	11.4	11.1	83.82	0.245
Viscose®	165	0.63	11.8	10.9	82.26	0.262
Tencel®	152	0.63	11.8	10.8	83.93	0.241
Modal®	155	0.58	11.8	10.3	82.55	0.267
Polyester	142	0.61	12.0	9.5	83.24	0.233

where: m is the mass per unit area, g m^{-2}; t is the thickness, mm; D_h is the horizontal fabric density, cm^{-1}; D_v is the vertical fabric density, cm^{-1}; P is the porosity, %; T_{cf} is thickness cover factor, g m^{-3}.

Comparing the knitting fabric thickness, influence of raw material is noticeable. The range of thickness is from 0.58 mm up to 0.64 mm, where knitted fabric made by cotton yarn has higher thickness for 10.3 % related to the fabric made by Modal® yarn. Results have shown the difference up to 14.5 % in knitted fabric cover factor, which ranged from 0.233 g m^{-3} for fabric made from polyester yarn to 0.267 g/m^3 for fabric made by Modal® yarn. Due to the difference in fibre density and knitted fabrics mass per unit area, range of knitted fabric porosity is from 82.26 % (for the Viscose® knitted fabric) to 83.93 % (for knitted fabric made by Tencel® yarn) which is a difference of only 2 %. Different mass per unit area and thickness of knitted fabrics made by different raw material influenced on the knitted fabric horizontal and vertical density (Table 2.).

The thermal resistance of the knitted fabrics is in the range of 0.019 m2 °C W-1 to 0.030 m2 °C W^{-1} (Table 3.). The smallest thermal resistance value has the polyester knitted fabric, while the largest thermal resistance value has the cotton knitted fabric. Thermal resistance of cotton fabric is higher for 36.7 % related to polyester knitted fabric.

Table 3. The knitted fabrics thermal resistance.

Sample	R_{ct} m^2 °C W^{-1}	SD, m^2 °C W^{-1}	CV, %
Cotton	0.030	0.003	11.3
Viscose®	0.022	0.002	8.3
Tencel®	0.023	0.003	12.2
Modal®	0.023	0.001	5.3
Polyester	0.019	0.001	3.9

Where: R_{ct}, is the knitted fabric thermal resistance in m^2 °C W^{-1}, SD is the knitted fabric thermal resistance standard deviation in m^2 °C W^{-1}, CV is the knitted fabric thermal resistance coefficient of variation in %.

Considering that thermal resistance describes resistance of human body heat passage through knitted fabric to the environment, it can be concluded that cotton knitted fabric will provide better thermal barrier then polyester fabric. Thermal resistance difference of Viscose® (13.6 %), Tencel® (17.4 %) and Modal® knitted fabric (17.4 %) related to polyester is smaller than cotton to polyester thermal resistance difference. Comparing thermal resistance of the knitted fabrics and their basic parameters, it is visible that polyester fabric has the lowest mass per unit area and thickness cover factor, which gives heat an easier pass through fabric, i.e. gives the smallest resistance heat to pass (Figure 3 and 4). Considering basic knitted fabric parameters, highest mass per unit area doesn't provide the greatest thickness, i.e. greatest thickness factor. It can be concluded that beside basic knitted fabric parameters, different yarn raw material influence on thermal resistance.

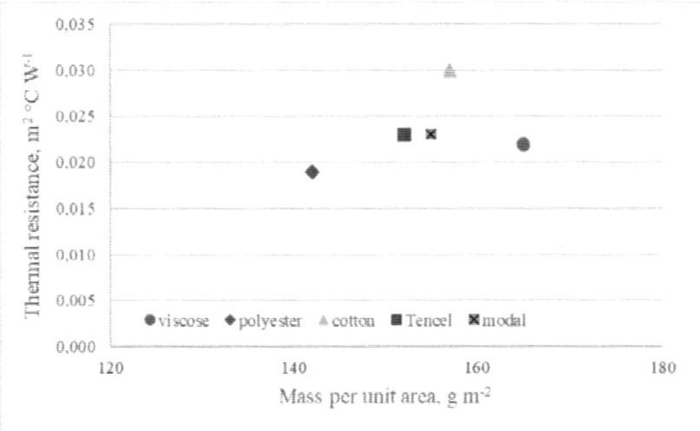

Figure 3. Influence of mass per unit area on knitted fabric thermal resistance.

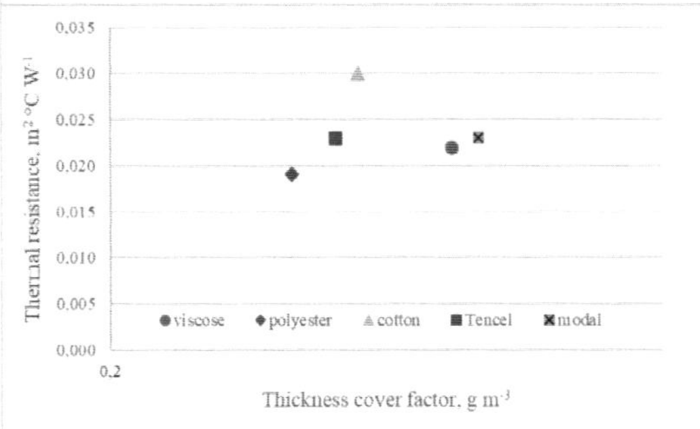

Figure 4. Influence of thickness cover factor on knitted fabric thermal resistance.

3. Conclusion

Based on the conducted research, it can be concluded that knitting with same yarn finesses and machine production parameters but different yarn raw materials do not provide knitted fabrics with the same constructional characteristic. Knitting with different raw materials namely, cotton, Viscose®, Tencel®, Modal® and polyester, gives mass per unit area in range of 142 g m^{-2} (knitted fabric made of polyester ring spun yarn) and 165 g m^{-2} (knitted fabric made of Viscose® ring spun yarn), which is different to up to 16 %. Beside mass per unit area, influence of raw material on knitted fabric is noticeable through thickness (from 0.58 mm for knitted fabric made by Modal® yarns up to 0.64 mm made by cotton yarns) and cover factor (from 0.233 g m^{-3} for fabric made by polyester yarn up to 0.267 g m^{-3} for fabrics made by Modal® yarn). The polyester has the lowest thermal resistance, while the cotton knitted fabric has the largest. Thermal resistance of cotton fabric is higher by 36.7 % related to polyester knitted fabric. Knitting with same yarn finesses and knitting machine production parameters but different yarn raw materials will give knitted fabrics different basic constructional parameters, which will influence on thermal resistance.

Acknowledgments
This work has been supported by the Croatian Science Foundation under the project (IP-2016-06-5278).

References

[1] Siejak V 2014 *The Fiber Year* PHP Fibers GmbH, Wuppertal, Germany
[2] Fanger P O 1972 *Thermal comfort – Analysis and applications in environmental engineering New York*, McGraw-Hill Book Company
[3] Hollies N R S, Custer A G, Morin C J and Howard M E 1979 A human perception analysis approach to clothing comfort, *Textile Research Journal* **49**(10), 557-564
[4] Oglakcioglu N, Celik P, Ute T B, Marmarali A and Kadoglu H 2009 Thermal Comfort Properties of Angora Rabbit/Cotton Fiber Blended Knitted Fabrics, *Textile Research Journal* **79**(10)pp. 888-94

[5] Sampath M B, Aruputharaj A, and Senthilkumar M and Nalankilli G 2012 Analysis of thermal comfort characteristics of moisture management finished knitted fabrics made from different yarns, *Journal of Industrial Textiles* **42** pp.19-33
[6] Azeem M, Hes L and Wiener J and Noman MT and Ali A and Mansoor T 2018 Comfort properties of nano-filament polyester fabrics: Thermo-physiological evaluation, *Industria textilă* **69**(4) 315-321
[7] Kim J O and Spivak S M 1994 Dynamic Moisture Vapor Transfer Trough Textiles – Part II: Further Techniques for Microclimate Moisture and Temperature Measurement, Textile Research Journal **64**(2), 112-121
[8] Heus R and Kistemaker L 1998 *Thermal comfort of summer clothes for construction workers*, Proceedings of 8th International Conference on Environmental Ergonomics, San Diego pp.273-276
[9] Tokura H and Jeong W S and Li X 1992 *The effect of clothing on thermoregulation and seasonal cold acclimation*, Proc. of the Fifth International Conference on Environmental Ergonomics, Maastricht: TNO-Institute of Perception pp 216-217
[10] Bakkevig M and Volla TT and Sandsund M 1994 *Sweat transport in double-and single-layer underwear* ed Ducharme M and Tikuisis P, Proc. of the Sixth International Conference on Environmental Ergonomics. Montebello: Defense and Civil Institute of Environmental Medicine pp 64-65
[11] Yasuda T, Miyama M and Yasuda H 1992 Dynamic Water Vapor and Heat Transport Through Layered Fabrics. Part II: Effect of Chemical Nature of Fibers, *Textile Research Journal* **62**(4) pp.227-235
[12] Dalidovič A S 1970 *Osnovy teorii vjazanija*, Moskva legkaja industrija
[13] Ray D 2017 *Biocomposites for High-Performance Applications: current barriers and future needs towards industrial development*, Woodhead Publishing
[14] Stana-Kleinschek K, Ribitsch V, Kreže T, Sfiligoj-Smole M and Peršin Z 2003, *Correlation of regenerated cellulose fibres morphology and surface free energy components,* Lenzinger Berichte, **82**(1), 83-95
[15] Huang J 2006 Sweating guarded hot plate test method, *Polymer Testing* **25**(5), 709–716
[16] Skenderi Z, Kopitar D, Ražić S E and Iveković G 2019 Study on Physical-mechanical Parameters of Ring-, Rotor- and Air-jet-spun Modal and Micro Modal Yarns, *Tekstilec* **62**(1), 42-53
[17] Skenderi Z, Kopitar D, Vrljičak Z and Iveković G 2018 Unevenness of air-jet spun yarn comparison with ring and rotor spun yarn made from micro modal fibers, *Tekstil* **67**(1-2) pp. 14-26
[18] Pavlović Ž, Lozančić M and Vrljičak Z 2017 Istraživanja strukture i vlačnih svojstava desno-desnih pletiva izrađenih pređama različitog sirovinskog sastava i procesa predenja pri jednakim parametrima pletenja, Tekstil: *Journal of Textile & Clothing* Technology **66** pp. 279-296

Physical-mechanical properties of thermal protective fabric subjected to textile care

I Schwarz[1, 3], S Brnada[1], A Kiš[2], S Kovačević[1]

[1]University of Zagreb, Faculty of Textile Technology, Department of textile design and management, Prilaz baruna Filipovica 28a, 10000 Zagreb, Croatia
[2]Čateks d.d., Quality Control Department, Zrinsko-Frankopanska 25, 40000 Čakovec, Croatia
[3]Author to whom any correspondence should be addressed

ivana.schwarz@ttf.hr

Abstract. Today, it is increasingly demanding, more complex and harder to achieve certain properties of fabric which are prescribed by standards for a particular purpose, and it is even more difficult to achieve the sustainability of these properties during use. In this paper, the most relevant physical-mechanical properties of thermal protective woven fabrics, made from meta-aramide yarns, after subjection to textile care, are investigated. This research is related to the objectives of the scientific project MF-WCOMPROTECET, which is based on the development of innovative multifunctional woven fabrics and composites for thermal protective clothing that will overcome the properties of the previous fabrics in the application.

1. Introduction

Protective fabrics are characterized by high quality properties and high numerous requirements placed on human body protection from the various external conditions present in different application areas. Development of protective textiles has been identified as a key market segment, due to its exceptional innovations, answering broader strategic, social, ecological and economic challenges and general benefit. With respect to market research and development requirements, there is a large number of projects funded in the segment of protective clothing [1-3]. MF-WCOMPROTECT (Multifunctional woven composites for thermal protective clothing) is one of them, which is funded by the Croatian Science Foundation (2018-2022). This project is focused on the fabric design process which will result in the creation of innovative, multi-layered, breathable and lightweight fabrics that will overcome the properties of the previous fabrics in the application for thermal protective clothing. Their protective properties will be based on the use of new high performance fibres but primarily on the construction of fabric parameters, forming multi-layer woven fabric that will provide properties of specific and strong protection, but simultaneously enhance wear comfort. Such compact fabric will have the function of a lightweight and thin composite and reflect the image of a protective high performance fabric. The compactness of such materials, taking into account the physical-mechanical properties, gives them an advantage over single-layer fabrics and well-known composite. Whereby it is reasonable to achieve properties of exceptionally good strength, abrasion resistance, durability and relevant external conditions, breathability, wear comfort and textile care [4-6]. In order to achieve the set objectives of the project, it is necessary to analyse the existing protective clothing products on the market with the aim of improving them through all segments of the specific requirements. This research continues on previously conducted research, which explored the influence of structural characteristics of woven fabric made from meta-aramide yarns on their thermal properties and properties associated with comfort [7].

Fabrics made from aramide yarns have high slitting and impact resistance, high strength, good chemical resistance and stability, high temperature resistance, high resistance to abrasion, acids and aggressive gases and are characterized by exceptional comfort and durability. The orientation of meta-aramid fibres allows molecule flexibility, providing exceptional textile properties for the use in making protective clothing for firefighters who are facing numerous hazards, especially those of heat, flame, and high temperature exposure [4].

Protective clothing, in addition to providing a high protection level, must be comfortable and easy to maintain. Washing, as a textile care process, is a cyclic hydrothermal treatment regulated by chemistry, mechanical agitation, temperature and time. One of the main importance in care is that washing has minimal negative impact on the protective fabrics properties. Negative influence of washing on woven fabrics physical and mechanical properties is decreasing breaking properties as well as the properties gained by surface treatments and increasing surface roughness [8]. Therefore, this paper will explore the influence of washing on the physical-mechanical properties of thermal protective woven fabrics.

2. Experimental

2.1. Materials and methods

The Samples used for this research are fabrics woven in different weaves from meta-aramide fibres and mixtures. Their specifications are shown in Table 1. Three samples have been compound from the same raw material (98% m-AR/2% HPCF) but in different weaves. The fourth sample K_AR_CV was woven in the same weave as K_AR but different raw material composition.

Table 1. Specification of tested samples.

Designation	Raw material composition	Weave
Kc_AR	98% m-AR / 2 % HPCF	Twill 2/2
K_AR	98% m-AR / 2 % HPCF	Twill 2/1
P_AR	98% m-AR / 2 % HPCF	Plain weave
K_AR_CV	68% m-AR / 30% viscose FR / 2 % HPCF	Twill 2/1

The structural and construction characteristics of the tested samples were determined by standard methods: warp and weft fineness (ISO 2060 (HRN F.S2.012)), fabric density (ISO 7211 (HRN F.S2.013)), fabric thickness (ISO 5084 (HRN F.S2.021)) and the fabric cover factor.

Determination of thermal properties of fabric samples was conducted according to the standard methods: EN ISO 9151:2016 – Method B (Determination of heat transmission on exposure to flame – the result of the test is the difference between the heat transfer index HTI24 – HTI12), ISO 17493:2016 (Test method for convective heat resistance using a hot air circulating oven intended to evaluate physical changes in a material at a given exposure temperature) and EN ISO 6942:2003 – Method B (Evaluation of materials and material assemblies when exposed to a source of radiant heat where protective effect of the materials is determined).

One of the most important indicators of textile material quality and the value of the finished product is its use durability, depending on the application. Therefore, the samples were subjected to textile care - washing process that entails 5 washing cycles with constant temperature of 60 °C. Relevant properties for thermal protective woven fabrics were tested before and after subjection to washing.

Breaking properties (breaking force and elongation at break) were determinated according to standard method (EN ISO 13934-1). Another very important parameter is the abrasion resistance, and to test it, the Martindale method was used in accordance with ISO 12947-2 (Determination of speciment breakdown). For determination of water repellence, standardised method EN ISO 4920 (Determination of resistance to surface wetting – Spray test) was applied, while for determination of oil repellence, EN ISO 14419 (Oli repellence – hydrocarbon resistance test) was used.

3. Results and discussion

Table 2 shows the structural parameters of tested samples of thermal protective woven fabrics. Cover factor were determined by analysing the image obtained by Dino-lite microscope, 60x magnification, by placing the sample on a plate with background light. From the Table 1 can been seen the differences in the weave construction, what have an impact on fabric thickness and cover factor. The results show that the thickness of tested samples does not differ significantly. Nevertheless, woven fabric in plain weave has the smallest thickness, as expected, while the fabric sample in the twill weave 2/1 has medium and the sample in twill 2/2 largest thickness. Cover factor of fabrics woven in twill 2/1 is higher than the fabric woven in twill 2/2 fabric, while the smallest cover factor has a fabric woven in plain weave resulting from its structure and slightly lower density.

Table 2. Structural parameters of tested woven fabric.

Properties		Sample			
		Kc_AR	K_AR	P_AR	K_AR_CV
Yarn fineness, Tex	warp	16.5	16.5	22.0	18.0
	weft	16.5	16.5	22.0	18.0
Fabric density, threads /cm	warp	36.0	36.0	26.0	31.0
	weft	23.0	23.0	20.0	26.0
Thickness, mm		0.56	0.54	0.52	0.55
Cover factor, %		93.8	99.6	97.1	99.8

The Table 3 shows the thermal characteristics of the tested fabric samples, such as the difference in the heat transfer indexes to the warming up of the sample at 24 °C and 12 °C (HTI24 - HTI12), the permeability of the thermal flux density (through the sample to calorimeter) (QC) and the heat transfer factor (TFQo).

Comparing fabric samples of identical structural characteristics and different raw material composition shows that the raw material composition affects all three parameters, i.e. the FR viscose component will reduce radiation resistance and accelerate heat transfer to flame, while the difference between heat transfer factors is negligible.

By comparing samples of the same raw material composition, but with different weaves, it can be concluded that fabric cover factor and thickness influence on rate of heat transfer of flame. Therefore, sample with smaller cover factor and highest thickness (with minimum number of interlacements) (twill 2/2) have a faster heat transfer of flame (5.70kW/m^2). In contrast, fabric woven in plain weave, because of maximum number of interlacements (resulting in highest compactness), has the highest resistance to radiative heat (5.53kW/m^2). Heat transfer factors in all samples also does not differ significantly.

Table 3. Thermal properties of tested woven fabric.

Properties	Sample			
	Kc_AR	K_AR	P_AR	K_AR_CV
HTI_{24} - HTI_{12}	1.57	1.53	1.57	1.70
Q_C, kW/m^2	5.70	5.56	5.53	5.89
TF Q_0	0.29	0.28	0.28	0.29

The results of mechanical properties in Table 4, shows significant differences between samples woven in different weaves. Fabric sample in the plain weave has the highest breaking force properties comparing to other samples due to its maximum interlacement. Twill 2/2 fabric (Kc_AR) has higher breaking force in warp than in weft direction, than the fabric woven in twill 2/1. Furthermore, the breaking force of Kc_AR is approximately three times lower in warp direction and almost two time lower in weft direction than the breaking forces of other fabrics. After the washing process, those relations are relatively detained in warp direction, while in the weft direction the values are almost equalized. Properties of elongation at break increases (for even 30%) after conducted washing process at all tested samples. Such changes after washing process occurs because of warp and weft crimp change due to fabric shrinkage/relaxation during washing. This results in lower tensile modules in the first elongation phase, which is reflected in the amounts of elongation at break.

The evidence that textile care (5 washing cycles) does not have negative impact on tested woven fabrics, is conducted test of abrasion resistance to wear, resulting in unchanged results regardless weave nor raw material.

The results of conducted tests of determination of water (resistance to surface wetting) and oil repellence shows that the samples Kc_AR and K_AR does not have resistance properties neither to wetting nor oil, due to non-applied surface treatments. At the same time, sample P_AR reduces its properties of surface wetting resistance for 10%, while the level of oil repellence was retained due to washing process. Sample K_AR_CV which is compound of 68% m-AR/30% viscose FR/2% HPCF under the subjection to washing reduce its properties for 30%.

Table 4. Physical-mechanical properties of tested woven fabric.

Properties			Sample			
			Kc_AR	K_AR	P_AR	K_AR_CV
Breaking force, N	Before washing	warp	393	1193	1228	1224
		weft	451	846	990	806
	After washing	warp	367	1117	1045	1241
		weft	559	597	511	857
Elongation at break, %	Before washing	warp	10.5	28.4	31.63	32.8
		weft	10.8	20.6	24.86	19.1
	After washing	warp	12.4	37.6	24.84	41.4
		weft	15.5	15.8	19.73	22.2
Abrasion resistance to wear, cycle/12kPa	Before washing		> 50 000	> 50 000	> 50 000	> 50 000
	After washing		> 50 000	> 50 000	> 50 000	> 50 000
Water repellence	Before washing		-	-	90-100	100
	After washing		-	-	90	70
Oil repellence	Before washing		-	-	6	5-6
	After washing		-	-	6	4-5

4. Conclusion

Conducted research gives an overview of the basic woven fabrics parameters intended for protective clothing, where high protection and comfort level must be obtained. From the obtained results, it can be concluded that the woven fabric cover factor is related to vertical porosity (affected by the weave), the fabric density and the yarn fineness effects the parameter of water vapour transfer rate and thus on the wearing comfort. The smaller cover factors of woven fabric is the faster heat transfer by flame. Woven fabric with more compact structure (which generally have less fabric thickness) will have the highest resistance to radiation heat. Furthermore, such fabrics must be easy to maintain and stabilize, with minimal negative impact on the relevant protective fabrics properties. Changes in tensile properties arising as a result of the washing process have no effect on the properties of abrasion resistance to wear. Furthermore, from the obtained results, we can conclude that raw material composition affects the decrease in water and oil repellence, where greater reductions were observed in samples with raw material composition of 68% m-AR/30% viscose FR/2% HPCF.

Acknowledgments
This work has been fully supported by the Croatian Science Foundation under the project number IP-2018-01-3170.

References

[1] European Commission, *Annex to the Proposal of the Regulation of the European Parliament and of the Council on Personal Protective Equipment*, Brussels
[2] European Commission, Lead Market Initiative (LMI). Available: http://s3platform.jrc.ec.europa.eu/documents/20182/117542/S2E_Fiche_LMI.pdf/137ddbea-582e-4674-a871-f7542f2a51dd. [Accessed: 18-Dec-2017].
[3] European Agency for Occupational Safety and Health, Directives 89/686, HRN EN 531 and HRN EN 15613. Available: https://osha.europa.eu/hr/legislation/directives/34. [Accessed: 24-April-2018].
[4] Scott R A 2005 *Textiles for protection* (Elsevier)
[5] Gokarneshan N 2009 *Fabric Structure and Design* (New Age International Publishers Limited)
[6] Hu J and Xin B 2008 *Structure and mechanics of woven fabrics. In: Structure and Mechanics of Textile Fibre Assemblies* (Woodhead Publishing) pp 48-83
[7] Kiš A and Brnada S and Kovačević S and Iskerka Pavlica B 2019 *Impact of structural fabric parameters on protective thermal properties of fabric and comfort* (Autex 2019 19th World Textile Conference, Ghent, Belgija) pp 1-3
[8] Truncyze D et al 2007 *The Influence of Washing on Fabric Wearing Properties* (Tekstil vol 56, 8) pp 493-498

Optical method for the determination of the roughness profile of woven fabrics

S Vassiliadis[1], D Matsouka[1,5], Q Watel[2] and H Jaouani[3,4]

[1] Department of Electrical and Electronics Engineering, University of West Attica, Thivon 250, Egaleo, 122 44 Athens, Greece
[2] ENSAIT, 2 Allée Louise et Victor Champier, 59056 Roubaix, France
[3] Higher School of Textile and Clothing Industry (ESITH), Route d'Eljadida, km 8, BP 7731 - Oulfa, Casablanca, Morocco
[4] Department of Renewable Energy and Dynamics Systems, University Hassan II, Casablanca 20000, Morocco.
[5] Author to whom any correspondence should be addressed

dmats@uniwa.gr

Abstract. Surface properties are important characteristics of textile fabrics. Apart from the tactile characteristics of the fabric, these properties can affect aspects of the fabric behaviour both in use and during further production processes. Fabric tactile characteristics and friction properties are just two of the properties connected to surface roughness. Typically surface roughness is determined using the KES-F system. This research paper proposes an innovative optical method for the measurement of fabric surface roughness based on image processing of photos of fabric cross-sections obtained through the stereoscopic microscope. Thirty parametric cotton woven fabrics were tested both in the warp and weft direction and the surface roughness for each fabric was obtained as a percentage of the void areas on the surface of the fabric.

1. Introduction

Fabric roughness is a property of fabrics connected with their surface properties. Roughness is a component of fabric "hand" and it affects surface friction properties, adhesion, and lamination and in more advanced applications, it could potentially influence properties such as flexible photovoltaic cell efficiency through light trapping [1]. While there exist a number of research papers that outline proposed methods for the determination of surface roughness, the only commercially available method currently in use is the Kawabata system (KES-F) [2-5]. The KES-F surface roughness method can determine the geometrical roughness and the coefficient of friction (μ) of the fabric. The measuring principle of the Kawabata system is based on the movement of a wire like probe over the surface of the fabric (Figure 1). The system translates the vertical movement of the sensor as it transverses the fabric surface into a proportional electric signal. Because of its design, the Kawabata system offers an averaged image of the surface roughness of the fabric. This is due to the relationship between the probe geometry and the usual range of numbers of yarns per unit length of woven fabric [2].

Figure 1. KES-F surface roughness probe.

Other methods for the determination of the surface roughness of woven fabrics as described in the relevant literature include the use of laser triangulation methods [4, 5]. These methods have not, as of the date of publication, resulted in the production of a commercially available device. Furthermore the use of lasers for the determination of the surface roughness of fabrics has its own limitations, regarding things like the acceptable specimen colours (white or light coloured fabrics can be difficult to measure) and the need for a very precise alignment of the specimen in the measuring position, which could prove difficulty due to the flexible and elastic nature of textiles.

The method described in this research paper is based on measuring the actual fabric geometry without the need for expensive, specialized equipment. The 3D profile of the fabric surface is captured on photographs taken through a stereoscopic microscope. The photographs are then analysed using an image analysis software. Thirty fabrics were analysed, and their surface roughness was calculated as a percentage of the indentations between warp yarns that appear on the surface of the fabrics.

2. Materials and Methods

2.1. Materials

This paper presents the study of the influence of yarn density on the surface roughness of woven fabrics using image processing of fabric cross-section photographs obtained through a stereoscopic microscope. The surface roughness of thirty fabrics, in total, was investigated. The fabrics were 100% cotton, plain weave with linear densities of Nec 40 and Nec 50 for warp and weft respectively. Five yarn densities were used for the warp direction, from 54 to 66 yarns per cm, with a step of 3 yarns/cm. For each of these warp densities four samples were produced with a weft yarn density of 20 to 45 yarns per cm with a step of 5 yarns/cm (Table 1).

Table 1. Characteristics of fabric samples.

i	Warp yarn density (yarns/cm)	Weft yarn density (yarns/cm)	Mass/unit area (g/m^2)
1	54	20	101.0
2	54	25	110.3
3	54	30	119.5
4	54	35	127.3
5	54	40	136.7
6	54	45	144.1
7	57	20	101.2
8	57	25	111.1

9	57	30	120.5
10	57	35	130.7
11	57	40	142.0
12	57	45	146.3
13	60	20	107.9
14	60	25	118.7
15	60	30	129.5
16	60	35	141.4
17	60	40	151.2
18	60	45	153.7
19	63	20	109.0
20	63	25	119.1
21	63	30	129.1
22	63	35	148.5
23	63	40	150.3
24	63	45	154.4
25	66	20	111.2
26	66	25	123.0
27	66	30	134.0
28	66	35	144.4
29	66	40	155.0
30	66	45	160.2

2.2. Methods

Fabric cross-sections were obtained for all the samples and photographed under the stereoscopic microscope. An example of the cross-section of the specimens can be seen in Figure 2.

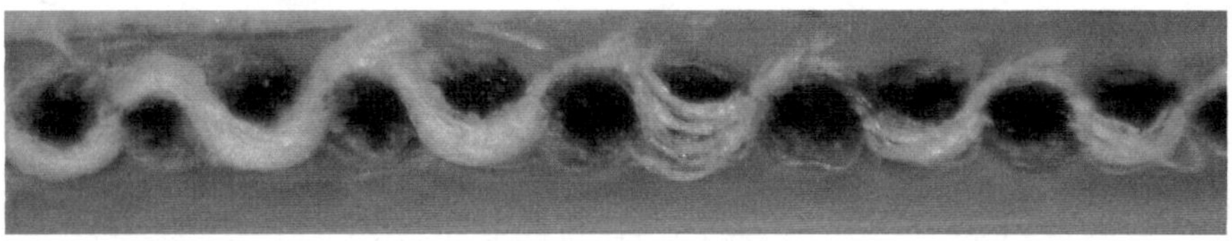

Figure 2. Specimen cross section.

The photographs were analysed using ImageJ image analysis software and the area that corresponded to the indentations created by the weft yarns was determined as a percentage of the total area of the photograph that corresponded to the fabric material. A visual representation of the definition of the void areas and the calculation of the corresponding area values can be seen in Figure 3. The photographs were cropped so that they only included the sample cross section and then the areas lacking material where outlined and the area was measured (i.e. the number of pixels was determined). The samples were analysed in both the warp and the weft directions.

IOP Conf. Series: Materials Science and Engineering **827** (2020) 012063 doi:10.1088/1757-899X/827/1/012063

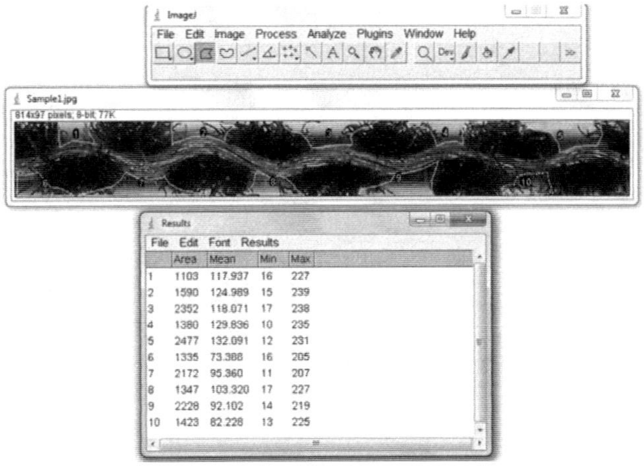

Figure 3. Measurement method.

3. Results – Discussion

The specimens examined under the stereoscopic microscope were prepared both in the warp and the weft directions. It should be noted that when the specimens were prepared along the warp direction, the weft yarns are prominent in the fabric cross section, while the warp yarns appear in a longitudinal form, while the reverse happens when the specimens are prepared in the weft direction.

The results obtained for the specimens prepared in the warp direction are presented in Figure 4 bellow. Figure 4 shows that the surface roughness of the specimens expressed as a percentage of the sum of the areas with no material (yarns) vs the total material in the specimen cross section is higher in the specimens with a lower warp density. Furthermore, when considering the same warp density, the surface roughness becomes lower with an increase in the weft yarn density.

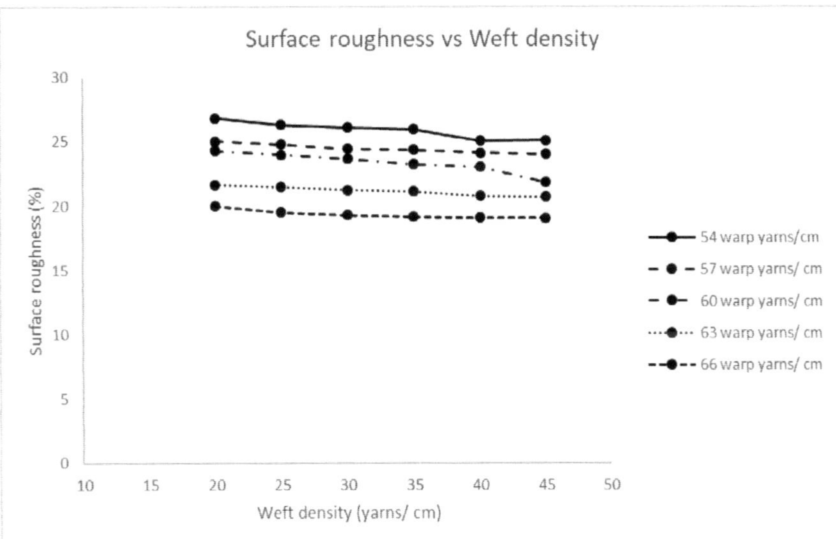

Figure 4. Surface roughness in warp direction specimens.

Figure 5 illustrates the results obtained for the specimens that where prepared in the weft direction. The Figure shows that the surface roughness of the specimens decreases as the warp density increases. Moreover, within the specimens with the same number of weft yarns per cm, surface roughness decreases as the weft yarn density increases.

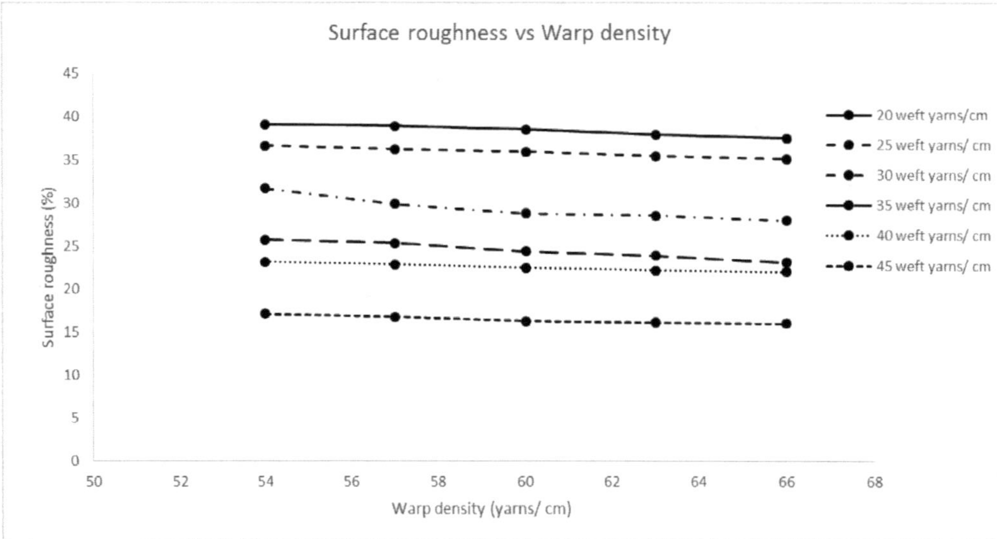

Figure 5. Surface roughness in weft direction specimens.

Regarding the results in both directions, it is important to note that while the method can be thought to be only measuring the dimensions of the yarns that are visible through the cross section, the effect of the yarns in the perpendicular direction is evident. Specifically the tension applied to the visible yarns and the deformation caused due to the presence of the other yarns. Moreover, it is significant to note that the values for the surface roughness that are obtained when examining either direction (warp/ weft) are not identical, which would is in step with the differences in fabric geometry in those directions. As materials fabrics are directional and quite similar to other products where the production direction has different properties than the direction that is perpendicular to the production direction e.g. paper.

4. Conclusions

The method outlined in this paper is presented as an alternative to the currently used methods that require expensive and specialized equipment. It is efficient as it provides actual measurement of the real geometry of the fabric and it can be applied to different fabric structures. The results obtained by using the method on a set of 30 fabrics to measure their roughness indicated a good agreement with what was described in theory of fabric structures.

References

[1] Lee S, Lee Y, Park J and Choi D 2014 *Nano Energy* **9** pp.88-93
[2] Akgun M, 2016 *J. Text. I.* **107** pp.1056-67
[3] Vassiliadis G S, Provatidis G C 2004 *Int. J. Clothing Sc. Tech.* **16** pp.445-57
[4] Mooneghi S A, Saharkhiz S, Varkiani S M H 2014 *J. Engineered Fibers Fabrics* **9** pp.1-18
[5] Ramgulam R B, Amirbayat J and Porat I 1993 *J. Text. I.* **84** pp.99-106.

IOP Conf. Series: Materials Science and Engineering **827** (2020) 012064 doi:10.1088/1757-899X/827/1/012064

Mathematical and numerical modelling of soiling effects of photovoltaic solar panels on their electrical performance

Marwane Rouway[1,3,6], Zoubair Boulahia[2], Nabil Chakhchaoui[3,4], Fraija Fouzia[1], Lhaj El Hachemi Omari[1], Omar Cherkaoui[3], Lieva Van Langenhove[5]

[1] Hassan II University, Faculty of Sciences Aïn Chock, Laboratory of Renewable Energy and Dynamic Systems, BP 5366 Maarif, 20100 Casablanca, Morocco
[2] Hassan II University, Faculty of Sciences Aïn Chock, Laboratory of Mechanics, BP 5366 Maarif, 20100 Casablanca, Morocco
[3] REMTEX Laboratory, Higher School of Textile and Clothing Industries (ESITH), Casablanca, Morocco
[4] BGIM Laboratory, Higher Normal School (ENS Q2), Hassan II University, Casablanca, Morocco
[5] Centre for Textile Science and Engineering, Department of Materials, Textiles and Chemical Engineering, Ghent University, Belgium
[6] Author to whom any correspondence should be addressed

marwanerouway@gmail.com

Abstract. With time, the photovoltaic solar panels naturally soiled from dust and other elements. These soils prevent light to reach solar cell and can reduce the long-term profitability. The study demonstrates the soiling effect on the electrical characteristics of panels, productivity and performance. A model simulation was developed in this direction in order to quantify and compare the performance of clean and dirty panels, the dust particles have been considered as spheres with an elliptical shadow on the panel. The results prove a power loss of 40% of a 0,224 mg/cm^2 dust deposition in comparison with a clean panel at the maximum operating point MPP. This study describes the identified critical parameters governing the fouling of the PV panels.
Keywords: photovoltaic panel; soiling effect; simulation; electrical performance

1. Introduction

The soiling effect of PV panels is a complicated and sophisticated problem since the phenomenon of dirt fouling is random and probabilistic. Many studies have observed the reduced power performance of PV module due to dust deposition and others environmental influences [1]. *Elminir et al.* confirmed the good reliance of dirt depositing, the transmittance and the power received [2]. *Hottel et al.* [3] proved a loss of 1% in solar radiation caused by dust deposition on glass tilted by an angle of 30° with a maximum degradation of 4.7%. *El Shobokshy et al.* have shown that the short-circuit current is decreased by 20% for a density of 28 g/m^2 of carbon nanoparticles [4,5]. *Mani and Pillai et al.* [6] have studied many climatic zones and their dust deposition level, and many recommended cleaning methods for each zone [7]. There is a theoretical model [8] for the simulation of a system with an opaque dust layer and a thickness varying from 0 μm to 5 μm, where 0 μm represents a cleaned module, and 5 μm represents a dirty module. This dust is formed of several thousand of small grains with different sizes placed on glass. The study investigates the influence of dust on the optical transmission of photovoltaic module compared to a clean module. According to the results an attenuation close to 90% has been detected [8]. There are different factors that affect the dust deposition, such as grain size, physical properties of module surfaces, type of glass, texture or antireflection, wind speed, and many other factors [9]. Also there are many types of pollutants (dust) that affect the performance of the modules, and various experiences have shown that each type of dust corresponds to a very specific voltage output as described in [10]. It is found that the most influential dust at the output voltage of the photovoltaic panel is ash with a reduction of 30%, and the least considerable type of dust is sand with a reduction of 4%. The study [11] revealed the impact of cement particles to be the most important, with a density of 73 g/m^2 of cement dust deposition resulting in an 80% decrease in PV short-circuit voltage, and the atmospheric dust with a diameter of 80 μm to 250 g/m^2, reduced the short-circuit current by 82%. Fine carbon particles (5 μm) had the most detrimental effect on PV efficiency. Also the finer particles have a greater impact than coarser particles on PV performance, for the same type of dust [11].

2. Mathematical model of soiling

2.1. Soiling modelling

The dust layer has a negative effect on the transmission of light through photovoltaic glazing surface, and it is studied by a mathematical model. For this, the different equations and the mathematical model are presented to calculate the particle transmittance. This model is based on a study by *Ahmed. Y et al.* [12]. The transmission of light at different angles of incidence for glass samples covered with dust particles will be calculated and compared with the mathematical model [12], where the transmittance for inclined angle of incidence $\tau_b(\theta)$ is:

$$\tau_b(\theta) = 1 - Q_e \cdot S_c(\theta) \quad (1)$$

With $S_c(\theta)$ is the total covered area, and Q_e is the efficiency extinction of light.

The beam radiation component I_b must be modified by the beam transmission factor $\tau_b(\theta)$. Therefore, the component due to dust particles $I_b(\theta)$:

$$I_b(\theta) = I_b \cdot \tau_b(\theta) \quad (2)$$

2.2. Simulation results of soiling modelling

The previously described mathematical model is validated by the model found in the article [12], as can be seen in figure 1. As the amount of dust installed on the surface of the panels decreases, beam light transmission reaches zero at very high angles of incidence. Furthermore, at high dust, the transmittance reaches zero at lower angles of incidence. We compared the radiation transmittance between the soil model described in [12] and our model. This comparison is made by taking two dust deposition densities of 0.224 mg/cm^2 and 0.02 mg/cm^2 as shown in Table 1. For the first soil density 0.224 mg/cm^2, the minimum relative error found is 0.88%, and increasing the inclination angle θ from 20° to 60°, the error reached a maximum value of 5.66. %. And for a second soil density 0.02 mg/cm^2, the minimum error is 0.2% for an inclination angle of 20°, and the maximum error is 1.09% for an angle equal to 78°.

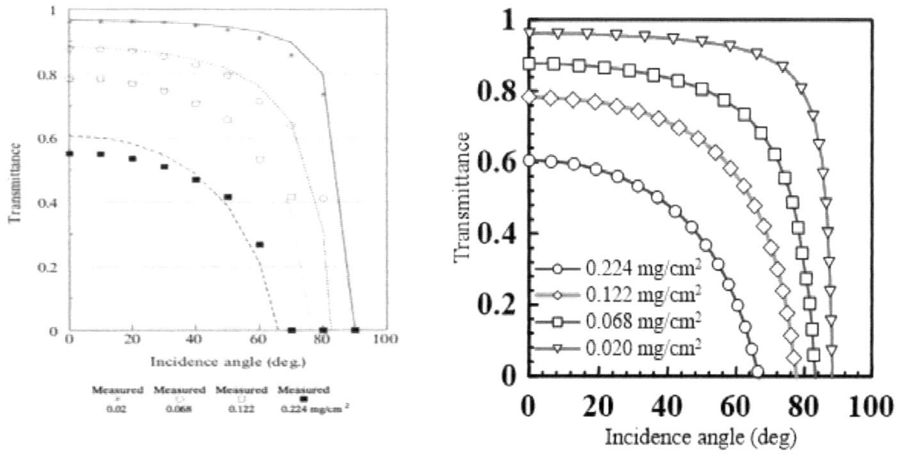

a) b)

Figure 1. Transmittance for different angles of incidence and dust masses, a) for model [12] and b) our calculated simulation.

The difference does not exceed 6%. This difference is due to uncertainty in calculation when reading the curves. From the comparison, it can be deduced that the calculation error is important when the value of the soil density is large.

Table 1. Comparison between calculated transmittance and that of the soiling model.

Density (mg/cm^2)		***θ*** **(°)**	**20**	**30**	**40**	**50**	**60**
0.224	τ (%)	Dust model	58.1	54.5	48.6	38.6	21.2
		Our results	58.6	55.4	50.03	39.1	20
		Relative error (%)	0.88	1.65	2.94	1.3	5.66
0.02	τ (%)	***θ*** **(°)**	**20**	**40**	**60**	**70**	**78**
		Dust model	96.5	95.7	93.3	89.9	82.1
		Our results	96.3	95.54	93.37	90.61	83
		Relative error (%)	0.2	0.16	0.07	0.78	1.09

3. Mathematical model of photovoltaic panel

3.1. Photovoltaic panel modelling

The mathematical model of solar cells is essential for the study of performance optimization operation. The photovoltaic module is represented by an equivalent circuit with experimental parameters from the current-voltage characteristic. These parameters are not usually measurable quantities, they must be calculated by systems of equations at different operating points given by the manufacturer or derived from the direct measurement on the module. This modeling is a crucial step and has led to diversification in the models proposed by the different researchers. Their differences are generally in the number of diodes, the finite or infinite shunt resistance, the ideality factor, as well as the numerical methods used to determine the various parameters. An electrical model of the photovoltaic cell has been presented; it is a real model with both series and parallel resistors (Shunt). This is the model on which builders rely by giving the technical characteristics of their solar cells. It is also considered satisfactory and even a reference for manufacturers to typically catalog solar modules [13]. It takes into account the voltage losses expressed by the R_s series resistor, and also the current leaks expressed by a parallel resistor R_{sh} as seen in figure 2:

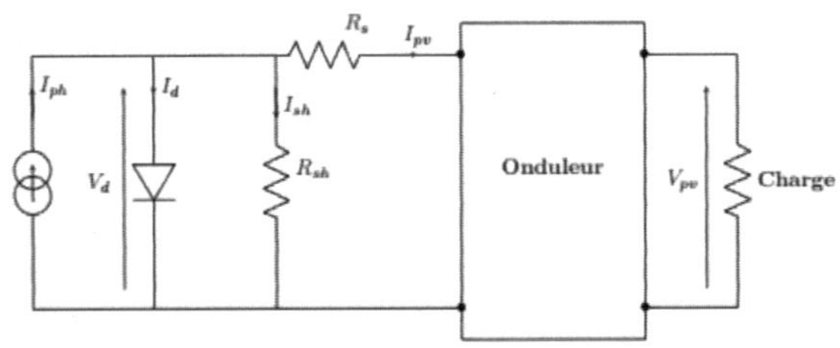

Figure 2. Equivalent circuit of a solar cell model with a diode.

3.2. Simulation and results of panel modelling

There are several factors influencing the photovoltaic system such temperature, irradiations, dust deposition, humidity and wind speed, which can impact the performance of the photovoltaic cell [14–18]. The Influence of dust is studied and shows that the dust deposition on panels affects all the electrical characteristics of the PV system; current, voltage and power. The figure 3 shows the evolution of the current as a function of the voltage of a clean panel with zero soiling density and under standard conditions STC (0 mg/cm^2, 25 °C and 1000 W/m^2), also the evolution of the dirty panel with a density of 0.224 mg/cm^2. The two curves kept the same shape of exponential decline, and the density of 0.224 mg/cm^2 had reduced the current from 17.5 A to 10 A at the peak of operating point (MPP) resulting in a power loss of 38.88 %. In addition, the figure 5 shows the evolution of the power according to the voltage, making also a comparison of the photovoltaic system in cleaned and dirty state. The dirt reduces the power from 10 kW to 6 kW at the maximum operating point MPP which gives a power loss of 40%.

Figure 3. a) Influence of dust on the current- voltage curve. b) Influence of dust on the power-voltage curve.

4. Conclusion

In this work, the problem of soiling and its effect on the panels has been presented as well as the various factors influencing the phenomenon of soiling. The dust accumulates on the panels due to gravitation, and glued when it mixes with water droplets already deposited on panels. It has been shown that the solar radiation received can be evaluated mathematically. This will help PV manufacturers to evaluate more precisely this radiation, as well as the predictive ability of existing solar models that can increase panel performance. In another perspective, the mathematical model performed could be improved if a statistical study is developed of the phenomenon of soiling. It can be concluded that soiling has a detrimental and serious effect on the power generation of PV plants, with a power loss of 40% with a dust deposition of 0,224 mg/cm^2.

References

[1] Sarver T, Al-Qaraghuli A and Kazmerski LL 2013. A comprehensive review of the impact of dust on the use of solar energy: History, investigations, results, literature, and mitigation approaches. *Renew Sustain Energy Rev* 2013,**22**, 698–733.

[2] Elminir HK, Ghitas AE, Hamid RH, El-Hussainy F, Beheary MM and Abdel-Moneim KM. 2006 Effect of dust on the transparent cover of solar collectors. *Energy Convers Manag* **47**, 3192–203

[3] HOTTEL HC 1942. Performance of flat-plate solar heat collectors. *Trans ASME,* **64**,91.

[4] El-Shobokshy M S and Hussein FM, 1993. Effect of dust with different physical properties on the performance of photovoltaic cells. *Solar energy*, **51**(6), pp.505-511.

[5] El-Shobokshy M S and Hussein F M, 1993. Degradation of photovoltaic cell performance due to dust deposition on to its surface. *Renewable Energy*, **3**(6-7), pp.585-590.

[6] Mani M and Pillai R, 2010. Impact of dust on solar photovoltaic (PV) performance: Research status, challenges and recommendations. *Renew Sustain Energy Rev*, **14**(9), pp.3124-3131.

[7] Berg R S, 1978. *Heliostat dust buildup and cleaning studies* (No. SAND78-0510). Sandia Labs., Albuquerque, N. Mex.(USA).

[8] Mesrouk, M. and ARAB, A.H., 2014. Effct de la poussière sur le rendement des modules photovoltaïques. Université Mouloud Mammeri de Tizi-Ouzou, 14. *Le 3 ème Séminaire International sur les Energies Nouvelles et Renouvelables*, Ghardaïa–Algérie

[9] Maghami M, Hizam H, Gomes C, Hajighorbani S and Rezaei N, 2015. Evaluation of the 2013 southeast asian haze on solar generation performance. *PloS one*, **10**(8).

[10] Khatib T, Kazem H, Sopian K, Buttinger F, Elmenreich W and Albusaidi AS, 2013,. Effect of Dust Deposition on the Performance of Multi-Crystalline Photovoltaic Modules Based on Experimental Measurements. *Int J Renew Energy Res IJRER* ;**3** pp.850–3.

[11] Sayigh A, Al-Jandal S and Ahmed H. Dust effect on solar flat surfaces devices in Kuwait 1985. *Proc. Workshop Phys. Non-Conv. Energy Sources Mater. Sci. Energy, ICTP Triest,* Italy, pp. 353–367.

[12] Al-Hasan AY 1998. A new correlation for direct beam solar radiation received by photovoltaic panel with sand dust accumulated on its surface. *Sol Energy*, **63**, pp.323–33.

[13] Wenham S R, Green M A and Watt M E, 1994. Applied photovoltaics centre for photovoltaic devices and systems. In *University of New South Wales*.

[14] Mekhilef S, Saidur R and Kamalisarvestani M, 2012 Effect of dust, humidity and air velocity on efficiency of photovoltaic cells. *Renew Sustain Energy Rev*,**16**,pp 2920–5..

[15] Adinoyi MJ, Said SAM 2013. Effect of dust accumulation on the power outputs of solar photovoltaic modules. *Renew Energy*,**60**,pp.633–6.

[16] Guan Y, Zhang H, Xiao B, Zhou Z, Yan X, 2017. In-situ investigation of the effect of dust deposition on the performance of polycrystalline silicon photovoltaic modules. *Renew Energy*;**101**,pp.1273–84.

[17] Dincer F, Meral ME.2010, Critical Factors that Affecting Efficiency of Solar Cells. *Smart Grid Renew Energy* **1**,pp.47–50.

[18] Meral ME, Dinçer F, 2011. A review of the factors affecting operation and efficiency of photovoltaic based electricity generation systems. *Renew Sustain Energy Rev*;**15**,pp.2176–84.

AI based forecasting in fast fashion industry: a review

El Hassan LAAZIZ[1, 2]
[1] Higher School of Textile and Clothing Industries (ESITH), Casablanca, Morocco
ESITH
[2] Author to whom any correspondence should be addressed

Laaziz02@gmail.com

Abstract. The textile and clothing sector is considered as one of the driving sectors of the Moroccan economy. It occupies a highly strategic position given the importance of its contribution in terms of growth and jobs. This sector, which is undergoing profound changes essentially linked to progressive dismantling of the Multi-Fibres arrangements (MFAs) that have governed trade long years, remains by far the largest export sector in the Moroccan industry. In this environment where competition will become more and more fierce, the ideal for a textile company is to produce exactly the products that its customers will buy. This ideal is unfortunately impossible, except in very special case where the company starts to manufacture from the receipt of the order of the client. Thus, in order to make the decisions relating to its proper functioning and to its durability, the company must rely on a reliable forecasting system. Such a system allows to achieve a competitive advantage by minimizing supply chain costs, avoiding shortages and minimizing financial losses due to the balance of unsold items.
The purpose of this article is to make a review of the literature on fashion retail sales forecasting methods. We have explored different kinds of analytical methods for fast fashion retail sales forecasting, their features and characteristics. From the reviewed literature we got a synthetic view of the pure statistical models, the pure Artificial Intelligence (AI) models, and the hybrid models.
Key words: Fast fashion, textile, forecasting, pure statistical methods, artificial intelligence, hybrid models, supply chain.

1. Introduction

The textile and clothing sector is considered as one of the driving sectors of the Moroccan economy. It occupies a highly strategic position given the importance of its contribution in terms of growth and jobs. This sector, which is undergoing profound changes essentially linked to progressive dismantling of the Multi-Fibres arrangements (MFAs) that have governed trade long years, remains by far the largest export sector in the Moroccan industry.

In this environment where competition will become more and more fierce, the ideal for a textile company is to produce exactly the products that its customers will buy. This ideal is unfortunately impossible, except in very special case where the company starts to manufacture from the receipt of the order of the client [1]. Thus, in order to make the decisions relating to its proper functioning and to its durability, the company must rely on a reliable forecasting system. Such a system allows to achieve a competitive advantage by minimizing supply chain costs, avoiding shortages and minimizing financial losses due to the balance of unsold items.

Sales forecasting is both necessary and difficult. It is necessary because it constitutes the point of a large number of business management tools: supply chain, retail logistics, production schedules, production plans and financing, marketing plans, cash flow plans, etc. [2]. It is difficult because it is out of our scope, whatever the quality of the methods adopted, to read the future with certainty as long as the parameters that come into play are numerous, complex and often unquantifiable, in any field whatsoever, and in particular that of textile-clothing which is distinguished by its unstable character and its vulnerability to disturbances from different sources (fashion phenomenon, climatic and economic variations, social events, etc.).

2. Literature review

We have conducted a review of fashion retail sales forecasting literature. We have depicted the characteristics of different kinds of analytical methods for fashion retail sales forecasting. We have also reported, through this review, the main findings related to real-world applications of the fashion retail sales forecasting models in the papers reviewed. From the reviewed literature, we got a summarized view of the pure statistical models, the pure Artificial Intelligence (AI) models, and the hybrid models. As in the industry sectors, the forecast of sales in fast fashion, has been mainly realized by the statistical forecasting methods such as: the moving averages, the linear regression, Bayesian analysis, the exponential smoothing (with & without trend) and double exponential smoothing.

To clarify the approach of multiple methods and techniques of sales forecasting, several classifications are commonly used: according to the term (short, medium, long-term methods), the sector of activity (industrial products, consumer durables, goods of great consumption, services, etc.), the technique used (econometric modeling, decomposition, exponential smoothing, Holt-Winters, Box-Jenkins, etc.), the nature of the data and their treatment (quantitative, qualitative methods), etc.

In *R. Maricourt* [2], one can find a classification in two approaches:

1. A pastist approach, based on the analysis of the past and the extrapolation of it into the future. Methods of this type, essentially quantitative, can themselves be subdivided into two great families based on different principles:
   a- Endogenous methods working only on directly related historical data to the phenomenon treated;
   b- Exogenous methods that are not content, like the previous ones, with a single causal variable, time, but by conducting a systematic analysis of one or several other explanatory variables.
2. A futuristic approach, often qualitative approach more directly connected to the exploration of the future while incorporating available elements of knowledge of the past. Futuristic methods seek to get as close as possible to men who are in the business of sales; is the questions, we simulate their behavior, we integrate them as much as possible in the elaboration of projections. These methods also give importance to abrupt changes in the environment, geo-political (revolutions, wars) or technological (discoveries scientists, development of new materials, etc.).

2.1. *Pure Statistical Methods*

In traditional fashion sales, forecasting is realized by the statistical methods such as moving average, weighted average, linear regression, Bayesian analysis, exponential smoothing, exponential smoothing with trend, double exponential smoothing, etc.

More sophisticated tools such as ARIMA and SARIMA are widely employed in sales forecasting for statistical time series analysis [3]. It is simple and easy to implement these methods and the results can be computed very quickly because they have a closed form expression for forecasting. Bayesian analysis is also very studied in the literature. *Green and Harrison* [4] apply a Bayesian approach to perform forecasting in case of a ladies dresses company. *Thomassey et al.* [5] use products segmentation to check the reliability of sales forecasting for new products. They stated that a wide range of product families and adequate segmentation criteria are necessary in order to achieve a reliable forecasting. They conclude that individual product's forecasting is less accurate than the product family and aggregated forecasting. *Mostard et al.* [6] proposed on a case study of an apparel company, a "top-flop" classification method and stated that it runs better than other methods. In addition, they found that the expert judgment methods is more effective, for a small aggregation of products, than the advance demand information method. The applicability of a hierarchical Bayesian approach for fashion demand forecasting was examined in *Yelland and Dong* [7]. The authors found that the approach yields superior quantitative results in comparison to other methods.

Here after in Table 1, a summary of most popularly used methods:

Table 1. The most popularly used forecasting methods.

Method	Paper	Findings
ARIMA and SARIMA	[3]	Closed form expression for forecasting, Easy to implement Results can be computed quickly
Bayesian approach	[4]	Efficient for mail order fast fashion company
	[5]	A wide range of products families and adequate segmentation criteria are conditions of forecasting precision. The individual product's forecasting is less accurate than product family and aggregated forecasting
Top flop method	[6]	It performs better than other methods The expert judgment methods outperform the advance demand information method for a small group of products.
Hierarchical Bayesian approach	[7]	Applicability of a Bayesian forecasting model for fashion demand forecasting. Yields superior quantitative results compared to many other methods.

Contribution on pure statistical methods are relatively rare in the literature over the last decades. This could be explained, in one hand, by the fact that the mathematical methods are enough fixed and stabilized. On the other hand, it is well known that these methods have some weaknesses:

- The choice of the suitable statistical methods is a difficult task that requires an "expert" knowledge.
- These methods do not output very reliable results compared to the methods such as AI methods.
- These methods may fail to achieve an enough good forecasting outcome because fashion sales are impacted by multiple variables such as the fashion trends and seasonality and exhibit a highly irregular pattern [8].

2.2. *AI based methods*

AI models came with strong advantages such us the capability to derive directly from the data "arbitrarily nonlinear" approximation functions. Here after a summary of this emerging methods that takes more and more advantage from computer science performance:

Table 2. The AI based forecasting approaches.

Method	area	Features and strengths
Artificial Neural Network (ANN)	Sufficient data	The AI forecasting models pioneer in fashion retail sales. Satisfactory results in various areas [9].
Fuzzy logic	Short term Sufficient data Color	Fuzzy theory sets is suggested by Zadeh [10] Multivariate model based on various product variables such as time, color, and size [11].

Evolutionary neural network (ENN) model	Low demand uncertainty and weak seasonal trends Short term	Search for the ideal network structure for a forecasting system, and then develop an ideal neural networks structure [12] Outperforms better than SARIMA for in case of products having low demand uncertainty and weak seasonal trends [12]
Extreme Learning Machine (ELM)	Price, size and color as significant factors	Outperforms back propagation neural networks based methods [13, 14]. Disadvantage: "unstable" as it can generate different outcome in each different run
Extended ELM Method (EELM)	Fast forecasting	- Overcomes the ELM drawback by computing the forecasting result by repeatedly running the ELM for multiple times.

2.3. Hybrid forecasting methods

The contributions on the hybrid models (i.e. models based on both pure statistical and AI methods) appear most frequently during last decade. As a consequence, we do believe that the topic is still worth exploring to produce more advanced hybrid models for fashion retail sales forecasting.

Hybrid forecasting methods use the strengths of different models together to form a new forecasting method. Many of these models are recognized to be more efficient than the AI models and the pure statistical models [6]. In the following a summary of these methods:

Table 3. Forecasting hybrid models.

Method	Paper	findings
Fuzzy Logic Based Hybrid Methods	[15] [16]	- Fuzzy hybrid model outperforms the conventional Holt-Winter Allows learning of the nonlinear explanatory variables influence.
Fuzzy logic and Neural Networks,	[17]	New system called the adaptive network based fuzzy inference system. Combines the learning capability of the neural networks and the generalization capability of the fuzzy logic technique
fuzzy logic, neural networks, and evolutionary	[18]	Versatility in processing the uncertain data
Neural Network Based Hybrid Methods	[19] [20]	Good results and forecasting for short and discontinuous time series Strength based on NN's ability of mapping the nonlinear relation between data output and inputs.
ANN and GM	[21]	Forecasting model with respect to color. Outperforms ANN, GM,Markov regime switching
ANN, Greymethod (GM) and autoregressive technique	[22].	Use neural network to establish a multivariable error model Uses the concept of "influence factors" Divides the "impact factors" into two distinct horizons (long term and short term).
ELM and NN	[23]	- It integrates an improved harmony search algorithm and an extreme learning machine & outperforms ARIMA
ELM and adaptive metrics	[24]	The inputs fix the amplitude changing and trend determination The over fitting of networks effect is reduced.

ELM and Grey relational analysis (GRA)	[25]	-	Outperforms competing models in fashion color forecasting

3. Conclusion and perspectives

In fast fashion domain, the global market, fueled by the explosion of new products and new technologies, especially the revolution in transportation and information technology has permitted manufacturers and retailers alike to establish international production and trade networks that cover vast geographical distances.

Through our literature review we found that majority of new contributions are focused on AI methods. However, AI (with a single method) models are not by themselves able to generate accurate results taking into account fashion sales variables. The contributions on the hybrid models appear most frequently during last decade. As a consequence, we do believe that the topic is still worth exploring to produce more advanced hybrid models for fashion retail sales forecasting. This was the main leitmotiv of this literature review.

In our future works, we will focus on the application of the most powerful AI methods mixed with the most popular pure statistical methods (Exponential smoothing and moving average) to real world applications in fast fashion industry.

References

[1] Courtois A, Martin-Bonnefous C, Pillet M 1999 *Gestion de production* (Les éditions d'organisation).
[2] De Maricourt R 1995 *la prévision des ventes* (Presses universitaires de France).
[3] Box G E, Jenkins G M and Reinsel G C 2008 *Time Series Analysis: Forecasting and Control, Wiley Series in Probability and Statistics*, John Wiley & Sons, Hoboken, NJ, USA, 4th edition.
[4] Green M and. Harrison P J 1973 Fashion forecasting for a mail order company using a bayesian approach *Operational Research Quarterly* **24**(2) pp. 193–205.
[5] Thomassey S, Happiette M and Castelain J M 2003 Mean-term textile sales forecasting using families and items classification *Studies in Informatics and Control* **12**(1) pp. 41– 52.
[6] Liu N, Ren S, Choi T M, Hui C L and Ng S F 2013 Sales forecasting for fashion retailing service industry: a review *Mathematical Problems in Engineering* **2013**.
[7] Yelland P M and Dong X J 2013 *Forecasting demand for fashion goods: a hierarchical Bayesian approach* in Handbook on Intelligent Fashion Forecasting Systems.
[8] Choi T M, Hui C L and Yu Y 2011 *Intelligent time series fast forecasting for fashion sales: a research agenda in* Proc. of the Int. Conf. on Machine Learning and Cybernetics (ICMLC '11), pp. 1010–4, Guilin, China.
[9] Olson D and Mossman C 2003 Neural network forecasts of Canadian stock returns using accounting ratios *International Journal of Forecasting* **19**(3) pp. 453–65.
[10] Zadeh L A 1965 Fuzzy sets *Information and Computation* **8** pp. 338–53.
[11] Sztandera L M, Frank C and Vemulapali B *Predicting women's apparel sales by soft computing* in Proc. of the 7th Int. Conf. on Artificial Intelligence and Soft Computing (ICAISC '04), pp. 1193–1198, Zakopane, Poland, June 2004.
[12] Au K F, Choi T M and Yu Y 2008 Fashion retail forecasting by evolutionary neural networks *International Journal of Production Economics* **114**(2) pp. 615–30.
[13] Huang G B, Zhu Q Y and Siew C K 2006 Extreme learning machine: theory and applications *Neurocomputing* **70**(1–3) pp. 489–501.
[14] Zhu Q Y, Qin A K, Suganthan P N and Huang G B 2005 Evolutionary extreme learning machine *Pattern Recognition* **38**(10) pp. 1759–63.
[15] Vroman P, Happiette M and Rabenasolo B 1998 Fuzzy adaptation of the Holt-Winter model for textile sales-forecasting *Journal of the Textile Institute* **89**(1) pp. 78–89.
[16] Thomassey S, Happiette M, Dewaele N and Castelain J M 2002 A short and mean term forecasting

system adapted to textile items' sales *Journal of the Textile Institute Part 1* **93**(3) pp. 95–104.

[17] Aksoy A, Ozturk N and Sucky E 2012 A decision support system for demand forecasting in the clothing industry *International Journal of Clothing Science and Technology* **24**(4) pp. 221–36.

[18] Thomassey S, Happiette M and Castelain J M 2005 A global forecasting support system adapted to textile distribution *International Journal of Production Economics* **96**(1) pp.81-95.

[19] Vroman P, Happiette M and Vasseur C 2001 A hybrid neural model for mean-term sales forecasting of textile items *Studies in Informatics and Control* **10**(2) pp. 149–68.

[20] Thomassey S and Happiette M 2007 A neural clustering and classification system for sales forecasting of new apparel items *Applied Soft Computing Journal* **7**(4) pp. 1177– 87.

[21] Choi T M, Hui C L, Ng C F and Yu Y 2012 Color trend forecasting of fashionable products with very few historical data *IEEE Transactions on Systems, Man and Cybernetics Part C* **42**(6) pp. 1003–10.

[22] Ni Y and Fan F 2011 A two-stage dynamic sales forecasting model for the fashion retail *Expert Systems with Applications* **38**(3) pp. 1529–36.

[23] Wong W K and Guo Z X 2010 A hybrid intelligent model for medium-term sales forecasting in fashion retail supply chains using extreme learning machine and harmony search algorithm *International Journal of Production Economics* **128**(2) pp. 614–24.

[24] Xia M, Zhang Y C, Weng L G and Ye X L 2012 Fashion retailing forecasting based on extreme learning machine with adaptive metrics of inputs *Knowledge-Based Systems* **36** pp. 253 9.

[25] Yu Y, Hui C L and Choi T M 2012 An empirical study of intelligent expert systems on forecasting of fashion color trend *Expert Systems with Applications* **39**(4) pp. 4383–9.

AUTHOR INDEX